Efficiency of Bank Filtration and Post-Treatment

Efficiency of Bank Filtration and Post-Treatment

Special Issue Editors

Thomas Grischek
Chittaranjan Ray

MDPI • Basel • Beijing • Wuhan • Barcelona • Belgrade

MDPI

Special Issue Editors

Thomas Grischek
University of Applied Sciences Dresden
Germany

Chittaranjan Ray
Nebraska Water Center University of Nebraska-Lincoln
USA

Editorial Office
MDPI
St. Alban-Anlage 66
4052 Basel, Switzerland

This is a reprint of articles from the Special Issue published online in the open access journal *Water* (ISSN 2073-4441) from 2018 to 2019 (available at: https://www.mdpi.com/journal/water/special_issues/Bank_Filtration)

For citation purposes, cite each article independently as indicated on the article page online and as indicated below:

LastName, A.A.; LastName, B.B.; LastName, C.C. Article Title. *Journal Name* **Year**, *Article Number, Page Range.*

ISBN 978-3-03921-305-4 (Pbk)
ISBN 978-3-03921-306-1 (PDF)

Cover image courtesy of Thomas Grischek.
RBF scheme with horizontal collector wells along the River Danube, Budapest, Hungary.

Contents

About the Special Issue Editors

Thomas Grischek Ph.D., Prof. Dr.-Ing. Thomas Grischek is a Professor of Water Sciences at the Department of Civil Engineering, University of Applied Sciences Dresden, Germany. Dr. Grischek has 28 years of professional and academic experience in groundwater management and water supply. His main research interests are natural water treatment techniques, such as riverbank filtration and artificial recharge, and the removal of iron, manganese, micropollutants, and pathogens. He has been involved in several national and international research projects, including the EU projects SAPH PANI and AquaNES. He has initiated and supported feasibility studies on riverbank filtration in India, Egypt, Thailand, and Vietnam. Dr. Grischek has more than 80 publications on water-related subjects in peer-reviewed scientific journals, book chapters, and international conference proceedings.

Chittaranjan Ray Ph.D., Prof. Chittaranjan Ray is the director of the Nebraska Water Center and Professor of Civil Engineering at the University of Nebraska, Lincoln, Nebraska, USA. Prof. Ray has nearly 35 years of research and consulting experience in the field of ground water, with special focus on ground water quality, riverbank filtration, and unsaturated zone flow and transport processes. He has been involved in several projects in the United States focused on the nitrate and pesticide contamination of ground water, riverbank filtration for water supply, and leaching assessment of contaminants from land surface to ground water. He has collaborated on riverbank filtration projects in the USA, India, South Korea, Egypt, and Malaysia. Dr. Ray has published nearly 100 journal papers and edited six books, and has presented research results at numerous conferences.

Preface to "Efficiency of Bank Filtration and Post-Treatment"

Riverbank filtration (RBF) has been used for many decades in Europe and the United States to provide drinking water to communities located on riverbanks. RBF is a well-proven water treatment step, which at numerous sites is part of a multi-barrier approach to drinking water supply. RBF schemes for the production of drinking water are increasingly challenged by new constituents of concern, such as organic micropollutants and pathogens in the source water and hydrological flow variations due to weather extremes. RBF and new technology components are integrated, and monitoring and operating regimes are adopted to further optimize water treatment in bank filtration schemes for these new requirements. This Special Issue presents results from the EU project AquaNES "Demonstrating synergies in combined natural and engineered processes for water treatment systems" (www.aquanes.eu). Additionally, papers from other research groups cover the efficiency of bank filtration and post-treatment, advantages and limitations of combining natural and engineered processes, parameter-specific assessment of removal rates during bank filtration, and the design and operation of RBF wells. The feasibility, design, and operation of RBF schemes under specific site conditions are highlighted for sites in the US, India, and South Korea.

From a sustainability point of view, RBF systems make better sense than full-scale treatment plants using surface water, since the energy and resource use in RBF are lower and little to no chemical residues are produced. RBF systems require less energy to operate and to deliver a unit amount of water than conventional surface water treatment systems.

At many sites worldwide, RBF provides good water quality which only requires disinfection as a post-treatment safety measure. There is a high potential for RBF application in countries or regions facing water supply problems—especially along large rivers in Asia, but also along the Nile River in Egypt. This Special Issue is intended to promote more applications of this natural, low-cost, and low-waste treatment step, with just disinfection at some sites, filtration for iron and manganese removal, or membrane filtration for the removal of pathogens and organic micropollutants.

Thomas Grischek, Chittaranjan Ray
Special Issue Editors

water

MDPI

Article

The AquaNES Project: Coupling Riverbank Filtration and Ultrafiltration in Drinking Water Treatment

Robert Haas [1],*, Ruediger Opitz [1], Thomas Grischek [2] and Philipp Otter [2]

[1]   DREWAG NETZ GmbH, 01067 Dresden, Germany; ruediger_opitz@drewag-netz.de
[2]   Dresden University of Applied Sciences, 01069 Dresden, Germany; grischek@htw-dresden.de (T.G.); otter@autarcon.com (P.O.)
*   Correspondence: robert_haas@drewag-netz.de; Tel.: +49-351-20585-4946

Received: 9 November 2018; Accepted: 18 December 2018; Published: 21 December 2018

Abstract: Natural water treatment techniques combined with engineered solutions were investigated at demonstration sites in Europe within the AquaNES project. Ultrafiltration is well-established in water treatment, but is not feasible for many water utilities due to its high operational costs compared to conventional treatment. These differences in cost are caused by membrane fouling and the associated cleaning required. This study aims to assess the economic and energetic operation factors based on studies of an out/in ultrafiltration treatment plant for river water and bank filtrate. The fouling potential of both raw water sources was investigated as well as the quality of the resulting water. In addition, the results show the potential utility of a combined approach utilizing bank filtration followed by ultrafiltration in drinking water treatment. In a separate consideration of the treatment process, the water quality does not fulfill the requirements of the German drinking water ordinance. A new method for the removal of dissolved manganese from the bank filtrate is presented by inline electrolysis. While this improves water quality, this also has a significant influence on fouling potential and, thus, on operating costs of ultrafiltration. These aspects lead to a fundamental decision for operators to choose between more costly ultrafiltration with enhanced microbiological safety compared to cost-effective but less stringent drinking water treatment via open filtration.

Keywords: river bank filtration; ultrafiltration; surface water treatment; energy efficiency; out/in membrane comparison; inline electrolysis

---

1. Introduction

According to the Federal Ministry for the Environment, most of the water resources in Germany are in a chemically and ecologically poor state [1]. In addition to chemical parameters, such as nitrate, anthropogenic inputs and microbiological pollution are increasingly affecting the raw water quality. Therefore, a robust and efficient water treatment is essential for a safe drinking water supply. Water treatment requirements are exacerbated by the increase in extreme weather events, such as flooding and prolonged droughts, as well as improved laboratory analytics and new legislation.

Riverbank filtration (RBF) is a robust and natural water treatment process. It has been used for drinking water treatment in Dresden, Germany for more than 140 years [2]. Surface water infiltrates due to a hydraulic gradient created by nearby wells. The underground passage between the river and the wells provides pre-treatment. Sorption, biodegradation, filtration, and mixing processes in the aquifer (Figure 1) result in a partial removal of heavy metals, organic compounds, bacteria, viruses, and protozoa as a function of residence time, flow path length, and hydrogeological/geochemical properties of the aquifer material [3].

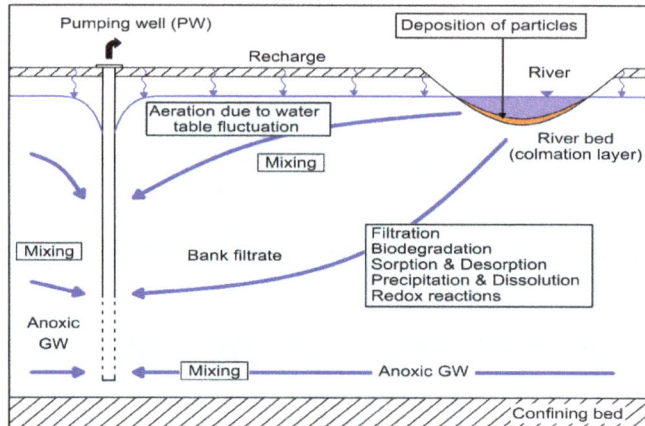

Figure 1. A schematic diagram of processes affecting water quality during riverbank filtration and groundwater (GW) [4].

At the same time, RBF is used for water quantity management. The interaction between surface water and groundwater (GW) prevents over-exploitation of the aquifer, salinisation by rising deep saline groundwater, and subsidence due to groundwater abstraction [4]. Against the background of the world's increasing demand for drinking water, RBF is also suitable as a low-cost treatment process with low technical requirements for developing countries [5].

Extreme events, such as floods, or increases in the capacity of the water intake lead to a shorter retention time of the bank filtrate in the exploited aquifer. This is a problem that can occur in the medium term, especially in conurbations with bank filtrate abstraction, such as the Rhine River region, or large cities, such as Berlin, Dresden, Budapest, and Poznan. With channeling rivers, more frequent weather extremes, and higher drinking water demand, abstraction of bank filtrate is becoming increasingly strained. One consequence may be breakthroughs of contaminants, which lead to microbiological and other pollution of the bank filtrate. A medium-term strategy for such scenarios is, therefore, advisable for water supply companies [6]. During such events, bank filtration should be coupled with an additional technical treatment stage.

Ultrafiltration is a treatment stage that reliably avoids microbiological breakthroughs. It is a pressure-based filtration through a membrane that cannot be passed by macromolecular compounds. Particles between 0.1 and 0.01 μm are retained regardless of the raw water quality (Figure 2). These include bacteria, viruses, protozoa, and suspended solids [7].

In Germany, there are more than 200 ultrafiltration plants for water treatment, mostly for reservoir water or groundwater [8]. Worldwide, ultrafiltration plants for drinking water treatment are increasingly used in arid areas or in the case of necessary use of water resources that are microbiologically polluted [9].

A disadvantage is the comparatively high energy requirement of ultrafiltration compared to other treatment techniques, such as sand filtration [10]. The energy consumption depends primarily on the fouling potential of the membrane, which is affected by the raw water quality and the cleaning processes. By combining bank filtration and ultrafiltration, the disadvantages of the respective treatment process can be reduced as both treatment steps complement each other.

The AquaNES project investigates various combinations of natural and technical water treatment processes for drinking water and wastewater treatment. The water company DREWAG works on an energetic, operational, and economical comparison considering the ultrafiltration of different source waters such as bank filtrates and untreated river water on a semi-technical scale. The focus is on the treatment performance and energy consumption of ultrafiltration membranes. In addition, the effect of the treatment of water for rinsing/cleaning by inline electrolysis on the fouling potential of ultrafiltration in terms of dissolved manganese and chlorine concentration in the backwash water is shown. Dissolved manganese and iron from the bank filtrate can lead to severe fouling of ultrafiltration membranes. By introducing oxygen into the backwashing process, the dissolved manganese can be oxidized and precipitate on the filtrate side of the membrane. Removal of such deposits requires additional chemical cleaning. Backwash water treatment by inline electrolysis can reduce this fouling and, thus, have a positive effect on the economic operation of an ultrafiltration plant.

This paper presents the first results of the semi-technical experiments and illustrates the advantages of the combination of bank filtration and ultrafiltration for water treatment compared to conventional treatment techniques.

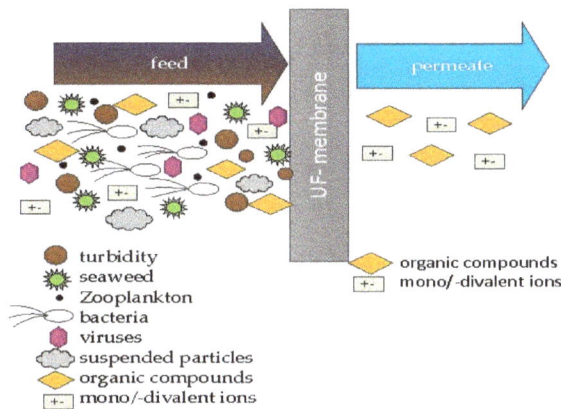

Figure 2. A schematic representation of the retention effect of ultrafiltration membranes.

2. Materials and Methods

2.1. Raw Water for Drinking Water Treatment and Feed Water for the Ultrafiltration Plant

The ultrafiltration plant was operated in the second largest waterworks in the city of Dresden, which is located at the Elbe River. This waterworks has a total capacity of 72,000 m³/day, and 111 vertical siphons wells and 36 wells with submersible pumps that extract bank filtrate from a depth of 5–8 m at a distance of 60–120 m to the river. After extraction, the water is aerated and filtered through activated carbon before it is disinfected with chlorine and distributed as drinking water (Figure 3, Scenario S01).

Besides bank filtrate (BF), river water (RW) is used as raw water for drinking water production. High drinking water demand requires direct abstraction of river water. The river water is first pre-treated by coagulation, sedimentation, and filtration through a multilayer filter. Subsequently, the water is artificially recharged into the aquifer via infiltration basins. The artificial recharge significantly increases the capacity of the waterworks. The infiltrated water is pumped together with the bank filtrate to the treatment stages aeration, activated carbon filtration, and disinfection (Figure 3, Scenario S02).

Both bank filtrate and untreated river water has been used as feed water for the ultrafiltration plant. From May 2018 to September 2018, river water was directly treated via ultrafiltration (Figure 3, Scenario S1). A robust feed pump with a capacity of 30 m³/h delivered river water directly into the storage tank of the membrane plant through a 435-m long supply pipe. Because the river water was not pre-treated, some sediment entered the storage tank. Bank filtrate was extracted from a well group of eight vertical filter wells with submersible pumps (Figure 3, Scenario S2). The wells are located at a distance of 80–110 m to the river with a depth of 6–8 m. Because of the riverbed and aquifer composition, some wells show an increased manganese concentration of 0.17 mg/L on average. Maximum concentrations up to 0.43 mg/L have been observed.

The bank filtrate was delivered via a 700-m long supply line into an overflow tank in the waterworks. From October 2017 to May 2018, a feed pump transported the bank filtrate to the ultrafiltration plant.

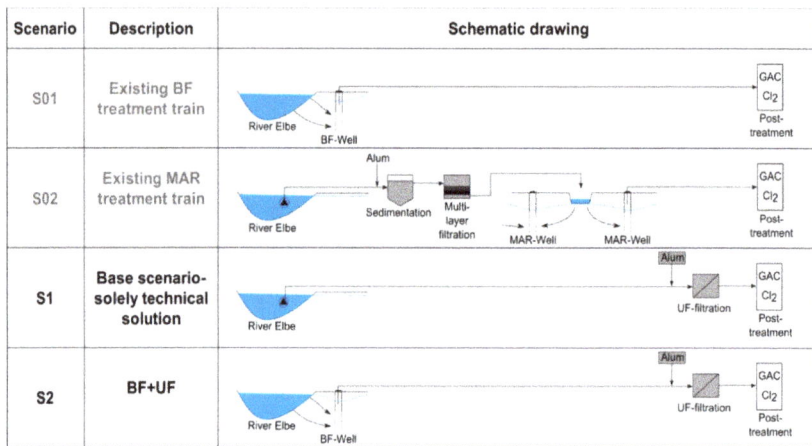

Scenario	Description	Schematic drawing
S01	Existing BF treatment train	
S02	Existing MAR treatment train	
S1	Base scenario-solely technical solution	
S2	BF+UF	

Figure 3. An overview of the treatment processes in the waterworks Dresden with Managed Aquifer Recharge (MAR) and the treatment scenarios S1 and S2 with an ultrafiltration plant (UF) and bank filtrate (BF).

2.2. Construction and Operation of the Ultrafiltration Plant

The maximum treatment capacity of the ultrafiltration plant was 20 m^3/h. The feed water tank with a volume of 1.9 m^3 was filled with BF or RW. The pumping was controlled by the water level. A permanent operation was guaranteed. From the feed tank, the water was pumped into the membrane unit (Q = 20 m^3/h, 3 bar operating pressure). A pre-filter with a pore size of 300 μm protected the membrane from coarse material and damage. The pre-filter was pressure-controlled and cleaned automatically when the pressure increased or after 2 min.

The first measuring point in the plant was placed after the pre-filter with online recording of the parameters temperature, turbidity, and conductivity. Subsequently, the feed water reached the membranes. There were two membrane lines, each consisting of three membrane modules (Figure 4). The inflow and filtration performance of the membrane line was controlled through flowmeters, pressure measurement, and control valves in front of the modules. Pressure directly before and after the membrane modules was recorded online, allowing for determination of the permeability and the transmembrane pressure (TMP). The permeate was stored in the permeate tank (3 m^3), and afterwards pumped (20 m^3/h) to one of the infiltration basins. The filtrate was also used for backwashing and chemical cleaning of the modules. All measurement and control data converged at the control panel.

The wastewater from backwashing without chemicals was discharged into the sedimentation basin of the waterworks. Chemical cleaning was carried out with a 50% hydrochloric acid, 35% sodium hydroxide solution, and 12% sodium hypochlorite solution. Depending on the cleaning program, setting parameters, such as temperature, concentration of the chemicals, and contact time were adjusted automatically via dosing pumps. The chemical wastewater was collected in the clean-in-place (CIP) container, neutralized, and discharged into a sewer after a safety residence time. Flowmeters recorded the volumes of the filtrate, backwash water flow, and chemical wastewater. The power consumption of the system was also measured. The ultrafiltration plant was fully automated.

Next to the membrane station, an AUTARCON electro-chlorination ECl$_2$ system (AUT, Figure 4) was installed for treatment of backwash water. A small portion of the UF filtrate passed through the electrolysis reactor, where chlorine ($\leq$1 mg/L) was produced from the natural chloride content of the water. This led to an increase in the redox potential of the water, allowing for the oxidation of dissolved Mn^{2+} to Mn^{4+} and its removal in the Green Sand Plus (GSP) filter unit.

The treated water was then stored in a 1 m^3 Intermediate Bulk Container (IBC) storage tank. Here, the redox potential was monitored in order to derive the required potential, which assures the oxidation of Mn^{2+}. The redox potential could be used as a control parameter in future applications.

The ECl$_2$ system was originally designed for a flow rate of up to 1000 L/h. Due to the low natural chloride concentration in the river water and bank filtrate, and the anticipated residual chlorine concentration of 0.2–0.5 mg/L, the flow rate was set to 400 L/h. This was sufficient to meet the demand for backwash water for both UF membrane modules. The GSP media was automatically backwashed every two days with a flow rate of 900 L/h for 20 min, including rinse using UF permeate. Backwash and rinse water were discharged.

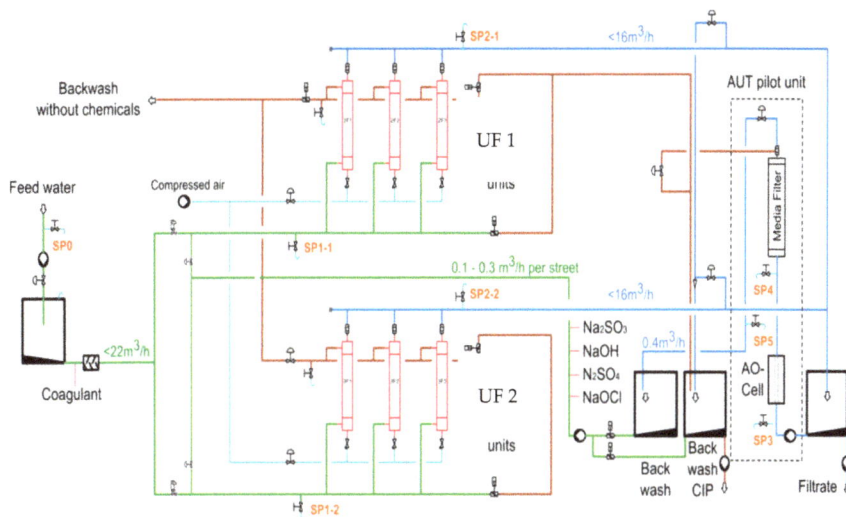

Figure 4. A flowsheet of the ultrafiltration membrane unit in combination with the AUTARCON electro-chlorination ECl$_2$ system (AUT) inline electrolysis unit. CIP, clean-in-place.

Sampling of the feed water and the two permeates (UF 1 and UF 2) was done weekly. The analyses were carried out in the accredited drinking water laboratory of DREWAG NETZ GmbH in accordance with the German Drinking Water Ordinancee.

Additionally, manganese and chlorine were analysed onsite using a handheld spectrophotometer (Aqualytic AL410, Dortmund, Germany). The chlorine was determined using the diethyl-p-phenylendiamin (DPD) method, where the free, bound, and total chlorine concentration is measured by the color reaction with diethyl-p-diphenylenediamine. According to wastewater regulations, the wastewater was tested monthly for arsenic, adsorbable organic halogens (AOX), and pH.

ATP measurements can detect living cells simply and easily and, thus, provide an onsite analysis of the effect for treatment steps or disinfections in addition to the legally required laboratory measurements. Set up for the measurement was carried out by measuring bioluminescence. An increase in light can be measured through enzymatic degradation of adenosine triphosphate (ATP) and adenosine monophosphate (AMP) using luciferase and pyruvate phosphate dikinase, which results in a biochemical reaction (bioluminescence).

The results are quantified as numerical Relative Light Units (RLU). If the number of micro-organisms in the water sample is high, the reactions are more intense and, therefore, the bioluminescence is high.

2.3. Ultrafiltration Membranes and Membrane Cleaning

Each membrane line was from a different membrane manufacturer. The first membrane line (UF 1) had modules with a 60 m^2 membrane area per module. The second line had modules with a 55.7 m^2 membrane area. The membrane fibres of both modules were made of polyvinylidene fluoride (PVDF) and operated in the filtration mode OUT-IN. The flux per membrane modules was 40–80 (L/m^2 h). The operating pressure of 1.5 bar was set in consultation with membrane manufacturers as the TMP limit.

The cleaning process of the membrane modules was adjusted depending on the filtration performance for the respective modules during operation in consultation with the manufacturers. Backwashing took place for bank filtrate every 90 min and for river water every 30 min.

UF 1 was cleaned with 60 s backwash as a combination of air and water with 5 m^3/h per module. Air was added in the filtration direction, water in the reverse direction. Afterwards, a 45 s forward flush was performed with 7 m^3/h feed in the direction of filtration.

UF 2 was first pre-cleaned with a 10 s air flushing from the feed side. Subsequently, the air-water backwashing from the filtrate took place in the flow direction as per UF 1 for 50 s. The cleaning was completed with a 15 s forward flush at 7 m/h in the filtration direction.

For manganese-containing feed water (bank filtrate), the backwash water was taken from the IBC tank of the inline electrolysis system.

The settings for the chemical cleaning process were different for the membrane line. Enhanced flux maintenance (EFM) and CIP (clean-in-place) were applied to UF 1, CEB (chemical enhanced backwash), and CIP in UF 2.

EFM is a purification process. Five hundred litres (500 L) of permeate were heated up to 32 °C and dosed with 1500 mg/L NaOH (35% solution) and 500 mg/L NaOCl (12% solution). The cleaning solution was pumped in the permanent filtration circuit of the membrane modules for 2400 s with 6 m^3/h. After the EFM, the membrane was rinsed for 90 s with permeate (15 m^3/h) and a forward flush for 30 s with feed water (7 m^3/h).

The dosage of 1750 mg/L NaOH and 500 mg/L NaOCl was done directly into the membrane with a flow rate of 6 m^3/h permeate in the CEB. This was followed by an exposure time of 1200 s with 3 min air intervals in the membrane. The chemical was rinsed out with filtrate (15 m^3/h) for 90 s. Afterwards, H$_2$SO$_4$ was added. After the exposure time and filtrate rinse like at NaOH, the TMP is reduced, which reflects a cleaning of the membrane. A forward flush for 15 s by 7 m^3/h completed the cleaning process. The strongest cleaning was the CIP. This was carried out separately with H$_2$SO$_4$ as well as with NaOH or NaOCl. The cleaning processes were identical to the EFM; however, the chemical exposure time of 7200 s was significantly longer.

3. Results

3.1. Ultrafiltration Efficiency

The laboratory results from the ultrafiltration permeate show an almost complete removal of turbidity and ATP, independent of the feed water quality. As expected, the dissolved organic carbon (DOC) concentration and the total hardness did not change. The efficiency of the pre-treatment by bank filtration is proven by low or negative findings for coliforms, *Escherichia coli*, *Clostridium perfringens*, and *Enterococci*. Differences between bank filtrate and river water with respect to the removal capacity of the membrane are visible in the parameters iron, manganese, and ultraviolet absorption coefficient at a wave-length of 436 nm (UVA-436).

The dissolved manganese from the bank filtrate completely passed through the membrane, indicating that Mn was 100% dissolved. The iron concentration was below the detection limit of 0.02 mg/L in the bank filtrate.

Almost complete removal of particulate iron and manganese was achieved during ultrafiltration of river water. The total organic carbon (TOC) in the river water (feed) was reduced by around 26% and the UVA-436 by around 14%. The TOC concentration is not usually determined in the bank filtrate, as it normally corresponds to the DOC concentration.

The differences between the removal performances of the membranes UF 1 and UF 2 were found to be negligible. A reduction in total hardness and potassium, calcium, and chloride concentrations was not determined. As an example, total hardness is shown in Table 1, where 98% of the ATP in the permeate was cell-free ATP (see Figure A1 in Appendix A).

Table 1. The water quality of feed water and permeate from the ultrafiltration plant (average values shown; for standard deviation and convergence see Table A1 in Appendix A).

Parameter	Feed	Permeate		Feed	Permeate	
	Elbe (n = 11)	UF 1 (n = 7)	UF 2 (n = 7)	BF (n = 12)	UF 1 (n = 12)	UF 2 (n = 12)
Colony counts at 22 °C (/mL)	1914	0.5	2	8	1	2
Colony counts at 36 °C (/mL)	1980	7	8	10	3	2
Coliform bacteria (1/100 mL)	3608	0	0	14	0	0
Escherichia coli (1/100 mL)	366	0	0	1	0	0
Clostridium perfringens (1/100 mL)	204	0	0	0	0	0
Enterococci (1/100 mL)	173	0	0	0	0	0
Turbidity (FNU)	10.2	0.1	0.1	0.5	0.1	0.1
TOC (mg/L)	7.2	5.1	5.2	-	-	-
DOC (mg/L)	5.1	4.9	5.0	2.3	2.3	2.3
UVA 436 (1/m)	0.7	0.60	0.6	0.1	0.1	0.1
Total iron (mg/L)	0.6	<0.02	<0.02	<0.02	<0.02	<0.02
Total manganese (mg/L)	0.3	<0.01	<0.01	0.1	0.1	0.1
Dissolved Mn (mg/L)	0.01	<0.01	<0.01	0.1	0.1	0.1
ATP total (RLU)	11,880	45	37	84	32	28
Total hardness (mg/L)	8	8	8	8.8	8.8	8.8
Smell at 23 °C (TON) (n = 4)	7.2	7.2	7.2	2.6	2.6	2.6

3.2. Fouling Potential

The fouling potential was assessed through the difference of the normalized transmembrane pressure (TMP) at 20 °C over the operating time. It is characterized by the TMP increase over the operating time and has a major impact on the amount and type of cleaning process as well as energy consumption. Figure 5 shows the fouling potential at a flux of 50 (L/m^2 h) for UF 1 as an example.

The TMP increase was 0.14 bar/day for RW and 0.005 bar/day for BF. A significant difference in the fouling potential of the different membrane manufacturers (UF 1, UF 2) was not found (see Table 2).

Table 2. A comparison of the fouling potential of both membrane types with differentiated flux.

Feed Water	UF 1 (Bar/Day)	UF 2 (Bar/Day)	Flux (L/m^2 h)
Elbe River water	0.078 (0.03 *)	0.077 (0.03 *)	40
Bank filtrate	0.005 (0.01 *)	0.005 (0.01 *)	60

* standard deviation.

Chemical cleaning was not required due to the low fouling potential of bank filtrate. In consultation with the membrane manufacturers, an EFM for UF 1 and a CEB for UF 2 was carried out once a week during filtration of river water. The aim was to delay the CIP cleaning and to reduce chemical wastewater. The results of chemical cleaning vary between the membranes for the river water according to the cleaning method. The cleaning performance for EFM at UF 1 is 0.7 bar (0.06 standard deviation) TMP reduction. The CEB cleaning performance for UF 2 is 0.4 bar (0.04 standard deviation)

TMP reduction. CIP completely purified the membrane depending on the flux and initial condition of the TMP.

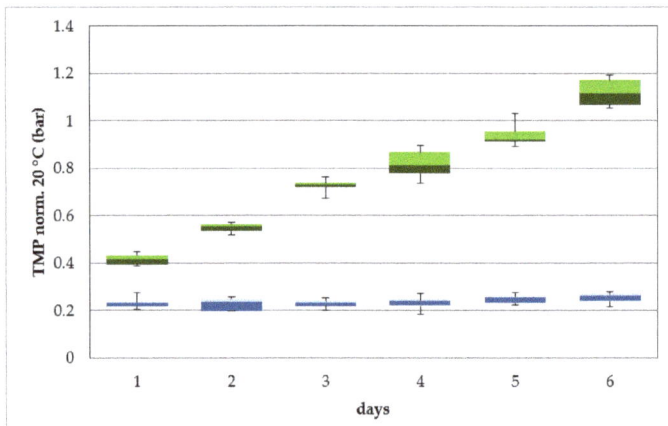

Figure 5. The increase in transmembrane pressure (TMP) for UF 1 during filtration of Elbe River water (green) and bank filtrate (blue). TMP was standardized over temperature. Flux was 50 L/m^2 h.

3.3. Energy Consumption

Depending on raw water quality and associated operation settings for cleaning, membrane pressure, and flux, the energy consumption was 0.23–0.18 kWh/m^3 for bank filtrate and 0.34–0.26 kWh/m^3 for river water (Figure 6).

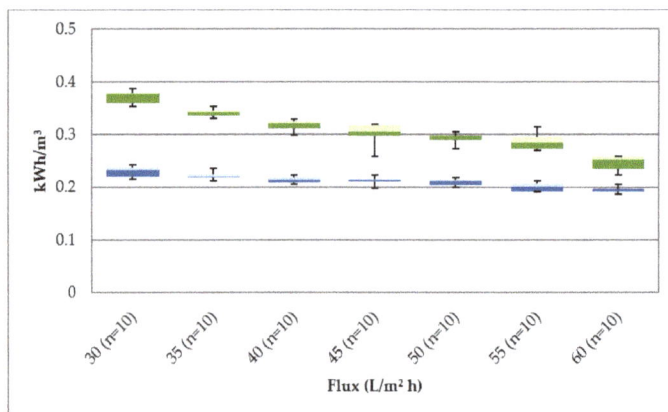

Figure 6. The energy consumption per m^3 filtrate produced from Elbe River water (green) and bank filtrate (blue) with identical operation settings of the membrane plant (UF 1 and UF 2).

The results show that the energy consumption per cubic meter of produced filtrate was 28% higher for the filtration of river water than for bank filtrate. The total energy consumption includes filtration, backwashing, and chemical backwashing.

A detailed analysis shows that membrane cleaning had a major impact on energy consumption for ultrafiltration of river water (Figure 7). In addition, the permeate production during treatment of river water was lower due to the cleaning cycles, which had a negative impact on the energy balance.

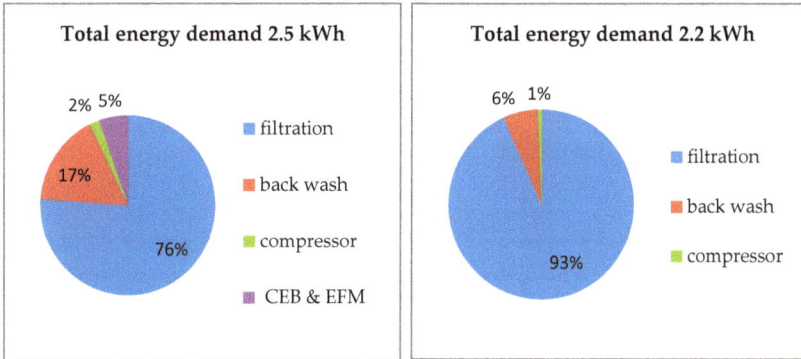

Figure 7. The percentage distribution of process-related energy demand for the filtration of Elbe River water with chemical cleaning (**left**) and for the filtration of bank filtrate without chemical cleaning (**right**). CEB, chemical enhanced backwash; EFM, enhanced flux maintenance.

The filtrate production at a flux of 50 L/m^2 h was 12.9 m^3/h for river water and 16.8 m^3/h for bank filtrate. It varied depending on the flux. Energy consumption for chemical cleaning showed a significant difference between EFM and CEB. CEB had a total energy consumption of 1.63 kWh, EFM 17.1 kWh. Seventy-five percent (75%) of the energy was needed to heat up the cleaning solution to 32 °C.

3.4. Effect of Inline Electrolysis on the Fouling Potential

Inline electrolysis (IEL) in combination with GSP filtration removed about 100% of the dissolved manganese from the bank filtrate (Table 3). On average, the produced chlorine concentration was 0.29 mg/L free and 0.39 mg/L total chlorine for the backwash water from the bank filtrate (n = 14). Manganese was completely removed by ultrafiltration because it was present in particulate form and not dissolved in river water (Table 4). The chlorine concentration was 0.2 mg/L free and 0.38 mg/L total chlorine (n = 4). The energy consumption for chlorine production and, therefore, manganese removal was on average 0.27 kWh/m^3. The fouling potential was reduced with manganese-free and slightly chlorinated backwash water from the IEL for both feed waters (Table 4).

Table 3. The removal performance of the inline electrolysis (IEL) for manganese with standard deviation.

Parameter	Before IEL	After IEL	n
Dissolved manganese (mg/L)	0.08	<0.01	19
Standard deviation	0.03	0.01	

Table 4. The fouling potential with and without backwash water from inline electrolysis (IEL) (n = 12).

Feed Water	without IEL	with IEL	Flux (L/m^2 h)
Elbe River water (bar/day)	0.14 (0.05 *)	0.12 (0.03 *)	50
Bank filtrate (bar/day)	0.013 (0.01 *)	0.005 (0.01 *)	80

* standard deviation.

4. Discussion

Microbiological particles, such as bacteria and other particulate matter, were efficiently removed by ultrafiltration independent from feed water quality.

Typically, all particles >0.02 μm should be removed by the membrane. Ions with mono or bivalent charge pass through the membrane. These include calcium, potassium, and chloride as well as pesticides and humic substances [11]. This statement was confirmed during this investigation. The results obtained through the monitoring of manganese concentration at different stages illustrate that dissolved manganese passes through the membrane. Particulate manganese from the river water, on the other hand, was completely removed by the membrane.

The parameter ATP was measured in addition to common microbiological parameters to better assess the removal performance of ultrafiltration concerning micro-organisms. The results illustrate that ultrafiltration reduced ATP and the number of micro-organisms (bacteria). A complete removal was found. The ATP in the permeate consisted of 98% cell-free ATP, which passed through the membrane. Cumming (2015) showed that free ATP was detected after ultrafiltration but not cell-bound [12]. The presented results confirm this statement (see Figure A1 in Appendix A).

It has to be noted that the ultrafiltration results were achieved without addition of activated carbon or flocculants before the membrane. The retention of humic substances, DOC, and reduction in UV-absorption (UVA) by ultrafiltration can be significantly improved by adding flocculants [13].

Furthermore, the requirements from the German Drinking Water Guideline were not met with ultrafiltration and surface water treatment without flocculation or activated carbon in two parameters, including UVA 436nm, which is an indicator for color and smell.

Humic substances that passed through the membrane were responsible for this, resulting in a slightly yellow color of water, which is measured by UVA.

On the other hand, the smell and taste of the surface water is not affected by the membrane [11]. In conclusion, exclusive treatment of Elbe River water using ultrafiltration without flocculation/activated carbon will not comply with German drinking water standards.

Microbiological and chemically relevant parameters for drinking water aside from UVA and color did comply with the German Drinking Water Guideline. Thus, in emergencies or regions without strict taste and color requirements, ultrafiltration can be an efficient treatment process for the provision of drinking water. In combination with flocculation or activated carbon, ultrafiltration effectively reduces DOC and UVA [14]. Disadvantages are increased operating costs through frequent cleaning and increased wastewater.

On the other hand, the results shown here indicate that bank filtration significantly reduces the number of micro-organisms but does not assure complete removal. Therefore, an efficient barrier against bacteria and viruses is essential, especially with regard to reduced residence times of bank filtrate during to floods or increased water abstraction. The advantages of the combination of both treatment processes are the production of safe drinking water independent from the raw water quality and the residence time of the bank filtrate in the aquifer as well as a more efficient operation of the ultrafiltration.

Higher content of particles in the river water leads to a significantly higher fouling potential than during ultrafiltration of bank filtrate.

This means that bank filtration acts as an efficient pre-treatment step for membrane filtration. The reduction of fouling indicators, such as bacteria, DOC, and UVA, as well as particulate matter minimizes the accumulation on the membrane. Bank filtration is comparable to slow sand filtration. Sand filters mainly remove biopolymers, proteins, and polysaccharides, which minimizes the fouling of membranes [15]. The result is an economically efficient operation compared to direct treatment of surface water. The low fouling potential of the membrane leads to longer filtration times and minimizes wastewater/backwashing. This reduces operating costs. A longer filtration time also leads to a more efficient use of energy in relation to energy consumption and filtrate production. This was demonstrated by the reduction in energy consumption for filtration of river water from 0.25 kWh/m^3 to 0.18 kWh/m^3 for filtration of bank filtrate at a flux of 60 L/m^2 h. The decrease in energy consumption with increasing flux was due to the optimized operating point of the raw water pump in terms of filtration volume and pressure.

The pilot plant's energy consumption is relatively high, with 0.18 kWh/m^3 at a flux of 60 L/m^2 h even for bank filtrate compared to other ultrafiltration plants with 0.05–0.2 kWh/m^3. One reason could be hydraulic losses due to control valves. Pressure filtration or open filters have an energy consumption of 0.03–0.1 kWh/m³ and are more energy-efficient [10,11]. After upscaling of the ultrafiltration system, a reduction in power consumption is expected. The pre-cleaning of the surface water by bank filtration also results in a reduced use of chemicals. In the case of surface water, chemical cleaning was required weekly. It is estimated that cleaning for bank filtrate is necessary every six months depending on the TMP and water quality. Operation costs are, therefore, reduced, as the life-span of the membrane is increased, which is mainly influenced by the use of chemicals [16].

Another advantage of ultrafiltration of bank filtrate is the possibility to waive another pre-treatment, such as activated carbon or flocculation. DOC and UVA and, thus, humic substances are significantly reduced by bank filtration. An additional fouling of the membrane is, therefore, avoided. The volume of wastewater is reduced and, thus, the filtration efficiency increased. A disadvantage of bank filtration is the potential increase in dissolved manganese and iron concentration during an anoxic aquifer passage, which has a negative impact on membrane filtration and fouling. Oxidation and precipitation of the solutes in the membrane should be avoided, as it can lead to severe fouling [17]. Chemically intensive cleanings are necessary in the case of fouling with manganese or iron to completely clean the membrane. One solution could be a pre-oxidation with potassium permanganate or by aeration. This leads to increased operation costs and can increase the fouling on the membrane in the case of an inaccurate reaction sequence [18]. In principle, this does not apply if the raw water has to be de-acidified.

The filtration system is designed in a way that as little water as possible is vented to prevent the precipitation of manganese or iron. Without air entrainment, the dissolved manganese passes through the membrane without causing fouling. A disadvantage of this approach is the combination of air-permeate backwash during the membrane operation out/in in which dissolved manganese from the bank filtrate would precipitate during the backwash on the filtrate side of the membrane and could lead to irreversible damage to the membrane. Inline-electrolysis produced exactly the backwashing volumes that were needed for the membrane system.

Manganese was oxidized by raising the redox potential and removed from the backwash water via a GSP filter. The additional low chlorine dose in the treated backwash water additionally purified the membrane, which led to a lower fouling potential. The reason for this is that, during each backwash, a short disinfection of the membrane fibres was achieved. Besides the protection of the membrane against manganese precipitation, the biological fouling was minimized. The same results have been seen during filtration of river water. A disadvantage of such a backwash process is that the permanent dosage of chlorine could reduce the life-span of the membrane. Whether such a low dosage of chlorine during a short reaction time of less than 50 s will permanently damage the membrane has not been investigated so far. A long-term study to determine this would be required.

It should also be noted that, aside from the membrane protection against manganese or iron fouling, the determination of the operation costs in terms of the power consumption of the inline-electrolysis is still outstanding compared to the use of chemicals, the fouling potential, and the cleaning performance without IEL.

One point of discussion is the retention of anthropogenic substances, such as drugs or microplastics. Microplastics are by definition completely retained by ultrafiltration membranes; however, nanoparticles are not removed [19]. A large number of drugs can pass through the membrane [20]. Studies show that bank filtration significantly removes pharmaceuticals such as antibiotics [21–23]. A retention of drugs can be improved with flocculation and/or activated carbon before ultrafiltration [24]. This supports the statement that drinking water treatment via ultrafiltration without flocculation of surface water is not effective for operational and economic reasons. However, in emergencies, it can maintain the drinking water supply under the aspect of preventing microbiological contamination. The combination of both treatment processes—bank filtration and ultrafiltration—has certain advantages and can offer a safe drinking water supply even under difficult boundary conditions.

5. Conclusions

Direct river water treatment using ultrafiltration was investigated in comparison to using bank filtrate as feed water. Neither bank filtration nor ultrafiltration treatment processes are suitable as single treatment steps for drinking water production under German law at the location of the Elbe waterworks. The combination of both techniques leads to an efficient and economically more feasible treatment for drinking water production. However, membrane filtration cannot compete with open sand filters in terms of energy efficiency. A decision between 100% retention and microbial safety for drinking water compared to a more cost-effective operation by means of open filtration must be made by each operator.

Dissolved iron and manganese in the bank filtrate can enhance fouling of ultrafiltration membranes. Dissolved manganese and iron can be oxidized by inline-electrolysis and then removed by filtration.

Treatment of bank filtrate using ultrafiltration without pre-treatment, such as flocculation, demonstrated the high performance of the membranes in terms of 100% removal of bacteria and turbidity with an energy consumption of 0.18 kWh/m^3 at a flux of 60 L/m^2 h (Q = 20.8 m^3/h).

The results of the research project AquaNES for the combination of natural and technical treatment processes are intended to serve as indicators for potential process design for water suppliers and constructors for drinking water treatment.

Limits and operational boundary conditions of ultrafiltration in the out/in operation have been demonstrated for the combination of bank filtration and ultrafiltration.

A reference is given for a meaningful use of such a combination in water treatment in the context of the achievement of requirements.

Author Contributions: R.H. reviewed the literature, analysed the energy and laboratory data, and prepared the draft of the publication; R.O. made important decisions in the project and contributed expertise in interpreting results, T.G. and P.O. supported the planning of experiments, produced additional laboratory and energy data, and contributed expertise in interpreting results. All co-authors reviewed and edited the draft.

Funding: All primary data was collected within the AquaNES project. This project has received funding from the European Union's Horizon 2020 Research and Innovation Program under grant no. 689450.

Acknowledgments: The authors gratefully acknowledge support from the DREWAG NETZ GmbH and its employees, especially in the waterworks Dresden-Elbwasserwerk (D. Vogt), water quality lab (J. Storm and his team), and from the University of Applied Sciences Dresden, namely R. Bartak, G. Orzechowski, and Y. Adomat during operation and sampling.

Conflicts of Interest: The authors declare no conflict of interest. The founding sponsors had no role in the design of the study, in the collection, analyses or interpretation of data; in the writing of the manuscript; or in the decision to publish results.

Appendix A

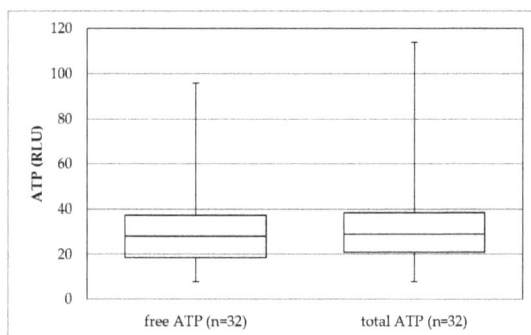

Figure A1. The results of the ATP measurements after ultrafiltration comparing free non-cell bound ATP (left) and total ATP (right).

Table A1. The standard deviation of the laboratory results for the UF membrane plant, convergence checked.

Parameter	Feed	Permeate		Feed	Permeate	
	Elbe (n = 11)	UF 1 (n = 7)	UF 2 (n = 7)	BF (n = 12)	UF 1 (n = 12)	UF 2 (n = 12)
Colony counts at 22 °C (/mL)	1064	0.4	0.7	1.5	2.5	5.8
Colony counts at 36 °C (/mL)	769	2.3	5.3	1.0	0.8	2.0
Coliform bacteria (1/100 mL)	1249	-	-	1	-	-
E. coli (1/100 mL)	495	-	-	-	-	-
Cl. perfringens (1/100 mL)	94	-	-	-	-	-
Enterococci (1/100 mL)	50	-	-	-	-	-
Turbidity (FNU)	6.0	0.08	0.03	0.01	0.1	0.03
TOC (mg/L)	0.2	0.3	0.4	-	-	-
DOC (mg/L)	0.3	0.3	0.2	0.1	0.2	0.2
UVA 436 (1/m)	0.07	0.05	0.04	0.005	0.03	0.03
Total iron (mg/L)	0.1	-	-	-	-	-
Total manganese (mg/L)	0.04	0.1	0.01	0.02	0.02	0.02
Dissolved Mn (mg/L)	-	0.1	0.01	0.02	0.02	0.02
ATP total (RLU)	5211	7.0	6.0	362	2.0	1.0
Total hardness (mg/L)	0.7	0.2	0.2	0.05	0.3	0.3
Smell at 23 °C (TON) (n = 4)	0.3	0.3	0.3	0.1	0.1	0.1

References

1. Arle, J.; Blondzik, K.; Claussen, U.; Duffek, A.; Grimm, S.; Hilliges, F.; Kirschbaum, B.; Kirst, I.; Koch, D.; Koschorreck, J.; et al. *Gewässer in Deutschland: Zustand und Bewertung*; Umweltbundesamt: Dessau-Roßlau, Germany, 2017; pp. 11–69. (In German)
2. Grischek, T.; Schoenheinz, D.; Eckert, P.; Ray, C. Sustainability of riverbank filtration—Examples from Germany. In *Groundwater Quality Sustainability*; Maloszewski, P., Witczak, S., Malina, G., Eds.; Taylor & Francis Group: London, UK, 2012; pp. 213–227.
3. Hiscock, K.M.; Grischek, T. Attenuation of groundwater pollution by bank filtration. *J. Hydrol.* **2001**, *266*, 139–144. [CrossRef]
4. Grischek, T.; Paufler, S. Prediction of iron release during riverbank filtration. *Water* **2017**, *9*, 317. [CrossRef]
5. Schiermeier, Q. The parched planet: Water on tap. *Nature* **2014**, *510*, 326–328. [CrossRef] [PubMed]
6. Schmoller, C.; Perfler, R. Uferfiltration—Stand der Technik und neue Herausforderung für die Weiterführung der Trinkwasseraufbereitung an Rhein und Donau. Master's Thesis, Universität für Bodenkultur, Vienna, Austria, 2014.
7. Cheryany, M. *Ultrafiltration and Microfiltration*; CRC Press: Boca Raton, FL, USA, 1998; Volume 2.
8. Lipp, P.; Baldauf, G. *Stand der Membrantechnik in der Trinkwasseraufbereitung in Deutschland*; DVGW Technologiezentrum Wasser (TZW): Karlsruhe, Germany, 2008.
9. American Water Works Association (AWWA). *Microfiltration and Ultrafiltration Membranes for Drinking Water*; American Water Works Association: Middleburg, VA, USA, 2008; Volume 100, pp. 84–97.
10. Elgg, J. Technischer Vergleich marktüblicher Filterverfahren. *Bädertechnik* **2010**, *10*, 631–650.
11. Baur, A.; Fritsch, P.; Hoch, W.; Merkl, G.; Rautenberg, J.; Weiß, M.; Wricke, B. *Mutschmann/Stimmelmayr Taschenbuch der Wasserversorgung*, 16th ed.; Springer Vieweg: Berlin, Germany, 2014; pp. 318–322.
12. Cumming, A. Assessing Biofiltration Pretreatment for Ultrafiltration Membrane Process. Ph.D. Thesis, University of Central Florida, Orlando, FL, USA, 2015.
13. Zularisam, A.W.; Ismaila, A.F.; Salimc, M.R.; Sakinaha, M.; Matsuurad, T. Application of coagulation ultrafiltration hybrid process for drinking water treatment: Optimization of operating conditions using experimental design. *Sep. Purif. Technol.* **2009**, *65*, 193–210. [CrossRef]
14. Lahoussine-Turcaud, V.; Wiesner, M.R.; Bottero, J.-Y.; Mallevialle, J. Coagulation pretreatment for ultrafiltration of a surface water. *J. Am. Water Works Assoc.* **1990**, *82*, 76–81. [CrossRef]
15. Zheng, X.; Ernst, M.; Jekel, M. Pilot-scale investigation on the removal of organic foulants in secondary effluent by slow sand filtration prior to ultrafiltration. *Water Res.* **2010**, *44*, 3203–3213. [CrossRef] [PubMed]

16. Regula, C.; Carretier, E.; Wyart, Y.; Gésan-Guiziou, G.; Vincent, A.; Boudot, D.; Moulin, P. Chemical cleaning/disinfection and ageing of organic UF membranes: A review. *Water Res.* **2014**, *56*, 325–365. [CrossRef] [PubMed]
17. Choo, K.H.; Choi, J. Iron and manganese removal and membrane fouling during UF in conjunction with prechlorination for drinking water treatment. *J. Membr. Sci.* **2005**, *267*, 18–26. [CrossRef]
18. Qu, F.; Du, X.; Liu, B.; Liang, H. Control of ultrafiltration membrane fouling caused by Microcystis cells with permanganate preoxidation: Significance of in situ formed manganese dioxide. *Chem. Eng. J.* **2015**, *279*, 56–65. [CrossRef]
19. Verschoor, A.J. *Towards a Definition of Microplastics*; RIVM Letter Report; National Institute for Public Health and Enviroment: Bilthoven, The Netherlands, 2015.
20. Boleda, M.R.; Gaceran, M.T.; Venura, F. Behavior of pharmaceuticals and drugs of abuse in a drinking water treatment plant (DWTP) using combined conventional and ultrafiltration and reverse osmosis (UF/RO) treatments. *Environ. Pollut.* **2011**, *159*, 1584–1591. [CrossRef] [PubMed]
21. Fritz, B. Uferfiltration und Wasseranreicherung. *wwt Wasserwirtschaft Wassertechnik* **2003**, *10–11*, 10–17.
22. Petrivic, M.; Lopez de Alda, M.; Cruz, S.; Postigo, C. Fate and removal of pharmaceuticals and illicit drugs in conventional and membrane bioreactor wastewater treatment plants and by riverbank filtration. *Phil. Trans. R. Soc. A Math. Phys. Eng. Sci.* **2009**, *367*, 3979–4003. [CrossRef] [PubMed]
23. Maeng, S.K. Multiple Objective Treatments Aspects of Bank Filtration. Ph.D. Thesis, Delft University of Technology, Delft, The Netherlands, 2010.
24. Shenga, C.; Nnanna, A.; Liu, Y.; Vargo, J.D. Removal of trace pharmaceuticals from water using coagulation and powdered activated carbon as pretreatment to ultrafiltration membrane system. *Sci. Total Environ.* **2016**, *550*, 1075–1083. [CrossRef] [PubMed]

water

MDPI

Article

Capillary Nanofiltration under Anoxic Conditions as Post-Treatment after Bank Filtration

Jeannette Jährig [1,*], Leo Vredenbregt [2,*], Daniel Wicke [1], Ulf Miehe [1] and Alexander Sperlich [3]

[1]   Kompetenzzentrum Wasser Berlin (KWB), Cicerostraße 24, 10709 Berlin, Germany;
      daniel.wicke@kompetenz-wasser.de (D.W.); ulf.miehe@kompetenz-wasser.de (U.M.)
[2]   Pentair X-Flow, Marssteden 50, 7547 TC Enschede, The Netherlands
[3]   Berliner Wasserbetriebe (BWB), Cicerostraße 24, 10709 Berlin, Germany; alexander.sperlich@bwb.de
*   Correspondence: jeannette.jaehrig@kompetenz-wasser.de (J.J.); leo.vredenbregt@pentair.com (L.V.);
     Tel.: +49-(0)30-5365-3843 (J.J.); +31-53-428-7284 (L.V.)

Received: 28 September 2018; Accepted: 1 November 2018; Published: 7 November 2018

Abstract: Bank filtration schemes for the production of drinking water are increasingly affected by constituents such as sulphate and organic micropollutants (OMP) in the source water. Within the European project AquaNES, the combination of bank filtration followed by capillary nanofiltration (capNF) is being demonstrated as a potential solution for these challenges at pilot scale. As the bank filtration process reliably reduces total organic carbon and dissolved organic carbon (DOC), biopolymers, algae and particles, membrane fouling is reduced resulting in long term operational stability of capNF systems. Iron and manganese fouling could be reduced with the possibility of anoxic operation of capNF. With the newly developed membrane module HF-TNF a good retention of sulphate (67–71%), selected micropollutants (e.g., EDTA: 84–92%) and hardness (41–55%) was achieved together with further removal of DOC (82–87%). Fouling and scaling could be handled with a good cleaning concept with acid and caustic. With the combination of bank filtration and capNF a possibility for treatment of anoxic well water without further pre-treatment was demonstrated and retention of selected current water pollutants was shown.

Keywords: decentralized capillary nanofiltration; anoxic; suboxic; organic micropollutants; bank filtrate; groundwater; sulphate

---

1. Introduction

The present work was developed within the European project AquaNES. The goal of the project is to demonstrate the benefits of the combination of natural treatment processes with engineered systems as sustainable adaptations to e.g., water scarcity, high nutrient loads and organic micropollutants (OMP) in the water cycle.

Bank filtration schemes for the production of drinking water are increasingly affected worldwide by e.g., OMP, pathogens, nitrate or sulphate in the source water, flood and low water or riverbed clogging [1–7]. Within AquaNES new technology components were integrated and monitoring and operating regimes were adopted to further optimize water treatment in bank filtration schemes for these new requirements.

In Berlin, drinking water is produced from 54% bank filtrate, 16% groundwater recharge and 30% ground water [8]. One challenge for the drinking water supply is the increasing sulphate concentration in raw water. Background concentrations for sulphate in ground water of Berlin are mostly >100 mg/L, in the inner city >360 mg/L [9]. Selected wells already exceed the limit for drinking water of 250 mg/L [10], reaching up to 900 mg/L at some eastern locations along the Havel River [11]. Sulphate in groundwater originates from different sources: leaching of dumps of building rubble and debris from the Second World War [12], domestic waste water [13] and oxidation of sulphide containing organic material [11].

The other source of sulphate in drinking water wells is the input by bank filtrate: higher sulphate concentration in the Spree River is caused by release of sulphate from dump sediments of abandoned open pit lignite mines upstream of Berlin. In the Spree River concentrations up to 320 mg/L sulphate [5] were measured. During subsurface passage of bank filtrate sulphate is either not affected [14] or only marginally attenuated under anaerobic conditions (sulphate reduction) [15]. Good sulphate removal from drinking water can be reached with technologies such as ion exchange, nanofiltration (NF), reverse osmosis and low pressure reverse osmosis, here retentions of >90% were observed [16,17]. Retention values depend on operational parameters, feed water quality and initial sulphate concentrations. Different agricultural residues were tested as raw materials to produce anion adsorbent, e.g., rice straw [18]. In addition, studies for adsorption of sulphate to synthesized zeolite [19] or kaolinite [20] were carried out.

The complexing agent ethylendiamintetraacetic acid (EDTA) is used in a wide range of applications, e.g., in photographic industries, cleaning agents, cosmetics and agriculture. EDTA is not easily biodegradable or adsorbable and can pass the drinking water processing steps [21,22]. EDTA enters the environment mostly via waste water (e.g., infiltration of waste water treatment plant effluent [23]). At water works, the main source of EDTA is leaching from former sewage irrigation fields and increasing concentrations are expected in the future. In addition, bank filtrate from a water body receiving wastewater treatment plant effluent is a (minor) source of EDTA. The health orientation value for drinking water is 10 µg/L, the measured values in drinking water are in this range [24].

Nanofiltration is a pressure driven membrane filtration process. The separation performance is located between ultrafiltration and reverse osmosis with a typical molecular weight cut-off (MWCO) between 200–1000 Da and a required feed pressure of 5–30 bar. Mono-valent ions and small molecules pass the membrane, whereas multi-valent ions and other molecules such as biopolymers and large organic micropollutants are retained. The negatively charged membrane surface leads to an increased retention of charged molecules. Nanofiltration is used for treatment of drinking water, surface water, urban and industrial waste water for e.g., softening, color removal and removal of turbidity, dissolved organic matter and microorganisms [25–28]. In case of industrial processes the membrane can be used for the separation of valuable components. Nanofiltration membranes are produced as spiral-wound, tubular or capillary modules, each of them has their own benefits: spiral-wound modules have the advantages of high packing density and low costs, tubular modules require less pre-treatment and capillary modules do not need expensive pre-treatment and can be backwashed [29].

For this study, a commercially available capillary membrane was upgraded with a new coating, enabling the retention of sulphate and selected OMPs at lower feed pressure compared to conventional sulphate removal technologies. The new developed membrane is applied for the first time to demonstrate the long-term stability for treatment of well water (ground water and bank filtrate) under anoxic to suboxic conditions. The main goal is to show the removal of selected compounds as well as the benefit of bank filtration to prevent biofouling of the membrane in comparison to direct surface water treatment via capNF.

Different studies show the possibility of anoxic operation of nanofiltration at higher iron concentrations up to 8 mg/L for the production of drinking water from ground water and bank filtrate [30–33]. In all former studies spiral wound membrane modules were used, which cannot be backwashed and are harder to clean. To overcome these disadvantages, a capillary module is applied in this study to demonstrate the benefits of better cleaning properties.

2. Materials and Methods

2.1. Feed Water Source

The water works Tiefwerder is one of nine water works supplying drinking water to the city of Berlin. A mixture of groundwater and bank filtrate is extracted by 55 wells along the river Havel with a well depth of 30–100 m. The water treatment consists only of aeration followed by rapid filtration to remove iron and manganese. Disinfection is not required but possible if necessary. The site was selected because of higher sulphate concentrations [12] and elevated EDTA concentrations in single wells [24].

Data shown in this paper were collected during two different periods of operation: during the first period drinking water was used as feed water whereas during the second period well water (ground water and bank filtrate) from a collecting line of one well field was used. The operation with drinking water as feed water was carried out to get a baseline for the operational conditions. For an intensive sampling campaign during the well water period it was possible to operate only wells with higher concentrations of sulphate and EDTA to increase feed concentrations of these contaminants.

2.2. Capillary Nanofiltration

The pilot plant was planned and built by Pentair X-Flow BV and is equipped with a new developed capNF module (HF-TNF). A schematic overview is shown in Figure 1. The filtration direction of the capillaries is inside-out. The plant is operated in feed and bleed mode, this means, during filtration a part of the concentrate is discharged and another part is recirculated over the membrane to increase the cross-flow velocity and therefore reduce concentration polarization and scaling (precipitations) on the feed side. Part of the permeate is stored in the permeate tank for backwash and cleaning and the excess permeate is drained in a nearby infiltration pond. All pumps are flow controlled and equipped with frequency converters. The membrane area is 40 m^2 and simulates one stage of a full-scale system. Operation with oxic, suboxic and anoxic feed water is possible. A chemical cabinet with four storage tanks and dosing pumps is available for flexible dosing of chemicals e.g., for membrane cleaning. The pilot plant is placed in a 40 ft. container and equipped with remote control for decentralized operation.

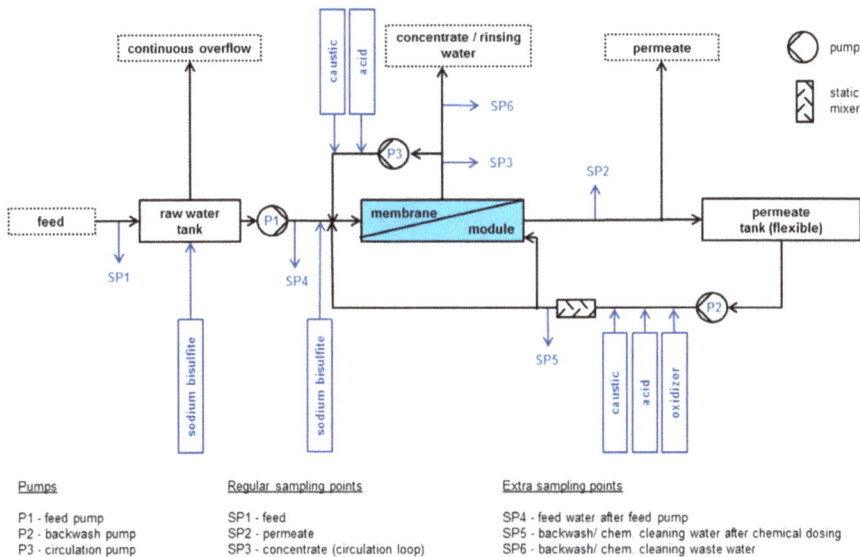

Figure 1. Settings and sampling points of the pilot plant.

One challenge for the treatment of anoxic source water is the operation under anoxic process conditions, as potential iron and manganese precipitation within the membrane material can result in irreversible loss of permeability. As permeate is used for hydraulic backwash cleaning after each filtration cycle, oxygen input needs to be prevented when tank levels change. Various measures were implemented to operate the capNF system under anoxic conditions: a flexible permeate tank (that prevents the entry of oxygen in permeate used for backwash), a continuously overflowing feed tank, oxidation-reduction potential (ORP) online measurements in feed, permeate and concentrate. Especially for removal of oxygen during start-up phase and after maintenance the option to dose sodium bi-sulphite to feed water was implemented. To prevent the influence of sodium bi-sulphite on water samples, the pilot plant was operated at least four days without sodium bi-sulphite before sampling. The plant is equipped with online measurements for operational parameters (flow, pressure, tank level, temperature) and for water quality parameter (conductivity, pH, ORP, color, turbidity, UV_{254}).

Different operational parameters were tested to investigate the efficiency of the membrane system including variation of:

- flux (15/22.5/27.5/30 L/m^2·h)
- recovery (rec) (50/75/85%)
- cross-flow velocity (cfv) (0.2/0.5/1.0 m/s)

2.3. Membrane

Based on the already existing capillary open NF-membrane HFW1000 (Pentair X-Flow, Enschede, The Netherlands) a new tighter membrane (HF-TNF) was developed and full-scale test modules were produced for the Berlin capNF-pilot of the AquaNES project. The new membrane was improved for high retention of sulphate and removal of specific OMPs (such as EDTA) by applying specific extra coatings using Layer-by-Layer (LbL) technology as demonstrated in the research project LbLBRANE (EC 7th Framework Programme, Grant Agreement no. 281047) [34]. The principle of LbL technology is shown in Figure 2.

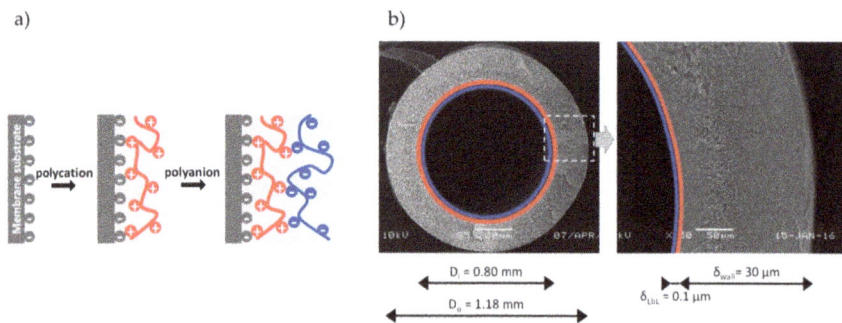

Figure 2. Principle of LbL technology: sequentially coating of a support membrane with charged polyelectrolytes (**a**), Microscopic pictures of a capillary fiber cross-cut showing coating layers including some characteristic dimensions (**b**) [35].

Both, the membrane fibers of the relatively open HFW1000 membrane and the HF-TNF, are manufactured from polyethersulfone (PES) and modified PES. The HF-TNF membrane has a negatively charged surface which ensures a strong binding with the positively charged polyelectrolytes to form coated layers. The MWCO of the HF-TNF is about 200–300 Da, estimated from retention measurements with organic substances with specific molecular weights such as polyethyleen glycol.

In total 11.376 capillary membrane fibers with an inner diameter of 0.8 mm and a length of 1.5 m are potted in a module resulting in a surface area of 40 m^2. The outside diameter of the module housing is 0.2 m. With the coated separation layer on the inside of the fibers the optimal filtration direction is inside-out which reduces the risk of abrasive wear. With outside-in filtration, the separating layer is on the outside and the membrane is vulnerable for outside influences such as the movement of fibers against each other causing damage to the separating layer. Moreover, with inside-out filtration concentration polarization can be controlled more precisely resulting in a better permeate quality.

The HF-TNF module can be operated with a maximum system pressure of 7.0 bar, a maximum TMP of 6.0 bar and a maximum backflush pressure of also 6.0 bar. The operation temperature can be 0–40 °C. During filtration a pH from 3–11 and during chemical cleaning a pH of 2–12 is possible. The membrane is also resistant to chlorine, allowing the application of 200 mg/L NaOCl for cleaning.

2.4. Online Data and Water Analysis

Online data (pressure, flow rate, level, pH, ORP, T, UV_{254}, color, turbidity) are logged every 20 s. Details on measured parameters, measuring locations and details of probes can be found in appendix, Table A2. In addition, several calculated values were monitored and evaluated such as salt retention, recovery, cross-flow velocity, permeability, resistance, transmembrane pressure (TMP), pressure drop and flux. Online data were analyzed using a software tool [36] that was developed within AquaNES with the free software R [37].

Regular sampling of feed, concentrate and permeate water was carried out every second week, during intensive sampling two times per week and complemented with occasional sampling of wastewater streams. Values for pH, temperature, ORP and oxygen were determined weekly using the portable multi-parameter probe SmarTROLL™ MP from In-Situ. All other chemical parameters were analyzed in the laboratory of Berliner Wasserbetriebe, including sulphate, DOC, iron species, manganese species, UV_{254}, color (436 nm), total hardness, Ca, Mg and conductivity. Furthermore, selected organic micropollutants were analyzed, e.g., gabapentin, acesulfame, vinyl chloride, EDTA, methyl tertiary butyl ether (MTBE), tertiary butyl alcohol (TBA), carbamazepine-10,11-trans dihydrodiol (CBZD), valsartan acid, formylaminoantipyrine (FAA) and phenylethyl-malonamide (PEMA)). All parameters, measuring methods and limits of quantification (LOQ) can be found in appendix, Table A3. For results <LOQ a value of $\frac{1}{2}$ LOQ was used for calculations and figures.

2.5. Calculation of Energy Consumption

Calculation of the energy consumption was performed according to Sethi et al. [38]. The main part of energy is consumed by pumps (feed pump, circulation pump, backwash pump). The energy demand of the feed pump (E_{filt}) during filtration is calculated as follows:

$$E_{filt} = \frac{p_{filt} \times \dot{V}_{filt}}{\eta_{filt}} \times t_{filt} \tag{1}$$

where p_{filt} is the pressure downstream of the feed pump during filtration; $\dot{V}_{filt}$ is the volumetric filtration flow rate, η_{filt} is the efficiency of feed pump during filtration and t_{filt} is the filtration time. An efficiency of 0.7 was estimated for all pumps. For calculation of the energy demand of the circulation pump the axial pressure drop over the module length (Δp_m) is used:

$$E_c = \frac{\Delta p_m \times \dot{V}_c}{\eta_c} \times t_{filt} \tag{2}$$

with:

$$\dot{V}_c = \frac{D_{cap} \times U_0 \times A_m}{4 \times L_{cap}} - \dot{V}_{filt} \tag{3}$$

where U_0 is the average cfv at the entrance of a capillary; D_{cap} is the diameter of the capillary and L_{cap} the length of a capillary; A_m is the total membrane surface and η_c is the efficiency of circulation pump during filtration.

The energy demand of the backwash and feed pumps was calculated according to Equation (1):

$$E_{bw} = \frac{p_{bw} \times \dot{V}_{bw}}{\eta_{bw}} \times t_{bw} \tag{4}$$

$$E_{ff} = \frac{p_{ff} \times \dot{V}_{ff}}{\eta_{ff}} \times t_{ff} \tag{5}$$

where p_{bw} is the pressure downstream the backwash pump; $\dot{V}_{bw}$ is the volumetric flow rate during backwash and η_{bw} is the efficiency of the backwash pump; p_{ff} is the pressure downstream of the feed pump during forward flush; $\dot{V}_{ff}$ is the volumetric flow rate of forward flush and η_{ff} is the efficiency of feed pump during forward flush. t_{bw} and t_{ff} are the times for backwash and forward flush, respectively.

The specific energy consumption per m^3 produced permeate was calculated as following:

$$Spec\ E_{total} = \frac{E_{filt} + E_c + E_{bw} + E_{ff}}{\dot{V}_{perm} \times t_{filt}} \tag{6}$$

where $\dot{V}_{perm}$ is the volumetric flow rate of permeate.

2.6. Chemical Cleaning

For the membrane cleaning, hydraulic and chemical procedures are available. The hydraulic cleaning consists of forward flush (with feed water), backwash (with permeate) and flush with permeate; backwash and forward flush can be carried out at the same time. The filtration cycle during the main trial period was one hour followed by a hydraulic cleaning. For optimizing of operation filtration cycles up to three days were tested.

The chemical cleaning was always started with a hydraulic cleaning. Chemicals were flushed in together with permeate and circulated on the feed side of the membrane. After circulation the chemicals were flushed out with permeate, followed by another hydraulic cleaning. Different chemicals were tested for cleaning: hydrochloric acid, sodium hydroxide and ascorbic acid, all chemicals and concentrations can be found in appendix Table A5. The cleaning interval was varied as well as the chemical concentrations, temperature and soaking/circulation time. To evaluate the success of the cleaning, online data for permeability, TMP and pressure drop of the membrane were recorded and analyzed.

3. Results and Discussion

3.1. Operational Parameters and Energy Consumption

During the first phase of operation with drinking water as feed different fluxes were tested (see Section 2.2). As a flux of 30 L/m^2·h exceeded the maximum allowed TMP for this membrane (5.5 bar), the highest flux applied was reduced to 27.5 L/m^2 h.

In Figure 3, the specific energy consumption for operation of the capNF system is shown for different operational phases with varying flux, recovery and cross-flow velocity. Main energy consumption is caused by the feed and circulation pumps, whereas the cleaning only had a minor influence. In Figure 4 the salt retention is displayed, calculated from the conductivity of feed water and permeate.

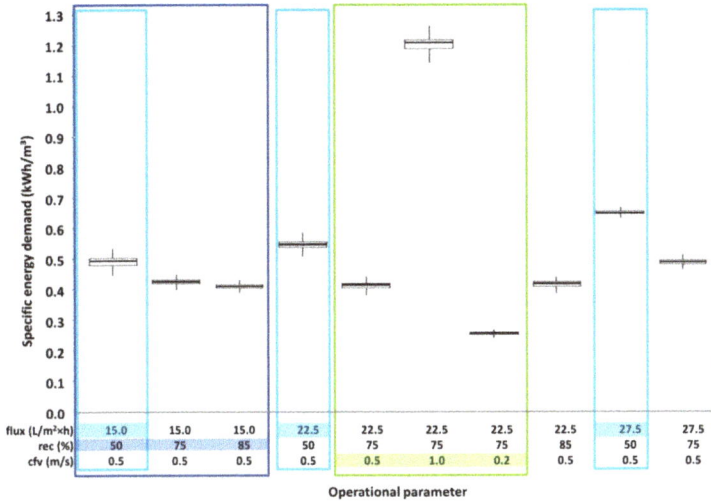

Figure 3. Specific energy demand (kWh/m³ permeate) for different operational parameter (drinking water as feed source), light blue: increase of flux at constant rec and cfv, dark blue: increase of recovery at constant flux and cfv, green: increase of cfv at constant flux and rec.

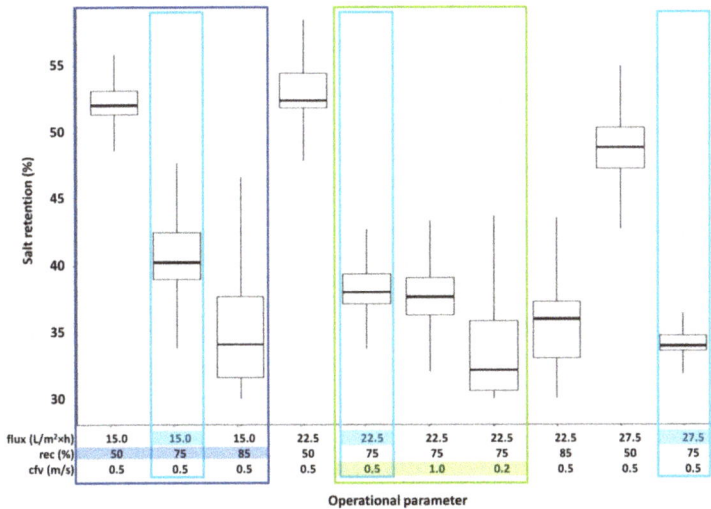

Figure 4. Salt retention (%) for different operational parameter (drinking water as feed source), light blue: increase of flux at constant rec and cfv, dark blue: increase of recovery at constant flux and cfv, green: increase of cfv at constant flux and rec.

Increase of flux from 15.0 L/m²·h to 22.5 L/m²·h (at constant rec and cfv, see light blue boxes in Figure 3) resulted in an only minor increase of the specific energy consumption, whereas further increase to 27.5 L/m²·h resulted in an increase by 31%. As salt retention only decreased slightly (5–15%, see Figure 4) for increasing flux, a flux of 22.5 L/m²·h was selected for long term operation.

Increase of recovery from 50% to 75% (at constant flux and cfv, see dark blue box in Figure 3) resulted in 23% lower specific energy consumption, with not much change with further increase to 85%. Salt retention decreased by about 23–35% for increasing recovery (Figure 4). Therefore, a recovery setting of 75% was preferred for long term operation.

Increase of cfv from 0.5 to 1.0 m/s (at constant flux and rec, see green box in Figure 3) resulted in an increase of specific energy consumption by almost 200% with similar salt retention. With a lower cfv of 0.2 m/s the energy consumption decreased by about 39% compared to 0.5 m/s, but with decreasing salt retention by about 16% (Figure 4). As a higher fouling rate is expected with a low cfv, a cfv setting of 0.5 m/s was selected for further operation.

Salt retention is automatically calculated using the online conductivity measurement values in feed and permeate. The behavior during flux variation is not consistent. Increase of flux from 15.0 L/m²·h to 22.5 L/m²·h and further to 27.5 L/m²·h (at constant rec and cfv, see light blue boxes in Figure 4) for a recovery of 75% resulted in a slight decrease of salt retention (2–4%), whereas for recovery of 50% and 85% no decrease or rather very slight increase of salt retention (0.5–2%) with the increase of flux from 15.0 L/m²·h to 22.5 L/m²·h and with further increase to 27.5 L/m²·h in slight decrease (2.5%) was observed.

Increase of recovery (at constant flux and cfv, see dark blue box in Figure 4) resulted in a clear decreasing salt retention (18% in total).

Between cfv of 0.5 and 1.0 m/s (at constant flux and rec, see green box in Figure 4) only a marginal difference of 0.5% is visible, but with further decrease to 0.2 m/s also the salt retention decreases by about 5.5%.

Due to retention of dissolved salts an interface directly at the membrane surface with higher salt concentration is formed (concentration polarization). This causes an increased concentration gradient and therefore an increased permeation of salts; the thicker the interface the higher the concentration gradient. The thickness of interface increases with increasing flux, increasing recovery and decreasing cfv. These effects could be observed with the available data. The thickness can be decreased e.g., with an increase of cfv with increase of turbulence, with the effect of less concentration polarization and higher salt retention. In this case with cfv of 0.2–1.0 m/s only laminar flow conditions occur with calculated Reynolds numbers of about 160–800.

In summary, specific energy consumption increased with increasing flux and cfv and decreasing rec (see Figure 3). For salt retention, retention decreases with increasing flux and recovery and decreasing cross-flow velocity (see Figure 4). Same behavior was found for the retention of other compounds such as sulphate and DOC (not shown). But it must be noted: the additional energy consumption of capNF is in the same range as the total specific energy consumption for drinking water supply in Germany (0.51 kWh/m³ [39]).

The contribution of the hydraulic cleaning to the total energy consumption is very low (about 0.5%) compared to the feed and circulation pumps, which are the main energy consumers. Therefore, an increase of filtration time and consequential reduction of hydraulic cleaning only has a minor effect on the overall energy consumption. A potential for improving the energy consumption could be a further reduction of cfv, mainly values between 0.2 and 0.5 m/s should be tested, to get a sufficient retention together with low energy consumption. However, the increase of scaling or fouling on the feed side has to be observed with decreasing cfv. Further increase of recovery could also be an option to reduce the energy consumption of the filtration.

For the switch to well water sodium bi-sulphite was dosed to the feed water to remove all oxygen from pipes and membrane module. This procedure was also applied before and after every rebuilding, air integrity tests or other maintenance.

Operation with well water started with a low flux (15 L/m² h), low recovery (50%) and filtration time of 60 min to get the direct comparison with start conditions with drinking water. During this period no increase of TMP could be observed. Subsequently, the flux was increased to 22.5 L/m²·h and the recovery set to 75%. With these settings a rapid increase of TMP was observed, the chemical cleaning could not control this increase and the flux and recovery had to be set back to start conditions until a suitable cleaning concept was found (see Section 3.3). With the improved cleaning concept the operation was possible with a flux of 22.5 L/m²·h and recovery of 75% without problems.

As every backwash causes an interruption of production as well as a consumption of permeate with increasing amount of waste water, it was tested to increase the filtration time from 60 min to 24 h and further up to 72 h. It was observed that the increase of TMP was slower at 24 h and 72 h filtration time compared to 60 min, which was not expected. The filtration with these increased intervals was possible without any problems. The reason for the slower increase of TMP was assumed to be caused by lower entrance of oxygen via the permeate outlet. Although a flexible tank is installed to prevent the entrance of oxygen, a small amount of oxygen can enter the permeate side of the membrane with every backwash due to decreasing tank level. As a consequence, a higher amount of iron and manganese precipitation could be formed with increasing backwash frequency. However, a lower backwash frequency also results in longer stagnation of water in the permeate tank, which could lead to microbiological growth. To prevent this, an exchange with fresh permeate should be considered.

3.2. Retention of Compounds

In Table 1 selected parameters for feed water, permeate and concentrate are shown (see Table A1 in appendix for more detailed version). During the tests with well water the feed temperature was between 11 and 14 °C, the pH about 7, the ORP about 100 mV and the oxygen concentration below LOQ.

Table 1. Water quality feed, permeate, and concentrate; feed: well water.

Parameter	Unit	Feed			Permeate *			Concentrate *		
Temperature	°C	12.8	±0.92	(n = 60)	13.0	±0.83	(n = 51)	13.1	±0.83	(n = 33)
pH	–	7.0	±0.21	(n = 60)	7.0	±0.16	(n = 51)	7.1	±0.16	(n = 33)
ORP	mV	−100	±26	(n = 17)	-80	±25	(n = 16)	n.d. **		
Oxygen	mg/l	<LOQ		(n = 16)	0.2	±0.14	(n = 12)	n.d. **		
Conductivity	µS/cm	961	±80	(n = 60)	744	±101	(n = 51)	1549	±93	(n = 51)
Color$_{436nm}$	1/m	0.3	±0.00	(n = 5)	<LOQ		(n = 2)	1.0	±0.14	(n = 2)
UV$_{254}$	1/m	11.6	±0.13	(n = 5)	1.0	±0.14	(n = 2)	39.3	±3.18	(n = 2)
DOC	mg/l	4.8	±0.15	(n = 5)	0.6	±0.02	(n = 2)	15.5	±1.06	(n = 2)

* flux = 22.5 L/m^2 h, rec = 75%, cfv = 0.5 m/s; ** n.d.: not determined.

In Figure 5, feed and permeate concentrations for selected parameters are shown. High retentions were found for UV$_{254}$, DOC and sulphate: UV$_{254}$ could be reduced by 85–89%, DOC by 82–87% (four permeate samples <LOQ (<0.5 mg/L)) and sulphate by 67–71%. Medium retentions were determined for hardness (41–55%), calcium (40–54%) and magnesium (50–64%), as well as for iron species (48–49%) and manganese species (42%). Only minor retention was observed for conductivity (22–32%) since monovalent ions are hardly retained by capNF.

In Figure 6, feed and permeate concentrations for selected OMPs are displayed. High retention of 84–92% was found for EDTA, medium retention of 44–58% for valsartan acid (similar for Gabapentin and FAA) and only minor or no retention for MTBE (also for TBA, acesulfame and vinyl chloride).

For iopamidol (LOQ = 0.02 µg/L), candesartan (LOQ = 0.01 µg/L) and olmesartan (LOQ = 0.01 µg/L) the permeate concentration was always <LOQ. As the feed concentration was less than 5 times of LOQ, the retention was not calculated. For CBZD, PEMA; primidone, gaba-lactam and carbamazepine retention could not be determined too, as the feed and permeate concentrations were only slightly above LOQ.

In Table 2, feed concentrations and calculated retentions of selected parameters are compared for the sampling campaigns using drinking water and well water as feed. If the permeate concentration was below limit of quantification (LOQ), a value of $\frac{1}{2}$ LOQ was used for calculation of the retention. In this case only feed concentration values of 5 times LOQ were used for calculation, otherwise no retention was calculated to prevent a high calculation failure.

The commercial available module HFW 1000 was developed for removal of color and natural organic matter; the new membrane HF-TNF was developed for improved retention of sulphate and selected OMPs. A comparison of both membranes can be found in Table 3.

About the same retention of UV$_{254}$ was observed with a slightly higher retention of DOC. A much better retention is shown for bivalent ions such as hardness (Mg^{2+} and Ca^{2+}).

Using capNF, the sulphate retention of about 67% could be sufficient for decentralized treatment of water from wells with higher concentrations of sulphate without using high pressure technology.

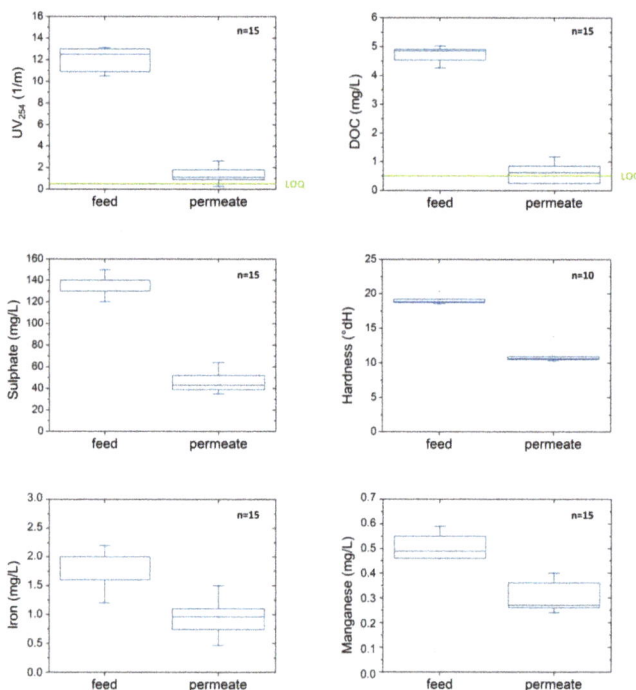

Figure 5. Feed and permeate concentrations of selected compounds; well water as feed; for permeate <LOQ the $\frac{1}{2}$ LOQ was used.

Table 2. Feed concentrations (average and standard deviations) and retentions (average and standard deviations) of selected parameters during drinking water and well water period (for permeate <LOQ the $\frac{1}{2}$ LOQ was used).

Parameter	Unit	Feed: Drinking Water [1]				Feed: Well Water [2]			
		Feed		Retention (%)		Feed		Retention (%)	
UV_{254}	1/m	8.7	±0.23	85	±6.5	(n = 9)	12.1	±1.01	89	±5.5	(n = 15)
DOC	mg/L	3.8	±0.21	82	±7.8	(n = 11)	4.7	±0.24	87	±6.4	(n = 15)
Sulphate	mg/L	149.1	±9.44	71	±8.5	(n = 11)	136.0	±7.37	67	±5.5	(n = 16)
Mg	mg/L	10.7	±0.43	64	±13.8	(n = 5)	11.7	±0.49	50	±6.0	(n = 10)
Hardness	°dH	20.5	±0.66	55	±14.7	(n = 5)	18.8	±0.51	41	±5.2	(n = 10)
Ca	mg/L	128.6	±4.93	54	±14.8	(n = 5)	115.1	±3.16	40	±5.0	(n =10)
Fe_{total}	mg/L	<LOQ		–		(n = 11)	1.8	±0.29	48	±8.7	(n = 15)
Fe^{2+}	mg/L	<LOQ		–		(n = 11)	1.6	±0.28	48	±7.7	(n = 15)
$Fe_{dissolved}$	mg/L	<LOQ		–		(n = 11)	1.8	±0.31	49	±8.0	(n = 14)
Mn_{total}	mg/L	<LOQ		–		(n = 11)	0.5	±0.05	42	±5.8	(n = 15)
$Mn_{dissolved}$	mg/L	<LOQ		–		(n = 11)	0.5	±0.05	42	±6.0	(n = 15)
Conductivity	µS/cm	948.0	±71.0	32	±5.8	(n = 11)	970.0	±78.3	22	±4.3	(n = 40)
Color	1/m	0.2	±0.04	–		(n = 5) [4]	0.3	±0.05	–		(n = 15) [4]
EDTA	µg/L	11.0	±0.00	84	±10.3	(n = 2)	12.3	±3.73	92	±1.1	(n = 5)
Gabapentin	µg/L	0.1	±0.04	55	±12.1	(n = 4)	0.05	±0.03	46	±10.6	(n = 8) [3]
Valsartan acid	µg/L	0.2	±0.06	58	±11.5	(n = 4)	0.1	±0.08	44	±10.2	(n = 8)
FAA	µg/L	0.05	±0.01	57	±13.9	(n = 4)	0.1	±0.01	41	±7.3	(n = 8) [3]
MTBE	µg/L	0.6	±0.32	14	±13.6	(n = 3)	9.2	±4.04	10	±9.2	(n = 10)
TBA	µg/L	<LOQ		–		(n = 3)	5.4	±2.87	8	±6.7	(n = 10)
Acesulfame	µg/L	0.7	±0.16	24	±9.5	(n = 4)	0.7	±0.08	−5	±2.1	(n = 8)
Vinyl chloride	µg/L	<LOQ		–		(n = 4)	1.6	±0.47	−13	±14.1	(n = 10)

[1] Variation of operational parameters (flux = 15.0–27.5 L/m² h, rec = 50–85%, cfv = 0.2–1.0 m/s); [2] No variation of operational parameters (flux = 22.5 L/m² h, rec = 75%, cfv = 0.5 m/s); [3] Feed and permeate values near LOQ; [4] Feed values near LOQ, permeate values <LOQ.

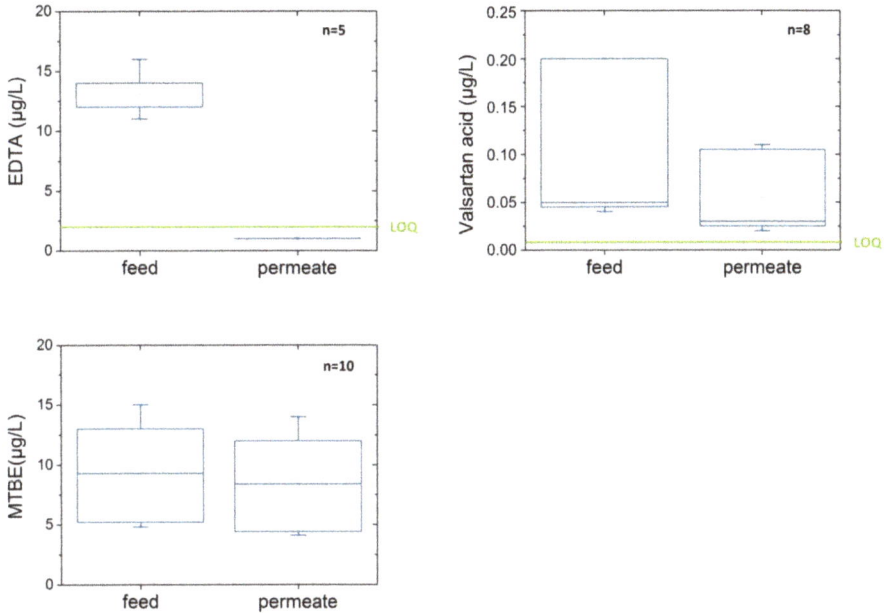

Figure 6. Feed and permeate concentrations of selected OMP; well water as feed; for permeate <LOQ the $\frac{1}{2}$ LOQ was used.

Table 3. Comparison of commercial available module HFW 1000 with new developed membrane HF-TNF using literature data [40,41] and data from pilot tests.

Parameter	Unit	HFW 1000	HF-TNF
MWCO	Da	~1000	~200–300
Flux	L/(m²·h)	10–20	22.5
Permeability	L/(m²·h·bar)	11.0–13.6	6.0–7.7
Resistance × 10¹³	1/m	2.65–3.23	4.6–6.3
TMP	bar	1.1–2.8	3.6–4.6
Pressure drop	bar	0.5–0.7	0.35–0.50
Sulphate removal	%	n.d. *	67 ± 5.5
DOC removal	%	70–80	87 ± 6.4
UV₂₅₄ removal	%	80–90	89 ± 5.5
Hardness removal	%	<20	41 ± 5.2

* n.d.: not determined; MWCO: molecular weight cut-off; TMP: transmembrane pressure; DOC: dissolved organic carbon.

For OMP, the retention depends on size, structure and charge of the molecule. The MWCO for the new HF-TNF membrane is 200–300 Da. For molecules with higher molecular weight (MW) such as EDTA (MW: 292.24 g/mol), FAA (231.251 g/mol), gabapentin (171.24 g/mol) and valsartan acid (266.08 g/mol) a medium/high retention was expected and could be confirmed by the laboratory results. Molecules with a lower MW such as MTBE (88.15 g/mol), TBA (74.12 g/mol) and vinyl chloride (62.50 g/mol) showed no or only minor retention, which was also expected. The weight of the molecules iopamidol (MW: 777.08 g/mol), candesartan (MW: 440.45/610.66 g/mol) and olmesartan (558.585 g/mol) is clearly above the MWCO of the membrane; a medium/high retention could be expected but could not be calculated due to low feed concentrations. Nevertheless, the permeate concentrations are always below LOQ. For CBZD (MW: 270.288 g/mol), PEMA

(MW: 206.24 g/mol); primidone (MW: 218.252 g/mol), gaba-lactam (MW: 153.225 g/mol) and carbamazepine (MW: 236.27 g/mol) the feed concentration is slightly >LOQ too, but retention could hardly or not be detected. The weight of all molecules is in the range or below the MWCO; therefore medium/low or no retention was expected but could also not be shown with these low feed concentrations. The MW and the structure of OMP can be found in appendix, Table A4.

The measured MTBE and TBA pollutions in well water of the water works are caused by a former fuel depot in an industrial area. The derived no-effect level (DNEL) for ground water for MTBE of 5 µg/L was defined by LAWA [42,43], only single wells show an elevated concentration; in drinking water the measured values are much lower. For TBA, a DNEL for ground water of 2 mg/L was defined by UBA [24,43], the measured values are far below DNEL. The pollution by vinyl chloride is caused by volatile chlorinated hydrocarbon contaminant plumes from a legacy from an industrial area. Because of carcinogenicity the limit value for drinking water is 0.5 µg/L [10]. Vinyl chloride is volatile and completely removed by common aeration of raw water [24,43].

For iron a reduction of 48% was observed, for manganese a reduction of 42%. In the present case the permeate concentrations are 0.97 ± 0.29 mg/L iron and 0.29 ± 0.06 mg/L manganese, this means a further treatment step is necessary to keep the limits for drinking water (iron: <0.2 mg/L; manganese <0.05 mg/L) [10]. Permeate could be fed back to the common drinking water treatment process of aeration and filtration to remove iron and manganese residues.

3.3. Hydraulical and Chemical Cleaning

Using drinking water as feed, no increase of TMP could be observed during all tested operational settings. Nevertheless, after each filtration cycle of 60 min a backwash was executed. Chemical cleaning was carried out before each change of operational parameters to set back the membrane starting conditions for the next test.

With well water as feed source and a flux of 15 L/m²·h, recovery of 50% and cross-flow velocity of 0.5 m/s, only a slight increase of TMP of about 0.15 bar during the first days was visible, therefore preventive weekly chemical cleaning was carried out. With a higher flux of 22.5 L/m²·h and higher recovery of 75% the TMP increased rapidly, which indicates a high fouling potential of the feed water. With hydraulic cleaning and chemical cleaning with HCl and NaOH it was not possible to restore the starting TMP. As tests with higher chemical concentrations and higher temperature during the cleaning process did not show any success, HCl was exchanged with ascorbic acid which resulted in a successful restoration of the TMP to start conditions (see Figure 7).

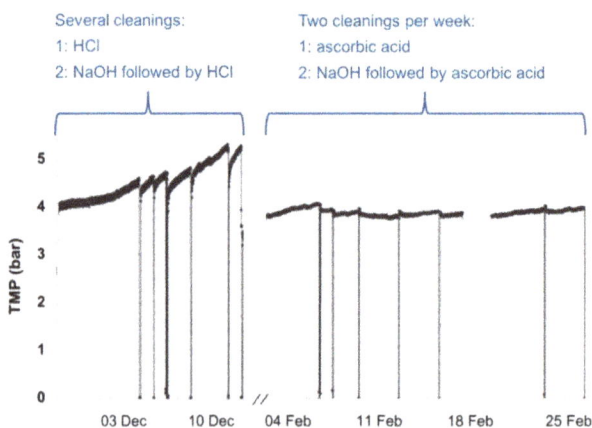

Figure 7. TMP trend and cleaning strategies; well water as feed water source, flux 22.5 L/m² h, rec 75%, cfv 0.5 m/s. each vertical line marks a chemical cleaning.

A second acid cleaning after caustic cleaning is necessary to prevent high pH values and precipitation of feed water compounds during the following first filtration cycle. With adaptation of the control software the use of two kinds of acid for the chemical cleaning process was possible. To reduce the consumption of the more expensive ascorbic acid, the second acid cleaning to lower the pH inside the module was carried out with HCl.

Main fouling was expected from precipitation of iron and manganese due to dissolved iron and manganese concentrations in well water. The common used cleaning concept with HCl and NaOH was not successful for well water as feed water. With the change from HCl to ascorbic acid for chemical cleaning the initial TMP could be restored. The reductive dissolution of iron from fouled membranes using ascorbic acid was already described in former studies [44,45].

However, with the use of ascorbic acid the amount and costs for chemical cleaning increase. The goal, to reduce the amount or replace ascorbic acid, is still in progress. As first successful step the second acid cleaning to lower the pH is now carried out with HCl instead of ascorbic acid, resulting in a reduction of ascorbic acid consumption by 50%. Tests with reduced ascorbic acid amounts and adjustment of pH with HCl during the first acid cleaning were also carried out, here first positive results could be observed but longer test duration is necessary. In further trials e.g., citric acid could be tested as cheaper alternative, but was found to be less effective in other studies [45]. Also, oxalic acid could be another substitute because of its reducing properties.

The backwash showed no additional positive effect on membrane performance, a reduction of backwash intervals was possible without any problems. The disadvantage of oxygen introduction to the backwash process is caused by insufficient exclusion from ambient air at the pilot plant. As each hydraulic cleaning cause an interruption of the production process and consumption of permeate, the increase of filtration time is recommended.

3.4. Online Probes

Several online probes are installed in feed, circulation, concentrate and permeate pipes. The measuring values are regularly compared with manual measurements and lab results. The s::can probes for turbidity, UV_{254} and colour showed iron/manganese fouling in bypass hoses and on measuring window. A regular (weekly) cleaning with HCl (5%) was necessary to produce usable results. The Endress + Hauser probes for conductivity (conductive measuring principle) showed no stable measuring values after switch to operation with well water, here also iron/manganese precipitation were found. After cleaning of probes with HCl (5%) the measured values decrease within a few hours, the evaluation of this data was not possible. It was decided to install new probes with inductive measuring principle (Endress + Hauser, Indumax CLS50D); this allows measurement without media contact. After installation of the new probes a stable measuring value was observed for conductivity with much less cleaning effort.

3.5. Concentrate Discharge

During operation of capNF about 25% of concentrate will be discarded continuously. The drainage or treatment of such waste water is a significant problem and cost factor in membrane processes. Different possibilities exist for handling of concentrate: in most cases it is discharged to the receiving water body, sporadically discharged to the sewer for further treatment [46]. However, other technologies for zero liquid discharge (ZLD) are currently under investigation. ZLD is a concept to avoid liquid waste in membrane processes with the possibility to reuse water and salts. In [47] thermal processes for concentrate (brine) treatment are described as not economical because of high investment costs and high energy consumption. To reduce costs, membrane processes for post-treatment of brine are tested to reach higher brine concentrations and thus reduce liquid waste prior thermal treatment. One challenge is the presence of compounds which cause fouling and scaling during membrane processes. These compounds must be removed via pre-treatment, depending on brine source and typical contaminants, promising technologies include chemical precipitation or coagulation, electrocoagulation, ion exchange or adsorption [47].

4. Conclusions

The results of the tests have shown that the operation of capillary nanofiltration under anoxic conditions is possible at a flux up to 22.5 L/m^2·h and a recovery up to 75%. Pre-treatment via bank filtration resulted in low organic fouling. The fouling caused by precipitation of iron and manganese could be removed with an optimized operation and cleaning concept using effective chemicals.

With the new developed HF-TNF membrane module good retentions of DOC (82–87%), sulphate (67–71%) and hardness (41–55%) could be demonstrated. In addition, micropollutants were retained depending on size and charge of the molecule, e.g., EDTA (84–92%).

Results indicate that a stable long-term operation of the membrane system is possible. With a filtration time of 24 h and backwash with forward flush after each cycle the operation was stable. A two-weekly chemical cleaning with ascorbic acid/HCl and caustic led to an effective reduction of slowly increasing TMP.

Author Contributions: Conceptualization, M.U. and S.A.; Methodology, V.L., M.U. and W.D.; Validation, J.J. and V.L.; Formal Analysis, V.L. and J.J.; Investigation, J.J. and V.L.; Data Curation, J.J.; Writing-Original Draft Preparation, J.J. and V.L.; Writing-Review & Editing, W.D., S.A. and M.U.; Visualization, J.J.; Supervision, M.U.; Project Administration, W.D.; Funding Acquisition, M.U. and S.A.

Funding: The AquaNES project has received funding from the European Union's Horizon 2020 research and innovation programme under grant agreement no. 689450.

Acknowledgments: We thank all the colleagues of Berliner Wasserbetriebe for providing the infrastructure, supporting the installation of the plant and carrying out the lab analysis. The operation of the pilot plant would not have been possible without the following persons: Henry Hamberg, Michael Rustler, Jan Schütz, Timo Hoff, Victoire Schellenberg, Charlotte Rohde and Carolin Flöter.

Conflicts of Interest: The authors declare no conflict of interest. The funders had no role in the design of the study; in the collection, analyses, or interpretation of data; in the writing of the manuscript, and in the decision to publish the results.

Abbreviations

capNF	capillary nanofiltration
CBZD	Carbamazepine-10,11-trans dihydrodiol
cfv	cross-flow velocity
DNEL	derived no-effect level
DOC	dissolved organic carbon
EDTA	Ethylendiamintetraacetic acid
FAA	Formylaminoantipyrine
LbL	Layer-by-Layer
LOQ	limit of quantification
MTBE	Methyl tertiary butyl ether
MW	molecular weight
MWCO	molecular weight cut-off
NF	nanofiltration
OMP	organic micropollutants
ORP	oxidation-reduction potential
PEMA	Phenylethyl-malonamide
PES	polyethersulfone
rec	recovery
TBA	Tertiary butyl alcohol
TMP	transmembrane pressure
UV$_{254}$	absorption of ultraviolet light (wave length: 254 nm)
ZLD	zero liquid discharge

Appendix A

Table A1. Water quality feed, permeate, and concentrate; feed: well water.

Parameter	Unit	Feed		Permeate				Concentrate			
				(a)		(b)		(a)		(b)	
Temperature	°C	12.8 [4]	±0.92	12.0 [2]	±0.12	13.0 [5]	±0.83	11.9 [2]	±0.15	13.1 [6]	±0.83
pH	–	7.0 [4]	±0.21	7.6 [2]	±0.31	7.0 [5]	±0.16	7.7 [2]	±0.34	7.1 [6]	±0.16
ORP	mV	−100 [8]	±26	–	–	−80 [9]	±25	–	–	–	–
Oxygen	mg/L	<LOQ		–	–	0.2 [7]	±0.14	–	–	–	–
Conductivity	μS/cm	961 [4]	±80	932 [2]	± 19	744 [5]	±101	1162 [2]	±38	1549 [5]	±93
Color$_{436nm}$	1/m	0.3 [1]	±0.00	<LOQ [2]		<LOQ [3]		0.5 [2]	±0.00	1.0 [3]	±0.14
UV$_{254}$	1/m	11.6 [1]	±0.13	4.3 [2]	±0.26	1.0 [3]	±0.14	19.0 [2]	±1.42	39.3 [3]	±3.18
DOC	mg/L	4.8 [1]	±0.15	1.9 [2]	±0.10	0.6 [3]	±0.02	7.4 [2]	±0.43	15.5 [3]	±1.06
Fe $_{total}$	mg/L	1.8 [1]	±0.23	1.6 [2]	±0.17	0.6 [3]	±0.21	2.6 [2]	±0.35	3.3 [3]	±1.70
Mn $_{total}$	mg/L	0.5 [1]	±0.02	0.4 [2]	±0.05	0.2 [3]	±0.00	0.6 [2]	±0.08	1.2 [3]	±0.14
Sulphate	mg/L	131 [1]	±0.92	73 [2]	±3.06	36 [3]	±0.71	200 [2]	±0.00	395 [3]	±7.07

(a) flux = 15 L/m^2 h, rec = 50%, cfv = 0.5 m/s; (b) flux = 22.5 L/m^2 h, rec = 75%, cfv = 0.5 m/s; [1] n = 5; [2] n = 3; [3] n = 2; [4] n = 60; [5] n = 51; [6] n = 33; [7] n = 12; [8] n = 17; [9] n = 16. ORP: oxidation-reduction potential; DOC: dissolved organic carbon.

Table A2. Online measuring probes for operational and water quality parameter; unit, measuring place, measuring probe, measuring method.

Parameter	Unit	Measuring Place	Measuring Probe	Measuring Method
Level	%	feed tank,	Endress + Hauser Liquiphant M	submerged gauge pressure sensor
		permeate tank	Endress + Hauser Prosonic M	reflection ultrasonic pulses
T	°C	feed	Endress + Hauser Omnigrad M TR10	resistance thermometer
Flow	m^3/h	feed, concentrate, circulation, backwash	Endress + Hauser Promag 10P25	magnetic induction
Pressure	bar	feed before/after restriction, feed, permeate, concentrate, backwash before/after restriction	Endress + Hauser Cerabar M	pressure-dependent change in capacitance
pH	-	feed, circulation	Endress + Hauser Memosens CPS16D	glass electrode with Ag/AgCl reference
ORP	mV	feed, permeate, concentrate (circulation)		Pt electrode with Ag/AgCl reference
T	°C	circulation		NTC 30kΩ
Conductivity	μS/cm	feed, permeate, concentrate (circulation)	Endress+Hauser Condumax CLS21D (old) Indumax CLS50D (new)	resistance between 2 electrodes (old) digital inductive (new)
Turbidity UV$_{254}$	NTU 1/m	feed, permeate	i::scan V1 Y04	ISO7027/EPA 180.1 multi-wavelength photometer with narrow band light source

Table A3. Analyzed parameter (manual and laboratory), unit, limit of quantification (LOQ), measuring method.

Parameter	Unit	LOQ	Measuring Method
Calcium	mg/L	0.8	DIN EN ISO 11885 (E22)
Color$_{436nm}$	1/m	0.2	DIN EN ISO 7887 (C01)
Conductivity	μS/cm	-	DIN EN 27888 (C08)
DOC	mg/L	0.5	DIN EN 1484 (H03)
Fe^{2+}	mg/L	0.03	DIN 38406-E01
Fe$_{dissolved}$	mg/L	0.03	DIN EN ISO 11885 (E22)
Fe$_{total}$	mg/L	0.03	DIN EN ISO 11885 (E22)
Hardness	°dH	-	DIN 38409-H06
Magnesium	mg/L	0.1	DIN EN ISO 11885 (E22)
Mn$_{dissolved}$	mg/L	0.01	DIN EN ISO 11885 (E22)
Mn$_{total}$	mg/L	0.01	DIN EN ISO 11885 (E22)
Sulphate	mg/L	6.0	DIN EN ISO 10304-1 (D20)
UV$_{254}$	1/m	0.5	DIN 38404-C03
Acid capacity	mmol/L	0.02	DIN 38409-H07-1/2
Base capacity	mmol/L	0.02	DIN 38409-H07-2
Calcite dissolving capacity	mmol/L	-	DIN 38404-C10-R3
Chloride	mg/L	5.0	DIN EN ISO 10304-1 (D20)
Hydrogen carbonate	mg/L	-	calculated using DIN 38409-7
Nitrate	mg/L	0.2	DIN EN ISO 10304-1 (D20)
Nitrate-N	mg/L	0.05	DIN EN ISO 10304-1 (D20)

Table A3. *Cont.*

Parameter	Unit	LOQ	Measuring Method
Potassium	mg/L	0.02	DIN EN ISO 11885 (E22)
Sodium	mg/L	0.5	DIN EN ISO 11885 (E22)
Acesulfame	µg/L	0.1	DIN 38407-F47
Candesartan	µg/L	0.01	DIN 38407-F47
Carbamazepine	µg/L	0.01	DIN 38407-F47
CBZD	µg/L	0.02	DIN 38407-F47
EDTA	µg/L	2.0	DIN EN ISO 16588 (P10)
FAA	µg/L	0.01	DIN 38407-F47
Gaba-lactam	µg/L	0.01	DIN 38407-F47
Gabapentin	µg/L	0.01	DIN 38407-F47
Iopamidol	µg/L	0.02	DIN 38407-F36
MTBE	µg/L	0.03	DIN 38407-F43
Olmesartan	µg/L	0.01	DIN 38407-F47
PEMA	µg/L	0.01	DIN 38407-F47
Primidone	µg/L	0.01	DIN 38407-F47
TBA	µg/L	1.0	DIN 38407-F43
Valsartan acid	µg/L	0.01	DIN 38407-F47
Vinyl chloride	µg/L	0.1	DIN EN ISO 10301 (F04)
pH	-	0.01	Std. Methods 4500-H+ EPA 150.2 (SmarTROLL™)
T	°C	−5.0	EPA 170.1 (SmarTROLL™)
ORP	mV	±1400	Std. Methods 2580 (SmarTROLL™)
Conductivity	µS/cm	5.0	Std. Methods 2510 EPA 120.1 (SmarTROLL™)
Oxygen	mg/L	0.1	EPA-approved In-Situ Methods 1002-8-2009 1003-8-2009 1004-8-2009 (SmarTROLL™)

Table A4. Analyzed organic micropollutants, molecular weight, structure.

Parameter	Molecular Weight (g/mol) [1]	Structure [1]	Parameter	Molecular Weight (g/mol) [1]	Structure [1]
Vinyl chloride	62.50		TBA	74.12	
MTBE	88.15		Gabalactam	153.225	
Gabapentin	171.24		Acesulfame	201.24	
PEMA	206.24		Primidone	218.252	
FAA	231.25		Carba-mazepine	236.27	
Valsartan acid	266.08		CBZD	270.29	

Table A4. *Cont.*

Parameter	Molecular Weight (g/mol) [1]	Structure [1]	Parameter	Molecular Weight (g/mol) [1]	Structure [1]
EDTA	292.24		Cande-sartan	440.45 (candesartan) 610.66 (candesartan- cilexetil)	
Olmesartan	446.51		Iopamidol	777.08	

[1] all data and pictures from [48].

Table A5. Cleaning chemicals.

Chemical	Concentration (wt. %)	Active stock Concentration (g/L)
HCl	25	280
NaOH	35	483
Ascorbic acid	20	200
NaHSO$_3$	39	522.6

References

1. Ascott, M.J.; Lapworth, D.J.; Gooddy, D.C.; Sage, R.C.; Karapanos, I. Impacts of extreme flooding on riverbank filtration water quality. *Sci. Total Environ.* **2016**, *554–555*, 89–101. [CrossRef] [PubMed]
2. Przybyłek, J.; Dragon, K.; Kaczmarek, P.M.J. Hydrogeological investigations of river bed clogging at a river bank filtration site along the River Warta, Poland. *Geologos* **2017**, *23*, 201–214. [CrossRef]
3. Kaczmarek, P.M.J. Hydraulic conductivity changes in river valley sediments caused by river bank filtration—An analysis of specific well capacity. *Geologos* **2017**, *23*, 123–129. [CrossRef]
4. Uhlmann, W.; Zimmermann, K. *Fallanalyse der Sulfatbelastung in der Spree 2014/2015*; Senatsverwaltung für Stadtentwicklung und Umwelt Berlin: Berlin, Germany, 2015.
5. SenStadtUm Berlin and MWE Brandenburg. *Sulfatgespräche der Länder Berlin und Brandenburg—Aktueller Sachstand und Maßnahmen zur Beherrschung der Bergbaulich Bedingten Stoffeinträge*; Senatsverwaltung für Stadtentwicklung und Umwelt des Landes Berlin; Ministerium für Wirtschaft und Energie des Landes Brandenburg: Berlin, Germany, 2016.
6. Galloway, J.N.; Aber, J.D.; Erisman, J.W.; Seitzinger, S.P.; Howarth, R.W.; Cowling, E.B.; Cosby, B.J. The Nitrogen Cascade. *BioScience* **2003**, *53*, 341–356. [CrossRef]
7. Heberer, T. Occurrence, fate, and removal of pharmaceutical residues in the aquatic environment: A review of recent research data. *Toxicol. Lett.* **2002**, *131*, 5–17. [CrossRef]
8. Destatis. *Trinkwasser Wird Überwiegend aus Grundwasser Gewonnen*; Statistisches Bundesamt Pressestelle: Wiesbaden, Germany, 2013.
9. SenUVK. *Grundwasser in BerlinVorkommen Nutzung Schutz Gefährdung*; Senatsverwaltung für Umwelt, Verkehr und Klimaschutz: Berlin, Germany, 2007.
10. Bundesgesundheitsamt. Verordnung über die Qualität von Wasser für den Menschlichen Gebrauch (Trinkwasserverordnung-TrinkwV). Available online: http://www.gesetze-im-internet.de/trinkwv_2001/BJNR095910001.html#BJNR095910001BJNG000201310 (accessed on 12 September 2018).
11. SenStadt Berlin. Qualität des oberflächennahen Grundwassers, Senat für Stadtentwicklung und Umwelt, Umweltatlas. Available online: www.stadtentwicklung.berlin.de/umwelt/umweltatlas/ka204.htm (accessed on 8 April 2018).
12. SenStadtUm. *Das Grundwasser in Berlin—Bedeutung, Probleme, Sanierungskonzeptionen*; Senatsverwaltung für Stadtentwicklung und Umwelt des Landes Berlin: Berlin, Germany, 1986.

13. Wurl, J. *Die geologischen, hydraulischen und hydrochemischen Verhältnisse in den südwestlichen Stadtbezirken von Berlin*; Fachbereich Geowiss., FU: Berlin, Germany, 1995; 3-89582-003-2 978-3-89582-003-8.

14. Schmidt, C.K.; Lange, F.T.; Brauch, H.-J.; Kühn, W. *Experiences with Riverbank Filtration and Infiltration in Germany*; DVGW-Water Technology Center (TZW): Karlsruhe, Germany, 2003.

15. Ziegler, D. *Untersuchung zur Nachhaltigen Wirkung der Uferfiltration im Wasserkreislauf Berlins*; Technische Universität Berlin: Berlin, Germany, 2001.

16. Darbi, A.; Viraraghavan, T.; Jin, Y.-C.; Braul, L.; Corkal, D. Sulfate Removal from Water. *Water Qual. Res. J. Can.* **2003**, *88*, 169–182. [CrossRef]

17. Lipp, P.; Gronki, T.; Lueke, J.; Lanfervoss, A.; Baldauf, G. Sulphate Removal from Ground Water—A Case Study. *gwf-Wasser/Abwasser* **2011**, *152*, 46–51.

18. Cao, W.; Dang, Z.; Zhou, X.-Q.; Yi, X.-Y.; Wu, P.-X.; Zhu, N.-W.; Lu, G.-N. Removal of sulphate from aqueous solution using modified rice straw: Preparation, characterization and adsorption performance. *Carbohydr. Polym.* **2011**, *85*, 571–577. [CrossRef]

19. Liu, H.; Li, F.; Zhang, G. Experimental Study on Adsorption Removal of Sulfate with Synthesized Zeolite Made from Fly Ash. In Proceedings of the 4th International Conference on Bioinformatics and Biomedical Engineering, Chengdu, China, 18–20 June 2010.

20. Rao, S.M.; Sridharan, A. Mechanism of Sulfate Adsorption by Kaolinite. *Clays Clay Miner.* **1984**, *32*, 414–418. [CrossRef]

21. Lindner, K.; Knepper, T.P.; Müller, J.; Karrenbrock, F.; Rörden, O.; Brauch, H.-J.; Sacher, F. *Entwicklung von Verfahren zur Bestimmung und Beurteilung der Trinkwassergängigkeit von Organischen Einzelstoffen*; Internationale Arbeitsgemeinschaft der Wasserwerke im Rheineinzugsgebiet: Köln, Germany, 2000.

22. Lindner, K.; Knepper, T.P.; Müller, J.; Karrenbrock, F.; Rörden, O.; Juchem, H.; Brauch, H.-J.; Sacher, F. *Bestimmung und Beurteilung der Mikrobiellen Abbaubarkeit von Organischen Einzelstoffen bei Umweltrelevanten Konzentrationen in Gewässern*; Internationale Arbeitsgemeinschaft der Wasserwerke im Rheineinzugsgebiet: Köln, Germany, 2003.

23. Wisotzky, F.; Jakschik, S.; Denzig, D. Herkunft und Dauer einer EDTA-Belastung im Grundwasser zweier Trinkwassereinzugsgebiete. *gwf-Wasser/Abwasser* **2010**, *3*, 278–284.

24. Krueger-Marondel, J. Senatsverwaltung für Umwelt, Verkehr und Klimaschutz, Berlin, Germany. Personal communication, 2018.

25. Samhaber, W.M. *Erfahrungen und Anwendungspotential der Nanofiltration*; Institut für Verfahrenstechnik, Johannes Kepler Universität Linz: Linz, Austria, 2006.

26. Cadotte, J.; Forester, R.; Kim, M.; Petersen, R.; Stocker, T. Nanofiltration membranes broaden the use of membrane separation technology. *Desalination* **1988**, *70*, 77–88. [CrossRef]

27. Raman, L.P.; Cheryna, M.; Rajagopalan, N. Consider nanofiltration for membrane separations. *Chem. Eng. Prog.* **1994**, *90*, 68–74.

28. Hilal, N.; Al-Zoubi, H.; Darwish, N.A.; Mohamma, A.W.; Abu Arabi, M. A comprehensive review of nanofiltration membranes: Treatment, pretreatment, modelling, and atomic force microscopy. *Desalination* **2004**, *170*, 281–308. [CrossRef]

29. Radier, R.G.J.; van Oers, C.W.; Steenbergen, A.; Wessling, M. Desalting a process cooling water using nanofiltration. *Sep. Purif. Technol.* **2001**, *22–23*, 159–168. [CrossRef]

30. Hiemstra, P.; van Paassen, J.; Rietman, B.; Verdouw, J. Aerobic versus Anaerobic Nanofiltration: Fouling of Membranes. In Proceedings of the AWWA Membrane Technology Conference, Long Beach, CA, USA, 28 February–3 March 1999; pp. 55–82.

31. Nederlof, M.M.; Kruithof, J.C.; Taylor, J.S.; van der Kooij, D.; Schippers, J.C. Comparison of NF/RO membrane performance in integrated membrane systems. *Desalination* **2000**, *131*, 257–269. [CrossRef]

32. Vrouwenvelder, J.S.; Kruithof, J.C.; Van Loosdrecht, M.C.M. Integrated approach for biofouling control. *Water Sci. Technol.* **2010**, *62*, 2477–2490. [CrossRef] [PubMed]

33. Beyer, F.; Rietman, B.M.; Zwijnenburg, A.; van den Brink, P.; Vrouwenvelder, J.S.; Jarzembowska, M.; Laurinonyte, J.; Stams, A.J.M.; Plugge, C.M. Long-term performance and fouling analysis of full-scale direct nanofiltration (NF) installations treating anoxic groundwater. *J. Membr. Sci.* **2014**, *468*, 339–348. [CrossRef]

34. Potreck, J. Enhanced membrane retention and mass transfer through smart surface modification. In Proceedings of the EuroMebrane 2015, Aachen, Germany, 6–10 September 2015.

35. Vriezekolk, E.; Pacak, A.; Potreck, J. Poster presentation: Layer-by-layer nanofiltration membranes; From fundamental research to large-scale membrane modules. In Proceedings of the Imagine Membrane, Azores, Portugal, 24–29 September 2017.

36. Rustler, M. *Aquanes Report*; Version v.0.5.0; Zenodo: Genève, Switzerland, 2018. [CrossRef]

37. R Development Core Team. R: A Language and Environment for Statistical Computing. 2017. Available online: www.R-project.org/ (accessed on 12 September 2018).

38. Sethi, S.; Wiesner, M.R. Cost Modeling and Estimation of Crossflow Membrane Filtration Processes. *Environ. Eng. Sci.* **2000**, *17*, 61–79. [CrossRef]

39. Arbeitsgemeinschaft Trinkwassertalsperren e. V.; Bundesverband der Energie und Wasserwirtschaft e. V.; Deutscher Bund der verbandlichen Wasserwirtschaft e. V.; Deutscher Verein des Gas und Wasserfaches e. V.—Technisch-wissenschaftlicher Verein; Deutsche Vereinigung für Wasserwirtschaft, Abwasser und Abfall e. V.; Verband kommunaler Unternehmen e. V. *Branchenbild der deutschen Wasserwirtschaft 2015*; wvgw Wirtschafts- und Verlagsgesellschaft Gas und Wasser: Bonn, Germany, 2015.

40. Keucken, A.; Wang, Y.; Tng, K.H.; Leslie, G.; Spanjer, T.; Köhler, S.J. Optimizing hollow fibre nanofiltration for organic matter rich lake water. *Water* **2016**, *8*, 430. [CrossRef]

41. Heidfors, I.; Vredenbregt, L.H.J.; Homes, A.; van Es, M.B. Pilot testing with hollow fiber nano filtration membranes for removal of NOM from surface water in Sweden. In Proceedings of the NOM 6 2015 (6th Specialist Conference on Natural Organic Matter in Drinking Water), Malmö, Sweden, 7–10 September 2015.

42. Dieter, H.H.; Frank, D.; Gihr, R.; Konietzka, R.; Moll, B.; Stockerl, R.; von der Trenck, T.; Schudoma, D.; Zedler, B.; Brodsky, J. Ableitung von Geringfügigkeitsschwellenwerten für das Grundwasser—Aktualisierte und Überarbeitete Fassung 2016. Available online: www.lawa.de/documents/Geringfuegigkeits_Bericht_Seite_001-028_6df.pdf (accessed on 12 September 2018).

43. Möller, K.; Kade, N.; Havermeier, L.; Paproth, F.; Burgschweiger, J.; Wittstock, E.; Günther, M.; Naumann, K.; Broll, J. *Wasserversorgungskonzept für Berlin und für das von den BWB Versorgte Umland (Entwicklung bis 2040)*; Berliner Wasserbetriebe: Berlin, Germany, 2008.

44. Zhang, Z.; Bligh, M.W.; Waite, T.D. Ascorbic acid-mediated reductive cleaning of iron-fouled membranes from submerged membrane bioreactors. *J. Membr. Sci.* **2015**, *477*, 194–202. [CrossRef]

45. Zhang, Z.; Bligh, M.W.; Wang, Y.; Leslie, G.L.; Bustamante, H.; Waite, T.D. Cleaning strategies for iron-fouled membranes from submerged membrane bioreactor treatment of wastewaters. *J. Membr. Sci.* **2015**, *475*, 9–21. [CrossRef]

46. Müller, U.; Baldauf, G.; Osmera, S.; Göttsche, R. *Erfassung und Bewertung von Nanofiltrations und Niederdruckumkehrosmoseanlagen in der Öffentlichen Wasserversorgung in Deutschland*; DVGW-Technologiezentrum Wasser Karlsruhe (TZW): Karlsruhe, Germany, 2009.

47. Semblante, G.U.; Lee, J.Z.; Lee, L.Y.; Ong, S.L.; Ng, H.Y. Brine pre-treatment technologies for zero liquid discharge systems. *Brine Pre-Treat. Technol. Zero Liq. Discharge Syst.* **2018**, *441*, 96–111. [CrossRef]

48. PubChem Substance and Compound databases. PubChem—Open Chemistry Database. Available online: https://pubchem.ncbi.nlm.nih.gov (accessed on 22 August 2018).

Article

Coupling Riverbank Filtration with Reverse Osmosis May Favor Short Distances between Wells and Riverbanks at RBF Sites on the River Danube in Hungary

Endre Salamon and Zoltán Goda *

Faculty of Water Sciences, National University of Public Service, 6500 Baja, Hungary; salamon.endre@uni-nke.hu
* Correspondence: goda.zoltan@uni-nke.hu; Tel.: +36-70-374-2674

Received: 28 November 2018; Accepted: 4 January 2019; Published: 10 January 2019

Abstract: Bank filtration and other managed aquifer recharge techniques have extensive application in drinking water production throughout the world. Although the quality of surface water improves during these natural processes, residence time in the aquifer and length of the flow paths are critical factors. A wide range of data is available on the physical–chemical processes and hydraulic conditions, but there is limited knowledge about the top layer of the porous media. An investigation was conducted on the hydraulic behavior and on the change of microbiological indicator parameters in the filter cake. The purpose of the experiment was to: (1) investigate if the reverse osmosis is sustainable when fed with only slow filtered water, and (2) show that a short travel distance can provide extensive pathogen removal and beneficial conditions for the reverse osmosis. A slow sand filter was operated over a one-year long period while changes in head loss and microbiological parameters were being monitored. Head loss and membrane permeability were monitored between 3 November 2016 and 24 October 2018 and microbiological sampling was performed from 19 July 2017 to 6 November 2018. The filtered water was fed to a reverse osmosis (RO) filter as the water above the sand filter had been spiked with dissolved iron. Results show that even a thin biofilm cake of 1–3 mm thickness can result in a significant (10–100%) reduction in microbiological activity in the infiltrate, while favorable short retention times and oxic conditions are maintained. Avoiding anoxic conditions, subsequent iron and manganese dissolution and precipitation is beneficial for membrane processes. Building on these results, it can be stated that when reverse osmosis is directly fed with slow filtered or bank filtered water, (1) a short distance from the surface water body is required to avoid dissolved iron and manganese from entering the groundwater and (2) proper pathogen rejection can be achieved even over short distances.

Keywords: bank filtration; biofilm; clogging; filter cake; pathogen barrier; pressure loss; slow sand filtration

---

1. Introduction

Due to the increasing strain on drinking water supplies and the energy demand for drinking water treatment, the combination of natural and engineered systems (CNES) gains more and more attention. While the currently reported annual volume of managed aquifer recharge (MAR) is only 1% of global groundwater use, in some countries it is considerably higher, especially where river bank filtration (RBF) is practiced, e.g., in Hungary, the Slovak Republic and Germany [1]. Although the benefits of these processes are well-known and they have been studied extensively over the last century, some unresolved site-specific issues and ambiguous scientific terminology remain. In Hungary, with the exception of the RBF sites on islands in Budapest, it is typical for smaller rivers that had elevated concentrations of dissolved iron and manganese in the portion of pumped land-side groundwater to require further water treatment [2].

The Fe and Mn issue is hydrogeology related, and it involves the distance between the surface water body (rivers in the case of RBF, infiltration trenches and basins in the case of MAR) and the abstraction point. If the abstraction point is far away (horizontally or vertically) from the surface water body, undesired flow from the background/land-side commonly increases, causing higher dissolved iron and manganese concentrations in the pumped water. On the other hand, if the flow path length is short, the withdrawal of surface water increases, but the short retention time reduces the efficiency of the porous media as a barrier and bioreactor.

A large number of publications are dealing with the question of how distance and travel time affect the removal capacity of MAR schemes [1]. Hydraulic conditions and physical and chemical changes have been extensively studied. Due to limitations concerning the calibration of geohydrochemical models, there is a lack of data on the exact spatial extent of water quality changes on the subsurface. It is understood from the slow sand filtration process that the highest (micro)biological activity is found at the topmost few centimeters of the filter. The flow velocity in slow sand filters (1–50 cm/h) [3,4] are comparable with those documented for BF sites (0.1–50 cm/h) [5,6].

Both sand filtration and bank filtration were studied extensively in the past regarding the rejection of potential microbiological hazards. In slow sand filters, 1–3 log removal (90–99.9%) was observed by different authors, both by colony counting and other methods [7,8]. RBF was reported to be capable of 2–5 log reduction for pathogen indicator parameters, such as *E. coli* and coliforms [9,10].

Consequently, with respect to the importance of the first few centimeters of the top layer, both RBF and MAR in general have some analogy with slow sand filtration and cake filtration. Since long residence times and flow path lengths can have an adverse effect on water quality, it raises the question about the extent that this thin boundary layer (or zone) can be utilized in order to avoid discharge from land-side or deeper groundwater.

Concerning RBF, a large number of scientific studies dealt with clogging and flow conditions in the riverbed, and various methods were applied and further developed to assess the permeability of the riverbed, i.e. the fragment approach and complex numerical flow simulation models [11,12]. The main difference between a pure slow sand filtration setup and RBF is the presence of the shear stress and sediment transport at the boundary coupled with an abstracted portion of land-side groundwater. With respect to MAR, the shear force at the boundary is usually negligible, depending on the specific layout and hydraulic conditions of the infiltration structures. Because of this, MAR is more closely related to slow sand filtration.

As part of the AquaNES "Demonstrating Synergies in Combined Natural and Engineered Processes for Water Treatment Systems" project, a slow sand filter was operated at the pilot water treatment plant (PWTP) of the National University of Public Service's Faculty of Water Science in Baja, Hungary. The aim of the study described here was to assess the efficiency of slow sand filtration when a thin, clogged, biologically active top layer is present. This way an assessment can be made about the removal of pollutants and pathogen retention at the porous media–water body boundary. Results are compared with the efficiency of long retention time systems. The investigation is focused on determining the extent of reduction in microbiological indicator parameters.

The main application of reverse osmosis is seawater and brackish water desalination in the present day. Even though there is considerable drinking water production by reverse osmosis worldwide, most scientific research focuses on the process itself and extensive case studies do not exist for bank filtration coupled with reverse osmosis. The main area of the present research is the rejection of pollutants, which are known to be unaffected by bank filtration [13]. Other research focuses on the energy efficiency [14] as reverse osmosis (RO) equipment used for freshwater desalination operates on a considerably lower pressure and energy demand than those for seawater desalination. A similar pilot scale investigation with single element RO was carried out at three different sites in the United States [15]. In developed countries, reverse osmosis of freshwater is typically used for wastewater reclamation and is usually located at the end of the process train either in wastewater treatment or drinking water treatment, for example in Orange County, California [16].

Although large water companies and some authorities possess reverse osmosis equipment for emergency water treatment, the long-term treatment of bank filtered water on reverse osmosis has never been studied in Hungary. Bank filtration supplies almost 50% of the country's drinking water demand and bank filtration sites are vulnerable to not just emerging micropollutants, but industrial and municipal wastewater discharge as well. Therefore, it is imperative to start investigations on the long-term sustainability of advanced treatment processes.

In order to investigate the impact of the bank filtration process on membrane filtration, a reverse osmosis (RO) filter was installed on the filtrate stream. This unique setup (despite RO never having been the first choice for treating fresh water in the past) was chosen because among all membranes, RO is the most sensitive to fouling. Because RO operating parameters respond very rapidly to fouling effects, it could be a good choice to investigate the impact of naturally treated water quality on engineered processes. If it can be proven that a thin boundary layer is adequate during MAR or even RBF to properly improve the water quality for membrane filtration, then more CNES systems with short retention times may be developed in the future.

2. Materials and Methods

2.1. Slow Sand Filter Setup

The experiment was carried out in a fully saturated open rapid sand filter structure (Figure 1). The grain size distribution of the 1 m thick sand layer shows mainly grains between 0.5–0.8 mm (11%) and 0.8–1.5 mm (87%) and larger than 2 mm (1.4%), with 0.6% of grains smaller than 0.5 mm. The supporting gravel layer consists of 2–5 mm pebbles (100%). A constant filtration rate of 50 mm/h was maintained by constant withdrawal of the filtrate with the help of a centrifugal pump (Q = 260 L/h, type FC 25-2C, SAER, Guastalla, Italy) with throttle control. Raw water was extracted directly from Sugovica, a branch of the Danube river on a daily basis. Daily change of the water depth above the sand layer was 1.0–1.5 m and a minimum coverage depth of 0.8 m was maintained at all times.

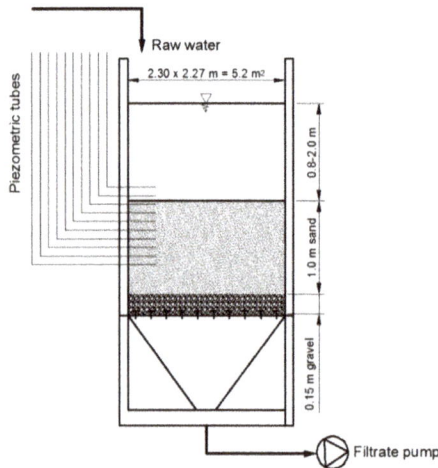

Figure 1. Slow sand filter layout.

Piezometric tubes (silicone rubber, nominal diameter = 8 mm) were installed at different depths in the sand layer. The distribution of the piezometers along the sand layer is detailed in Table 1. The inlets of the tubes were covered with geotextile in order to prevent the sand grains from entering and clogging the tubes. A vacuum pump was installed on a joint collector headspace of the piezometer tubes. Since this system has a closed headspace, once the vacuum pump is turned off, the differences between water levels in the piezometric tubes reflect the pressure differences between the tube inlets. These differences were subtracted from the calculated hydrostatic pressure in order to obtain the pressure distribution. The purpose of the two inlets above the sand layer is to signal if any obstruction, clogging or biofilm growth takes place on the protective cover at the inlet.

Table 1. Piezometer distribution.

Piezometer No.	Inlet Depth from Sand Layer Top (cm)
1	105
2	82
3	62
4	43
5	21.5
6	5.5
7	8 cm above sand layer
8	37.5 cm above sand layer

Differences in piezometric levels were registered daily in order to monitor the clogging of the top layer. In order to read the level differences between the piezometers, manual reading was applied. Before reading the piezometers, the water raised to eye-level with the vacuum pump and all air bubbles were removed from the tubing. After the levels stabilized (5–15 min), the differences between each piezometer were registered.

2.2. Reverse Osmosis Unit

Filtrate from the slow sand filter was fed to small RO filter, which had a pressure booster and a 5 μm cartridge filter of its own. The RO unit contained cross-flow, spiral wound composite (thin film) membranes (manufacturer: General Electric, catalog number: AK2540TM, Boston, MA, USA) in two vessels (Figure 2).

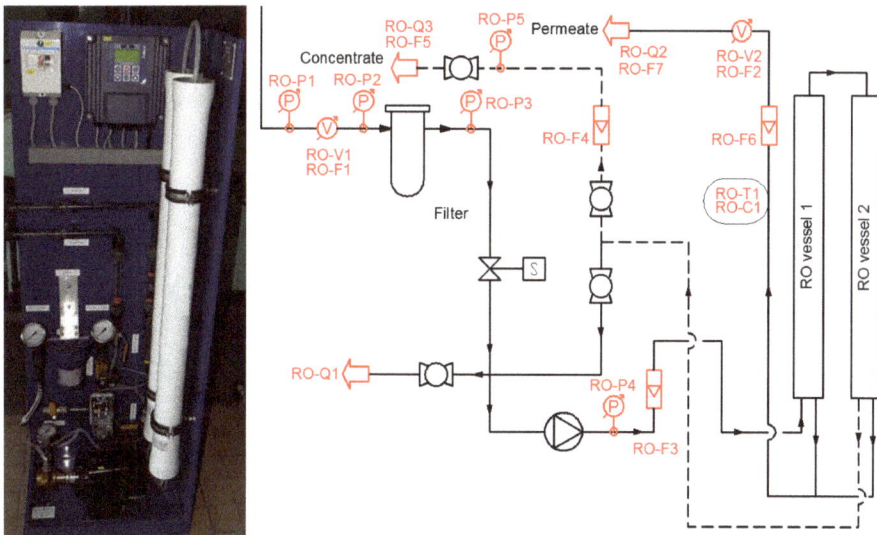

Figure 2. Small scale reverse osmosis (RO) unit (measurement points: F = flowrate; C = conductivity; P = pressure; Q = sampling; T = temperature; V = volume).

At the beginning of each filtration cycle, the recovery ratio was set to 50–60% and the initial throttle on the concentrate outlet and on the concentrate recycle was left undisturbed. This way, as the filtration went on, the permeability slowly declined due to membrane fouling. Daily average feed, concentrate and permeate quantities were determined based on installed water meters. Transmembrane pressure was calculated based on the average of feed and concentrate pressure.

2.3. Sampling and Analysis

Weekly and bi-weekly sampling for microbiological parameters was conducted according to the Hungarian MSZ EN ISO 19458:2007, sampling for microbiological testing standard [17] between November 2016 and October 2018. The samples were sent to the laboratory of Budapest Waterworks (BUWW) for analysis (Table 2). Parallel samples were taken on the same day and the colony count was determined on site in the PWTP laboratory as well.

Table 2. Measured microbiological parameters and applied standard methods.

Parameter	Standard Method of Analysis
Colony count, 22 °C	MSZ EN ISO 6222:2000
Colony count, 37 °C	MSZ EN ISO 6222:2000
Coliforms	MSZ EN ISO 9308-1:2015
E. coli	MSZ EN ISO 9308-1:2015
Enterococcus	MSZ EN ISO 7899-2:2000
Clostridium (anaerobic sulfate reducers)	MSZ EN 26461-2:1994
Pseudomonas aeruginosa	MSZ EN ISO 16266:2008

Two samples were taken at each sampling, one from the raw water above the sand filter and one from the filtered water. The sample from the raw water was taken by a thoroughly rinsed metal bowl dipped in the open water above the filter. Samples from the filtrate were taken from a sampling tap installed on the filtrate pipe. The tap and the pipeline were rinsed with 1% hypochlorite solution before and after every sampling.

3. Results

3.1. Clogging and Pressure Loss Development

Based on the measured piezometric level differences at different depths in the sand layer of the filter, the pressure distribution (or hydraulic gradient) was calculated along the filter in the form of a Michau (or Lindquist) diagram (Figure 3). The difference shown at a depth of 150 cm above the bottom of the filter indicates different water levels above the surface of the filter (water column of 1.0–1.5 m).

As expected, the pressure distribution shows the characteristics of a slow sand filter, with the steepest gradient present in the top 5 cm of the sand layer and no observable deviation from the linear gradient below. A deviance from the linear hydrostatic pressure can be attributed to the biofilm development on the geotextile covering the outlets of the piezometer tubes. This affected only the tubes near the top of the sand. Because of this, the equalization of the pressure differences in the piezometers takes a longer time.

The gradient in the top layer plotted as a function of elapsed time shows an increasing linear tendency (Figure 4). The calculated p-value of the F-test performed to investigate the linear correlation is 1.6×10^{-5}. At a 0.05 confidence level, it can be rejected that there is no linear correlation. The error of the slope is 3.86 ± 0.53 cm/d. In spite of the large number of influencing factors, the increase of the gradient indicates an effective rejection of suspended solids and microbial growth.

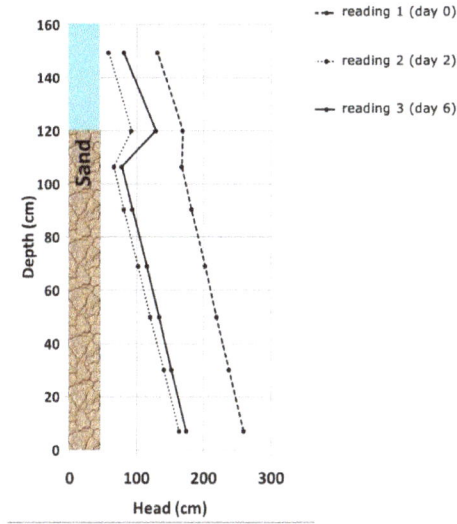

Figure 3. Measured pressure distribution in filter.

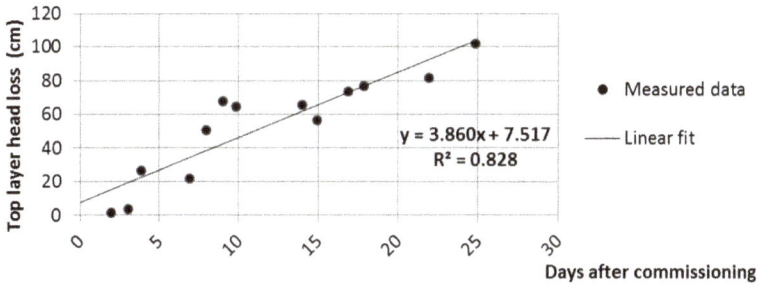

Figure 4. The relationship between the top layer and the days after commissioning.

The gradient and the resistance build-up are subject to many influencing factors apart from the microbiological growth and biofilm development. Combined, these factors can cause large random variations in the hydraulic gradient and the development of the resistance against filtration. Such factors include:

- sedimentation of small particles (e.g., in stagnant river areas),
- erosion of the top boundary layer (e.g., during floods or dredging),
- three-phase flow (e.g., gas bubbles from biological activity),
- non-zero order biomass growth (fluctuations in substrate concentration, temperature, pH, etc.),
- changes in viscosity,
- precipitation and dissolution of carbonates, iron-hydroxides and other compounds at the boundary.

Under the conditions of the experiment, the rate of the resistance build-up was 4 cm/d (cm stands for the measured difference in the piezometric levels). The developed biofilm and boundary layer could be observed with the naked eye after 1–2 weeks of operation (Figure 5a–c). Its thickness was 2–3 mm.

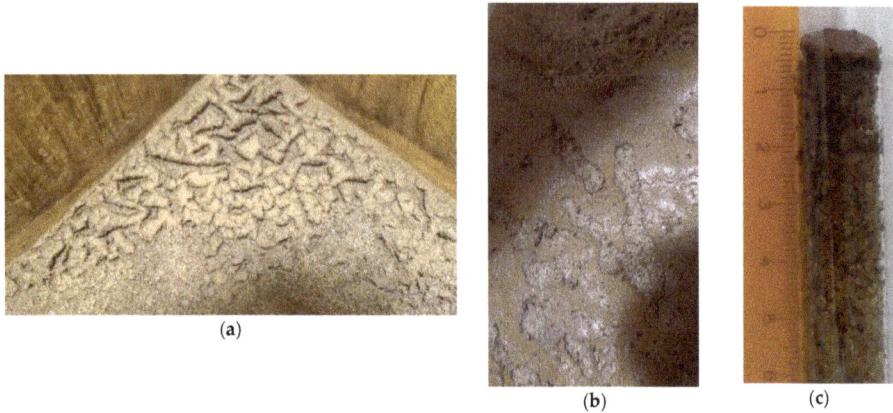

Figure 5. Clogged layer at the porous media–water boundary. (**a**) Dried state; (**b**) wet state; (**c**) approximate thickness of 2–3 mm.

3.2. Changes in Microbiological Indicator Parameters

A box plot was created for the measured values of each parameter in the raw and in the filtered water (Figure 6a,b). Some results for coliforms and *E. coli* were omitted, because four samples could not be analyzed for *E. coli* by the BUWW laboratory due to the unexpectedly high interference from all other microbiological growth. In case of *E. coli* and *Enterococcus*, only 1–3 non-zero results were obtained and only the minimum (0), the maximum and the average are shown on the graph. The number of samples is also given after the name of the parameters.

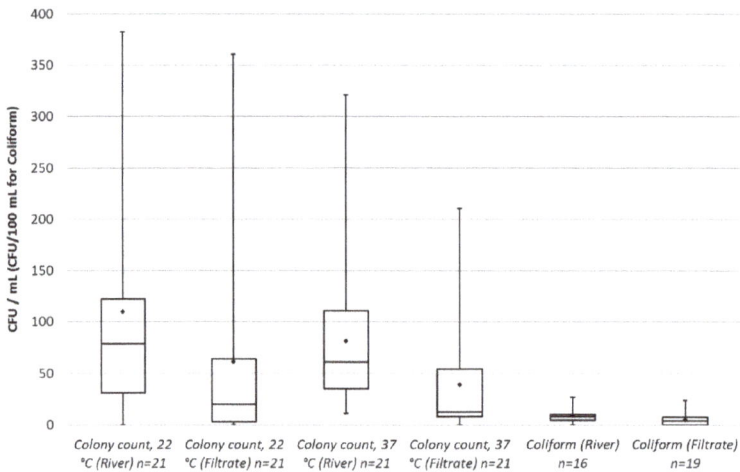

(**a**)

Figure 6. *Cont.*

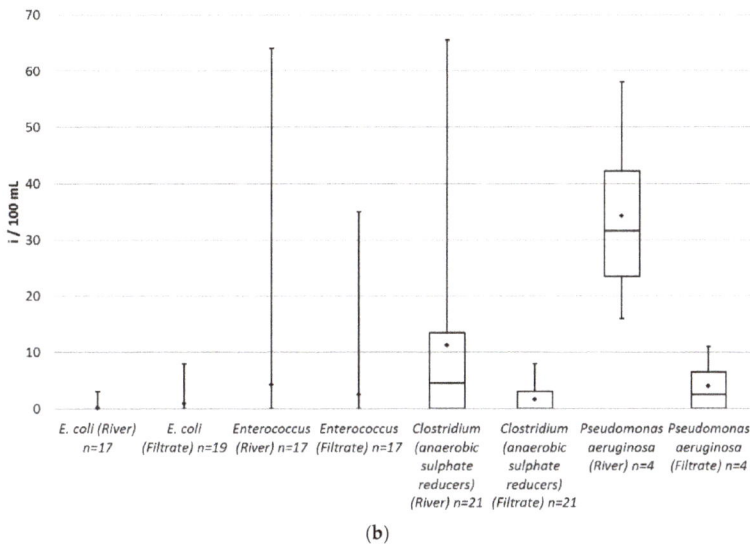

(b)

Figure 6. Values of microbiological indicator parameters. CFU = colony forming units, i = individuum. (**a**) Parameters measured as CFU, (**b**) Parameters measured as individual entities.

All parameters, except *E. coli*, have a lower average and median in the filtrate compared to river water. *E. coli* was present in only one river and two filtrate samples: 3 i/100 mL (individuum/100 mL), 6 i/100 mL and 3 i/100 mL respectively. Therefore, there is insufficient data to evaluate the change during the filtration for *E. coli*. The reduction in percentage was calculated for every parameter based on the raw water samples with non-zero values (Table 3).

Table 3. Reduction of microbiological indicator parameters.

Parameter	Reduction in %			Number of Non-Zero River Water Samples	Number of Filtered Water Samples with Zero Result (100% Reduction)
	Minimum	Average	Maximum		
Colony count, 22 °C	15	43	100	16	2
Colony count, 37 °C	10	39	100	18	2
Coliforms	20	41	100	10	3
Enterococcus	45	68	100	3	1
Clostridium	16	80	100	15	8
Pseudomonas aeruginosa	57	81	100	4	2

Twenty samples were analyzed for each parameter, except for *Enterococcus* (17 samples) and *Pseudomonas aeruginosa* (four samples). Both the raw water and the filtered water showed high variation in the measured concentrations. Based on the average reductions, the box diagrams and the non-zero samples from river, the most effective reduction is achieved regarding *Pseudomonas*, *Clostridium* and colony count. Coliform and *E. coli* numbers were close to zero at all times. This is due to the clean state of the source water and to the lack of any significant fecal contamination and wastewater discharge during the sampling period.

A paired *t*-test was carried out on all parameters except *E. coli*, which has too much zero measured data (Table 4). At a 0.05 significance level, it can be rejected that there is no difference between samples from the river and the filtered water for all parameters, except for coliforms and *Enterococcus*.

Table 4. Paired *t*-test results.

	Colony Count, 22 °C	Colony Count, 37 °C	Coliforms	*Enterococcus*	*Clostridium*	*Pseudomonas Aeruginosa*
Mean difference	49.00	41.95	2.63	1.76	9.67	30.25
Standard error	23.47	18.00	1.67	1.75	3.41	10.87
t-test *p*-value	0.049	0.030	0.132	0.327	0.010	0.050

3.3. Reverse Osmosis Results

The RO unit has been operated with slow filtered water produced purely from surface water from 1 February 2017 to 10 July 2017. The river water was spiked with groundwater extracted from a nearby well during a second test period from 12 July 2017 to 2 November 2017. The groundwater contained dissolved iron in concentrations of 5–10 mg/L and manganese in concentrations of 2–5 mg/L. Before spiking, dissolved iron concentrations were below 100 µg/L and manganese was below the detection limit (10 µg/L) above the filter. Dissolved iron concentration increased to 150–200 µg/L and manganese to 300 µg/L above the filter after spiking.

In order to push the slow filtration system more toward anoxic conditions, fresh riverbed material from the Danube was placed on top of the sand in a 10 cm thick layer. Even after this, 1–4 mg/L dissolved oxygen could be observed in the filtrate, but dissolved iron concentration in the filtrate increased to 200–250 µg/L. As shown in Figure 7, the long-term permeability rate increased from 0.011 LMH/bar/day to 0.017 LMH/bar/day. LMH reprsenets the membrane flux and it stands for "liter per square meter per hour", $L \cdot m^{-2} \cdot h^{-1}$. Divided by the transmembrane pressure and the elapsed time, it gives the average rate of permeability decrease.

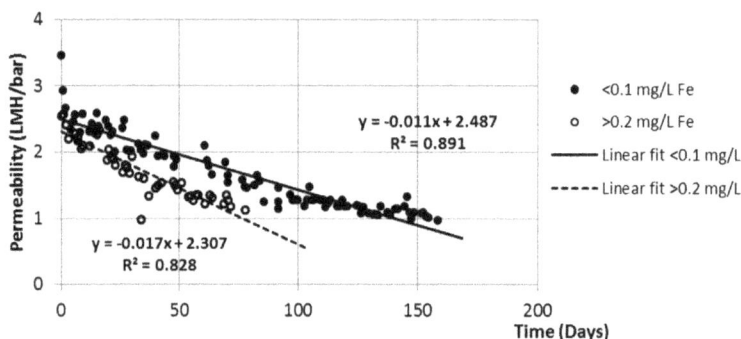

Figure 7. Relationship between permeability with 100 µg/L and 200 µg/L dissolved iron and time (LMH = $L \cdot m^{-2} \cdot h^{-1}$).

Apart from the random errors due to the manual reading of pressure gauges and flow meters, the permeability shows a clear decreasing tendency, as expected. A sudden drop can be observed when the filtration is started after each chemical cleaning. After the first 1–2 days, the plotted linear tendency is present for a long duration (50–150 days). Even though the permeability decreased to one-third of its original value, it was possible to restore it.

4. Discussion

It has to be noted that indicator parameters are not well suited for the quantitative assessment of pathogens or microbial activity. These standard methods are primarily designed as safety indicators for distinguishing between the negative and positive (non-zero) samples. In order to assess microbiological activity and the biomass, other emerging methods are available such as ATP or RNA analysis, measurement of certain enzymes, etc. [18].

Due to the low number of bacteria in the river water, only less than 1 log removal values could be observed during the time of the experiment. The good quality of the river water can be explained by no wastewater discharge nor any polluting activity in the vicinity and the location of the water intake structure at a stagnant zone in the river. Most suspended solids from the main stream settle before they reach the inlet of the water intake.

Because of these low initial values, a strong reduction could only be observed for three parameters: colony count at different temperatures and *Pseudomonas*. Although colony count has the largest range in reduction (10–100%), it gives the best proof of reduction in the number of microorganisms since it had the most positive samples in the river water. An even better reduction was achieved regarding *Pseudomonas*, however, two out of four samples from the river were found to be free of *Pseudomonas*.

The results obtained in this experiment for the reduction of microbiological indicator parameters are well within the range of previously reported values [7]. For *Pseudomonas* and total bacteria, 60% removal was obtained in small scale filter columns with 1 m thick sand layer and 0.15 m/h loading rate. It was also found that *Pseudomonas* species dominate the bacterial biofilm [19]. During the interpretation of the results, it has to be taken into account that even though there is an overall reduction in microbiological activity due to substrate consumption, the developed biofilm itself can detach from the solid surfaces of the filter and may cause higher results in the samples. Small slow sand filters with a thin 15 cm fine sand layer were reported to reject coliforms with 90–97% efficiency [20].

Due to the very low pathogen content of surface water involved in this demonstration, it is problematic to compare results with similar studies. A similar investigation on the application of reverse osmosis directly after bank filtration described in "Desalination and Water Purification Research and Development Program Report No. 122" [15], where coliforms ranged between 200–10,000 colony forming units (CFU)/100 mL in three investigated river sites. The report also mentions 0.052–1.62 mg/L of iron and 0.05–0.617 mg/L of manganese in the bank filtered water used for reverse osmosis membrane feed from the Ohio, Missouri and Raccoon rivers. Total rejection of coliforms on the reverse osmosis was also reported [15].

The feed water quality requirement for the membranes involved in the current demonstration are less than 0.1 mg/L of total iron and manganese and less than 100 CFU/mL according to the manufacturer's suggestions.

The transport of slowly biodegradable organic matter in the surface water should be taken into account when the short distance to the extraction point is considered. Such materials are humic acids, lignin and another plant residues, which can be found in surface waters used for bank filtration. The biologically active layer and the subsequent adsorption processes during slow filtration are capable of reducing the concentration of such components [21]. If iron and manganese precipitates in the active layer due to the oxic conditions, it can be beneficial not only to the adsorption of the colloid material, but for the immobilization of microorganisms as well [22], as long as severe clogging can be avoided.

Since the experiment was conducted under circumstances characteristic for pure slow sand filtration, its results support the hypothesis that the thin boundary layer and a short (1.2 m) flow through a porous media can be responsible for as much as 80% or more reduction in the common microbiological indicator parameters. Compared with common flow path lengths at Hungarian bank filtration sites of 10–50 m and the high iron and manganese concentration in land-side groundwater, it is important to define the ideal flow path length during the design. This requires a cost–benefit analysis, especially when iron and manganese sensitive membrane technologies are to be installed.

5. Conclusions

In general, it is customary to increase the distance of the extraction element from the surface water boundary for safety considerations. In case of combined systems, when new post-treatment technologies such as ultrafiltration, nanofiltration or reverse osmosis with high pathogen removal efficiency are applied, shorter flow path lengths could be an advantage to prevent increased concentration of dissolved iron and manganese. In the case of RBF, there are severe uncertainties and safety considerations for the distance of the extraction point from the river. The most important is riverbed erosion, which heavily puts the extent of the biologically active zone into question. On the other hand, artificial structures such as filter dams with geotextile support may be adequate to provide proper conditions for short travel distances.

It was found that as long as the bank filtered water fulfills the required quality standards set by the manufacturer, the permeability of the reverse osmosis membrane can be sustained for a long time, even for 50–100 days. The quality of the pumped water as a mixture of bank filtrate and land-side groundwater is site specific and in Hungary, especially at the lower Danube region, tends to contain more than 0.2 mg/L of dissolved iron due to the groundwater flow. In order to make the long-term operation of reverse osmosis possible, short distances between the river bank and the extraction point will help to maintain oxic conditions and minimize the iron input from groundwater flow. Reverse osmosis is known to be capable of rejecting pathogens almost completely. Even if a higher number of bacteria break through the bank filtration because of the shorter distances, the risks should not be higher than for classic water treatment processes.

In this way, the natural system (the biologically active zone) can be utilized not only to decrease turbidity and reject a significant number of pathogens, but to reduce the biofouling potential of the membrane by the removal of easily degradable substrates which otherwise contribute to biofilm growth. Membrane filtration techniques have a high efficiency in pathogen rejection, but are sensitive to fouling caused by iron or manganese precipitation. When membrane filtration is applied, the decrease in the portion of land-side groundwater is of first priority and the pathogen rejection can be safely carried out with the combination of short filtration by RBF or other MAR schemes and membrane treatment.

Author Contributions: All authors collaborated in this work. Writing—Original Draft Preparation, E.S.; Writing—Review and Editing, Z.G.

Funding: All primary data was collected within the AquaNES project. This project has received funding from the European Union's Horizon 2020 Research and Innovation Program under grant no. 689450. www.aquanes-h2020. eu.

Acknowledgments: The authors would like to express their gratitude to the staff operating the pilot water treatment plant: Ilona Dalkó, Kitti Tafner and Tamás Papp in particular. The authors also acknowledge the helpful comments from T. Grischek during final paper preparation.

Conflicts of Interest: The authors declare no conflict of interest.

References

1. Dillon, P.; Stuyfzand, P.; Grischek, T.; Lluria, M.; Pyne, R.D.G.; Jain, R.C.; Bear, J.; Schwarz, J.; Wang, W.; Fernandez, E.; et al. Sixty years of global progress in managed aquifer recharge. *Hydrogeol. J.* **2018**, 1–30. [CrossRef]

2. Salamon, E. Investigation of slow filtration followed by reverse osmosis (Lassú szűrés–fordított ozmózisos ivóvíz tisztítási technológiai sor vizsgálata). In Proceedings of the Membrane Technology Water Industry Day Conference, Budapest, Hungary, 9 November 2017.

3. Collins, M.R.; Eighmy, T.T.; Fenstermacher, J.M.; Spanos, S.K. Removing Natural Organic Matter by Conventional Slow Sand Filtration. *J. Am. Water Works Assoc.* **1992**, *84*, 80–90. [CrossRef]

4. Jenkins, M.W.; Tiwari, S.K.; Darby, J. Bacterial, viral and turbidity removal by intermittent slow sand filtration for household use in developing countries: Experimental investigation and modeling. *Water Res.* **2011**, *45*, 6227–6239. [CrossRef] [PubMed]

5. Hoffmann, A.; Gunkel, G. Bank filtration in the sandy littoral zone of Lake Tegel (Berlin): Structure and dynamics of the biological active filter zone and clogging processes. *Limnologica* **2011**, *41*, 10–19. [CrossRef]

6. Ray, C.; Grischek, T.; Schubert, J.; Wang, J.Z.; Speth, T.F. A Perspective of Riverbank Filtration. *J. Am. Water Works Assoc.* **2002**, *94*, 149–160. [CrossRef]

7. Seeger, E.M.; Braeckevelt, M.; Reiche, N.; Müller, J.A.; Kästner, M. Removal of pathogen indicators from secondary effluent using slow sand filtration: Optimization approaches. *Ecol. Eng.* **2016**, *95*, 635–644. [CrossRef]

8. Pfannes, K.R.; Langenbach, K.M.W.; Pilloni, G.; Stührmann, T.; Euringer, K.; Lueders, T.; Neu, T.R.; Müller, J.A.; Kästner, M.; Meckenstock, R.U. Selective elimination of bacterial faecal indicators in the Schmutzdecke of slow sand filtration columns. *Appl. Microbiol. Biotechnol.* **2015**, *99*, 10323–10332. [CrossRef] [PubMed]

9. Weiss, W.J.; Bouwer, E.J.; Aboytes, R.; LeChevallier, M.W.; O'Melia, C.R.; Le, B.T.; Schwab, K.J. Riverbank filtration for control of microorganisms: Results from field monitoring. *Water Res.* **2005**, *39*, 1990–2001. [CrossRef] [PubMed]

10. Van Driezum, I.H.; Chik, A.H.S.; Jakwerth, S.; Lindner, G.; Farnleitner, A.H.; Sommer, R.; Blaschke, A.P.; Kirschner, A.K.T. Spatiotemporal analysis of bacterial biomass and activity to understand surface and groundwater interactions in a highly dynamic riverbank filtration system. *Sci. Total Environ.* **2018**, *627*, 450–461. [CrossRef] [PubMed]

11. Grischek, T.; Bartak, R. Riverbed Clogging and Sustainability of Riverbank Filtration. *Water* **2016**, *8*, 604. [CrossRef]

12. des Tombe, B.F.; Bakker, M.; Schaars, F.; van der Made, K.-J. Estimating Travel Time in Bank Filtration Systems from a Numerical Model Based on DTS Measurements. *Ground Water* **2018**, *56*, 288–299. [CrossRef] [PubMed]

13. Albergamo, V.; Blankert, B.; Cornelissen, E.R.; Hofs, B.; Knibbe, W.-J.; van der Meer, W.; de Voogt, P. Removal of polar organic micropollutants by pilot-scale reverse osmosis drinking water treatment. *Water Res.* **2019**, *148*, 535–545. [CrossRef] [PubMed]

14. Davies, P.A. A solar-powered reverse osmosis system for high recovery of freshwater from saline groundwater. *Desalination* **2011**, *271*, 72–79. [CrossRef]

15. Gooters, S. *The Role of Riverbank Filtration in Reducing the Costs of Impaired Water Desalination*; Desalination and Water Purification Research and Development Program Report No. 122; Department of the Interior, Bureau of Reclamation, Water Treatment Engineering and Research Group: Denver, CO, USA, 2006.

16. Argo, D.R. Use of Lime Clarification and Reverse Osmosis in Water Reclamation. *J. Water Pollut. Control Fed.* **1984**, *56*, 1238–1246.

17. *MSZ EN ISO 19458:2007 Water Quality—Sampling for Microbiological Analysis (ISO 19458:2006) 21*; International Organization for Standardization: Geneva, Switzerland, 2006.

18. Chan, S.; Pullerits, K.; Riechelmann, J.; Persson, K.M.; Rådström, P.; Paul, C.J. Monitoring biofilm function in new and matured full-scale slow sand filters using flow cytometric histogram image comparison (CHIC). *Water Res.* **2018**, *138*, 27–36. [CrossRef] [PubMed]

19. Calvo-Bado, L.A.; Pettitt, T.R.; Parsons, N.; Petch, G.M.; Morgan, J.A.W.; Whipps, J.M. Spatial and Temporal Analysis of the Microbial Community in Slow Sand Filters Used for Treating Horticultural Irrigation Water. *Appl. Environ. Microbiol.* **2003**, *69*, 2116–2125. [CrossRef] [PubMed]

20. Guchi, E.; Leta, S.; Boelee, E. Efficiency of slow sand filtration in removing bacteria and turbidity from drinking water in rural communities of central Ethiopia. *Afr. J. Microbiol. Res.* **2014**, *8*. submitted.

21. Mo, J.; Yang, Q.; Zhang, N.; Zhang, W.; Zheng, Y.; Zhang, Z. A review on agro-industrial waste (AIW) derived adsorbents for water and wastewater treatment. *J. Environ. Manag.* **2018**, *227*, 395–405. [CrossRef] [PubMed]

22. Bouabidi, Z.; El-Naas, M.; Zhang, Z. Immobilization of microbial cells for the biotreatment of wastewater: A review. *Environ. Chem. Lett.* **2018**, 1–17. [CrossRef]

water

MDPI

Article

Combination of River Bank Filtration and Solar-driven Electro-Chlorination Assuring Safe Drinking Water Supply for River Bound Communities in India

Philipp Otter [1,*], Pradyut Malakar [2], Cornelius Sandhu [3], Thomas Grischek [3], Sudhir Kumar Sharma [4], Prakash Chandra Kimothi [4], Gabriele Nüske [5], Martin Wagner [5], Alexander Goldmaier [1] and Florian Benz [1]

1   AUTARCON GmbH, D-34117 Kassel, Germany; goldmaier@autarcon.com (A.G.); benz@autarcon.com (F.B.)
2   International Centre for Ecological Engineering, University of Kalyani, Kalyani, West Bengal 741235, India; pradyutmalakar2@gmail.com
3   Division of Water Sciences, University of Applied Sciences Dresden, D-01069 Dresden, Germany; cornelius.sandhu@htw-dresden.de (C.S.); thomas.grischek@htw-dresden.de (T.G.)
4   Uttarakhand State Water Supply and Sewerage Organization, Uttarakhand Jal Sansthan (UJS), Dehradun 263139, India; sudhirksharma10@yahoo.com (S.K.S.); pckimothi@gmail.com (P.C.K.)
5   Technologiezentrum Wasser (TZW) Karlsruhe, D-01326 Dresden, Germany; gabriele.nueske@tzw.de (G.N.); martin.wagner@tzw.de (M.W.)
*   Correspondence: otter@autarcon.com; Tel.: +49-561-5061 868 92

Received: 30 October 2018; Accepted: 2 January 2019; Published: 11 January 2019

Abstract: The supply of safe drinking water in rural developing areas is still a matter of concern, especially if surface water, shallow wells, and wells with non-watertight headworks are sources for drinking water. Continuously changing raw water conditions, flood and extreme rainfall events, anthropogenic pollution, and lacking electricity supply in developing regions require new and adapted solutions to treat and render water safe for distribution. This paper presents the findings of a pilot test conducted in Uttarakhand, India, where a river bank filtration (RBF) well was combined with a solar-driven and online-monitored electro-chlorination system, treating fecal-contaminated Ganga River water. While the RBF well provided nearly turbidity- and pathogen-free water as well as buffered fluctuations in source water qualities, the electro-chlorination system provided disinfection based on the inline conversion of chloride to hypochlorous acid. The conducted sampling campaigns provided complete disinfection (>6.7 log) and the adequate supply of residual disinfectant (0.27 ± 0.17 mg/L). The system could be further optimized to local conditions and allows the supply of microbial-safe water for river bound communities, even during monsoon periods and under the low natural chloride regimes typical for this region.

Keywords: electro-chlorination; smart villages; disinfection; river bank filtration; rural water supply; online monitoring

---

1. Introduction

The Millennium Development Goal to halve the number of people without access to improved water sources was achieved in 2015—five years ahead of schedule. By that, 2.6 billion people gained access to improved water sources. However, there is substantial evidence that improved sources of drinking water, including piped water, can contain fecal contamination and studies estimate that 1.8–2.0 billion people drink such water [1–4]. Every year 502,000 deaths are caused by diarrheal diseases that can be attributed to the consumption of unsafe water [4]. Especially rural communities are prone to having no access to safe drinking water. Lack of infrastructure, technical expertise, user compliance, as well as the lack of supply of chemicals and electricity have been identified as reasons for the failure of rural water treatment and supply systems [5]. Point-of-use (PoU) treatment approaches are often considered as alternatives and have shown to reduce the risk of diarrheal infections by 40% [6]. However, the effectiveness of PoU disinfection (including chlorination) depends highly on the comprehension and willingness of the households to apply the treatment systems correctly, especially under varying source water conditions [6–8]. Turbidity impedes the application of chlorine and other disinfection methods. In that case additional filtration is required, increasing the complexity and costs for PoU treatment. In the end, the responsibility for safe water supply is passed on to the end user and the educational and motivational efforts required for establishing a reliable application of PoU may not pay off.

The here presented combination of river bank filtration (RBF) and solar-driven electro-chlorination (ECl$_2$) could be a feasible option for the decentralized treatment of surface water in river bound communities. Reported data show that RBF can effectively remove many major water pollutants and micro-pollutants, including particulates, colloids, algae, pathogens, organic as well as inorganic compounds, microcystins, and heavy metals [9,10]. Log reductions for total coliforms of 5.5–6.1 and for bacteriophages of >4.4 were reported by [11,12]. Total organic carbon (TOC) removal rates of 60% are possible [13]. Whereas conventional treatment methods, like coagulation-filtration, can reduce the disinfection by-product (DBP) formation potential by 25% [13], the reduction can reach 50%–80% using RBF, without any waste sludge produced. Furthermore, RBF is able to attenuate temperature peaks and can provide protection against shock loads. Although inorganic contamination is less likely found in bank filtrate [9], oxygen may be depleted during the passage of the water through the bank. Under anoxic conditions, iron, manganese, and even arsenic can re-mobilize and enter bank filtration wells [14]. During the planning process of RBF abstraction sites such aspects have to be considered and recommendations for safe management of RBF sites in India were published [15].

In the US, RBF has received log-credits for pathogen removal and is mainly used for the removal of suspended solids. The sites are often designed with shorter travel times compared to Europe, where RBF has been widely applied for more than 130 years to produce drinking water along the Rhine, Elbe, Danube, and Seine rivers. Furthermore, in developing regions, the interest in RBF is increasing and the feasibility of its application has been evaluated under different hydrological and hydrogeological conditions (e.g., in India [16], Egypt [17], and Thailand [18]).

However, the application of RBF wells alone does not assure long-term microbial-safe water. Despite the cited removal rates, monitoring campaigns and risk assessment studies have repeatedly shown the presence of total coliforms and Escherichia coli (E. coli) in RBF wells, even at greater distances (48–190 m) to the river bank [12,19]. In Haridwar (northern India), where the pilot site for this project is located, such incidents could be related to the seepage of fecal contamination in the direct vicinity of the wells [19]. Further, recontamination may occur also during distribution and storage, justifying further disinfection. Here, chlorine, in contrast to, for example, UV-treatment or ultrafiltration (UF), has a long proven record of rendering water safe during storage and distribution—if handled correctly [20,21]. In rural communities; however, chlorination systems have failed for the same reasons as stated above, as they require constant availability of chemicals, skilled personnel capable in evaluating the chlorine demand of the water, and strict compliance with existing guidelines. Furthermore, the application of chlorine compounds is challenged by the formation of DBPs if applied in unfavorable source water conditions. Even though the risks for microbial contamination usually exceed the adverse side effects of chlorination [22,23], guideline values for chlorine dose and inorganic and organic DBP concentrations exist (Table 1).

The inline-electrolytic production of chlorine (ECl$_2$) could pose a feasible alternative towards the dosing of chlorine. Here, gaseous chlorine is produced directly at the anode of an electrolytic cell from the chloride dissolved in the water that is to be treated (Equation (1)). The chlorine gas rapidly dissociates in water to hypochlorous acid, being chemically the same oxidizing agent as in conventional chlorine dosing systems (Equation (2)). The chlorine gas production is accompanied by a decrease of pH (Equation (3)) and the evolution of hydrogen gas at the cathode (Equation (4)) [24].

Table 1. Selected guideline values concerning the chlorination of drinking water.

Parameter	Germany	EU	WHO	India IS 10500
Free Available Chlorine (FAC) [mg/L]	1.2 [a]/0.1–0.3 [b]		>0.5	0.2/1.0
Bromate [µg/L]	10	10	10	-
Chlorate [µg/L]	200 [d]	250 [e]	700	-
Chlorite [µg/L]	200	250 [e]	700	-
Trihalomethanes (THM) [µg/L]	10 [b]/50 [c]	100	60–300	-
Bromoform [µg/L]				100
Dibromochloromethane [µg/L]				100
Bromodichloromethane [µg/L]				60
Chloroform [µg/L]				200

[a] During treatment. [b] At the end of treatment. [c] Point of use. [d] Valid by the time of pilot test; was reduced to 70 µg/L in December 2017. [e] As currently proposed [25].

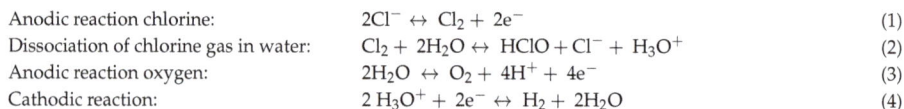

Anodic reaction chlorine:	$2Cl^- \leftrightarrow Cl_2 + 2e^-$	(1)
Dissociation of chlorine gas in water:	$Cl_2 + 2H_2O \leftrightarrow HClO + Cl^- + H_3O^+$	(2)
Anodic reaction oxygen:	$2H_2O \leftrightarrow O_2 + 4H^+ + 4e^-$	(3)
Cathodic reaction:	$2\,H_3O^+ + 2e^- \leftrightarrow H_2 + 2H_2O$	(4)

To power this process a DC voltage is applied to dimension stable (DSA) titanium electrodes coated with platinum group metals. Studies have shown that coatings comprising iridium and or ruthenium oxides (MOX electrodes) produce consistently higher chlorine output compared to platinum coatings [26,27]. In comparison to the manifold in literature-described boron-doped diamond (BDD) electrodes, MOX electrodes are less prone to produce DBPs, especially considering chlorate and perchlorate [28,29].

To control the production of chlorine, fundamental knowledge about the functional interrelationship between chloride concentration, current, current density, electrode material, temperature, and source water quality is required and has become available only very recently [30]. Systematical evaluation on the effectiveness of the produced disinfecting agents and the potential formation of disinfection by-products (DBP) has shown that the application of inline-electrolysis is comparable to the application of hypochlorous acid [24,27,28,30]. However, uncertainty towards the long-term operability, the effectiveness under very low chloride regimes, and elevated hardness levels persists [27,28].

For the first time a combination of RBF and solar-driven inline-electrolysis was tested in a long term trial in northern India. The intention of this combination between natural and engineered solutions (cNES) was to merge the above-mentioned benefits of RBF for surface water treatment with the benefits of residual chlorination, eliminating the above-mentioned drawbacks of chemical dosing. The here presented data summarize the findings of two intensive sampling periods conducted within a two and a half year pilot trial. The main target was to evaluate the pathogen removal and residual disinfection capacity. The first eight-month sampling campaign lasted from March–November and included one monsoon season (July–September). The second sampling campaign lasted for two weeks and was conducted after system optimization. Further, the formation of DBPs and energy efficiency of this water treatment approach was evaluated and suggestions for long-term operation and maintenance requirements were derived.

2. Materials and Methods

2.1. Bank Filtration at the Haridwar Site

The test was conducted in Haridwar, India, where 68% of the drinking water supply for the city is produced by RBF [15]. The used large diameter well (IW #18) is situated on Pant Dweep Island, located between the Upper Ganga Canal and the Ganga River (Figure 1). The distance to the nearest canal bank is 115m. The siting of the well (IW #18) on an island and the significant natural gradient of the water table result in a high proportion of bank filtrate in the abstracted water. The water table in the well varies between 6 and 8.5 m bgl.

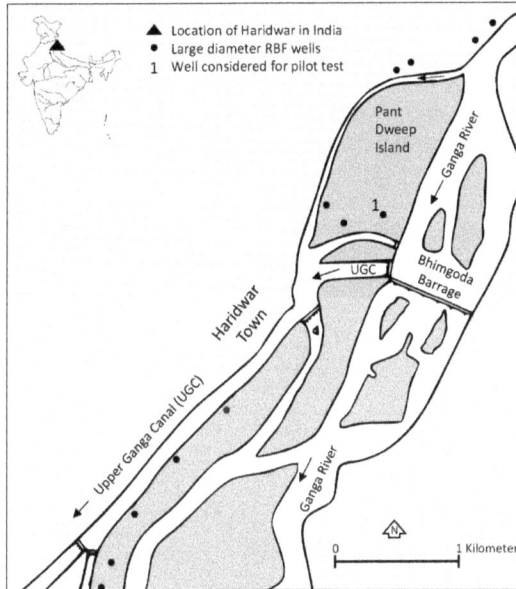

Figure 1. Location of the large diameter well on Pant Dweep Island at Haridwar (after [12]).

Studies conducted at two monitoring wells, starting in 2005, revealed that the bank filtrate contains dissolved organic carbon (DOC) of less than 1 mg/L under aerobic conditions. Trace metals were found to be below the Indian Standard IS 10500 (1991) limit [12]. The abstracted water from all the RBF wells in Haridwar only require disinfection and thus are well suited for the conduction of the pilot test.

2.2. Inline Electrolysis

The inline electrolytic chlorination unit tested during this trial was originally designed for surface water filtration and disinfection. The ECl_2 cell stack in this pilot had a total surface area of 600 cm^2 and was operated with a maximum current of 5 A, which resulted in a maximum current density of 8.5 mA/cm^2. The cells polarity was inverted every 60 min to remove potentially-forming calcareous deposits from the cathode. At very low chloride concentrations the chlorine production efficiency may not be sufficient to meet the chlorine demand of the water. In that case, the station automatically reduces its flow rate. This works well; however, it also reduces the treatment capacity of the station and thus the economic feasibility. In prior studies conducted with good source water conditions, 10 mg/L of natural chloride in the water has been identified as the minimum chloride concentration for flow rates up to 100 L/h [26]. In the here described pilot test an average treatment capacity of 200 L/h was anticipated and the natural chloride concentration of the bank filtrate was only 14 ± 2 mg/L. Due to that, the pilot station was equipped with an automated NaCl dosing system, which would start to dose NaCl solution into the feed water tank under the following conditions:

(a) $ORP_{\text{drinking water tank}} < ORP_{\text{target}}$ and

(b) $Voltage_{\text{ECl2 cell}}/Current_{\text{ECl2 cell}} \geq 3$.

If these requirements were met the system would start dosing NaCl solution until either the ratio of 3 or the target oxidation reduction potential (ORP) was reached. At a constant cell voltage of 12 V, a current of 4 A would be required to stop dosing and the water would then contain a chloride concentration of about 50 mg/L. As target ORP a value of 720 mV in the first and 700 mV in the second pilot phase were set. With that a Free Available Chlorine (FAC) conentration of 0.2–0.5 mg/L was anticipated.

For the first trial, the treatment system was equipped with a 9-inch pressurized vessel containing Activated Filtration Media (AFM) to remove potential turbidity still present in the bank filtrate. The filter was automatically backwashed after three days, independent of the quantity of water that had passed through the filter. During the second short test phase, a second filter was installed after the ECl_2 cell to remove calcareous deposits that were released from the cathode after polarity inversion and had slightly increased the turbidity in the final storage water tank.

A submersible pump lifted the bank filtrate into a 2 m³ feed tank. The water was then pumped by an internal system pump through Filter 1 and the electrolytic reactor and into a 1 m³ final water storage tank (Figure 2). The chlorine production capacity of the cell was increased by adjusting ECl_2 cell current and the flow rate, which was measured with a GEMÜ 850 flow sensor.

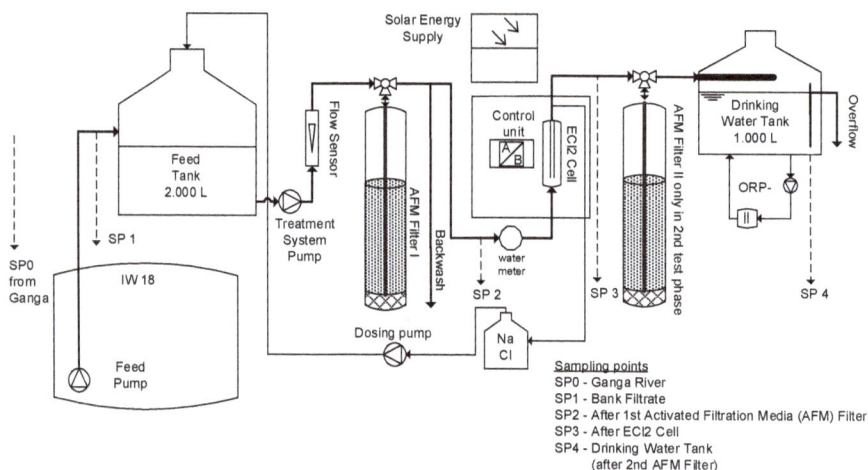

Figure 2. Pilot system setting at well IW #18, Haridwar.

2.3. Sampling, Water Analysis, and Monitoring

For water quality analysis, random samples were taken one to two times per week at five sampling points (SP0 Ganga River, SP1 bank filtrate, SP2 after AFM filtration, SP3 directly after inline electrolysis, SP4 in final drinking water storage water tank) (Figure 2). Electric conductivity (Hach CDC 101), dissolved oxygen (LDO 101), and pH (Hach PHC 101) were measured with a Hach Multimeter HQ40d (Düsseldorf, Germany). The ORP in SP1 (bank filtrate) and SP4 (drinking water) was measured directly with the pilot system using a Jumo tecLine Rd electrode (Fulda, Germany). For parameters shown in Table 2, an Aqualytic AL410 (Dortmund, Germany) handheld photometer was used. Analysis of immediate parameters and parameters in Table 2 were done on site.

Table 2. Parameters and methods used for analysis with the AL410 photometer.

Parameter	Wavelength in nm	Method	Range
Free Available Chlorine (FAC)	530	100: DPD1	0.01–6
Total Chlorine	530	100: DPD3	0.01–6
NH_4-N in mg/L	610	60: Indophenole	0.02–1
Cl^- in mg/L *	530	90: Silver nitrate/turbidity	0.5–25
Total Hardness in mg/L $CaCO_3$	560	200: Metalphthalein	2–50

* Chloride was additionally determined through titration based on APHA Method 4500-Cl^-A.

Pathogens, DBPs, and UV-absorption (UVA, wavelength 254 nm) were analyzed in laboratories of Uttarakhand Jal Sansthan, the state water supply agency, or at TZW Dresden, Germany. Total coliforms and *E. coli* were monitored following DIN EN ISO 9308-2 using Colilert© Quanti-Tray© from IDEXX Laboratories, Inc. (Westbrook, CT, USA) with a 24 h incubation time. Samples containing chlorine were quenched using thiosulfate directly after sampling. Turbidity was measured in Nephelometric Turbidity Units (NTU) with a Turb 430 IR/T from WTW (Weilheim, Germany) following DIN EN ISO 7027 (Nephelometric Turbidity Unit). Operational parameters, such as electrolytic cell current, flow rate, power consumption of the pump, and filtration intervals, were monitored using a system integrated Supervisory Control and Data Acquisition (SCADA) system. The chlorine demand was determined based on [31] by determining the difference between FAC directly after chlorine production and after 30 min at SP3. Combined chlorine mainly caused by reaction with nitrogen compounds was determined as the difference between total chlorine and FAC at SP3 and SP4. UVA-254 was determined using a Lambda 25 PerkinElmer (Waltham, MA, USA) following DIN 38-404-C3. Inorganic DBP analysis (chlorate, chlorite, perchlorate, bromate) for the first trial period was done for a duration of four months following DIN EN ISO 10304-4 and TZW lab method, using an ICS 3000 by Thermo Fischer Scientific (Waltham, MA, USA) having a detection limit of 1 µg/L. In the second short time trial, random samples were analyzed for Trihalomethanes (THMs) following DIN EN ISO 10301, using a 7890A GC/MS by Agilent Technologies (Santa Clara, CA, USA) with a detection limit of 0.1 µg/L. DBP samples were transported to Germany. In order to reduce the number of samples in the first test phase, all samples of a sampling week (generally 2–3) were mixed, conserved, and then analyzed.

The specific energy demand per m^3 of drinking water produced was calculated using SCADA data, summarizing the produced water and the power required for running the system. Here, the energy consumption of the pump lifting the bank filtrate, the pump pushing the water through the system, the electrolytic cell, and the power supply for the control and online monitoring units, was evaluated.

2.4. Solar Energy Supply System

Due to the potentially non-existing or unreliable electricity supply in future target regions of the here tested system, the supply with solar photovoltaic (PV) electricity only was evaluated. Planned or unplanned electricity shortages are permissible when ECl_2 is applied, as the residual disinfectant assures safe water conditions during water storage. This is one main advantage compared to alternative disinfection processes based on, for example, UV radiation, whereby power supply has to be always guaranteed if water is supplied for 24 h per day. For sizing an adequate solar PV system, different combinations of the photovoltaic (PV) generator size and battery capacities were subject to a sensitivity analysis using Homer [32]. The established model hereby considered the technical parameters given in Table 3.

Figure 3a shows the clearness indicies and the global horizontal radiation values at the pilot site. The acutal solar PV generator installed in Haridwar is shown in Figure 3b.

Following the below presented results of the sensitivity analysis, the solar energy supply system installed at the pilot site comprised a 900 Wp PV generator and 2 × 96 Ah (C10) valve-regluated lead-acid (VRLA solar batteries).

Table 3. Technical parameters for sensitive analysis conducted with Homer.

Parameter	Value
Solar radiation	
Average monthly clearness indices for Haridwar in kWh/m^2d *	See Figure 1
Photovoltaic (PV) panel	
Slope, Azimuth	20, 0° West of South
Nominal operational temperature and temperature coefficient	47 °C, −0.5%/°C
PV module Efficiency	14%
PV generator size considered for sensitivity analysis (24 V)	0.6, 0.8, 0.9, and 1.0 kW
Batteries	
Minimum state of charge (SOC)	40%
Battery capacity (C10) considered for sensitivity analysis (24 V)	50, 96, 144 Ah
Load	
Load, day-to-day variability, time-step-to-time-step variability	70 W, 5%, 5%
Operational duration **	22 h per day
Constraints	
Maximum annual capacity shortage, operational reserve	1% (88 h/a), 10% hourly load

* These data were obtained from the NASA Langley Research Center Atmospheric Science Data Center Surface Meteorological and Solar Energy (SSE) web portal, supported by the NASA LaRC POWER Project, and are based on 30-year average meteorological and solar monthly and annual climatology (January 1984–December 2013) averages. ** A 24-h per day supply of solar-generated electricty is economically not feasible in regions with distinct rainy seasons.

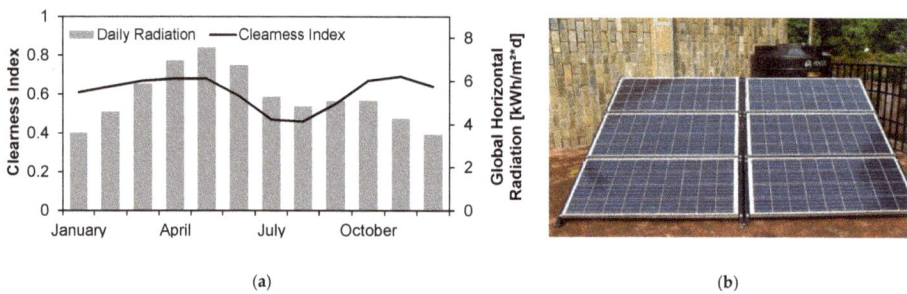

(a)   (b)

Figure 3. Clearness indices and monthly global horizontal radiation values (**a**); and the installed solar system (**b**).

3. Results and Discussion

3.1. System Operation Haridwar

The evaluated trial phase lasted 244 days, including downtimes of in total 24 days. During most of the time the station operated without technical problems and allowed continuous sampling. Reasons for downtimes were, for example, low water levels in the well, infrequent cleaning of PV modules with subsequent power failure, pump failure. Despite the polarity inversion, the operation in Haridwar was challenged from time-to-time by sudden growth of calcareous deposits on the electrolytic cell following elevated levels of hardness in the source water. Those are believed to have initiated crystallization at the cell allowing fast build-up of deposits. Deposits needed to be manually removed from the cell using acid.

3.2. Water Quality Parameters Haridwar

Major water quality parameters of the Ganga River water and bank filtrate of the first sampling period are summarized in Table 4.

Table 4. Water quality of the Ganga River water and bank filtrate.

Parameter	Mean ± SD Ganga River	Mean ± SD Bank Filtrate	n Ganga	n Bank Filtrate
Pathogens				
Total coliforms (MPN/100 mL)	$1.07 \times 10^6 \pm 1.89 \times 10^6$	$8.87 \times 10^1 \pm 1.20 \times 10^2$	10	30
E. coli (MPN/100 mL)	$2.34 \times 10^4 \pm 6.34 \times 10^4$	$1.09 \times 10^1 \pm 1.79 \times 10^1$	13	26
Chemical parameters				
Hardness (mg/L)	92 ± 39	207 ± 40	19	41
Chloride * (titrated) (mg/L)	7 ± 3	14 ± 2	15	20
Ammonium NH_4-N (mg/L)	0.15 ± 0.07	0.23 ± 0.18	34	41
Physico-chemical parameters				
Electrical conductivity (µS/cm)	156 ± 17	403 ± 31	15	38
pH	7.8 ± 0.3	7.5 ± 0.1	23	43
Temperature (°C)	25.5 ± 2.6	24.8 ± 1.1	21	41
ORP [mV]	ND	476 ± 58	ND	38
Ultraviolet absorbance (UVA-254) (1/m)	51.6 ± 28.1	0.8 ± 0.9	14	
Total organic carbon (TOC) (mg/L)	ND	1.96 ± 0.49 **	ND	16

* titrated, ** mix values, MPN-most propable number, ND-not determined.

3.3. Turbidity

Turbidity in the Ganga River averaged to 501 ± 243 NTU and was reduced to 0.55 ± 0.63 NTU in the bank filtrate (Figure 4), underlining the role of bank filtration as a barrier for particle ingress. The AFM filter further reduced the turbidity down to 0.30 ± 0.34 NTU, and by that substantially improved water quality prior to the chlorination. This value could be nearly maintained during the passage in the electrolysis cell but increased to 0.70 ± 1.16 NTU with outliers and 0.40 ± 0.38 NTU without outliers, which were caused by filter-breakthrough. The slight but constant increase of turbidity was traced back to calcareous deposits released from the electrolytic cell after polarity inversion. In the second short term trial, a second media filter was installed to remove those deposits.

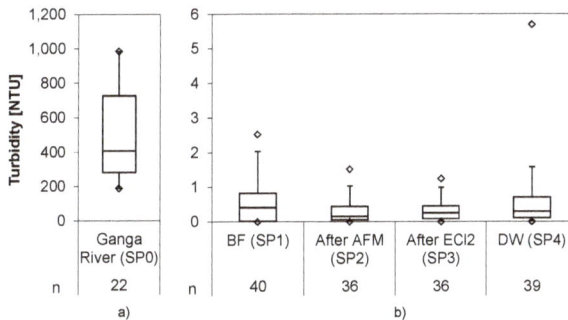

Figure 4. Turbidity values along water treatment.

3.4. Disinfectant Production

Figure 5 shows the ORP values of the bank filtrate and the final drinking water as well as the FAC and total chlorine in the drinking water storage tank.

The ORP increased from 476 ± 58 mV in the bank filtrate to 720 ± 85 mV in the drinking water tank. FAC and total chlorine values reached 0.27 ± 0.17 mg/L and 0.30 ± 0.16 mg/L, respectively. Despite the fact that increased ORP values indicate the presence of chlorine, no direct correlation between both values could be drawn. As reasons for that, the slow reaction time of the ORP sensors in combination with a relatively small storage volume for the tested flow rates were identified. Whenever the ORP sensor indicated a low reading, the system automatically increased the chlorine production and reduced the flow rate. On the other hand, whenever the ORP sensor signaled a high reading, the system automatically decreased the chlorine production and increased the flow rate. Due to the small volume in the drinking water tank, an increase or decrease of chlorine concentration was not detected quickly enough by the ORP sensor. As a consequence, the chlorine concentration oscillated while the system tried to maintain the target ORP value. Because of this oscillation, the chlorine value fell from time-to-time below the minimum target value of 0.2 mg/L and reached high chlorine values of around 1 mg/L. Disabling the automatic flow rate adaptation or using a larger drinking water storage tank would compensate for the delay of the ORP sensors in adjusting to changing chlorine levels, and thus could stabilize the chlorine concentration. The control mechanism was adapted in the second short term trial by removing the automatic adaptation of the flow rate, which proved to produce more constant results.

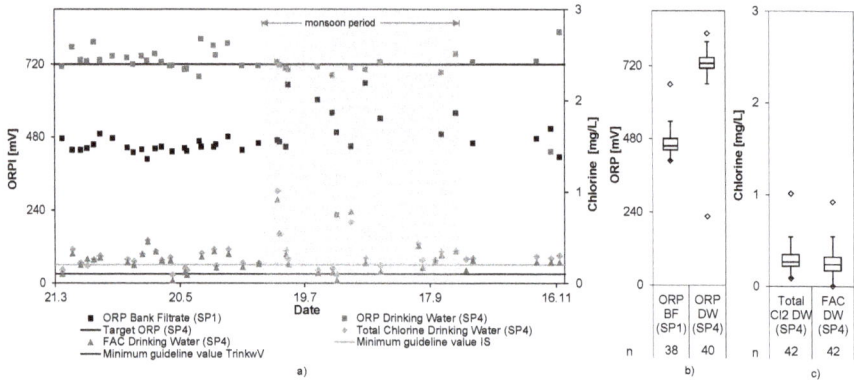

Figure 5. ORP of bank filtrate (SP1), ECl$_2$ effluent (SP4), and chlorine (FAC and total) in ECl$_2$ effluent (SP4) (a–c).

3.5. Pathogens

The analytical results of total coliforms and *E. coli*, as indicator pathogens, for the first trial period are shown in Figures 6 and 7.

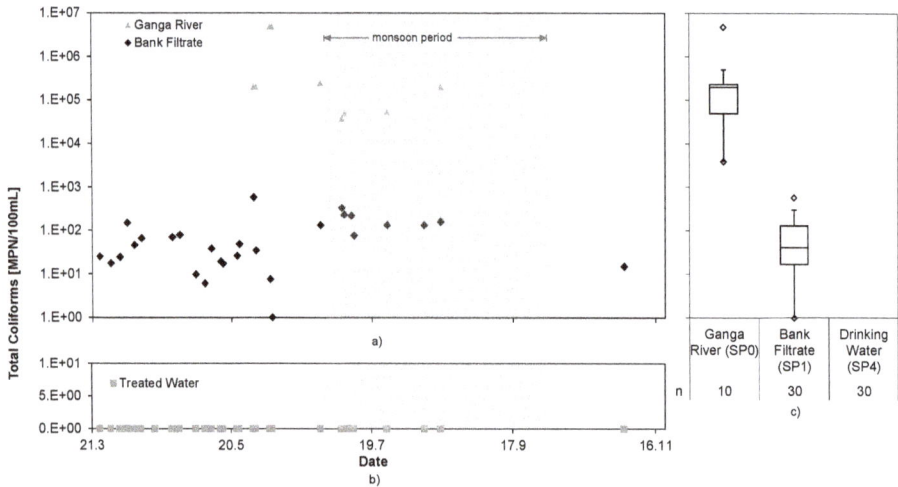

Figure 6. Total coliforms in the Ganga River water and bank filtrate (**a**) and drinking water (**b**), and statistical interpretation (**c**); trial period March–November.

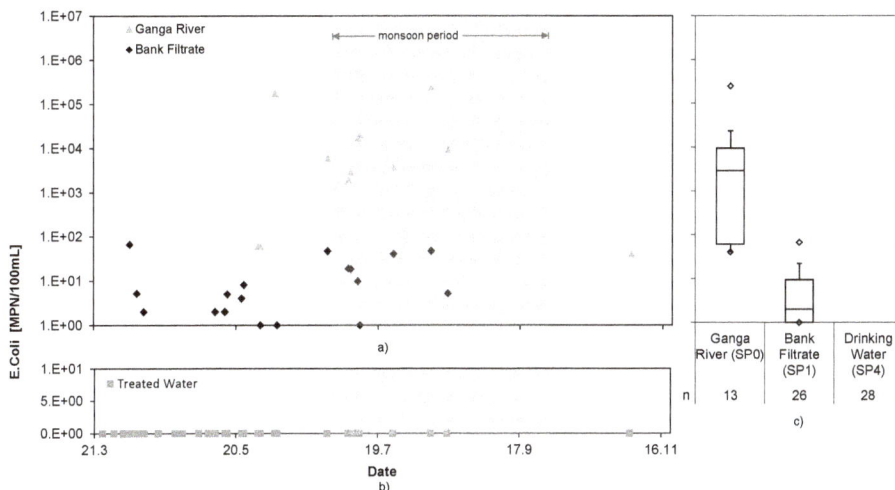

Figure 7. *E. coli* in the Ganga River water and bank filtrate (**a**) and drinking water (**b**), and statistical interpretation (**c**); trial period March–November.

The bank filtration achieved a $\log_{10}$ reduction of 3.9 and 3.6 for total coliforms and *E. coli*, respectively. It is assumed that the peak values in the bank filtrate did not originate from the Ganga River water, but rather came from seepage into the well from above, as described in [19]. However, the ECl_2 system completely removed still present fecal indicators and water could be kept microbially safe at all times. The maximum log reduction of the RBF + ECl_2 cNES achieved was >6.7 for total coliforms and >5.4 for *E. coli*. It can be assumed that even higher log reductions could be reached, considering presence of FAC in the treated water.

3.6. Chlorine Demand

During the trial period the chlorine demand (ΔFAC) was very low, with 0.03 ± 0.03 mg/L on average, but peaked to 0.33 mg/L. Combined chlorine, formed directly at SP3 and SP4, were 0.09 ± 0.08 mg/L and 0.02 ± 0.02 mg/L, respectively (Figure 8).

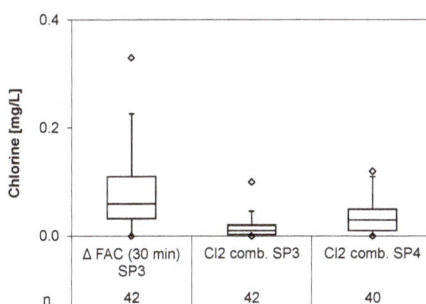

Figure 8. Combined chlorine formed at SP3 and SP4 as well as chlorine demand.

The low average values resulted from the low concentrations of ammonium and organics in the bank filtrate (see Table 4) and do not indicate any critical potential for organic disinfection by-product formation. However, as there are substantial fluctuations of organic and nitrogen compounds in the well water, an automated adaption of the chlorine production process is required to compensate for changing chlorine demand and combined chlorine formation. Even though there was no good correlation between ORP and chlorine concentration, the ORP can indicate insufficient supply of disinfectant and can; therefore, next to cell current and flow rate, be consideredas an additional parameter to control chlorine production.

3.7. Electrical Conductivity and Chloride Concentration

The correlation between electrical conductivity and chloride concentration is shown in Figure 9.

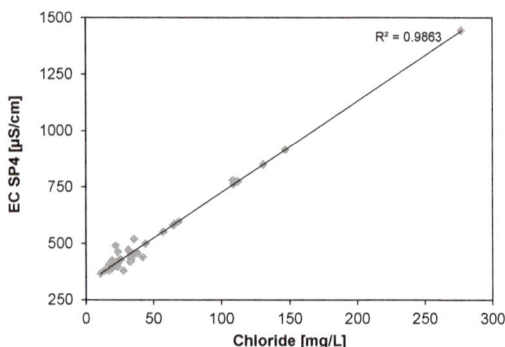

Figure 9. Correlation between conductivity and chloride concentration in treated water (SP4).

The scatterplot indicates that, especially at higher chloride concentrations, the effect of the chloride on the conductivity prevails towards other ions. In order to limit the NaCl consumption, the chloride concentration was supposed to be kept below 50 mg/L. The incidents when higher concentrations occurred could be tracked back to either a nearly empty feed tank, into which the dosing pump dosed too much chloride for the water available, or to calcareous deposits that were formed on the cell. Those deposits have hampered the ability to reach a voltage/current ratio of ≤3.

3.8. Inorganic Disinfection By-Products

The concentrations of the mixed samples for chloride and chlorate in SP3 (directly after the electrolysis cell) are shown in Figure 10.

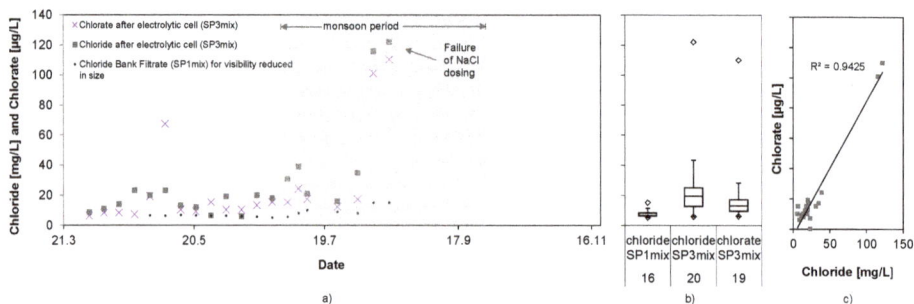

Figure 10. Chloride and chlorate concentrations during field test (**a**,**b**), and their correlation (**c**).

Even after the long storage period of several weeks until analysis in Germany, the chlorate concentrations reached only 22 ± 29 µg/L and is not of concern considering WHO and German guideline values (Table 1). Uncertainty towards the maximum chlorate values exists due to the mixing of two to three random samples into one sample per week, as the concentration of samples with higher concentration might have been lowered with samples of lower concentration. However, the correlation between chloride concentration and chlorate production (Figure 10c) show that higher chloride concentrations are required to reach elevated levels of chlorate. The two maximum chlorate concentrations above 100 μg/L went along with an excess of chloride added into the feed tank. Chlorite and perchlorate were always below the detection limit of 1 μg/L and are; therefore, not of concern when water is disinfected by means of inline-electrolysis with the here applied MOX-electrodes.

3.9. Hardness

Figure 11 shows the total hardness values measured during the first system trial.

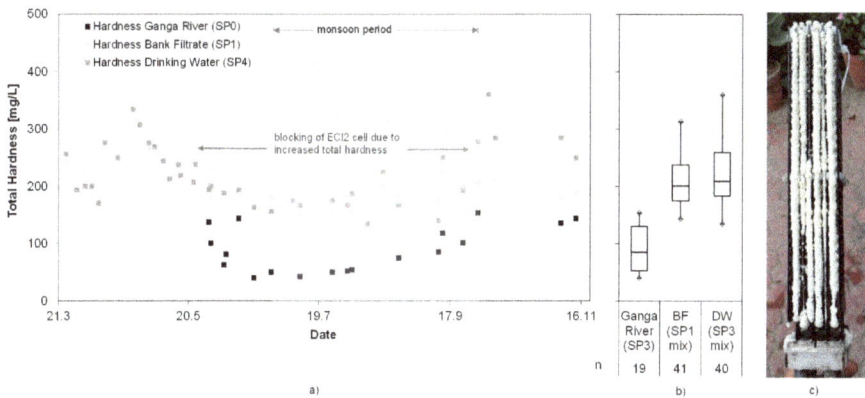

Figure 11. Total hardness values in SP0 and SP1 during the first long term test (**a,b**); and calcareous deposits on the ECl$_2$-cell (**c**).

The hardness values were fluctuating throughout the test period in the Ganga River (92 ± 39 mg/L) and the bank filtrate (207 ± 40 mg/L). Whereas the values in the bank filtrate ranged around an unproblematic 200 mg/L during monsoon, the levels reached ~300 mg/L before and after the monsoon season. Those values have shown to be problematic for system operation, as spontaneous growth of calcareous deposits on the ECl$_2$ cell intermittently reduced chlorine production efficiency and required extra maintenance.

3.10. Second Optimized Test Phase

In the second short term test the automatic flow rate adaption was disabled and constant flow rates of 160, 220, and 280 L/h were established This was giving the ORP sensor sufficient time to detect changing chlorine concentrations and allowed the ability to test the system's reaction on changing chlorine demand in SP4, adjusting the cell current only. The main results of the second short term test are presented in Figure 12.

Despite the very short duration of this second pilot trial, it was sufficient to show that the fluctuation around the set target ORP of 700 mV could be reduced to an acceptable level. The constant flow rates permitted an adequate utilization of the ORP reading for controlling the chlorine production and keeping the concentration in the desired range of 0.26 ± 0.04 mg/L. The presence of FAC in this concentration range was represented by elevated ORP values of ~700 mV. It can be assumed that a larger drinking water storage tank (SP4) would have had a similar effect. Pathogens could still be completely removed through ECl$_2$.

Further, the effect of the second AFM filter, placed behind the electrolytic cell to remove calcareous deposits, and the THM concentrations are shown in Figure 13b.

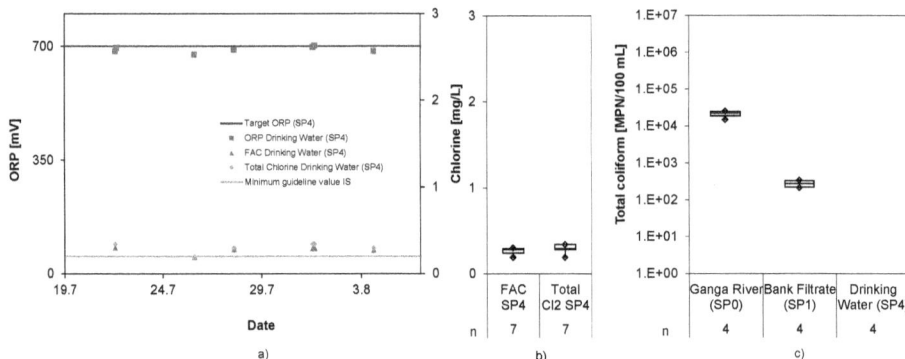

Figure 12. ORP and chlorine concentration in SP4 (**a**,**b**); and pathogenic contamination in short term test with optimized system setting (**c**).

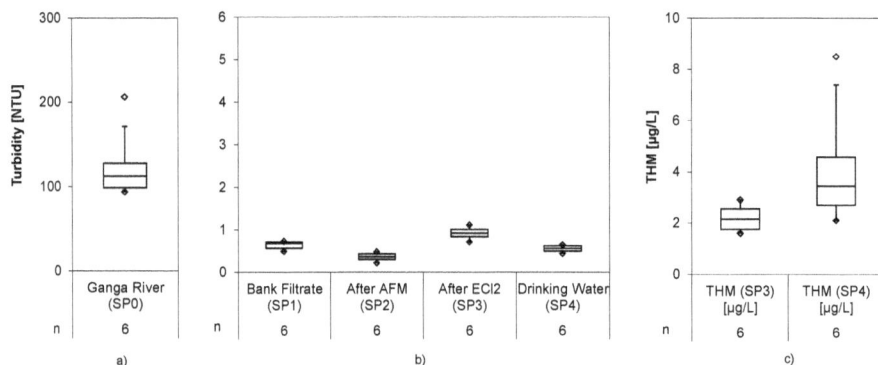

Figure 13. Turbidity removal (**a**,**b**) and THM formation (**c**) during short term test with optimized system setting.

The second filter reduced the turbidity down to 0.55 ± 0.08 NTU, after it had increased to 0.92 ± 0.13 NTU behind the electrolytic cell, improving overall water quality. The THM analysis showed concentrations of 2.2 ± 0.5 µg/L in SP3 after 30 min and 4.2 ± 2.1 µg/L in SP4. Those low concentrations were expected due to the low TOC and dissolved organic carbon DOC content of the water, allowing full compliance even with strict guideline values for DBPs (Table 1).

3.11. Energy Demand and Solar Energy Supply

During the first trial period a water volume of 1037 m^3 was treated and the total electricity demand was summed up to 271 kWh without and 412 kWh with bank filtrate pumping. The average flow rate through the system, including off times (e.g., at night, during maintenance or repair) in Haridwar, was 180 L/h. This resulted in an average power demand of 46 W without and 70 W with bank filtrate pumping, and a per m^3 energy consumption of 0.4 kWh (Table 5).

Table 5. Energy demand of the tested ECl$_2$ system.

Energy Demand in [kWh/m^3]	Water Pumping through System incl. AFM Filtration	Inline Electro-lysis	Auxiliary and Online Monitoring	Total Demand Water Treatment	Pumping of Bank Filtrate (BF)	Total Demand incl. Pumping
RBF AFM ECl$_2$ station in Haridwar	0.06	0.19	0.02	0.26	0.14	0.40

The power requirement of 70 W was used for the sensitivity analysis using different sized PV generators and different battery capacities considering the parameters mentioned in Table 3. The analysis shows that at least a 96 Ah battery system with minimum 800 Wp are required to power the water treatment system for 22 h per day and a permitted capacity shortage of 1%. The results are summarized in Table 6.

Table 6. Results of sensitive analysis based on shown PV battery combinations.

PV Generator Size and Battery Capacity [W], [Ah]	Total Prod. [kWh/a]	Total Cons. [kWh/a]	Input Battery [kWh/a]	Output Battery [kWh/a]	Excess Electricity [kWh/a], [%]	Unmet Load [kWh/a], [%]	Capacity Shortage [kWh/a] [%]
600, 96	928	553	334	278	319, 34.4%	9.07, 1.6%	10.3, 1.8%
600, 144	928	560	343	285	310, 33.4%	1.90, 0.3%	2.13, 0.4%
800, 96	1237	559	332	277	622, 50.1%	3.10, 0.6%	3.51, 0.6%
800, 144	1237	561	335	279	619, 50.3%	0.80, 0.1%	0.89, 0.2%
900, 96*	1392	560	330	275	776, 55.8%	2.13, 0.4%	2.38, 0.4%
900, 144	1392	562	332	277	774, 55.6%	0.48, 0.1%	0.53, 0.1%
1000, 96	1546	561	339	274	930, 60.2%	1.58, 0.2%	1.78, 0.3%
1000, 144	1546	562	330	275	929, 60.1%	0.16, <0.1%	0.19, <0.1%

* installed solar PV-battery system combination.

The simulation shows, that with the installed solar PV-battery combination of 900 Wp and 2×96 Ah, the required electricity can be supplied nearly throught the year. Only a few hours of electrictiy shortages including a 7-h long power cut in the middle of the monsoon in August occured. The total capacity shorage summed up to 0.4%. Whether this is permissible in a real case scenario and whether this could be compensated by, for example, an increase of the water storage capacity depends on local conditions. The simulated power production and SoC is shown in Figure 14a,b, respectively.

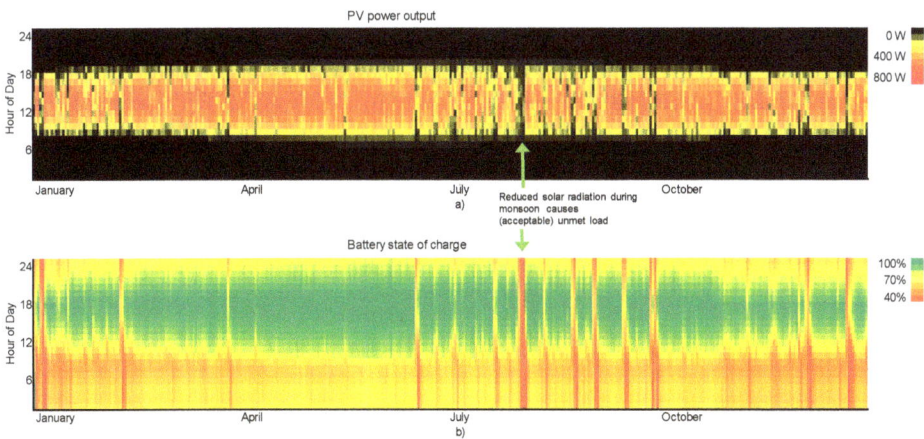

Figure 14. Power output from 900 Wp PV panel (**a**) and corresponding State of Charge (SoC) (**b**).

During the trial the station was continuously running on the given solar PV system, as long as the modules were cleaned from dust frequently.

4. Conclusions

The presented results of a first long term and a second short term trial of a RBF ECl_2 combination in India show that the tested system poses a feasible alternative for decentralized and safe drinking water supply for river bound communities in developing countries. RBF serves as a very efficient pre-treatment step, substantially reducing pathogens, turbidity, and DBP precursors. The installed AFM filter is capable of further reducing the already low turbidity values and the ECl_2 system completely removes all still-present indicator pathogens, and supplies sufficient residual disinfectant for safe water distribution. The station complies with given water regulations concerning indicator pathogen and chlorine concentrations. Additionally, the production of DBPs is of no concern and stays well below the given guideline values. The first test period revealed some optimization potential of the control algorithm and the system setting, which was successfully implemented for the second trial. After that, the system reacted reliably to changing source water and operating conditions by keeping the residual disinfectant at a constant level. The used ORP sensor is able to indicate "sufficient" or "insufficient" disinfectant once it is given sufficient reaction time. For more accurate and faster readings, the application of chlorine probes may be an alternative to ORP probes.

Clogging of the electrolytic cell, due to increased levels of hardness, remains an operational challenge of the ECl_2, which needs to be specifically addressed. With polarity inversion alone, and total hardness levels above 200 mg/L, the operation of an ECl_2 system currently reduces the maintenance intervals to about once every two months. The operation of the ECl_2 system at total hardness values above 300 mg/L is currently not advisable. With the implementation of an additional probe to measure electrical conductivity, cell overgrowth could be detected by monitoring current and voltage of the cell and comparing them with the actual conductivity of the water.

After this trial, the system is mature enough to be implemented in a real scenario and under favorable operational conditions and source water quality, and it should be possible to reduce the maintenance intervals of the station to six months. For this, the implemented SCADA system will play an important role. An increase of the treatment capacity is straight forward by increasing the ECl_2-cell size and the solar energy supply accordingly.

Author Contributions: P.O., A.G., and F.B. developed the pilot station; P.O. reviewed the literature, analyzed the data, and prepared the draft of the publication; F.B. dimensioned the solar energy supply system; P.M., S.K.S., and P.C.K. took care of the infrastructure for the pilot study; P.M., G.N., and M.W. analyzed water samples and contributed expertise in interpreting results; and C.S. and T.G. supported the planning and design of the experiments and the preparation of the draft. All co-authors reviewed and edited the draft.

Funding: All primary data was collected within the AquaNES project. This project has received funding from the European Union's Horizon 2020 Research and Innovation Program, under grant no. 689450

Acknowledgments: This study was based on the excellent support from Uttarakhand Jal Sansthan (UJS), which provided the pilot locations and supported in-system construction and operation, and the work performed by many staff members of UJS. The authors also gratefully acknowledge support from S. Hertel, for supporting literature review, and from F. Kowalczyk, F. Bauer, R. Rajoriya, F. Naumann, Binod Das, and P. Patwal during sampling, water analysis, and system maintenance.

Conflicts of Interest: The authors declare no conflict of interest.

References

1. Onda, K.; LoBuglio, J.; Bartram, J. Global access to safe water: Accounting for water quality and the resulting impact on MDG progress. *Int. J. Environ. Res. Public Health* **2012**, *9*, 880–894. [CrossRef] [PubMed]
2. Bain, R.; Cronk, R.; Hossain, R.; Bonjour, S.; Onda, K.; Wright, J.; Yang, H.; Slaymaker, T.; Hunter, P.; Prüss-Ustün, A.; et al. Global assessment of exposure to faecal contamination through drinking water based on a systematic review. *Trop. Med. Int. Health* **2014**, *19*, 917–927. [CrossRef] [PubMed]
3. Bain, R.; Cronk, R.; Wright, J.; Yang, H.; Slaymaker, T.; Bartram, J. Fecal contamination of drinking-water in low- and middle-income countries: A systematic review and meta-analysis. *PLoS Med.* **2014**, *11*, e1001644. [CrossRef] [PubMed]
4. World Health Organization (WHO). Drinking Water Key Facts. 2018. Available online: http://www.who.int/en/news-room/fact-sheets/detail/drinking-water (accessed on 2 October 2018).
5. Hossain, M.A.; Sengupta, M.K.; Ahamed, S.; Rahman, M.M.; Mondal, D.; Lodh, D.; Das, B.; Nayak, B.; Roy, B.K.; Mukherjee, A.; et al. Ineffectiveness and Poor Reliability of Arsenic Removal Plants in West Bengal, India. *Environ. Sci. Technol.* **2005**, *39*, 4300–4306. [CrossRef] [PubMed]
6. Sobsey, M.D.; Handzel, T.; Venczel, L. Chlorination and safe storage of household drinking water in developing countries to reduce waterborne disease. *Water Sci. Technol. A J. Int. Assoc. Water Pollut. Res.* **2003**, *47*, 221–228. [CrossRef]
7. Montgomery, M.A.; Elimelech, M. Water and Sanitation in Developing Countries: Including Health in the Equation. *Environ. Sci. Technol.* **2007**, *41*, 17–24. [CrossRef] [PubMed]

8. Roberts, L.; Chartier, Y.; Chartier, O.; Malenga, G.; Toole, M.; Rodka, H. Keeping clean water clean in a Malawi refugee camp: A randomized intervention trial. *Bull. World Health Organ.* **2001**, *79*, 280–287. [PubMed]
9. Lorenzen, G.; Sprenger, C.; Taute, T.; Pekdeger, A.; Mittal, A.; Massmann, G. Assessment of the potential for bank filtration in a water-stressed megacity (Delhi, India). *Environ. Earth Sci.* **2010**, *61*, 1419–1434. [CrossRef]
10. Dash, R.R.; Bhanu Prakash, E.V.P.; Kumar, P.; Mehrotra, I.; Sandhu, C.; Grischek, T. River bank filtration in Haridwar, India: Removal of turbidity, organics and bacteria. *Hydrogeol. J.* **2010**, *18*, 973–983. [CrossRef]
11. Weiss, W.J.; Bouwer, E.J.; Aboytes, R.; LeChevallier, M.W.; O'Melia, C.R.; Le, B.T.; Schwab, K.J. Riverbank filtration for control of microorganisms: Results from field monitoring. *Water Res.* **2005**, *39*, 1990–2001. [CrossRef]
12. Sandhu, C.; Grischek, T.; Kumar, P.; Ray, C. Potential for Riverbank filtration in India. *Clean Technol. Environ. Policy* **2011**, *13*, 295–316. [CrossRef]
13. Wang, J.; Smith, J.; Dooley, L. (Eds.) *Evaluation of Riverbank Infiltration as a Process for Removing Particles and DBP Precursors*; American Water Works Association: Denver, CO, USA, 1996.
14. Grischek, T.; Paufler, S. Prediction of Iron Release during Riverbank Filtration. *Water* **2017**, *9*, 317. [CrossRef]
15. Sandhu, C.; Grischek, T. Riverbank filtration in India—Using ecosystem services to safeguard human health. *Water Sci. Technol. Water Supply* **2012**, *12*, 783–790. [CrossRef]
16. Wintgens, T.; Nattorp, A.; Elango, L.; Asolekar, S.R. Natural Water Treatment Systems for Safe and Sustainable Water Supply in the Indian Context: Saph Pani. *Water Intell. Online* **2016**, *15*, 9781780408392. [CrossRef]
17. Ghodeif, K.; Grischek, T.; Bartak, R.; Wahaab, R.; Herlitzius, J. Potential of river bank filtration (RBF) in Egypt. *Environ. Earth Sci.* **2016**, *75*, 255. [CrossRef]
18. Pholkern, K.; Srisuk, K.; Grischek, T.; Soares, M.; Schäfer, S.; Archwichai, L.; Saraphirom, P.; Pavelic, P.; Wirojanagud, W. Riverbed clogging experiments at potential river bank filtration sites along the Ping River, Chiang Mai, Thailand. *Environ. Earth Sci.* **2015**, *73*, 7699–7709. [CrossRef]
19. Bartak, R.; Page, D.; Sandhu, C.; Grischek, T.; Saini, B.; Mehrotra, I.; Jain, C.K.; Ghosh, N.C. Application of risk-based assessment and management to riverbank filtration sites in India. *J. Water Health* **2015**, *13*, 174–189. [CrossRef] [PubMed]
20. Hashmi, I.; Farooq, S.; Qaiser, S. Chlorination and water quality monitoring within a public drinking water supply in Rawalpindi Cantt (Westridge and Tench) area, Pakistan. *Environ. Monit. Assess.* **2009**, *158*, 393–403. [CrossRef] [PubMed]
21. Clasen, T.; Haller, L.; Walker, D.; Bartram, J.; Cairncross, S. Cost-effectiveness of water quality interventions for preventing diarrhoeal disease in developing countries. *J. Water Health* **2007**, *5*, 599–608. [CrossRef] [PubMed]
22. Morris, R.D.; Audet, A.M.; Angelillo, I.F.; Chalmers, T.C.; Mosteller, F. Chlorination, chlorination by-products, and cancer: A meta-analysis. *Am. J. Public Health* **1992**, *82*, 955–963. [CrossRef]
23. World Health Organization (WHO). *Guidelines for Drinking-Water Quality*; World Health Organization: Geneva, Switzerland, 2017.
24. Kraft, A. Electrochemical Water Disinfection: A Short Review. *Platin. Met. Rev.* **2008**, *52*, 177–185. [CrossRef]
25. European Commission. *Proposal for a Directive of the European Parliament and of the Council on the Quality of Water Intended for Human Consumption (Recast)*; European Commission: Brussels, Belgium, 2018.
26. Otter, P. Experimental Determination of the Optimization Potential of an Energy Autarkic Drinking Water Purification System Under the Consideration of Local Conditions in the Rubber Trapper Reserve "RESEX do Rio Ouro Preto" Located in the Brazilian State of Rondônia. Master's Thesis, University of Kassel, Kassel, Germany, 2010.
27. Kraft, A.; Blaschke, M.; Kreysig, D.; Sandt, B.; Schröder, F.; Rennau, J. Electrochemical water disinfection. Part II: Hypochlorite production from potable water, chlorine consumption and the problem of calcareous deposits. *J. Appl. Electrochem.* **1999**, *29*, 895–902. [CrossRef]
28. Schmidt, W. *Untersuchungen zur Desinfektionswirkung und Sicherheit der In-line-Elektrolyse von Chlor als umweltschonendes Verfahren für die Desinfektion von Trinkwasser—In-line-Elektrolyse für die Trinkwasserdesinfektion (Investigations on the Disinfecting Effectiveness and Safety of the In-Line Electrolysis of Chlorine as an Environmentally Friendly Process for the Disinfection of Drinking Water In-Line Electrolysis for Drinking Water Disinfection)*; DBU Report; Deutsche Bundesstiftung Umwelt: Osnabrück, Germany, 2012.

29. Haaken, D.; Dittmar, T.; Schmalz, V.; Worch, E. Influence of operating conditions and wastewater-specific parameters on the electrochemical bulk disinfection of biologically treated sewage at boron-doped diamond (BDD) electrodes. *Desalin. Water Treat.* **2012**, *46*, 160–167. [CrossRef]
30. Kraft, A.; Stadelmann, M.; Blaschke, M.; Kreysig, D.; Sandt, B.; Schröder, F.; Rennau, J. Electrochemical water disinfection Part I: Hypochlorite production from very dilute chloride solutions. *J. Appl. Electrochem.* **1999**, *29*, 859–866. [CrossRef]
31. Deutscher Verein des Gas und Wasserfaches. *DVGW W 296 (A): Trihalogenmethanbildung—Vermindern, Vermeiden und Ermittlung des Bildungspotentials (Trihalogenmethane Formation—Reduction, Avoidance and Determination of Formation Potential)*; Wirtschafts- und Verlagsgesellschaft Gas und Wasser mbH: Bonn, Germany, 2014.
32. Lilienthal, P.; Lambert, T.; Gilman, P. *Homer: The Micropower Optimization Model*; Homer Energy, LLC.: Boulder, CO, USA, 2009.

water

MDPI

Case Report

Operational Strategies and Adaptation of RBF Well Construction to Cope with Climate Change Effects at Budapest, Hungary

Zsuzsanna Nagy-Kovács *, Balázs László, Elek Simon and Ernő Fleit

Budapest Waterworks Ltd., H-1138 Budapest, Hungary; balazs.laszlo@vizmuvek.hu (B.L.); elek.simon@vizmuvek.hu (E.S.); erno.fleit@vizmuvek.hu (E.F.)
* Correspondence: zsuzsanna.nagy-kovacs@vizmuvek.hu; Tel.: +36-30-439-4566

Received: 17 September 2018; Accepted: 22 November 2018; Published: 28 November 2018

Abstract: The objective of this paper is to give an overview on the Hungarian experience of river bank filtration (RBF) systems. The study addresses the conflict, which arises between the stochastic character of river water quantity and quality, and the required standard of drinking-water supply. Trends in water levels, flow, and water quality are discussed, along with technical measures and operational rules that were developed for implementation of RBF systems. This paper also provides an overview of the average lifespan of the wells and operational strategies. The emerging reconstruction and reconditioning needs are highlighted, and existing alternatives are presented. Large-scale infrastructural elements, such as the Danube-based RBF systems, have to be adapted to a changing environment. The increasing frequency of floods and droughts stresses the need to implement climate-adapted RBF systems and related operational strategies. Operational strategies which were developed by the Budapest Waterworks to deal with extreme hydrological scenarios are presented.

Keywords: river bank filtration; hydrological trends; sustainable water production; well structure remodeling

---

1. Introduction

The impact of climate change on the hydrological cycle poses a serious challenge to the water industry and society as a whole [1]. Basin flooding and low-water periods of the Danube river are major stress factors for river bank filtration (RBF) systems. Many existing RBF systems can only be operated within a certain range of river-water level [2]. The past century has seen increasing low-water periods for the Danube, whereby the mean water level has reduced by more than 1 meter. Regional interferences (construction of reservoirs in the upstream section) and global/regional climate change (increase of annual water temperatures and critically low water levels) require the development of novel well scheme operation methods [3]. RBF-based water supply systems are exposed to stochastic processes affected by climatic change [4]. Safety and security requirements of water supply contradict these natural processes [5]. Coming up with solutions for these contradictions is a primary focus for the waterworks. Climatic changes have a profound effect on river water levels [6]. More frequent flooding affects not only the flow conditions at RBF systems, but also the quality of the bank filtrate and abstracted water [7]. In addition to this, water demands are fluctuating, land-use patterns, and especially agricultural activities are changing, and energy demands and costs are increasing. However, water quality effects are somewhat tempered in the case of RBF as they are damped, compared to surface water abstraction [8].

During RBF, a significant removal of pollutants occurs in the river bed and the aquifer zone nearest to the river bed, where microbiological activity is high. Due to the changes in river water levels, the wetted zone can be increased by up to 50% under favorable conditions [9,10]. In extreme low-water periods, the microbiologically active layer decreases in size, resulting in lower quality of the bank filtrate. Given the potential of the Danube for water supply via RBF in particular, Danubian countries must have a long-term strategy that aims to mitigate problems rooted in global climate change and its aquatic consequences (droughts, floods, etc.) [11,12].

RBF systems can effectively remove pathogenic microorganisms, suspended solids, algae and their toxins, dissolved organic matter, ammonium, disinfection by-product precursors, etc. [13]. The removal efficiency greatly depends on the flow regime of the river; therefore, RBF systems are also sensitive to climate change, although mainly the increased frequencies of extreme water levels [14]. Low-water periods (Figure 1) can cause problems in terms of both water quantity and quality. Long rainy periods can lead to a deterioration of water quality, especially in times of flooding and strong erosion. Under high-flow conditions, the infiltration of river water into the aquifer increases, with a simultaneous increase in the flux of oxygen and organic carbon (3–4 times higher rates), which microorganisms in the active zone (acclimatized to lower flow rates) are not able to fully process. When low-water periods occur, microbiological non-compliances increase, particularly when the temperature of river-bank filtrate is above 15 °C [15]. In addition to this, a significantly higher number of pathogens is measured in river water during floods. In such cases, water disinfection becomes particularly important. However, compared to surface and groundwater extraction and treatment, RBF is considered to be a more sustainable alternative and less sensitive to climate change [16].

The operation of RBF systems under changing environmental conditions requires a careful analysis of future operational strategies, including the assessment of reconstruction needs at existing RBF systems in Budapest [17].

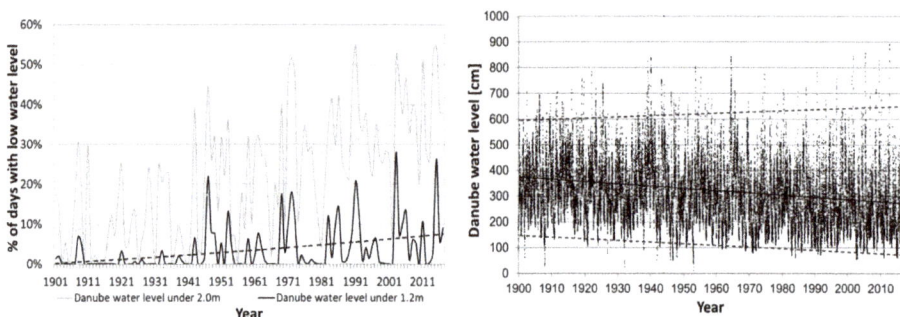

Figure 1. Percentage of days when Danube water levels were below 2 m and below 1.2 m (1901–2018) (**left**); long-term trend of the Danube water level (1646.5 km, Budapest) (**right**).

The treatment capacity of RBF systems is determined by the following processes:

- Hydrodynamic processes (convective—dispersive transport—transport, mixing, and dilution);
- Mechanical processes (natural filtration);
- Microbiological transformations (biodegradation by microorganisms);
- Physical–chemical and chemical processes (sorption, precipitation, redox processes).

RBF systems are affected by environmental and technical factors [4,7]. Water temperature influences the water's viscosity and thus also the infiltration and flow velocity, resulting in higher abstracted water volumes at higher water temperatures [4]. Temperature also affects biochemical processes and microbial activities, which ultimately affects water quality [8]. The input of waste water into a river determines the quality of the bank filtrate. For example, wells located upstream of the city of Budapest provide excellent water quality, whilst the ones located downstream require iron and manganese removal due to anoxic conditions [3].

Directions of water flow into or out of the aquifer are governed by the water levels of the river and the land-side groundwater [10]. Water-level fluctuations have a strong influence on aquifer saturation, the size of the biologically active zone, and the quantity of biofilms. These fluctuations determine the characteristics of the transportation and flow into the unsaturated zone, as this zone has lower removal capability than the saturated one. In the case of high water levels, river water also enters a previous non-wetted zone, resulting in deteriorated water quality [11]. RBF is an ultra-slow filtration process whereby the transport rate is only about 0.1–0.25 m/day at the bank surface. (For comparison, this value is 5 m/day at slow sand filtration, and 240 m/day at rapid sand filtration.)

River-bed regulations can have a negative effect on both the volume and quality of bank filtrate, as evidenced by an example from the Colorado River [6]. In the Hungarian Danube section, gravel dredging has regularly been done to improve navigation conditions. This causes the width of the infiltration zone to decrease, and its depth to change also. River regulations resulted in the decrease of river-bed load transport, and no further coarse fractions were sedimented. River training dams perpendicular to the flow direction also resulted in a siltation of the river bed. This induced anoxic conditions and, consequently, deterioration of water quality by the dissolution of iron and manganese.

The hydrogeological properties of the aquifer zone near the river, such as depth, particle size, transport conditions, hydraulic conductivity (K value), mineral composition, and pollution state cannot be influenced by the operator [17]. However, technical parameters can only be managed within certain limits, such as the type of water intake, well distance from the river, produced water volumes, and pump operation.

In the following, based on an overview of the Budapest experience with RBF systems, effects of climate change and changes in water quality will be discussed together with technical measures and operational strategies developed for adaptation of the RBF systems. Furthermore, an overview of construction and lifespans of RBF wells, as well as adaptation measures will be provided.

2. RBF Site Description

2.1. Characteristics of the RBF Site

The Szentendre Island is situated in the center of Hungary, upstream of Budapest. Its length is 30.85 km, and the average width is 2.3 km, with a maximal width of 3.5 km. Its area is 55.73 km^2. Its height above sea level varies between 100.0–123.5 m. The average yearly precipitation is 500–600 mm. The climate represents typical lowland continental conditions. There are no permanent surface water bodies on the island, and temporal wetlands are dispersed over its area. The mean annual air temperature is 11 °C.

2.2. Hydrogeology

The site is a separate geographical, hydrological, and ecological unit. From the northernmost tip to the central area, the Szentendre Island is formed by fluvial layers of upper Oligocene clay. On the southern part of the island, Miocene sandy and clay formations can be found. Tertiary sediments are covered by gravel of 7–9 m thickness. The gravel composition of the island is considered to be relatively diverse; however, it can be described as being predominantly sandy and locally intercepted with clay. Fluvial sand is found in the top 3–5 m of the gravel layer.

The layers of the active watershed area above the clayey substratum are gravelly, sandy, and silty. As the cover is predominantly clay or silt, the gravel layer can be considered as being semi-protected. The groundwater level follows the fluctuations in the Danube water level, and is not influenced by changes of the relief; hence, on the higher part of the island, the groundwater level reaches a depth of 8–10 m below the surface. Water communicating with the river flow decreases with distance from the river banks. Normally, groundwater flows toward the southern part of the island, while closer to the river bank, the main direction is perpendicular to the river flow. The water balance of the island is determined by the Danube water level. Abstraction wells considerably influence the natural groundwater flow and level. Abstracted water is predominantly bank filtrate; however, production can decrease the groundwater level in the inner parts of the island. The aquifer beneath the island is recharged when the river water level is above 2 m [3].

2.3. Historical Background

The construction of RBF wells on Szentendre Island started at the end of the 19th century, when large-capacity shaft wells were constructed on the southern parts of the island (Figure 2). As the population of the capital grew and the relative water consumption increased, drilled wells were installed around the middle of the 20th century. This well type proved to be suitable for the utilization of shallow aquifers. Finally, in the 1970s, mainly horizontal collector wells were constructed on the northern parts of the island.

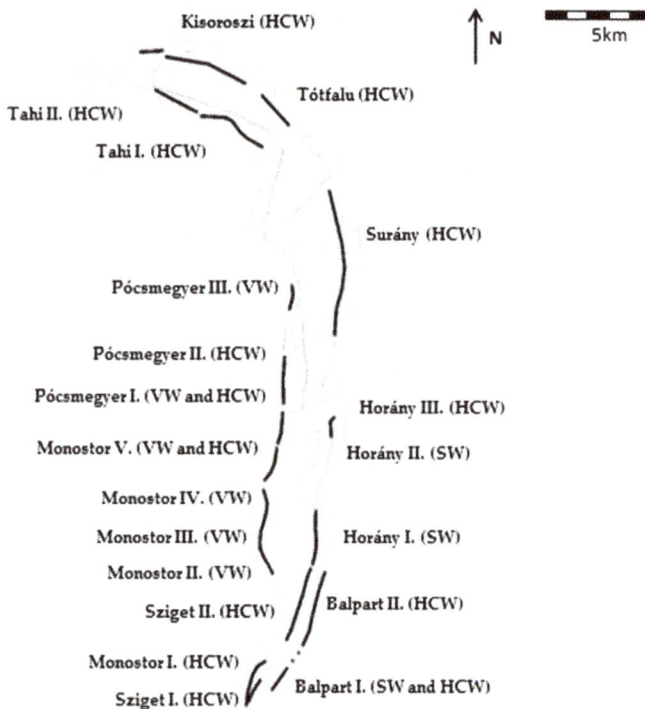

Figure 2. Location and names of well groups of the Budapest Waterworks at Szentendre Island. HCW: horizontal collector well, VW: vertical well, SW: shaft well.

Today, Budapest Waterworks operates 756 RBF wells to supply 1.89 million inhabitants. The wells are predominantly located on Szentendre Island and Csepel Island. The maximum capacity of the RBF systems of Budapest Waterworks is 1.0 million m^3/day, and the average supply is about 456,000 m^3/day. Compared to the average discharge of the Danube River at Budapest of 200 million m^3/day, the water abstraction via bank filtration from the river is, on average, only 0.23%. Figure 3 shows the volume of water fed into the distribution system from 1950 until today. Water consumption followed a continuously increasing trend until 1990, but the economic transition brought changes that influenced both the industrial and public water demands. Decreased industrial activity, an increase in water fees, and the availability of water-saving household devices brought about a change in consumption patterns. The dropping trend in consumption levelled out by the mid-90s, and by the first decade of the 21th century the average consumption settled at 120 L/day/capita.

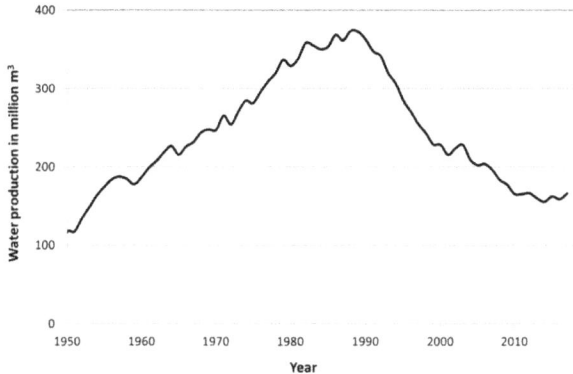

Figure 3. Annual water production from river bank filtration (RBF) schemes in Budapest, 1950–2016.

3. Discussion of Flow Conditions and Operational Strategies

The volume of potentially produced drinking water and its quality is largely governed by the already discussed environmental-hydrological factors. The waterworks has limited options to control the listed RBF processes. Infiltration rates and filtration velocity can be controlled by the abstraction rates of the wells. Water quality can also be controlled to a certain degree by controlling the hydraulic conditions of the RBF systems by adjusted well operation.

The present operational strategy has been developed throughout several decades, and continuously adapted to ever-changing requirements and various hydrological scenarios. As an operational rule of thumb, the Budapest Waterworks distinguishes three different operation modes, which are the normal, low-water, and flood conditions, where different challenges are faced and have to be mitigated.

3.1. Normal Conditions

Normal conditions are defined by the Danube water level in the range of 120–550 cm (Budapest gauging station, 1646.5 km). In this situation, no risk of well inundation is expected and the water production capacity of the wells can meet consumption needs, according to past experiences and modelling results.

The highest water demand in Budapest was recorded in the eighties, and since then, the demand has seen a steady decline (Figure 3). As a result of this, the available nominal water production capacity of the wells is significantly higher than the average water consumption. That would provide the necessary safety from an operational point, but seasonal changes of water consumption should not be neglected. In between the minimum levels of consumption in winter and maximum in summer, a 50% difference in consumption pattern can be regularly observed. The available water production capacity is also a continuously changing value, largely driven by the Danube water level and water temperature at any given water depression level. The lowest production capacity is observed at concurrently low water levels and low water temperatures, whilst the maximum capacity is expected to occur at high water levels and temperatures. The actual water production capacity is limited by the number of non-operational wells (due to maintenance, reconstruction, or water quality problems). Under "average" conditions, one might falsely conclude that the Budapest Waterworks has unnecessary well capacities. However, it first has to be stated that the maximum water demand must be met at any time. In this situation, the most critical engineering task is to make excluded wells functional again in the shortest time possible. A viable option would be to operate the not required wells at low capacity, thus optimizing the hydraulic conditions in the large-scale collector pipe and distribution systems.

The ecosystem services provided by the RBF systems rely largely, but not entirely, on microbiological processes. This makes the complete shutdown of the wells risky, as the full operation of the microbiological system involved in microbiological water treatment processes requires time to regenerate and to start up. During such a regeneration period, problems in water quality may occur. Depending on the duration of the idle phase and other factors, the full regeneration of RBF processes can take multiple weeks or even months. From an operational perspective, this situation should be avoided. Consequently, the practice which is commonly followed is that instead of completely switched-off periods, some limited operation is maintained at wells or well groups.

This practice is, of course, limited by the minimum pumping capacity of the built-in pumps. In some extreme periods, there are instances whereby water production cannot be decreased further unless certain wells stop operating—a factor which has its own risks, considering the time requirement to restart. For such cases, the following compromise was developed by the Budapest Waterworks. Based on their location, certain well groups (8–22 wells per group) are designated to be part of the solution. These groups have a minimum inter-well distance, and all the wells within each group are individually operated with submersible pumps (not siphon wells). Based on the requirement, every second well is shut down for a period of 24 h, or occasionally, 48 h. Afterwards, an intermittent operational cycle starts, the operating ones are switched off, and the temporarily idled are restarted. As a result of this intermittent operational mode, in the neighborhood of any switched-off well, there are two operational wells. Hence, some water transport and microbiologically active zones are sustained. Based on the operational experiences, it is stated that the water quality of the intermittently operated wells remains comparable to normal-operation well groups [18].

3.2. Low-Water Conditions

Low-water conditions are defined as being when the Danube water level is below 120 cm (Budapest gauging station). These periods are becoming more frequent, and their duration is increasingly longer (Figure 1). In these situations, the water production capacity decreases, and it can approach the water demands during critical periods. Therefore, these periods are considered as critical.

For the management of these critical periods, Budapest Waterworks has introduced the term, "nominal capacity". Nominal capacity is defined as water volume that can be produced at a given Danube water level and temperature, at which the value of the drawdown (depression) is 200 cm. This number is arbitrarily selected and based on several decades of operational experience, and it is highly site- and system-specific.

During the peak water production period in the 1980s, depression values of 5 m were frequently observed. In this early period, water quality was not such a high priority as it has recently become. High depression levels resulted in deteriorating water quality, as well as well-clogging by sand intrusion, which physically damaged the filter layers and the structure of the well.

From the beginning of the 90s, the decreasing water demand resulted in lower water production, and thereby lower depression in wells. It was observed that at a depression level of about 200 cm, no sand intrusion occurred, and the water quality improved. This was the basis of the determination of nominal capacity using this specific value. The continuous maintenance of this value requires careful monitoring and process control.

In the case of extreme and long low-water periods that usually occur in August, the depression limit cannot be maintained, and the depression value can even reach 300 cm. In this scenario, intensive water-quality measurements are commenced, and frequent surveillance of the wells is conducted. When problems arise, emergency actions are taken, such as additional disinfection measures, and the spatial distribution of the production is modified.

During low-water periods, wells must be operated with higher depression levels to ensure the required water supply, which considerably increases the risk of microbiological noncompliance. In extreme cases, this might lead to cease of production in some wells. Therefore, wells that are still in operation have to maintain production with an elevated production load (depression), resulting in increased water quality-related risks.

Mitigation of this challenge is managed by ensuring the necessary reserve water production capacity to avoid overloading the operational wells. A precondition to this is that surplus wells should always be maintained. This is complemented by the reconstruction of the existing wells, where shaft wells are transferred into horizontal wells (Ranney wells). Provision of surplus production capacities will make the RBF system more robust and resilient to climate change and hydrological extremes.

3.3. Flood Conditions

Flood conditions are defined as Danube water levels exceeding 550 cm. At this water level, some of the wells and gravitational collector channels become partly or fully covered by water. At flood conditions, the highest operation risk is the intrusion of surface water into the well. Should this occur, the well must be taken out of service unconditionally and immediately. Intrusion risks can be decreased by appropriate maintenance of the technical elements of the wells, with particular attention to the integrity of the well shaft and the water-impermeable layer above the filter zones (well sealing). This requires continuous monitoring and control of the wells, as well as the implementation of a reconstruction program.

The novel elements of the reconstruction program started by the Budapest Waterworks were partly initiated as a result of new flood patterns observed at the Danube. The highest measured flood level was 891 cm in 2013, whereby the highest recorded water levels were exceeded on two occasions in the past 15 years. Counteracting the floods requires the reconstruction of wells' pump houses and service appliances, along with the relevant elevation of electric equipment. It may seem like a paradox, but in the case of floods, the available reserve wells are also needed, as a result of the sudden closure of intruded wells. Having the appropriate number of reserve wells is of paramount importance in terms of water safety and security.

4. Well Construction Issues and Adaptation Measures

The water production capacities of the horizontal RBF wells are largely determined by the characteristics of the filtration layers, the diameter and length of the horizontal laterals, and their numbers in a given filtration layer.

According to the facts demonstrated in Sections 3.2 and 3.3, the well load increases in cases of both hydrological extremes, but as a consequence of different causes. Additionally, these modified operational conditions result in damaging of the well structure by inducing filter clogging, which shortens the well life-cycle. To prevent or even avoid such events, three different interventions are available which are presented in the following.

4.1. Adapted Filter Screen Design to Avoid Sand Intrusion at High Pumping Rates

Within the RBF system, two velocity ranges can be distinguished along the path travelled by the water. In the first zone, flow velocities are near to the edge of the shoreline till the beginning of the effective zone of depression around the well. In the second zone, velocities are increased from the edge of the zone of depression till the inlet points, resulting in increasing kinetic energies near to the filters of the constructed wells. Entering into the horizontal lateral, the velocity of the inflowing water suddenly decreases, resulting in sedimentation of suspended particles. This phenomenon causes a slow accumulation of sediments.

Clogging of the filter depends on many factors, such as the natural sediment composition around the filter (screen), the volume and quality of the water, operational procedures, and the duration between two cleaning periods. As the operational time goes on, the clogging increases and the volume of produced water decreases. As a result, producing the same volume of water requires higher depression levels. It is concluded that the water production capacity at a given horizontal well provides the same volume of water referring to the same filtration surfaces, irrespective of the technical design of the horizontal laterals of the well.

Budapest Waterworks operates 217 horizontal collector wells (Figure 4). At the beginning, the wall of the laterals was made of slotted carbon steel filters. Laterals were horizontally extruded into the aquifer. The walls of the laterals were slotted, resulting in filter voids of approximately 14% of the total surface of the lateral.

Recently, a new design was introduced to horizontal laterals, where the bridge-slotted filters are made of stainless-steel. The size of the surface formed on both sides of the so-called "ear" provides a useful permeable surface of the bridge. After perforation, the plate is formed by a longitudinal welding tube. One end of the tube is cushioned to provide lateral flow. The open inflow area of bridge-slotted filters is approximately 21% of the total surface.

Due to the geometry of the bridge-slotted filter, the direction of velocity changes in the inflow, and thus significantly reduces the inflow of solids from the well's environment. While slotted filters can hold a grain diameter above 6 mm, the bridge-slotted filter provides effective protection against particle diameters down to 3 mm due to its two-sided slit surface.

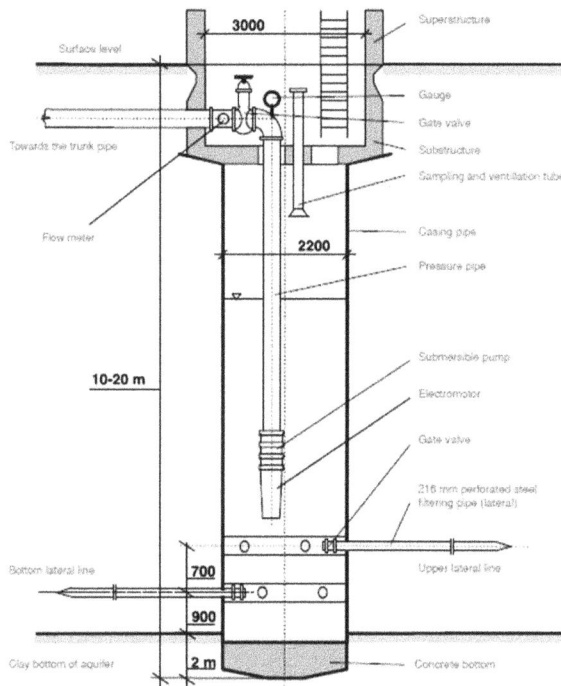

Figure 4. Schematic view of a horizontal collector well.

The deviation of the pipe direction is influenced by the bending stiffness of the tube and the structure of the soil. Technically, the bending stiffness can vary, which is the product of the second torque and modulus of elasticity of the tube. Secondary torque depends on the pipe wall thickness and the fourth power of the material from the distance to the center of gravity of the tube.

The material properties do not change because the bridge design does not involve any material loss. In the case of a bridge-slotted filter, the material distance is increased by extruding the bridge outward: 21% of the material moves away from the center of gravity by forming a bridge of 3 mm thickness. Thus, the bending stiffness of the slotted pipe is smaller than that of a bridge filter tube of the same diameter, with a higher risk of reversing under pressure.

4.2. Optimal Well Rehabilitation/Filter Cleaning Frequencies

Rehabilitation frequencies are based on operational practice and have been determined empirically. In general, the typical cleaning frequency is 7–8 years. Based on practical knowledge, the cycle cannot exceed 10 years. This time length results in massive cementation on the laterals that cannot be removed, with the exception of carbon steel laterals; however, there the cost of removing cementation is extremely high too. As the deterioration of the filter layer is a function of the well operation (velocity of the bank filtrate, volume of produced water), a production load-based cleaning cycle has been developed. The operational data and the amount of produced water have been examined and compared with the nominal capacity for each horizontal well, and the rehabilitation frequencies were also determined (Table 1).

Table 1. Horizontal well-filter cleaning frequencies.

Ratio of Production Load and Nominal Capacity in %		Cleaning Frequency in Years
≥80		6
<80	≥65	7
<65	≥50	8
<50	≥30	9
<30		10

Besides the production load, the following two criteria have been considered: wells that are in operation only at peak demands have a 10-year cycle; strategically important wells have a 7-year cleaning frequency. By applying these criteria and considering the number of wells, the average cleaning cycle has been determined at 7.88 years, which means 29 wells need to be reconditioned annually. This program has to be revised every other year, as operational needs and circumstances might change. Table 2 shows the determined and applied cleaning frequency distribution.

Table 2. Distribution of well rehabilitation frequencies at Budapest.

Cleaning Frequency in Years	Number of Wells
6	11
7	102
8	23
9	9
10	80
Total	225

4.3. Shaft Well Transformation into Horizontal Wells

The Sziget I. well group with nine shaft wells is located at the southernmost point of the Szentendre Island on the right bank of the Danube, between 1657–1658 km (Figure 2). The wells were constructed between 1897–1899. Considering that they are situated close to the network inlet point, their strategic importance is obvious as they are less prone to raw water infiltration in the collecting pipes during floods. By redesigning these shaft wells, higher production capacities can be achieved.

At the bottom of the reinforced concrete, where cylindrical shaft wells do not reach the impermeable layer, they hang on an "I" girder supported by two redwood stilts embedded into the substratum. Therefore, the shaft well and the filtering bell can be considered as a reinforced structure.

A gravel layer—or a casing shoe—can be found at the bottom of the well, 1.5–2 m above the bottom of the aquifer. This hinders the intrusion of the sandy-gravel aquifer material into the filtering bell. This solution is widely used under such conditions. In theory, a larger filtering area can be achieved by tapping the bottom of the well than applying a casing reaching down to the bottom of the aquifer.

At the lower part of the reinforced concrete shaft, or the so-called "bell", there is a slotted iron cylinder which is 5 m in diameter and 3–6 m in length. The 2.8 m high cylindrical structure above the bell was constructed according to contemporary concrete technology. Instead of the mesh structure applied nowadays or the reinforcing ironing, six steel rods placed at every 1.5 m ran along the shaft casing, fixed to the narrowing of the filtering bell. At every meter, a flat steel ring also consolidated the vertical steel rods. This was the complete reinforcement of the well shaft.

The wells are protected by a cast-iron cover on the top of a stone-paved protection cone that was raised above the maximal flood-water level. The vent hole and the sampling tube can be found on the well cover.

The aim of the shaft well redesign is to renew the well structure. The technology involves structural transformation, where a DN 2200 well shaft is installed into a DN 3000 shaft well. The laterals are extruded beneath the previous well screen, but above the bottom of the aquifer. The distance between the bottom of the aquifer and the bottom edge of the screen is 0.7–1.5 m, so the laterals have to fit within this range. The length of the stainless-steel laterals is 30 m, uniformly with a diameter of 200 mm. The laterals have a bridge-slotted structure which has been described in detail above, except for the first 5 m of the laterals—situated on the well-shaft side—that has a plain surface. For all wells, one lateral is perpendicular to the river, and two laterals are placed at 60° and 120° to the Danube. The new well-shafts are raised above the maximal flood level, and the casing is closed by cast-iron covers, but buildups are also feasible. After reconstruction, the same water quantity can be produced, while the operation and the water quality can be improved.

5. Conclusions

From an operational aspect, there are existing solutions that can be adapted at RBF systems to cope with extreme hydrological conditions. The first option for assuring the requested water supply in quantity and quality is the construction of a water treatment plant (which was not necessary so far). The second option is to maintain reserve well capacities, ensuring appropriate water quality. Having considered the characteristics of the existing water reserve capacities, as well as the historical background and the financial feasibility, Budapest Waterworks has chosen the second option.

Due to the particularities of RBF, appropriate water quality can be assured even under extreme hydrological conditions by avoiding well overloading. The necessary water volume is produced by operating more wells, thereby optimizing the load across all wells. This also explains the importance of having sufficient reserve wells available. The conditions require to keep these wells in operation, as the microbiological processes might need several weeks, or even months to be established.

Experience based on several decades of research and practical application led to the conclusion that reserve well capacities should fulfil at least 30% of the total production capacity in the case of extreme hydrological conditions. This figure is based on the particularities of Budapest Waterworks operations. In a broader and more general perspective, a decision on adaptation measures should be made based on investment costs of new wells, maintenance and cleaning interval costs of existing ones, operational costs, public health and technical risks (failures), and changing hydrological conditions.

Author Contributions: Z.N.K., B.L. and E.F. reviewed previous literature and prepared the article draft. E.S. supervised and provided documents related to Section 4.

Funding: The article was prepared within the AquaNES project to synthetize all available information on well operation. This project has received funding from the European Union's Horizon 2020 Research and Innovation Program under grant no. 689450.

Acknowledgments: The authors acknowledge the helpful comments of anonymous reviewers and support from T. Grischek during final paper revision.

Conflicts of Interest: The authors declare no conflict of interest. The funding sponsors had no role in the design of the study; in the collection, analyses, or interpretation of data; in the writing of the manuscript, and in the decision to publish the results.

References

1. Haines, A.; Kovats, R.S.; Cambell-Lendrum, D.; Corvalan, C. Climate change and human health: Impacts, vulnerability and public health. *Public Health* **2016**, *120*, 585–596. [CrossRef] [PubMed]
2. Gollnitz, W.D.; Whitteberry, B.L.; Vogt, J.A. Riverbank filtration: induced infiltration and groundwater quality. *J. Am. Water Works Assoc.* **2000**, *96*, 98–110. [CrossRef]
3. Davidesz, J.; Debreczeny, L. Long-Term Sustainability of RBF Systems from Aspects of Availability and Capacity. Presented at the MAVÍZ Conference, Sopron, Hungary, 11–12 June 2009. (In Hungarian)
4. Merkel, W.; Leuchs, W.; Oldenkirchen, G. Challenges of Global Climate Change for the Water Supply and Distribution in Germany: Experience Report. In *Handlungsfelder und Forschungsbedarf, Proceedings of Mülheimer Wassertechnisches Seminar, Mülheim, Germany, 20 November 2007*; IWW Wasser Zentrum: Mülheim, Germany, 2007; Volume 46, pp. 1–16.
5. Schubert, J. *How Does It Work? Field Studies on Riverbank Filtration, Proceedings of the International Riverbank Filtration Conference, Dusseldorf, Germany, 2–4 November 2000*; Julich, W., Schubert, J., Eds.; Internationale Arbeitsgemeinschaft der Wasserwerke im Rheineinzugsgebiet (IAWR): Düsseldorf, Germany, 2000; pp. 41–55.
6. Christensen, N.S.; Wood, A.W.; Voisin, N.; Lettenmaier, D.P.; Palmer, R.N. The effects of cimate change on the hydrology and water resources of the colorado river basin. *Clim. Chang.* **2004**, *62*, 337–363. [CrossRef]
7. Ascott, M.J.; Lapworth, D.J.; Sage, R.C.; Karapanos, I. Impact of extreme flooding on riverbank filtration water quality. *Sci. Total Environ.* **2016**, *554–555*, 89–101. [CrossRef] [PubMed]
8. Chiaudani, A.; Di Curzio, D.; Palmucci, W.; Pasculli, A.; Polemio, M.; Rusi, S. Fractal approaches on long time-series to surface-water/groundwater relationship assessment: A Central Italy alluvial plain case study. *Water* **2017**, *9*, 850. [CrossRef]
9. Kuriqi, A.; Ardiciouglu, M.; Muceu, Y. Investigation of seepage effect on river dike's stability under steady state and transient conditions. *Pollack Periodica* **2016**, *11*, 87–104. [CrossRef]
10. Kuriqi, A.; Ardiciouglu, M.; Muceu, Y. Investigation of the hydraulic regime at the middle part of the Loire River in context of floods and low flow events. *Pollack Periodica* **2018**, *13*, 145–156. [CrossRef]
11. Sandhu, C.; Grischek, T.; Musche, F.; Macheleidt, W.; Heisler, A.; Handschak, J.; Patwal, P.S.; Kimothi, P.C. Measures to mitigate direct flood risks at riverbank filtration sites with a focus on India. *Sustain. Wat. Res. Manag.* **2018**, *2*, 237–249. [CrossRef]
12. Martin, H.K.; Fuchs, M.P. Runoff conditions in the upper Danube basin under an ensemble climate change scenarios. *J. Hydrol.* **2012**, *424–425*, 264–277.
13. Ray, C.; Grischek, T.; Schubert, J.; Wang, J.Z.; Speth, T.F. A perspective of riverbank filtration. *J. Am. Water Works Assoc.* **2002**, *94*, 149–160. [CrossRef]
14. Schubert, J. Significance of Hydrologic Aspects on RBF Performance. In *Riverbank Filtration Hydrology -Impacts on System Capacity and Water Quality*; Hubbs, S.A., Ed.; Springer: Dordrecht, The Netherlands, 2006; pp. 1–20.
15. Doussan, C.; Ledoux, E.; Detay, M. River-groundwater exchanges, bank filtration, and groundwater quality: Ammonium behavior. *J. Environ. Qual.* **1998**, *27*, 1418–1427. [CrossRef]
16. Schubert, J. Hydraulic aspects of riverbank filtration—field studies. *J. Hydrol.* **2002**, *266*, 145–161. [CrossRef]
17. Vanek, V. Heterogeneity of Groundwater-Surface Water Ecotones. In *Groundwater/Surface Water Ecotones: Biological and Hydrological Interactions and Management Options*; Gibert, J., Mathieu, J., Fournier, F., Eds.; Cambridge University Press: Cambridge, UK, 1997; pp. 151–161.
18. László, B. *Assessment of River Bank Filtration in Changing Environmental and Operational Circumstances*; WQ III Program, Final Report, Internal Unpublished Document (in Hungarian). Budapest Waterworks: Budapest, Hungary, 2013.

![water logo] *water*

MDPI

Article

Water Quality Changes during Riverbank Filtration in Budapest, Hungary

Zsuzsanna Nagy-Kovács [1],*, János Davidesz [1], Katalin Czihat-Mártonné [1], Gábor Till [1], Ernő Fleit [1] and Thomas Grischek [2]

[1]   Budapest Waterworks Ltd, Budapest H-1138, Hungary; janos.davidesz@vizmuvek.hu (J.D.);
     katalin.martonne@vizmuvek.hu (K.C.-M.); gabor.till@vizmuvek.hu (G.T.); erno.fleit@vizmuvek.hu (E.F.)
[2]   Division of Water Sciences, University of Applied Sciences Dresden, D-01069 Dresden, Germany;
     thomas.grischek@htw-dresden.de
*   Correspondence: zsuzsanna.nagy-kovacs@vizmuvek.hu

Received: 19 October 2018; Accepted: 6 February 2019; Published: 11 February 2019

Abstract: The paper gives an overview on the changes in water quality during riverbank filtration (RBF) in Budapest. As water from the Danube River is of high quality, no problems occur during regular operation of RBF systems. Additionally, water quality improved through the past three decades due to the implementation of communal wastewater treatment plants and the decline of extensive use of artificial fertilizers in agriculture. Algae counts are used as tracer indicators to identify input of surface water into wells and to make decisions regarding shutdowns during floods. RBF systems have a high buffering capacity and resistance against accidental spills of contaminants in the river, which was proven during the red mud spill in October 2010. The removal rate of microorganisms was between 1.5 log and 3.5 log efficiency and is in the same order as for other RBF sites worldwide.

Keywords: riverbank filtration; water quality; organic carbon; nitrate; heavy metals; microorganisms

---

1. Introduction

Riverbank filtration (RBF) is a widely used natural water treatment process where, by definition, at least 50% of treated water must originate from surface water. It has been observed that the surface water source, the hydrogeological characteristics of the aquifer, the protected watershed area and the particularities of production play an important role on the quality of the produced water [1–3].

RBF offers many advantages concerning improvement of water quality. This type of ecosystem service is used in many watersheds globally, including India, China, USA and Germany. RBF along the Danube River has been used for water supply in Budapest for over 150 years [4]. Due to the high quality of the Danube River water and favorable hydrogeological conditions at the Szentendre Island upstream of the city of Budapest, no post-treatment except disinfection is required after RBF processes. This unique situation enables us to study long-term trends in the characteristics of water quality parameters.

The focus is set on basic water quality parameters to describe microbiologically mediated reactions, physical sorption and mixing processes during RBF and the resulting attenuation of pollutants. Spatial changes in redox potential conditions were studied in RBF processes in Berlin, where it has been found that temperature variation strongly influenced the efficacy of microbial removal processes [5].

The vulnerability of RBF processes to climate change has been discussed in prior studies concentrating on oxic/aerated and anoxic conditions of the aquifer layers [6]. Redox conditions are profoundly affected by both microbiologically mediated pathways of nitrogen transformations (nitrification and denitrification) and physicochemical sorption processes and phase equilibria [7]. Local redox conditions, however, can only be indirectly controlled in the aquifer by the operator (i.e., pumping rates). In spite of the improvements in traditional water quality parameters, concerns arise regarding the microbial parameters of the Danube both upstream [8] and downstream of Budapest [9] related to the increasing incidence and severity of extremities.

The aim of this study is to give an overview regarding the efficiency of RBF processes. The basic concept is to analyze physical, chemical, microbiological and biological parameters and highlight existing connections. Challenges include seasonal variations in river water quality, floods, droughts, industrial and agricultural pollutant input variations. Therefore, it is important to consider water quality parameters which can be determined at a high number, high frequency and at low cost. Also, it is important to determine how these measurements can improve the level of service by faster and established interventions, lower disinfectant concentration and effective operational strategies.

2. Materials and Methods

2.1. Site Description

As the efficiency of RBF is site specific and the water quality changes are affected by many other factors besides source water quality, e.g., water level changes, travel times of bank filtrate, pumping regime of wells, etc., a large dataset is required to be able to determine reliable operational methodology. In this paper, data from the period 2006 to 2017 from a total of up to 756 wells were overviewed to assess changes in water quality. The maximum capacity of the RBF systems of Budapest Waterworks is 1.0 million m^3/day; the recent average supply is about 456,000 m^3/day. Compared to the average discharge of the Danube River in Budapest, which stands at 200 million m^3/day, only 0.23% of the water is extracted from the river discharge via bank filtration. A unique situation occurs in Budapest whereby there is no riverbed clogging observed [10] and no distinct clogging layer exists in the riverbed affecting water quality. This may be due to the high river flow velocity of 0.8–1.6 m/s, the depth of the river and the related shear forces. At such levels of flow velocity, fine particles do not settle, only coarse sand and gravel do [11]. At many other RBF sites worldwide, clogging profoundly affects the infiltration rates of river water and results in a highly active biological layer in the riverbed which often notably contributes to water quality changes, especially considering oxygen consumption and attenuation of organic compounds [11–15].

Budapest Waterworks operates 756 RBF wells to supply water to 1.89 million inhabitants. The wells are predominantly located on Szentendre Island and Csepel Island (Figure 1). A detailed description of well types and operation procedures are to be found in Nagy-Kovács et al. [16].

2.2. Groundwater Flow Modeling

Travel times have been determined by ground water flow modeling using the MODFLOW software (USGS, Reston, VA, USA). The modeling served to determine the travel time of water particles arriving in the well. The ratio between river bank filtrate and ground water was investigated. The original proportion of the produced water is primarily controlled by the actual level of the Danube and the rate of drawdown. Calculations were carried out based on a 2-m average Danube level and a drawdown of 2 m. Later, separate well capacities for different Danube water levels were also determined in 2012 [17]. Table 1 gives an overview of the distances and travel times between the Danube River and the production wells.

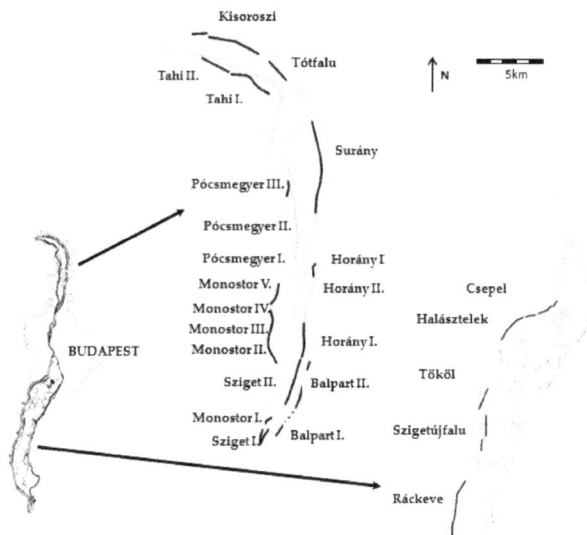

Figure 1. Location and names of well groups of the Budapest Waterworks at Szentendre Island (upstream the capital) and Csepel Island (downstream the capital).

Table 1. Distance and travel time between the Danube River and RBF wells in Budapest.

Well/Well Group	Type of Wells	Distance between the Riverbank and Wells (m)	Thickness of Aquifer (m)	Travel Time of Bank Filtrate (days)
Kisoroszi	HW	40–370	9–13	13–17
Tótfalu	HW	120	7–10	14–22
Tahi I.	HW	60	5–8	3–6
Tahi II.	HW	200–230	5–8	12–20
Surány well 1–7	HW	60–120	7–12	6–8
Surány well 8–14	HW	190–228	7–12	22–25
Surány well 15–20	HW	255–410	7–12	68–98
Horány I.	VW	19	7–9	4–6
Horány II.	VW	19	7–9	6–8
Horány III.	HW	85–245	7–9	9–11
Pócsmegyer I.	VW, HW	90–140	5–12	8–11
Pócsmegyer II.	HW	0–60	5–12	2–4
Pócsmegyer III.	VW	30	5–12	2–4
Monostor I	HW	30–270	5–13	11–12
Monostor II.	VW	70	5–13	9–12
Monostor III.	VW	70	5–13	7–9
Monostor IV.	VW	40	5–13	4–6
Monostor V.	VW, HW	40–80	5–13	9–13
Sziget I.	HW	10–35	5–13	2–5
Sziget II.	HW	10–45	5–13	2–5
Balpart I.	HW, SW	51–100	5–10	5–10
Balpart II.	HW	62–203	5–10	2–5
Csepel	HW	7–27	9–17	2–5
Halásztelek	HW	24–113	9–17	2–25
Tököl	VW	55–70	5–13	5–20
Szigetújfalu	HW	374–860	5–13	100–220
Ráckeve	HW	60–117	8–15	15–20

HW—horizontal (collector) well, VW—drilled (vertical) well, SW—shaft well.

2.3. RBF Monitoring Network and Samples

Water sampling from the river and the wells was carried out following the Hungarian guidelines and standards [18]. Analyses are carried out systematically and adjusted to changing circumstances to ensure safe and secure water supply and to gain data for optimal operation of the RBF systems. The Danube River was sampled at least weekly either on Szentendre Island or Csepel Island. Every well in operation was sampled regularly at least twice a year from its sampling tap. Some siphon systems were sampled at their collecting pipes.

All analytical methods for the determination of discussed parameters are provided in the Supplementary Material.

The minimum, median and maximum values were prepared. Due to the large number of data, short events such as floods or spills would not affect the median values which are used to determine removal rates for different groups of RBF wells.

For physical and chemical parameters, mean removal rates have been determined for the whole time period. Lowest and highest removal rates were calculated using the mean concentration in the river water and the maximum and the minimum concentration in the bank filtrate, respectively. As the sampled well water is a mixture of bank filtrate of different age depending on the location of infiltration in the riverbed, the depth of the flow path in the aquifer and the pumping rate of the well, pairing of data is not useful. In no case, data from the same date of sampling can be compared as the water sampled from the well is days to weeks old and has nothing to do with the river water quality at the sampling date. As RBF acts as a buffer for water quality, it is feasible to use the mean concentration in river water and to compare with minimum and maximum concentrations observed in the well water. For microbiological parameters, the mean logarithmic removal rate was calculated as the difference between the logarithm of the average cell count in Danube River water and the logarithm of the average cell count in the bank filtrate.

The available dataset has been discussed in five different Chapters (Sections 3.1–3.5). For all chapters, a table has been prepared to better demonstrate the results that have been analyzed during this study. Median values are in bold presented in the first row, the range of each parameter with minimum and maximum values is given in the second row, and *n* represents the total number of samples for the parameter in italics. The dataset is formed by culminating results from the monitoring plan determined by Hungarian regulations, operation-related experiments, sampling during extreme hydrological events and a major accidental pollution event in the Danube River basin. Due to this fact, the number of samples (*n*) varies for each parameter, as authors chose to present all reliable data from the period 2006–2017.

3. Results and Discussion

3.1. Physical Parameters and Selected Cations and Anions

This chapter summarizes the results of physical parameters, cations and anions measured on a regular basis. The temperature of the river water ranges from -1.4 to $26.3\,°C$ with a median of $13.2\,°C$ (Table 2). Due to the short distance between the riverbank and the wells and the heat capacity of the aquifer material, only a low buffering effect was observed—the temperature of bank filtrate ranges from 0.7 to $21.0\,°C$ with a median of $11.9\,°C$. As the aquifer thickness is only 5–17 m (Table 1), the buffering effect is lower in Budapest compared to other RBF sites with larger aquifer thickness, e.g., in Torgau with 50–60 m thickness [14,19]. The electrical conductivity (EC) of river water and bank filtrate varies from 283 to $652\,µS/cm$ and from 303 to $1809\,µS/cm$, respectively. The mean EC is very similar for the Danube River water and well groups on Szentendre Island, indicating a high portion of bank filtrate, which has been calculated from groundwater flow modeling as 60–80% [17]. The maximum EC values in bank filtrate, which are significantly higher than those of the river water are only observed on Csepel Island, where industrial and agricultural activities even outside of the well head protection zones are still affecting the water quality in some wells. The mean turbidity of bank filtrate is very low

(0.05–0.07 NTU) compared to river water and the removal of particles is not a function of travel time. Slightly increased turbidity values on Csepel Island are related to iron and manganese precipitates.

Table 2. Physical parameters and selected cations and anions, median (min–max) values of Danube River water and bank filtrate with different travel time (t), Budapest, 2006–2017.

Parameter	Unit	Danube River Water	Bank Filtrate t < 10 days	Bank Filtrate t = 10–25 days	Bank Filtrate t > 50 days	Effect
T	°C	13.2 −1.4–26.3 (n = 1476)	12.4 −0.7–21 (n = 4744)	11.7 4–20.8 (n = 4451)	11.8 7–16.6 (n = 753)	Buffering effect
EC	µS/cm	387 283–652 (n = 1010)	454 303–1617 (n = 3255)	475 336–1809 (n = 3677)	731 411–1379 (n = 663)	Increase with travel time
Turbidity	NTU	12.5 0.24–213 (n = 962)	0.07 <0.1–149 (n = 4152)	0.08 <0.1–44.5 (n = 3951)	0.05 <0.1–1.99 (n = 698)	Removal up to 99%
pH	-	8.03 6.86–8.95 (n = 1009)	7.52 7–7.95 (n = 2080)	7.5 6.11–7.92 (n = 2298)	7.4 7.05–7.7 (n = 375)	Decrease, no effect of travel time
Alk	mmol/L	3.2 2.4–4.0 (n = 140)	3.8 2.5–7.6 (n = 845)	3.1 2.9–7.7 (n = 1524)	5.5 3.8–8.2 (n = 268)	Slight increase for t < 50 days
Ca^{2+}	mg/L	54.3 40.3–80.3 (n = 352)	66.6 24.7–196 (n = 1799)	72.1 47.6–214 (n = 1605)	92.1 64–160 (n = 269)	Increase with travel time
Mg^{2+}	mg/L	13.3 9.1–20.6 (n = 354)	16.1 5.2–59.9 (n = 1799)	17.2 11.5–73.7 (n = 1604)	40.4 15.5–96 (n = 269)	Slight increase for t < 50 days
Na^+	mg/L	13.3 6–30.3 (n = 352)	14.3 8.0–75.3 (n = 1804)	14.6 8.7–78.7 (n = 1604)	19.7 11.3–45.9 (n = 269)	Slight increase for t < 50 days
K^+	mg/L	2.7 1.6–4.6 (n = 352)	2.7 1.5–11 (n = 1778)	2.6 1.0–36.0 (n = 1593)	2.6 1.4–10.8 (n = 268)	No change in median
NH_4^+	mg/L	0.07 <0.04–1.1 (n = 1046)	<0.04 (n = 3273)	<0.04 (n = 3628)	<0.04 (n = 656)	Slight decrease
Hardness	mg/L CaO	113 77–194 (n = 668)	142 91–374 (n = 464)	147 101–429 (n = 864)	215 135–427 (n = 150)	Increase with travel time
HCO_3^-	mg/L	189 140–256 (n = 223)	232 171–464 (n = 523)	256 183–469 (n = 937)	342 232–500 (n = 170)	Slight increase for t < 50 days
Cl^-	mg/L	19.0 8.9–48 (n = 978)	21.3 7.9–218 (n = 3305)	21.3 11.0–220 (n = 3460)	33.2 9.7–77.8 (n = 632)	Slight increase for t < 50 days
PO_4^{3-}	mg/L	0.123 <0.1–0.55 (n = 579)	<0.1 <0.1–0.46 (n = 1489)	<0.1 <0.1–0.378 (n = 1031)	<0.1 <0.1–0.106 (n = 151)	Removal 0–18% at least
F^-	mg/L	<0.2 <0.2–0.4 (n = 260)	<0.2 <0.2–0.9 (n = 1363)	<0.2 <0.2–0.9 (n = 899)	<0.2 <0.2–0.2 (n = 150)	No change
B	mg/L	0.028 <0.005–1.594 (n = 242)	0.031 0.008–0.313 (n = 152)	0.022 0.009–0.099 (n = 218)	0.035 0.014–0.079 (n = 36)	Buffering
Si	mg/L	2.82 2.29–3.21 (n = 11)	4.06 2.81–8.45 (n = 25)	4.43 2.97–12.1 (n = 30)	8.42 6.66–10.6 (n = 10)	Increase with travel time
CN^-	µg/L	No data	<10 <10–20.4 (n = 123)	<10 <10–18.8 (n = 258)	<10 <10–14.5 (n = 35)	No river data

Despite the long-term operation of most of the wells, there is still a dissolution of carbonates in the aquifer, resulting in an increase in hardness (Ca and Mg) compared to river water. Sodium and potassium concentrations in river water and bank filtrate are within the same range.

As for sodium, chloride concentrations in river water and bank filtrate are within the same range, indicating a high portion of bank filtrate. Fluoride, boron, silicon and cyanide concentrations are at low levels both in river water and bank filtrate and do not pose any risk for the water supply. Phosphate levels in river water are below the LOD (limit of determination), but are surprisingly also found in the bank filtrate.

Seasonal temperature changes in river water and bank filtrate and flow-related changes in EC can be used to estimate travel times of the bank filtrate [14,20]. Figure 2 shows the temperature data for the Danube River water and well group Balpart II with a short travel time in the range of 2–5 days. It can be seen how the seasonal temperature variation of the river influences the river bank filtrate as the buffering effect of the aquifer is considered low. Also, a change in the water temperature variation can be seen starting from August of 2012 that is linked to the fact that the production rate decreased by 30% at the site.

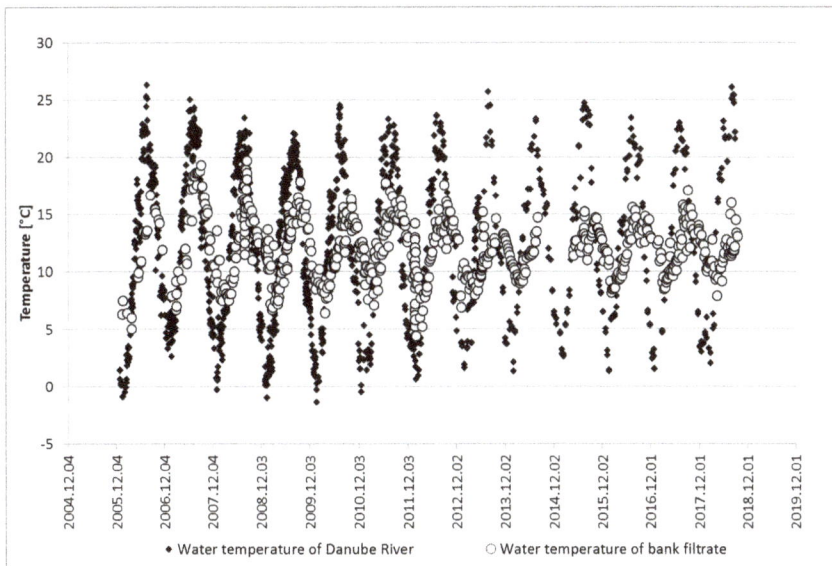

Figure 2. Seasonal temperature variation in Danube River water and bank filtrate, well group Balpart II, 2006–2018.

3.2. Redox-Related Parameters

The removal of organic compounds (mainly natural humic and fulvic acids) is of high relevance for the required post-treatment (to supply microbiologically stable drinking water) and especially for disinfection with regards to the potential formation of disinfection by-products. The total organic carbon (TOC) concentration in the Danube River water ranges from 1.6 to 10.0 mg/L (Table 3) and has a median similar to that of the Rhine River at Düsseldorf [21] and about half of the concentration in the Elbe River [14,19]. Despite the low input concentration, the removal rate for TOC is relatively high, ranging from 11 to 75%. The TOC removal is higher if the travel time is longer. The removal of organic aromatic compounds causing UV absorption ranges between 0 and 92%. The median specific UV-absorbance (calculated as UV_{254}/TOC) is 2.61 L/(m·mg) for Danube River water and 1.93–2.45 L/(m·mg) for bank filtrate. At the RBF site Torgau, Elbe River, Germany, the specific UV-absorbance of the river water was 2.94 L/(m·mg), increasing to 3.17 L/(m·mg) in the riverbed, where easily biodegradable and less UV-active organic compounds were removed, and decreased along the >200 m long flow path to 2.54 L/(m·mg) due to further attenuation of UV-active compounds [14]. In Budapest, a similar removal indicates high removal of UV-active compounds.

Table 3. Redox related parameters, median (min–max) values, Danube River water and bank filtrate with different travel time (t), Budapest, 2006–2017.

Parameter	Unit	Danube River Water	Bank Filtrate t < 10 days	Bank Filtrate t = 10–25 days	Bank Filtrate t > 50 days	Effect
TOC	mg/L	2.8 1.6–10.0 (*n* = 585)	1.1 0.7–2.5 (*n* = 319)	1.1 0.7–2.1 (*n* = 200)	0.9 0.7–1.3 (*n* = 10)	Removal 11–75%
UV_{254}	m^{-1}	7.4 1–29.6 (*n* = 426)	2.7 0.7–7.5 (*n* = 2617)	2.4 0.6–12.7 (*n* = 2771)	1.65 0.55–6.9 (*n* = 500)	Removal 0–92%
COD	mg/L	2.8 0.3–16.2 (*n* = 1039)	0.7 <0.2–2.6 (*n* = 3401)	0.7 <0.2–2.5 (*n* = 3639)	0.44 <0.2–2.4 (*n* = 657)	Removal 7–93%
DO	mg/L	>7 near to saturation	2.7 0.2–11.1 (*n* = 173)	3.1 0.2–11.1 (*n* = 97)	2.3 0.4–4.4 (*n* = 13)	No river data set available
NO_3^-	mg/L	8.1 2.9–17.8 (*n* = 1092)	8.4 <1–126 (*n* = 3531)	7.1 <1–144 (*n* = 3815)	22.9 2.2–89.6 (*n* = 661)	No effect at short travel times
NO_2^-	mg/L	0.05 <0.03–0.91 (*n* = 1052)	<0.03 <0.03–0.82 (*n* = 3296)	<0.03 <0.03–0.53 (*n* = 3649)	<0.03 <0.03–0.19 (*n* = 656)	Decreasing effect
Mn	µg/L	33.5 1.9–415 (*n* = 757)	1.3 <1–1752 (*n* = 3358)	1.9 <1–3255 (*n* = 3722)	3.9 <1–135 (*n* = 657)	Removal 0–97%
Fe	µg/L	201 <5–3600 (*n* = 1052)	5.8 <5–1670 (*n* = 918)	7.2 <5–3540 (*n* = 1215)	6.5 <5–82.2 (*n* = 257)	Removal 0–98%
SO_4^{2-}	mg/L	31.8 18.2–138.3 (*n* = 307)	38.4 20.7–438.8 (*n* = 2016)	45.5 20.7–491.9 (*n* = 1904)	103 18.4–351.6 (*n* = 319)	Slight increase

The chemical oxygen demand (COD) removal rates during RBF were about 7–93% but cannot be used as the TOC to assess the removal of organic compounds as it cumulatively removes all oxidizable compounds present in the water, including inorganic constituents such as ammonium and iron.

Both the dissolved oxygen concentration in the river water and the consumption in the aquifer by biological processes are strongly temperature dependent. The median dissolved oxygen (DO) concentration decreased from 8 mg/L in river water to 2.3–3.1 mg/L in bank filtrate. Considering the minimum values, it becomes obvious that RBF at Budapest is predominantly operated under oxic conditions; however, anoxic conditions occur during summer months. On Szentendre Island, no increase in Mn and Fe concentration was observed, whereas on Csepel Island, there was an increase in Mn and Fe to 15 µg/L and 43 µg/L, respectively, requiring post-treatment. The aquifer material on Csepel Island is of smaller grain size. Here, the river water quality is affected by effluents from the city, probably resulting in a higher portion of biodegradable organic carbon. Sulphate concentrations in river water and bank filtrate are very similar, indicating no sulphate reduction during RBF.

Although the temperature dependency of biological nitrate removal (denitrification) during RBF has been reported for many sites [22,23], a strong effect on dissolved organic carbon (DOC) removal has been rarely observed as DOC transport is a complex process and is also affected by temperature-related adsorption and desorption interactions [19,23]. The Danube and the bank filtrate at Budapest have very similar DOC and TOC concentrations. Also, it has been observed that these concentrations have decreased due to the improving raw water quality and decreased production rate. In general, the median TOC concentration in bank filtrate was 1.0 mg/L in summer and 1.6 mg/L in winter (data 2006–2017, n = 529). It has to be mentioned that TOC still includes organic particles and is only at the same level as DOC for the bank filtrate, not for river water. Elevated nitrate concentrations under longer travel times (t > 50) are related to the higher groundwater portion.

3.3. Metals

Metal concentrations are listed in alphabetical order of the chemical symbol of the element. Metal concentrations were below the limit determined by the Hungarian regulations [24,25] at the RBF sites in Budapest, for the river water and the bank filtrate (Table 4). All concentrations are below the limits of the drinking water quality guideline [24]. The removal of heavy metals follows the sequence: Cr > Zn > Cu > Pb > Ni > Cd > Co > Al. The range of removal is in agreement with a general removal of 20–70% for Cd, Co, Cu, Ni, Zn under oxic conditions according to [26].

Table 4. Median (min–max) metal concentrations in Danube river water and bank filtrate (BF) with different travel time (t), 2006–2017.

Parameter	Unit	Danube River Water	Bank Filtrate t < 10 days	Bank Filtrate t = 10–25 days	Bank Filtrate t > 50 days	Effect
Al	µg/L	157 5.2–4261 (n = 163)	<5 <5–107 (n = 192)	<5 <5–1114 (n = 267)	<5 <5–289 (n = 43)	Nearly complete removal
Sb	µg/L	<0.5 <0.5–1.28 (n = 159)	<0.5 <0.5–2.5 (n = 158)	<0.5 <0.5–2.8 (n = 229)	<0.5 <0.5–0.9 (n = 38)	Results below LOD
As	µg/L	1.8 1.0–7.7 (n = 163)	1.7 <1–3.7 (n = 325)	1.5 <1–6 (n = 521)	<1 <1–6.9 (n = 88)	Little change, around LOD
Ba	µg/L	35.1 <5–185 (n = 266)	40.2 <5–126 (n = 172)	42.9 19–126 (n = 257)	47.7 24.3–102 (n = 43)	Slight increase
Bi	µg/L	<0.2 (n = 12)	<0.2 <0.2–0.8 (n = 33)	<0.2 <0.2–0.8 (n = 33)	<0.2 (n = 10)	Results below LOD
Cd	µg/L	0.2 <0.2–0.8 (n = 266)	<0.2 <0.2–0.677 (n = 192)	0.2 <0.2–0.5 (n = 266)	<0.2 (n = 43)	Slight decrease, results close to LOD
Cr	µg/L	1.2 <1–23.3 (n = 266)	1.0 <1–6.6 (n = 192)	1.0 <1–21.9 (n = 267)	1.2 <1–3.1 (n = 45)	Removal 0–16%, results close to LOD
Co	µg/L	0.25 <0.2–2.4 (n = 266)	<0.2 <0.2–2.0 (n = 172)	<0.2 <0.2–0.4 (n = 257)	<0.2 <0.2–1.4 (n = 43)	Removal 0–20% (at least)
Cu	mg/L	<0.005 <0.005–3.34 (n = 266)	<0.005 <0.005–1.003 (n = 192)	<0.005 <0.005–0.059 (n = 266)	<0.005 <0.005–0.067 (n = 43)	Results below LOD
Pb	µg/L	1 <0.5–13.5 (n = 163)	<0.5 <0.5–5.7 (n = 193)	<0.5 <0.5–6 (n = 266)	<0.5 <0.5–1.2 (n = 43)	Results for BF below LOD
Li	µg/L	3 <0.1–11 (n = 11)	7 <0.1–15.6 (n = 26)	6.2 3.3–17.6 (n = 25)	14.4 10.9–18.9 (n = 10)	Increase
Hg	µg/L	<0.05 <0.05–0.65 (n = 116)	<0.05 <0.05–1.2 (n = 175)	<0.05 <0.05–0.8 (n = 263)	<0.05 (n = 43)	Results below LOD
Mo	µg/L	1.0 <1–23 (n = 265)	<1 <1–7.6 (n = 172)	<1 <1–20.4 (n = 258)	<1 <1–2.1 (n = 45)	Results for BF below LOD

Table 4. Cont.

Parameter	Unit	Danube River Water	Bank Filtrate t < 10 days	Bank Filtrate t = 10–25 days	Bank Filtrate t > 50 days	Effect
Ni	µg/L	1.672 <1–105 (n = 267)	<1 <1–28.6 (n = 192)	<1 <1–89.9 (n = 267)	<1 <1–4.46 (n = 45)	Results for BF below LOD
Se	µg/L	<1 <1–1.0 (n = 163)	<1 <1–3.6 (n = 191)	<1 <1–3.6 (n = 265)	1.6 <1–4.2 (n = 43)	Little change
Ag	µg/L	<1 <1–4 (n = 266)	<1 <1–2.3 (n = 178)	<1 <1–4.4 (n = 258)	<1 <1–0.839 (n = 43)	Results below LOD
Sr	mg/L	0.4 0.24–0.32 (n = 3)	0.33 0.23–0.95 (n = 20)	0.3 0.26–0.46 (n = 23)	0.58 0.46–0.78 (n = 8)	River data not sufficient
Zn	µg/L	12.1 <5–222 (n = 267)	5.8 <5–119.5 (n = 172)	6.2 <5–175 (n = 257)	11.5 <5–166 (n = 43)	Removal 0–59%

In October 2010, at Kolontár beside Torna Creek, located 120 km upstream from the Danube River, the dam of a reservoir containing red mud deposits from a bauxite processing company broke, and a so-called "red mud spill" happened. The mud directly affected Torna Creek, a tertiary tributary of the Danube River, ultimately discharging into the Danube River. In Budapest, no change in color was observed in the Danube; however, a slight increase in pH and increased concentrations of aluminium and molybdenum were determined. Figures 3 and 4 show the aluminium and molybdenum concentrations in Danube River water and adjacent RBF wells as a consequence of the spill. An increase in aluminium concentration in the river lasting about four days was observed. No water quality changes were detected in the RBF wells during a continuous monitoring of two years following the accidental pollution. This result clearly indicates the robustness and buffering capacity of RBF against industrial spills.

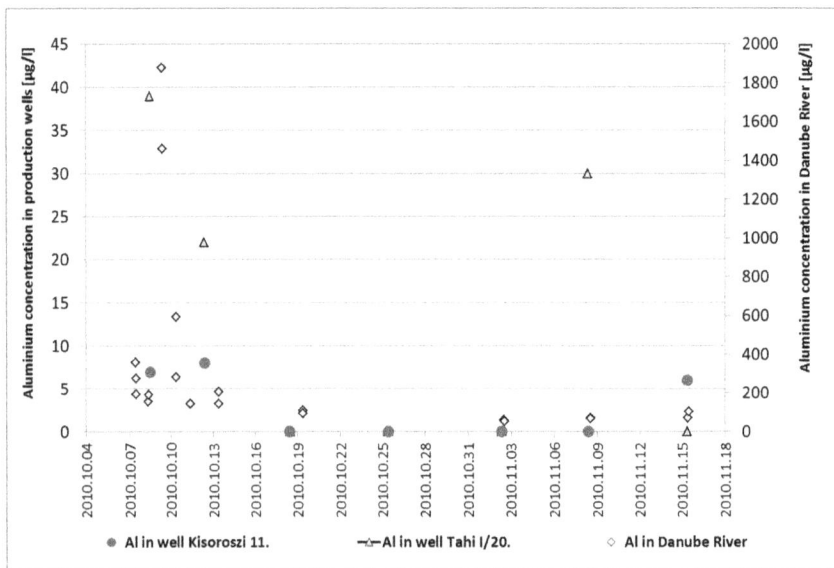

Figure 3. Aluminium concentrations in Danube River water and bank filtrate in production wells during the red mud spill in October 2010. Lower LOD is 5 µg/L for aluminium.

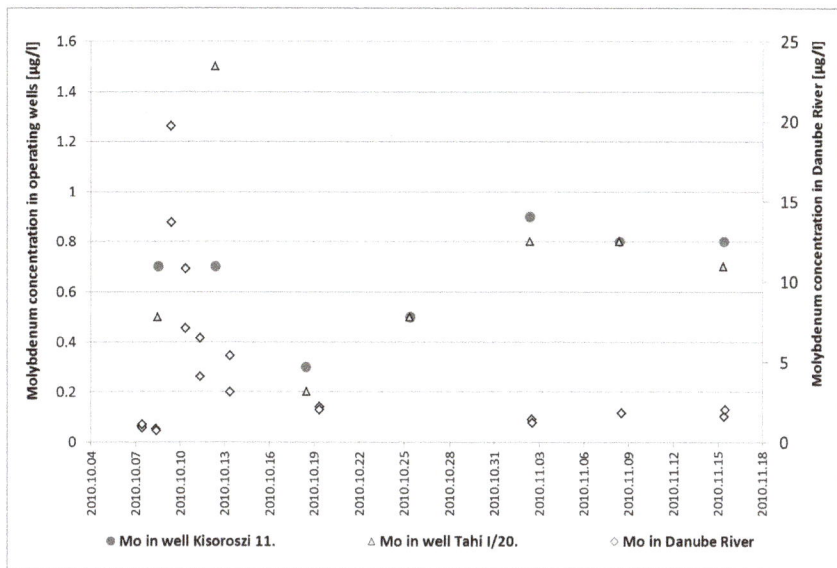

Figure 4. Molybdenum concentrations in Danube River water and bank filtrate in production wells during the red mud spill in October 2010. Lower LOD is 1 µg/L for molybdenum.

3.4. Microbiological Parameters

The microbiological pollution of the Danube River is more prominent in the downstream regions of large European rivers [2]. High bacterial counts in river water are typically observed during floods if direct wastewater effluents discharge into the river [27]. Sporadic microbiological non-compliances in the RBF wells occur very rarely and only as an effect of floods, independent from the bacteria numbers in the river water. In the case of any observed bacterial breakthrough into a well, the well operation is modified as described in Nagy-Kovács et al. [15]. Out of 9153 samples from RBF wells, only 13 samples showed a positive result for total coliforms. *Escherichia coli* (*E. coli*) was never detected in any of the wells. The heterotrophic plate counts (HPC) at 22 °C for bank filtrate were found to range from 0 to 300 counts per mL in 96% of all samples. Based on the experience of the water quality control laboratory operated by the Budapest Waterworks, all values below 400 c/mL are categorized not critical. The Hungarian drinking water guideline [24] does not define a limit for this as it may vary from site to site but requires certain measures if any sudden change in the number occurs.

The mean log removal rates for total coliforms, enterococci and *Pseudomonas aeruginosa* are 2.8, 1.9 and 1.8 respectively (Table 5). *Escherichia coli* and *Clostridium perfringens* are not included in Table 6 as there are no Danube River measurements for these parameters. Despite the short travel times at Budapest, these removal rates are only slightly lower than ranges found at other RBF sites [27–29]. We must highlight that even during higher bacterial loads of the river water, the bank filtrate was of high quality. If the highest total coliform count (TCC) value in river water is taken as input, a log removal of 3.5 (99.95%) would be calculated. Thus, the high efficiency of RBF to remove particles (turbidity) and potential pathogens is proven also for the Budapest RBF systems.

Table 5. Microbiological parameters, median (min–max), Danube River water and bank filtrate with different travel time (t), 2006–2017.

Parameter	Unit	Danube River	Bank Filtrate t < 10 days	Bank Filtrate t = 10–25 days	Bank Filtrate t > 50 days	Log Removal
HPC 22	c/mL	480 0–30,000 (*n = 894*)	0 0–30,000 (*n = 4381*)	0 0–60,000 (*n = 4068*)	0 0–26,000 (*n = 706*)	2.7
HPC 37	c/mL	220 0–18,000 (*n = 728*)	0 0–40,000 (*n = 3275*)	0 0–50,000 (*n = 1709*)	0 0–16,000 (*n = 215*)	2.3
TCC	c/100 mL	660 0–1600 (*n = 890*)	0 0–102 (*n = 4371*)	0 0–500 (*n = 4071*)	0 0–7 (*n = 711*)	2.8
Enterococci	c/100 mL	75 12–360 (*n = 245*)	0 0–160 (*n = 2851*)	0 0–3 (*n = 987*)	0 0–0 (*n = 142*)	1.9
Pseudomonas aeruginosa	c/100 mL	70 2–2800 (*n = 440*)	0 0–160 (*n = 2846*)	0 0–80 (*n = 996*)	0 0–2 (*n = 141*)	1.8

3.5. Biological Parameters

The measurement of biological parameters is required by the Hungarian legislation [24]. The dataset is unique, as there are only few similar studies at RBF sites known from other countries. For the majority of the samples, no positive results were obtained—see median values in Table 6. As for the parameters of protozoa, other protozoa, other worms and iron- and manganese bacteria results are not listed in Table 6 as no river water results are monitored in accordance with Hungarian regulations. For example, out of 1074 samples for protozoa in bank filtrate, only 31 samples were positive.

Table 6. Biological parameters, median (min–max), Danube River water and bank filtrate with different travel time (t), 2006–2017.

Parameter	Unit	Danube River Water	Bank Filtrate t < 10 days	Bank Filtrate t = 10–25 days	Bank Filtrate t > 50 days	Removal
Algae	c/L	24 1,727,200–66,739,440 (*n = 797*)	0 0–1,464,542 (*n = 2544*)	0 0–10,494 (*n = 2204*)	0 0–26 (*n = 356*)	100%
Protozoa	c/L	No measurement	0 0–360 (*n = 446*)	0 0–1503 (*n = 543*)	0 0–1 (*n = 85*)	No river data
Other protozoa	c/L	No measurement	0 0–50 (*n = 1059*)	0 0–8 (*n = 963*)	0 0–288 (*n = 194*)	No river data
Nematodes	c/L	0 0–0 (*n = 21*)	0 0–8 (*n = 1358*)	0 0–12 (*n = 1381*)	0 0–3 (*n = 256*)	-
Other worms	c/L	No measurement	0 0–28 (*n = 1150*)	0 0–6 (*n = 1055*)	0 0–1 (*n = 211*)	No river data
Amoebae	c/L	0 0–1 (*n = 23*)	0 0–42 (*n = 1358*)	0 0–8 (*n = 1381*)	0 0–2 (*n = 256*)	-
Fungi	c/L	0 0–0 (*n = 23*)	0 0–18 (*n = 1358*)	0 0–2 (*n = 1381*)	0 0–0 (*n = 256*)	-
Fe/Mn bact.	c/L	No measurement	97 0–238,440 (*n = 1358*)	97 0–8,253,349 (*n = 1381*)	24 0–7610 (*n = 256*)	No river data

An observed increase in algae numbers in bank filtrate during floods indicates surface water entering into the properly sealed wells via preferential flow paths in the subsurface (Figure 5). All well heads and pump houses are located above the highest observed flood levels or protected against surface water entrance. Additionally, a well reconstruction program to rehabilitate the well structure and sealing to prevent by-passes of surface water is nearly complete. Algae are used as a primary indicator to check if a well is affected (under risk) by direct surface water input. This is preferred against the use of bacteriological methods as they require more time for analysis.

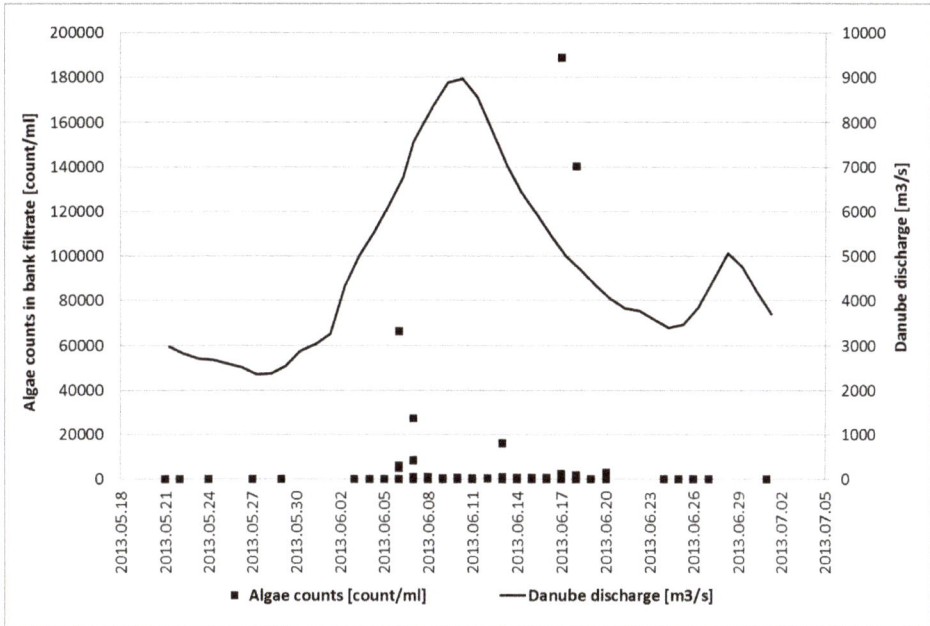

Figure 5. Algae counts in bank filtrate and Danube discharge values during a flood event in 2013.

4. Conclusions

The present study gives an overview of the Danube River water quality and the bank filtrate of related well groups in Budapest. Three different categories were determined for bank filtrates of the operating well groups based on travel times. The large amount of data enabled us to draw robust conclusions on the effect of RBF for the majority of analyzed parameters. Additionally, the effect of different environmental factors on the RBF systems was considered through multiple examples. Results in this case reassured the premise that this natural water treatment process is effective, even under extreme weather conditions.

The analyzed results serve as output of a monitoring procedure principally serving to support water safety and security objectives. Therefore, it should be regarded as an adaptive system that requires continuous updating. A well-defined monitoring program could serve to further investigate the particular characteristics and behavior of RBF systems in future. Considering that the majority of water resources centered around RBF originate in Hungary, the results of such research would serve to increase water safety at a national level.

Supplementary Materials: The following are available online at http://www.mdpi.com/2073-4441/11/2/302/s1, Table S1: Parameters and analytical methods.

Author Contributions: Z.N.-K. and T.G. prepared the article draft. J.D. prepared hydrogeological data. K.C.-M. was responsible for water quality data collection. E.F. helped in the literature review. G.T. participated in project coordination. All co-authors reviewed and edited the article draft.

Funding: All primary data were collected within the AquaNES project. This project has received funding from the European Union's Horizon 2020 Research and Innovation Program under grant no. 689450.

Acknowledgments: This study was based on the excellent work performed by many staff members of the Budapest Waterworks, from water sampling, analysis up to data management.

Conflicts of Interest: The authors declare no conflict of interest. The funding sponsors had no role in the design of the study; in the collection, analyses, or interpretation of data; in the writing of the manuscript, and in the decision to publish the results.

References

1. Ray, C.; Melin, G.; Linsky, R.B. (Eds.) *Riverbank Filtration: Improving Source Water Quality*; Kluwer Academic Publishers: Dordrecht, The Netherlands, 2002.
2. Barreto, S.; Bártfai, B.; Engloner, A.; Liptay, Á.Z.; Madarász, T.; Vargha, M. Water in Hungary, Status overview for the National Water Programme of the Hungarian Academy of Science. (Document of the National Water Programme). Budapest, Hungary, 2017. Available online: https://mta.hu (accessed on 11 February 2019).
3. Davidesz, J.; Debreczeny, L. Long-Term Sustainability of RBF Systems from Aspects of Availability and Capacity. In Proceedings of the MAVÍZ Conference, Sopron, Hungary, 11–12 June 2009. (In Hungarian)
4. László, F. The Hungarian experience with riverbank filtration. In *Riverbank Filtration: The Future Is NOW! Proceedings of the 2nd International Riverbank Filtration Conference, Cincinnati, OH, USA, 16–19 September 2003*; Melin, G., Ed.; National Water Research Institute: Fountain Valley, CA, USA, 2003.
5. Massmann, G.; Nogetzig, A.; Taute, T.; Pekdeger, A. Seasonal and spatial distribution of redox zone during lake bank filtration in Berlin, Germany. *Environ. Geol.* **2008**, *54*, 53–65. [CrossRef]
6. Sprenger, C.; Lorenzen, G.; Hülshoff, I.; Grützmacher, G.; Ronghang, M.; Pekdeger, A. Vulnerability of bank filtration systems to climate change. *Sci. Total Environ.* **2011**, *409*, 655–663. [CrossRef] [PubMed]
7. Bertelkamp, C.; Verliefde, A.R.D.; Schoutteten, K.; VanHaecke, L.; Vand Bussche, J.; Singhal, N.; van der Hoek, J.P. The effect of redox conditions and adaptation time on organic micropollutant removal during river bank filtration: A laboratory scale column study. *Sci. Total Environ.* **2016**, *544*, 309–318. [CrossRef] [PubMed]
8. Kischner, A.-K.-T.; Reischer, G.H.; Kakwerth, S.; Savio, D.; Ixenmaier, S.; Toth, E.; Sommer, R.; Mach, R.L.; Linke, R.; Eiler, A. Multiparametric monitoring of microbial faecal pollution reveals the dominance of human contamination along the whole Danube River. *Water Res.* **2017**, *124*, 543–555. [CrossRef] [PubMed]
9. Kolarevic, S.; Knezevic-Vukcevic, J.; Paunovic, M.; Tmovic, J.; Gagic, Z.; Vukovic-Gacic, B. The anthropogenic impact on water quality of the river Danube in Serbia: Microbiological analysis and genotoxicity monitoring. *Arch. Biol. Sci.* **2011**, *63*, 1209–1217. [CrossRef]
10. AquaNES Deliverable 1.6—Advantages and Limitations, Impact of BF Design, Recommendations for Operators. 2019. Available online: http://www.aquanes-h2020.eu (accessed on 10 February 2019).
11. Benedek, P. New trends in water quality management. *Water Sci. Technol.* **1982**, *14*, 47–82. [CrossRef]
12. Grischek, T.; Macheleidt, W.; Nestler, W. River bed specifics and their effect on bank filtration efficiency. In *Management of Aquifer Recharge for Sustainability*; Dillon, P., Ed.; Balkema Publ.: Lisse, The Netherlands, 2002; pp. 59–64.
13. Grischek, T.; Bartak, R. Riverbed clogging and sustainability of riverbank filtration. *Water* **2016**, *8*, 604. [CrossRef]
14. Grischek, T. Management of RBF along the Elbe River. Ph.D. Thesis, Dresden University of Technology, Department of Water Sciences, Dresden, Germany, 2003. (In German)
15. Nagy-Kovács, Z.; László, B.; Fleit, E.; Czichat-Mártonné, K.; Till, G.; Börnick, H.; Adomat, Y.; Grischek, T. Behaviour of organic micropollutants during riverbank filtration at Budapest, Hungary. *Water* **2018**, *10*, 1861. [CrossRef]
16. Nagy-Kovács, Z.; László, B.; Simon, E.; Fleit, E. Large scale, long-term operational experiences to sustain secure RBF and well structure remodeling. *Water* **2018**, *10*, 1751. [CrossRef]
17. Molnár, Z. *Determination of the Production Well Capacities by Modelling*; Internal Unpublished Document; Budapest Waterworks: Budapest, Hungary, 2013. (In Hungarian)

18. Ministry for Environment and Water. *Ministerial Decree No. 21/2002. (IV. 25.) on the Operation of Water Supply Systems*; Ministry for Environment and Water: Budapest, Hungary, 2002. Available online: https://net.jogtar.hu (accessed on 11 February 2019). (In Hungarian)
19. Schoenheinz, D.; Grischek, T. Behavior of dissolved organic carbon during bank filtration under extreme climate conditions. In *Riverbank Filtration for Water Security in Desert Countries*; Ray, C., Shamrukh, M., Eds.; Springer Science + Business Media B.V.: Dordrecht, The Netherlands, 2011; pp. 51–67.
20. Hoehn, E.; Cirpka, O.A. Assessing residence times of hyporheic ground water in two alluvial flood plains of the Southern Alps using water temperature and tracers. *Hydrol. Earth Syst. Sci.* **2006**, *10*, 553–563. [CrossRef]
21. Schmidt, C.K.; Lange, F.T.; Brauch, H.J.; Kühn, W. Experiences with riverbank filtration and infiltration in Germany. In *Proceedings of the Int. Symp. Artificial Recharge of Groundwater, Daejon, Korea, 14 November 2003*; DVGW-Water Technology Center (TZW): Karlsruhe, Germany, 2003.
22. Grischek, T.; Hiscock, K.M.; Metschies, T.; Dennis, P.; Nestler, W. Factors affecting denitrification during infiltration of river water into a sand and gravel aquifer in Saxony. *Water Res.* **1998**, *32*, 450–460. [CrossRef]
23. Henzler, A.F.; Greskowiak, J.; Massmann, G. Seasonality of temperatures and redox zonations during bank filtration—A modeling approach. *J. Hydrol.* **2016**, *535*, 282–292. [CrossRef]
24. Ministry for Environment and Water-Ministry of Health-Ministry of Agriculture and Regional Development. *Hungarian Drinking Water Guideline: Governmental Decree No. 201/2001. (X. 25) on Water Quality Standards and Monitoring of Drinking Water Quality*; Ministry for Environment and Water-Ministry of Health-Ministry of Agriculture and Regional Development: Budapest, Hungary, 2001. Available online: https://net.jogtar.hu (accessed on 11 February 2019). (In Hungarian)
25. Ministry for Environment and Water. *Ministerial Decree No. 6/2009 (IV. 14) on the Limits and Measurement Methods Necessary for the Protection of Ground and Ground Water against Contaminations*; Ministry for Environment and Water: Budapest, Hungary, 2009. Available online: https://net.jogtar.hu (accessed on 11 February 2019). (In Hungarian)
26. Stuyfzand, P.J. Fate of pollutants during artificial recharge and bank filtration in the Netherlands. In *Artificial Recharge of Groundwater, Proceedings of the Third International Symposium on Artificial Recharge of Ground Water, Amsterdam, Netherlands, 21–25 September 1998*; Peters, J.H., Ed.; Balkema: Rotterdam, The Netherlands, 1998; pp. 119–125.
27. Partinoudi, V.; Collins, M.R. Assessing RBF reduction/removal mechanisms for microbial and organic DBP precursors. *J. AWWA* **2007**, *99*, 61–71. [CrossRef]
28. Hijnen, W.A.M.; Brouwer-Hanzens, A.J.; Charles, K.J.; Medema, G.J. Transport of MS2 phage, Escherichia coli, Clostridium parvum, and Giardia intestinalis in a gravel and a sandy soil. *Environ. Sci. Technol.* **2005**, *39*, 7860–7868. [CrossRef] [PubMed]
29. Sandhu, C.; Grischek, T. Riverbank filtration in India—Using ecosystem services to safeguard human health. *Wat. Sci. Technol.* **2012**, *12*, 783–790. [CrossRef]

water

MDPI

Article

Behavior of Organic Micropollutants During River Bank Filtration in Budapest, Hungary

Zsuzsanna Nagy-Kovács [1,*], Balázs László [1], Ernő Fleit [1], Katalin Czihat-Mártonné [1], Gábor Till [1], Hilmar Börnick [2], Yasmin Adomat [3] and Thomas Grischek [3]

[1] Budapest Waterworks Ltd, Váci út 23-27, 1134 Budapast, Hungary; balazs.laszlo@vizmuvek.hu (B.L.);
    erno.fleit@vizmuvek.hu (E.F.); katalin.martonne@vizmuvek.hu (K.C.-M.); gabor.till@vizmuvek.hu (G.T.)
[2] Institute for Water Chemistry, Technische Universität Dresden, 01062 Dresden, Germany;
    Hilmar.boernick@tu-dresden.de
[3] Division of Water Sciences, University of Applied Sciences Dresden, Friedrich-List-Platz 1,
    01069 Dresden, Germany; yasmin.adomat@htw-dresden.de (Y.A.); thomas.grischek@htw-dresden.de (T.G.)
* Correspondence: zsuzsanna.nagy-kovacs@vizmuvek.hu; Tel.: +36-30-439-4566

Received: 26 October 2018; Accepted: 12 December 2018; Published: 14 December 2018

Abstract: This paper summarizes results from a half-year sampling campaign in Budapest, when Danube River water and bank filtrate were analyzed for 36 emerging micropollutants. Twelve micropollutants were detected regularly in both river water and bank filtrate. Bisphenol A, carbamazepine, and sulfamethoxazole showed low removal (<20%) during bank filtration on Szentendre Island and Csepel island, whereas 1H-benzotriazole, tolyltriazole, diclofenac, cefepime, iomeprol, metazachlor, and acesulfame showed medium to high removal rates of up to 78%. The concentration range in bank filtrate was much lower compared to river water, proving the equilibration effect of bank filtration for water quality.

Keywords: river bank filtration; attenuation; organic micropollutants; pharmaceuticals

1. Introduction

Organic micropollutants from various sources are present in most European surface water bodies [1,2]. Not all micropollutants can be completely removed during drinking water treatment using common techniques such as flocculation, filtration, and activated carbon filtration [3]. River bank filtration (RBF) is known to have a high efficiency in removing organic micropollutants, mainly depending on their biodegradability and adsorption properties [4,5]. Furthermore, attenuation of organic micropollutants is dependent on redox conditions during RBF. Whereas many compounds are better degraded under oxic conditions, there are other compounds which are only (partly) attenuated under anoxic conditions [6]. Authors even suggest operating sequential RBF systems to take advantage of both redox conditions [7,8].

Additionally, an important aspect is that the removal rates of different micropollutants cannot be transferred from one site to another, therefore, it is important to investigate site specific characteristics [9] for a river bank filtration site.

Due to the development of analytical methods, the number of compounds identified in source water is continuously increasing. Some of these compounds are defined as emerging pollutants, which are potentially hazardous compounds with limited available information about their possible effects on humans and aquatic organisms. They comprise of pharmaceuticals, hormones, perfluorinated compounds (PFCs), corrosion inhibitors, algal toxins, or pesticide transformation products [2,10]. It is of major interest to a water company to identify relevant micropollutants and indicators to assess the water quality, taking into account cost issues for regular monitoring. As there are no defined limit values for many emerging pollutants in the Hungarian and German drinking water guidelines, the water companies themselves have to define parameters, which should be included in their water quality monitoring programs. Additionally, knowledge about the behavior of emerging pollutants during RBF is a pre-requisite to eventually adjust the post-treatment accordingly. Recently, the combination of bank filtration and engineered post-treatment systems (e.g., ultrafiltration, nanofiltration, reverse osmosis, electro chlorination) has been investigated in the EU project AquaNES [11].

On one hand, the aim of the presented study was to improve knowledge on the occurrence, range, and behavior of typical emerging pollutants in the Danube River and RBF wells upstream and downstream of the Hungarian capital Budapest. It was also an important aspect to identify relevant pollutants to be included in future monitoring of the wells operated by Budapest Waterworks.

2. Materials and Methods

Budapest Waterworks supply 1.89 million inhabitants based on two large and several small RBF systems. For the study, sampling was focused on the two large RBF systems along the Danube River, on Szentendre Island upstream of the city and Csepel Island downstream of the city (Figure 1). The Danube River has been sampled from the shores at both sites to see the impact of the city on source water quality. The location of the wells on the islands is favorable for RBF, resulting in high portions of bank filtrate [12]. The sampling point upstream of Budapest (W_1) is fed by two separate well groups, Kisoroszi and Tótfalu (Figure 1), and the sampling point downstream of Budapest (W_2) is fed by the Ráckeve well group. For the Kisoroszi well group, the average pumping rate is 80.696 m^3/day and the ratio of bank filtrate is 70%; for the Tótfalu well group these values are 13.090 m^3/day and 91%; and for the Ráckeve well group 90.925 m^3/day and 70%.

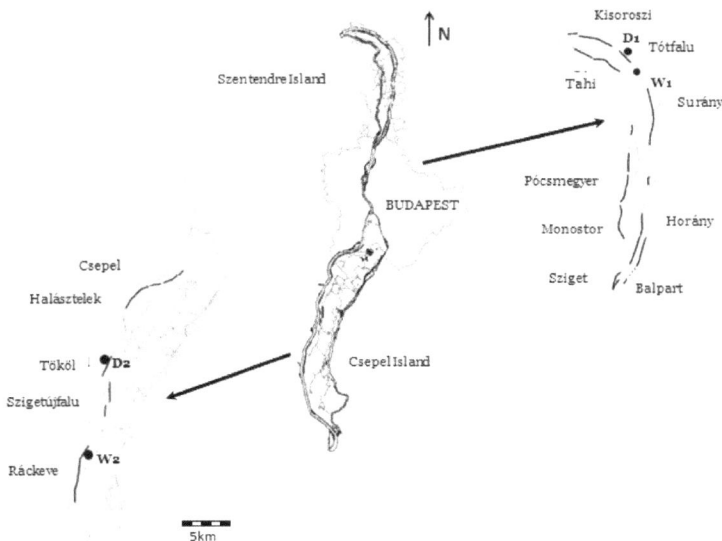

Figure 1. Danube River water and bank filtrate sampling points with the names of well groups on Szentendre Island and Csepel Island in Budapest. D_1: Danube River water sampling point upstream of Budapest. W_1: Bank filtrate sampling point upstream of Budapest. D_2: Danube River water sampling point downstream of Budapest. W_2: Bank filtrate sampling point downstream of Budapest.

On Szentendre Island, samples were taken from a collecting point as a mixture of bank filtrate from two different well groups. On Csepel Island, the collector pipe fed by several horizontal collector wells was selected for sampling. The bank filtrates were untreated at both locations. Samples of

Danube River water and bank filtrate were taken from October 2017 until March 2018. Sampling on Szentendre Island was performed monthly and sampling on Csepel Island was performed weekly, as downstream of the city water quality was expected to be more prone to pollution.

Sampling was done wearing single-use rubber gloves to prevent any contamination of the sample. Glass vial 1 (30 mL) was rinsed with the sampling water three times and emptied. A second glass vial, vial 2 (30 mL), was rinsed two times and half-filled. From vial 2, a volume of 5 mL was taken and transferred to vial 1. Next, 250 µL of an internal standard was added using a Hamilton microsyringe. The vial was closed and shaken. The spiked sample was taken with a one-way syringe and filtered through a 0.2 µm membrane filter (Chromafil Xtra RC-20/25, Macherey-Nagel Germany) and filled into a vial (ND 13) after the first 1 mL was wasted. Internal standards were stored at 2 °C–6 °C until usage and samples were stored at −18 °C until analysis. Before analysis, the samples were defrosted and analyzed without further preparation.

The analysis of 36 target compounds was carried out at the Institute for Water Chemistry, TU Dresden, using a UHPLC Shimadzu Nexera X2 coupled with a Sciex Q6500+ mass detector. Separation was realized on a porous silica column Phenomenex Luna Omega polar C18 (100 × 2.1 mm) with a particle size of 1.6 µm. For all determinations, the UHPLC was operated in gradient mode with a flow rate of 0.60 mL/min and a mobile phase of (A) water and (B) acetonitrile, both acidified with 0.02% formic acid. After an isocratic step for 1 min, a linear gradient was applied from 5% B to 98% B within 9 min. An isocratic step followed for 0.2 min, then, within 1.1 min, a linear gradient was applied again from 98% to 5% B. The column temperature was 40 °C. The mass spectrometer was operated in both positive and negative ion, multiple reaction-monitoring mode (MRM) using nitrogen as the collision gas. Quantification was accomplished using an internal standard method. In the case of compounds without an appropriate isotope-labelled internal standard, an external calibration method was applied. Instrument calibration was performed by analyzing standards at 0.1, 0.5, 1, 5, 10, 50, 100, 500, 1000, 5000, and 10,000 ng/L. The limit of quantification (LOQ) was set at a signal-to-rate ratio (S/N) $\geq$ 10. To prove that the instrument was properly calibrated throughout the analysis, a calibration verification standard was analyzed every 10 samples. Also, blank samples were analyzed between each compound to verify that the measured levels were not an artefact. Data acquisition was accomplished by MultiQuant™ Software (Sciex, version 1.62). Table 1 shows the list of compounds, including their range of quantification.

Table 1. List of analyzed emerging pollutants with range of quantification and MRM transitions.

Analyte	Range of Quantification in LOQ−10,000 ng/L	Quantifier MRM Transition $Q_1 \rightarrow Q_3$ (m/z)	Qualifier MRM Transition $Q_1 \rightarrow Q_3$ (m/z)
Industrial Chemicals			
1H-benzotriazole	50−10000	120→65	120→92
Bisphenol A	5−10000	233→138	233→215
Tolyltriazole	10−10000	134→77	134→79
Herbicides, Pesticides and Transformation Products			
Dimethachlor-ESA	1−10000	300→120	300→80
Dimethachlor-OA	10−10000	250→178	250→130
Dimethoate	10−10000	230→199	230→125
Diuron	10−10000	230→199	230→125
Imidacloprid	5−10000	256→209	256→175
Irgarol	1−10000	254→198	254→108
Isoproturon	1−10000	208→72	208→175
Metazachlor-ESA	5−10000	322→121	322→148
Metazachlor-OA	1−10000	271→67	271→65
Metolachlor-ESA	5−10000	328→120	328→80
Metolachlor-OA	5−10000	278→206	278→174
Nicosulfuron	5−10000	410→182	410→213
Terbuthylazine-2-hydroxy	1−10000	210→97	210→154
Terbutryn	5−10000	142→186	142→91

Table 1. Cont.

Analyte	Range of Quantification in LOQ−10,000 ng/L	Quantifier MRM Transition $Q_1 \rightarrow Q_3$ (m/z)	Qualifier MRM Transition $Q_1 \rightarrow Q_3$ (m/z)
Food Additives			
Acesulfame	1−10000	162→82	162→78
Pharmaceuticals and X-ray Contrast Agents			
Bezafibrate	10−10000	362→316	362→139
Carbamazepine	1−10000	237→194	237→179
Cefepime	50−10000	481→396	481→324
Cefotaxime	50−10000	456→396	456→167
Cefuroxime	50−10000	447→386	447→342
Clarithromycin	10−10000	748→590	748→158
Clindamycin	5−10000	425→126	427→126
Diclofenac	50−10000	294→250	294→252
Erythromycin	5−10000	734→576	734→158
Fluoxetin	10−10000	310→148	310→44
Gabapentin	50−10000	172→154	172→137
Ibuprofen	5−10000	205→161	205→159
Iomeprol	50−10000	778→687	778→405
Metoprolol	5−10000	268→116	268→133
Naproxen	10−10000	229→185	229→169
Paracetamol	5−10000	152→110	152→93
Roxithromycin	50−10000	837→679	837→158
Sulfamethoxazole	1−10000	254→156	154→108

3. Results

Out of the comprehensive list given in Table 1, 12 micropollutants representing each group were detected nearly regularly. The micropollutants bezafibrate, clarithromycin, clindamycin, erythromycin, gabapentin, ibuprofen, metoprolol, naproxen, paracetamol, dimethachlor-ESA, dimethachlor-OA, igarol, imidacloprid, isoproturon, nicosulfuron, metazachlor-OA, terbutylazine-2-hydroxy, and terbutryn were only found in Danube River water but either not found, or found at very low levels, in bank filtrate, thus the attenuation rate is nearly 100%. Azithromycin, cefotaxime, cefuroxime, dimethoate, diuron, fluoxetine, and roxithromycin were not detected in any samples.

The minimum, median, and maximum concentrations of these 12 compounds found in Danube River water and in bank filtrate (BF) are comprised in Tables 2 and 3, respectively.

Table 2. Minimum, median, and maximum concentrations in ng/L of most prominent compounds in the Danube River Water at sampling points on Szentendre Island and Csepel Island.

Compound	Danube River Water (Szentendre) n = 6			Danube River Water (Csepel) n = 24		
	Minimum	Median	Maximum	Minimum	Median	Maximum
1H-Benzotriazole	181	272	345	183	256	338
Bisphenol A	15	33	124	14	86	990
Tolyltriazole	84	121	172	86	142	255
Carbamazepine	19	30	40	19	31	54
Cefepime	194	358	532	135	394	680
Diclofenac	70	153	442	59	154	418
Iomeprol	106	131	161	68	122	272
Sulfamethoxazole	6	14	17	7	13	45
Metolachlor-ESA	33	113	162	24	85	163
Metolachlor-OA	6	31	49	7	23	53
Metazachlor-ESA	52	180	359	31	152	1142
Acesulfame	102	219	343	115	266	512

Table 3. Minimum, median, and maximum concentrations in ng/L of most prominent compounds in bank filtrate at sampling points on Szentendre Island and Csepel Island.

Compound	Bank Filtrate (Szentendre) $n = 6$			Bank Filtrate (Csepel) $n = 24$		
	Minimum	Median	Maximum	Minimum	Median	Maximum
1H-Benzotriazole	70	85	92	125	146	200
Bisphenol A	19	51	98	30	105	2381
Tolyltriazole	32	63	73	64	88	118
Carbamazepine	18	24	24	20	29	43
Cefepime	57	193	301	123	248	546
Diclofenac	36	103	144	13	87	231
Iomeprol	<LOQ	<LOQ	<LOQ	<LOQ	<LOQ	<LOQ
Sulfamethoxazole	9	13	18	6	9	16
Metolachlor-ESA	29	43	70	34	57	83
Metolachlor-OA	11	38	88	9	17	26
Metazachlor-ESA	25	40	273	28	125	686
Acesulfame	112	131	134	145	195	258

value lower then limit of quantification (LOQ).

It can be seen from the mean results of the Danube River water samples that the two locations have no considerable differences.

For each compound, only a mean removal rate has been determined, based on median concentrations for river water and bank filtrate for each site (Table 4). Data pairs were not suitable for this case, considering that well water samples are mixed samples of bank filtrate with different travel times. Samples from river water and bank filtrate taken on the same day are not related to each other. Thus, calculated negative removal rates could result from a higher concentration of a micropollutant before the start of the sampling campaign or on certain days during the sampling campaign, or when river water was not sampled or was sampled from another source (e.g. land-side groundwater).

Table 4. Median removal rates of the most prominent compounds in the River Danube water and bank filtrate at sampling points on Szentendre Island and Csepel Island.

Compound	Removal Rates in % (Szentendre)	Removal Rates in % (Csepel)
1H-benzotriazole	69	43
bisphenol A	−54	−22
tolyltriazole	48	38
carbamazepine	20	4
cefepime	46	37
diclofenac	32	44
iomeprol	bank filtrate concentrations below LOQ	
sulfamethoxazole	9	30
metolachlor-ESA	62	33
metolachlor-OA	−20	25
metazachlor-ESA	78	18
acesulfame	40	27

3.1. Industrial Products

All three compounds of the group of industrial chemicals were detected in river water and bank filtrate. The highest median values were found for 1H-benzotriazole followed by tolyltriazole and bisphenol A in both river water and bank filtrate (Figure 2).

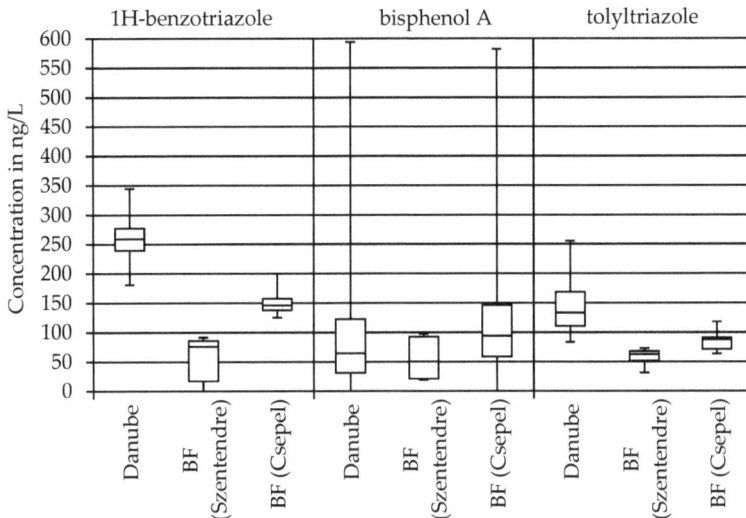

Figure 2. Boxplots representing concentrations of industrial chemicals in the Danube River water and bank filtrate.

For better visibility, two outliers for bisphenol A are not shown in Figure 2: 989 ng/L in Danube River water at Csepel (26 February 2018) and 2381 ng/L in bank filtrate at Csepel (12 March 2018). Concentrations were similar in river water samples at both sites, thus data for the Danube have been combined to have a higher number of samples ($n = 30$) as input concentration.

During the sampling campaign, bisphenol A was detected in all water samples (Figure 3). The obtained bisphenol A levels varied from 4 to 2381 ng/L. The highest concentrations were observed during spring season on Szentendre Island and differed significantly from the determined bisphenol A levels in autumn and winter. These variations may result from environmental factors such as precipitation and temperature or different usage patterns of bisphenol A related products.

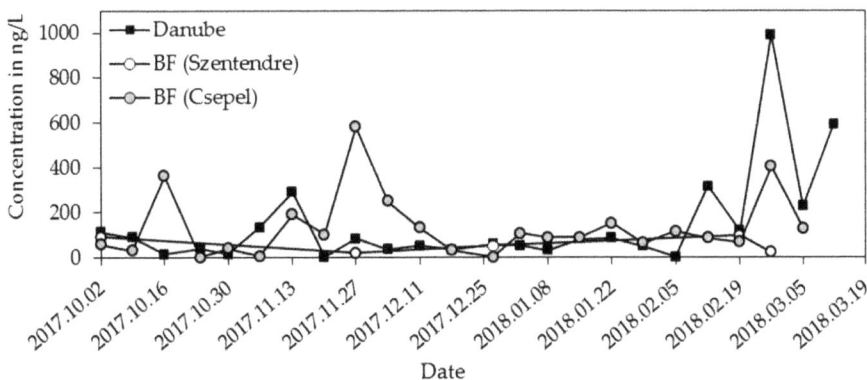

Figure 3. Seasonal fluctuation of bisphenol A concentration in Danube River water (squares) and bank filtrate (circles) (outlier of 2381 ng/L in BF (Csepel) on 26 February 2018, not shown).

3.2. Pharmaceuticals and X-ray Contrast Agents

Out of 19 monitored pharmaceuticals, five compounds were found to be regularly present in both Danube River water and bank filtrate. The cephalosporin antibiotic cefepime and the analgesic diclofenac were the most frequently detected pharmaceuticals, followed by the X-ray contrast agent iomeprol, the antiepileptic carbamazepine, and the antibiotic sulfamethoxazole (Figure 4). The median concentration of cefepime from all 30 Danube River water samples was 376 ng/L, of diclofenac was 154 ng/L, and of iomeprol was 126 ng/L. The levels of pharmaceutical residue in the bank filtrate were in all cases lower than those detected in the river water. Cefepime, diclofenac, and iomeprol concentrations decreased by 57%, 62%, and 96%, respectively. Iomeprol was found at much lower concentrations in river water compared to other European rivers with RBF sites, such as the Elbe River, with median concentrations from 500 to 800 ng/L since 2015 [13].

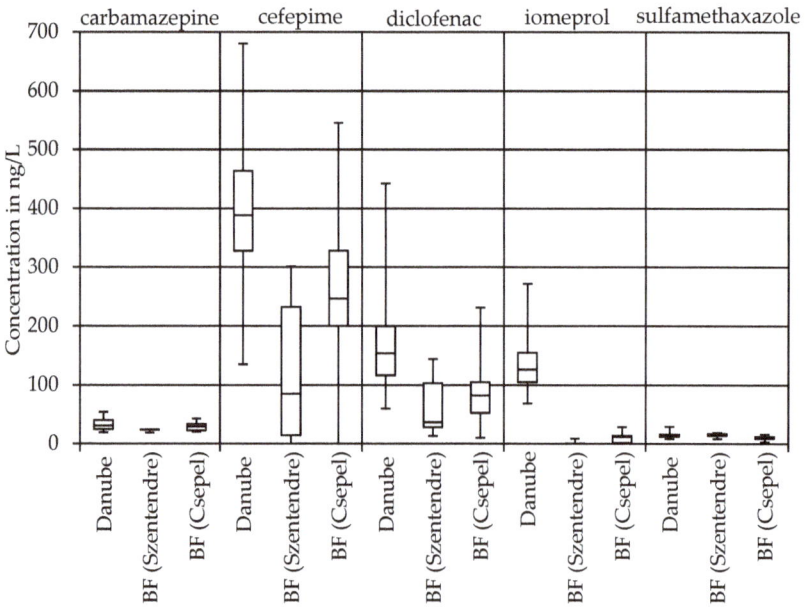

Figure 4. Boxplots representing concentrations of pharmaceuticals and X-ray contrast media in the Danube River water and bank filtrate.

3.3. Herbicides, Pesticides, and Transformation Products

Out of the 14 analyzed herbicides, pesticides, and transformation products, the metabolites of metazachlor and metolachlor were most frequently measured. The highest concentrations were determined for metazachlor ethane sulfonic acid (ESA), metolachlor oxanilic acid (OA), and metolachlor-ESA (Figure 5).

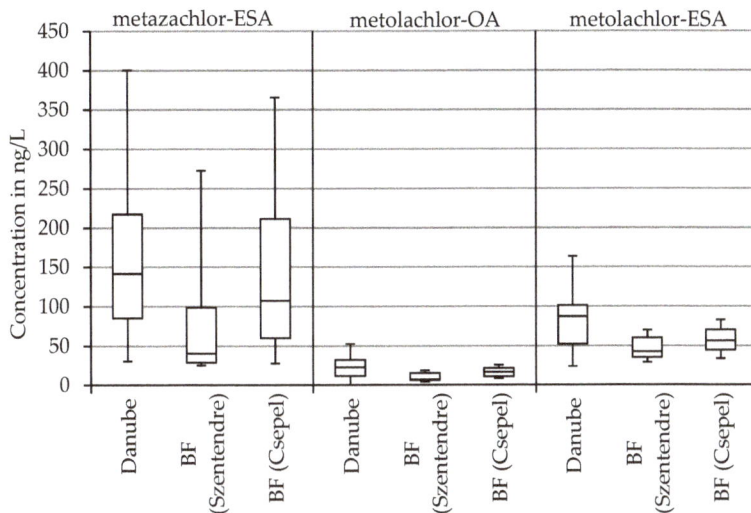

Figure 5. Boxplots representing concentrations of herbicides, pesticides, and transformation products in the Danube River water and bank filtrate.

The two highest levels of metazachlor-ESA are not shown in Figure 5. They were detected in December 2017 (1141 ng/L) in the Danube River water and in February 2018 (685 ng/L) in the bank filtrate on Csepel Island (Figure 6).

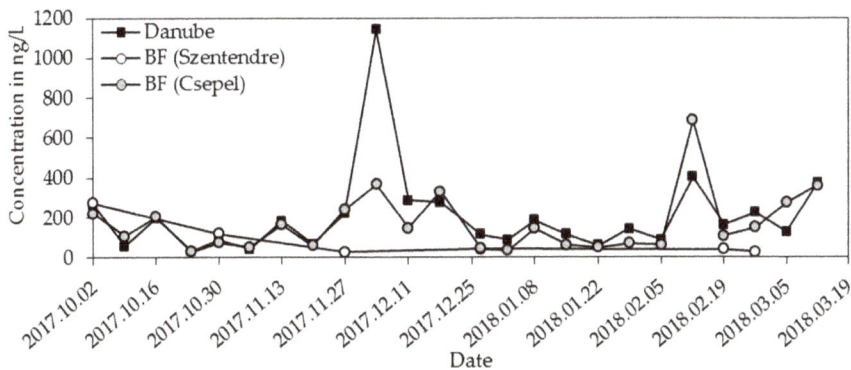

Figure 6. Seasonal fluctuation of metazachlor-ESA concentrations in the Danube River water (squares) and bank filtrate (circles).

3.4. Food Additives

The artificial sweetener acesulfame was detected in all water samples (Figure 7). The concentration in river water ranged from 102 to 512 ng/L ($n = 30$) and in bank filtrate from 112 to 258 ng/L ($n = 30$). The highest concentrations in river water were found after Christmas 2017 and during the low flow period starting end of February 2018 (Figure 8).

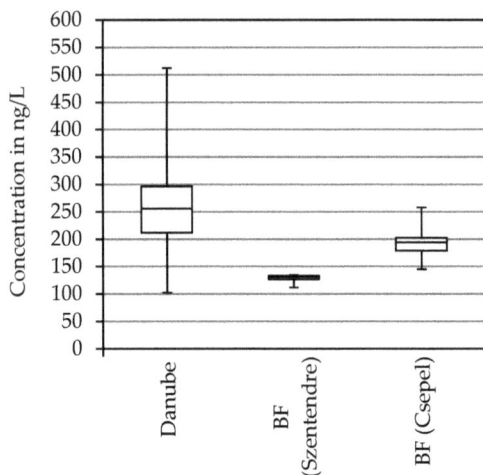

Figure 7. Boxplots representing concentrations of acesulfame in the Danube River water and bank filtrate.

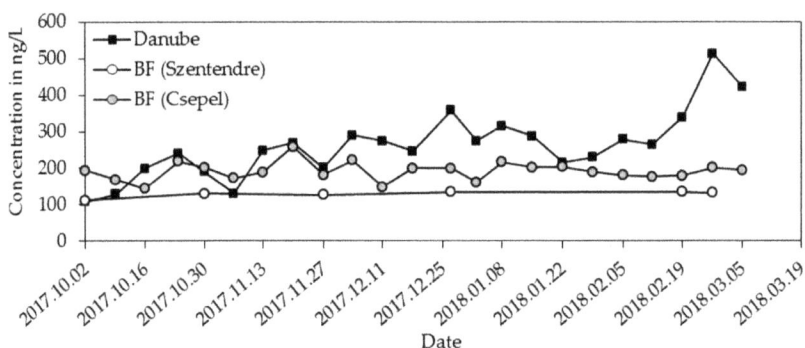

Figure 8. Seasonal fluctuation of acesulfame concentration in the Danube River water (squares) and bank filtrate (circles).

4. Discussion

The levels of 1H-benzotriazole and tolyltriazole are used as complexing agents (e.g., corrosion inhibitors) or for silver protection in dishwashing agents. Benzotriazoles can undergo several processes during RBF, such as biodegradation and retardation [14]. The concentration range of benzotriazoles in the Danube was determined to be between 130–300 ng/L [2]. During the demonstrated measurement campaign, they were present in the Danube River water at considerably higher concentrations than in the bank filtrate. Mean removal rates for the Szentendre and Csepel sites were 68% and 43% for 1H-benzotriazole and 48% and 37% for tolyltriazole, respectively.

Bisphenol A is used as an intermediate in the production of polycarbonate plastics and epoxy resins, unsaturated polyester-styrene resins, and as flame retardants in products like food and drink storage containers or protective coatings for metal cans [14–17]. Microbiological biodegradation has been investigated and according to the findings, bisphenol A is eliminated during bank filtration [18]. That is in contradiction with the results of the present study, where negative removal rates have been obtained. To exclude the effect of temporary spills of bisphenol A in the river on bank filtrate quality, preparation of mixed river water samples based on hourly sampling over at least one week would be an option.

Pharmaceuticals and X-ray contrast agents belong to the most predominant group of compounds in the aquatic environment, due to the high input quantities, their resistance to degradation (persistence or pseudo-persistence), and polar character-limiting attenuation by adsorption. Pharmaceuticals and their metabolites are mainly released into waterbodies via waste water effluents because they are not well attenuated in the human body and in the sewage treatment plant [2,19,20]. Considering the fact that most pharmaceuticals are very polar, they are hardly removed in conventional wastewater treatment plants. These properties also hinder the removal in drinking water treatment processes. As a result, residual concentrations are often found even in drinking water [9].

Carbamazepine is a medication used primarily in the treatment of epilepsy and neuropathic pain. It is reported that carbamazepine is a persistent compound, with relatively stable concentrations throughout bank filtration [21]. Measurement results are similar, with removal rates 20% and 4% for the sites upstream and downstream from the capital, respectively.

Cefepime is a fourth-generation cephalosporin antibiotic that has a broad spectrum of activity against both Gram-negative and Gram-positive bacteria. Cefepime was found to be partly removed by 46% and 37% during RBF at Budapest.

Diclofenac is a potent nonsteroidal anti-inflammatory drug (NSAID), taken or applied to reduce inflammation and as an analgesic reducing pain. Due to its wide use, it is a well observed pharmaceutical micropollutant. High concentrations were found in river waters but not in groundwater, suggesting that it is eliminated effectively. Also, it has been described that even in wells with relatively short travel times, its concentration decreased sharply [21]. At 32% and 43% removal rates, diclofenac can be considered to be relatively degradable.

Iomeprol is an iodinated X-ray contrast agent. According to Schittko et al., iomeprol was significantly removed during BF under anoxic conditions [22]. This was also the case at both sites in Budapest, where concentrations of iomeprol in bank filtrate were below LOQ.

Sulfamethoxazole is an antibiotic, effective in the treatment against Gram-negative and Gram-positive bacterial infections. It is considered to be a rather persistent pollutant, showing relatively stable concentrations through the subsurface water passage. [21] With removal rates at 9% and 30%, it proved to be fairly persistent in the present study as well.

The removal rates for sulfamethoxazole and diclofenac were higher on Csepel Island, which is assumed to be a result of longer flow paths and travel times. On average, the distance between the wells on Szentendre Island from the bank of the Danube River is 103 m, whereas it is 156 m on Csepel Island. From the data of this study, it is not yet possible to assess if the distance between the wells and the river bank or the travel time is more responsible for the attenuation, because the wells have different pumping rates and are not all continuously operated.

The concentration range of cefepime, diclofenac, and iomeprol was less than a factor of 2, whereas the discharge of the Danube River has been changing by a factor of 3.33 during the sampling period. For diclofenac and iomeprol, much lower fluctuations were found in bank filtrate, proving the buffering effect of RBF.

The median concentration of cefepime in Danube River water was calculated to 376 ng/L, which is low compared to findings from the Somes River, Romania [23].

Metazachlor and metolachlor are widely used herbicides, applied predominantly to maize crops and rape. Depending on their stability, they undergo decomposition processes. Therefore, not only active ingredients but their metabolites also occur as emerging contaminants. Metazachlor and metolachlor also have short half-lives in soils (5–30 days), therefore, they quickly degrade to oxanilic acid (OA), ethane sulfonic acid (ESA), and derivates [24,25]. Those transformation products are only weakly adsorbed onto soil, resulting in a high mobility. As a consequence, the OA and ESA metazachlor and metolachlor derivates are among the most frequent and concentrated water pollutants [25]. The maximum concentrations of metazachlor-ESA were determined during winter months. This may indicate a more frequent agricultural usage of parent herbicides during winter [26] but is not supported by results from common monitoring of herbicides, for which no peaks were observed in winter. As of metolachlor-ESA and metazachlor-ESA removal rates were higher at Szentendre with 62% and 78%, respectively, while for the Csepel site lower values were determined, at 33% and 18%, respectively. For metolachlor-OA results were inconclusive at the Szentendre site with a negative removal rate, while at Csepel it was 25%.

Acesulfame is one of the most used artificial sweeteners. It is passing in wastewater treatment plants and thus typically found in waste water affected river water. Other sweeteners, such as cyclamate or saccharine, are usually degraded during wastewater treatment [27]. Therefore, acesulfame is a favorable indicator for human sewage and could be used to estimate the portion of bank filtrate in the abstracted water from the RBF wells. Assuming no attenuation during RBF and no occurrence in natural groundwater, the median concentration of 143 ng/L in the wells on Csepel Island and 266 ng/L in the Danube River water at Csepel would indicate a portion of bank filtrate of 73%, which is within the range found from groundwater flow modeling [12].

5. Summary

Out of the 36 micropollutants that have been analyzed, 12 were present in almost all the samples. In the case of eight compounds, the median concentrations were lower in the Szentendre Island bank filtrate samples. Diclofenac, sulfamethoxazole, and metolachlor-OA results were lower in the Csepel Island bank filtrate samples. The results of bisphenol A showed considerable seasonal variations.

The median concentrations of iomeprol were below the limit of detection for both sites.

The results for the herbicides, pesticides, and transformation product groups showed considerable differences between the results originating from Szentendre Island and Csepel Island. It would be interesting to further investigate the concentration of micropollutants in the ground water [21].

This study presents the first measurement campaign of the Budapest Waterworks within the AquaNES project. Results give an overview about the occurrence of micropollutants, which are not yet monitored regularly, in the Danube River water and its bank filtrate at Budapest. Most of the analyzed micropollutants have no determined method nor defined limits in any Hungarian or European regulation. The applied methods are not yet accredited, and accordingly, measurement results are of an informative nature. In general, it can be declared that persistent micropollutants in the river water and bank filtrate are well below the concentrations of contaminants found in other alimentations. Nevertheless, this issue is of high priority for all waterworks that are operating RBF systems to assure safe drinking water.

Author Contributions: Z.N.-K., B.L., E.F., Y.A. and T.G. reviewed previous literature and prepared the article draft. K.C.-M. was responsible for the sampling. H.B. performed the analyses and researched the use of the compounds. G.T. participated in project coordination. All co-authors reviewed and edited the article draft.

Funding: All primary data was collected within the AquaNES project. This project has received funding from the European Union´s Horizon 2020 Research and Innovation Program under grant No. 689450. The financing of UHPLC-MS/MS system was supported by European Fund for Regional Development and by the Free State of Saxony.

Acknowledgments: This work was performed in cooperation between Budapest Waterworks and the Division of Water Sciences at the University of Applied Sciences, Dresden.

Conflicts of Interest: The authors declare no conflict of interest. The funding sponsors had no role in the design of the study; in the collection, analyses, or interpretation of data; in the writing of the manuscript, and in the decision to publish the results.

References

1. Loos, R.; Gawlik, B.M.; Locoro, G.; Rimaviciute, E.; Contini, S.; Bidoglio, G. EU-wide survey of polar organic persistent pollutants in European river waters. *Environ. Pollut.* **2009**, *157*, 561–568. [CrossRef] [PubMed]
2. Liška, I.; Wagner, F.; Sengl, M.; Deutsch, K.; Slobodník, J. Joint Danube Survey 3 | ICPDR – International Commission for the Protection of the Danube River. Available online: https://www.icpdr.org/main/activities-projects/jds3 (accessed on 12 December 2018).

3. Kim, M.-K.; Zoh, K.D. Occurrence and removals of micropollutants in water environment. *Eng. Res.* **2016**, *21*, 319–332. [CrossRef]

4. Gutiérrez, J.P.; van Halem, D.; Rietveld, L. River bank filtration for the treatment of highly turbid Colombian rivers. *Drinking Water Eng. Sci.* **2017**, *10*, 13–26. [CrossRef]

5. Hiscock, K.M.; Grischek, T. Attenuation of groundwater pollution by bank filtration. *J. Hydrol.* **2001**, *266*, 139–144. [CrossRef]

6. Schmidt, C.K.; Lange, F.T.; Brauch, H.-J. Assessing the impact of different redox conditions and residence times on the fate of organic micropollutants during river bank filtration. In *Proceedings of the 4th International Conference on Pharmaceuticals and Endocrine Disrupting Chemicals in Water*; National Ground Water Association: Minneapolis, MN, USA, 2004; pp. 195–205.

7. Regnery, J.; Wing, A.D.; Kautz, J.; Drewes, J.E. Introducing sequential managed aquifer recharge technology (SMART)-From laboratory to full-scale application. *Chemosphere* **2016**, *154*, 8–16. [CrossRef] [PubMed]

8. Hellauer, K.; Mergel, D.; Ruhl, A.S.; Filter, J.; Hübner, U.; Jekel, M.; Drewes, J.E. Advancing Sequential Managed Aquifer Recharge Technology (SMART) using different intermediate oxidation processes. *Water* **2017**, *9*, 221. [CrossRef]

9. Storck, F.R.; Sacher, F.; Brauch, H.-J. Hazardous and emerging substances in drinking water resources in the Danube River Basin. In *Danube River Basin*; Liska, I., Ed.; Springer: Berlin, Germany, 2015; pp. 251–270.

10. Greskowiak, J.; Hamann, E.; Burke, V.; Massmann, G. The uncertainty of biodegradation rate constants of emerging organic compounds in soil and groundwater-A compilation of literature values for 82 substances. *Water Res.* **2017**, *126*, 122–133. [CrossRef] [PubMed]

11. AquaNES. Available online: http://www.aquanes-h2020.eu (accessed on 27 September 2018).

12. Molnár, Z. *Determination of the Production Well Capacities by Modelling (in Hungarian)*; Budapest Waterworks: Budapest, Hungary, 2013. (in Hungarian)

13. AWE. *AWE: Arbeitsgemeinschaft der Wasserversorger im Einzugsgebiet der Elbe, Gütebericht 2016–2017*; Water Quality Report of Waterworks in the Elbe River Catchment: Torgau, Germany, 2018.

14. Calvo-Flores, F.G.; Isac-Garcéa, J.; Dobado, J.A. *Emerging Pollutants: Origin, Structure and Properties*; Wiley-VCH: Weinheim, Germany, 2018.

15. Milanović, M.; Sudji, J.; Grujić Letić, N.; Radonić, J.; Turk Sekulić, M.; Vojinović Miloradov, M.; Milić, N. Seasonal variations of bisphenol A in the Danube by the Novi Sad municipality. *J. Serb. Chem. Soc.* **2016**, *80*, 333–345. [CrossRef]

16. Arnold, S.M.; Clark, K.E.; Staples, C.A.; Klecka, G.M.; Dimond, S.S.; Caspers, N.; Hentges, S.G. Relevance of drinking water as a source of human exposure to bisphenol A. *J. Exposure Sci. Environ. Epidemiol.* **2013**, *23*, 137–144. [CrossRef] [PubMed]

17. Technical University of Denmark. *Benzotriazole and Tolyltriazole-Evaluation of Health Hazards and Proposal of Health Based Quality Criteria for Soil and Drinking Water, Toxicology and Risk Assessment*; Technical University of Denmark: Lyngby, Denmark, 2013.

18. Ray, C. *Riverbank Filtration: Understanding Contaminant Biogeochemistry and Pathogen Removal*; Kluwer Academic Publishers: Dordrecht, The Netherlands, 2002.

19. Radović, T.; Grujić, S.; Dujaković, N.; Radišić, M.; Vasiljević, T.; Petković, A.; Boreli-Zdravkovic, Đ.; Dimkić, M.; Laušević, M. Pharmaceutical residues in the Danube River basin in Serbia–a two-year survey. *Water Sci. Technol.* **2012**, *66*, 659–665. [CrossRef] [PubMed]

20. Sacher, F.; Metziger, M.; Wenz, M.; Gabriel, S.; Brauch, H.-J. Arzneimittelrückstände in Grund- und Oberflächenwässern. In *Spurenstoffe in Gewässern*; Track, T., Kreysa, G., Eds.; Wiley-VCH: Weinheim, Germany, 2006; pp. 97–106. ISBN 3-527-31017-7.

21. Hollender, J.; Huntscha, S. River Bank Filtration of Micropollutants, 2014. Available online: https://www.dora.lib4ri.ch/eawag/islandora/object/eawag%3A11880/datastream/PDF/view (accessed on 12 December 2018).

22. Schittko, S.; Putschew, A.; Jekel, M. Bank filtration: A suitable process for the removal of iodinated X-ray contrast media? *Water Sci. Technol.* **2004**, *50*, 261–268. [CrossRef] [PubMed]

23. Soran, M.-L.; Lung, I.; Opriş, O.; Floare-Avram, V.; Coman, C. Determination of antibiotics in surface water by solid-phase extraction and high-performance liquid chromatography with diode array and mass spectrometry detection. *Anal. Lett.* **2017**, *50*, 1209–1218. [CrossRef]

24. Hvězdová, M.; Kosubová, P.; Košíková, M.; Scherr, K.E.; Šimek, Z.; Brodský, L.; Šudoma, M.; Škulcová, L.; Sáňka, M.; Svobodová, M.; et al. Currently and recently used pesticides in Central European arable soils. *Sci. Total Environ.* **2018**, *613*, 361–370. [CrossRef] [PubMed]

25. Lewis, K.A.; Tzilivakis, J.; Warner, D.J.; Green, A. An international database for pesticide risk assessments and management. *Human Ecol. Risk Assess. Int. J.* **2016**, *22*, 1050–1064. [CrossRef]

26. Mai, C.; Theobald, N.; Lammel, G.; Hühnerfuss, H. Spatial, seasonal and vertical distributions of currently-used pesticides in the marine boundary layer of the North Sea. *Atmos. Environ.* **2013**, *75*, 92–102. [CrossRef]

27. Lange, F.T.; Scheurer, M.; Brauch, H.-J. Artificial sweeteners-a recently recognized class of emerging environmental contaminants: A review. *Anal. Bioanal.Chem.* **2012**, *403*, 2503–2518. [CrossRef] [PubMed]

water

MDPI

Article

Removal of Natural Organic Matter and Organic Micropollutants during Riverbank Filtration in Krajkowo, Poland

Krzysztof Dragon [1,*], Józef Górski [1], Roksana Kruć [1], Dariusz Drożdżyński [2] and Thomas Grischek [3]

[1]   Department of Hydrogeology and Water Protection, Institute of Geology,
       Adam Mickiewicz University in Poznań, ul. Bogumiła Krygowskiego 12, 61-680 Poznań, Poland;
       gorski@amu.edu.pl (J.G.); roksana.kruc@amu.edu.pl (R.K.)
[2]   Department of Pesticide Residue Research, Plant Protection Institute–National Research Institute,
       ul. Władysława Węgorka 20, 60-318 Poznań, Poland; d_drozdzynski@o2.pl
[3]   Division of Water Sciences, University of Applied Sciences Dresden, Friedrich-List-Platz 1,
       01069 Dresden, Germany; thomas.grischek@htw-dresden.de
*   Correspondence: smok@amu.edu.pl; Tel.: +48-618-296-058

Received: 19 September 2018; Accepted: 11 October 2018; Published: 16 October 2018

Abstract: The aim of this article is to evaluate the removal of natural organic matter and micropollutants at a riverbank filtration site in Krajkowo, Poland, and its dependence on the distance between the wells and the river and related travel times. A high reduction in dissolved organic carbon (40–42%), chemical oxygen demand (65–70%), and colour (42–47%) was found in the riverbank filtration wells at a distance of 60–80 m from the river. A lower reduction in dissolved organic carbon (26%), chemical oxygen demand (42%), and colour (33%) was observed in a horizontal well. At greater distances of the wells from the river, the removal of pharmaceutical residues and pesticides was in the range of 52–66% and 55–66%, respectively. The highest removal of pharmaceutical residues and pesticides was found in a well located 250 m from the river and no micropollutants were detected in a well located 680 m from the river. The results provide evidence of the high efficacy of riverbank filtration for contaminant removal.

Keywords: riverbank filtration; removal efficacy; dissolved organic carbon (DOC); pesticides; pharmaceutical residues

1. Introduction

Alluvial aquifers supply a significant amount of drinking water in many countries because they offer easy access to groundwater and usually have feasible hydraulic properties. One method used for increasing quantities of groundwater in alluvial aquifers is riverbank filtration (RBF). RBF is a good alternative to the direct supply of surface water because the passage of water through the aquifer improves water quality. First in the riverbed and then in the aquifer, the water undergoes combined physical, biological, and chemical processes such as dissolution, sorption, redox processes, and biodegradation [1]. Additionally, mixing with ambient groundwater usually occurs to some degree [2,3].

Among the multiple benefits of RBF, the removal of natural organic matter (NOM), which is usually present in surface waters at relatively high concentrations, is significant. During RBF, an effective removal of dissolved organic carbon (DOC) of more than 50% can be achieved [4,5]. The significant reduction in chemical oxygen demand (COD) is also important [6]. It has been documented that the effective reduction of high molecular weight organic fractions is achieved during RBF, but with a lower removal of low molecular weight fractions [7,8]. This finding is important for further water treatment due to the formation of by-products during water chlorination [9].

The nature of the RBF system results in the quality of extracted water being dependent on surface water quality. Pollution of rivers is observed in many European countries due to agricultural activities in the catchment area [10,11], and wastewater effluents [12]. Water pollution by nitrates is common around the world [13,14], but in recent years, the pollution of surface water by pesticides has become increasingly problematic [15–17]. Other emerging contaminants in surface water are pharmaceutical residues [18,19]. Due to the high vulnerability of RBF systems to contamination by source surface water, it is crucial to determine organic micropollutant removal rates to properly manage RBF systems.

The main goals of the present article are: (1) the determination of the changes in water chemistry during passage through the aquifer in relation to the seasonal surface water chemistry fluctuations; (2) the investigation of the occurrence and behaviour of selected pesticides and pharmaceuticals; and (3) the investigation of removal efficacy of RBF depending on the distance of the wells from the river. For the present investigation, the Krajkowo site was selected, where an RBF system of vertical wells exists as well as a horizontal well (HW), with drains located below the river bottom.

2. Materials and Methods

2.1. Site Description

The Krajkowo well field supplies water to Poznań City and is located 30 km south of the city on Krajkowo Island (52°12′47″N 16°56′49″E) in the Warta River valley (Figure 1). The wells are located in the region where two main groundwater bodies overlap—The Wielkopolska Burried Valley (WBV) aquifer and the Warszawa-Berlin Ice Marginal Valley (WBIMV) aquifer. The well field is located in the region where the sediments forming these aquifers overlap, thereby providing good conditions for water exploitation (water-bearing sediments with a thickness of 30–40 m).

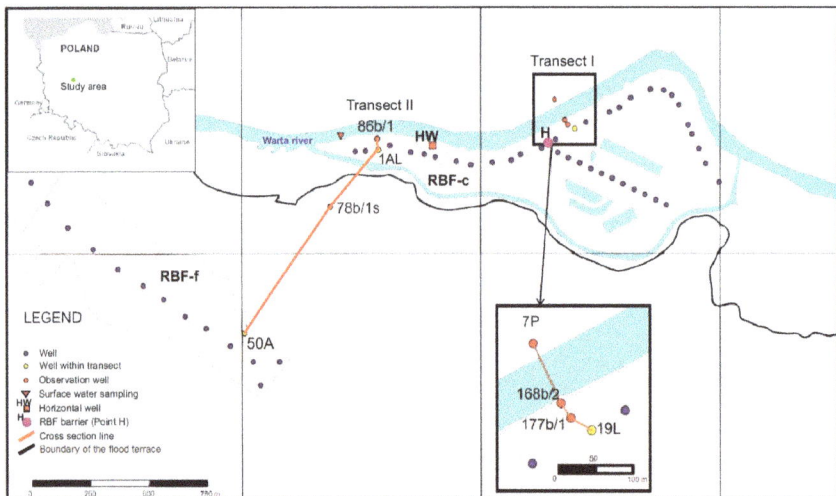

Figure 1. Map of the study area. RBF: riverbank filtration; RBF-c: wells on the flood terrace; RBF-f: wells on the higher terrace; HW: horizontal well.

The lithology of the upper aquifer (WBIMV) is dominated by fine and medium sands of fluvial origin (to a depth of 10 m) and by coarse sands and gravels of fluvio-glacial origin in the deeper portions (to a depth of 20 m) (Figure 2). The deepest aquifer (WBV) is also composed of fine and medium fluvial sands in the upper part (to a depth of 25–30 m) and by coarse fluvio-glacial sands and gravels in the deepest part of the aquifer. Unconfined aquifer conditions dominate the study area, whereas in small regions aquitard composed of glacial tills are present between the WBV and WBIMV aquifers. The static water level is approximately 3–5 m below the ground surface.

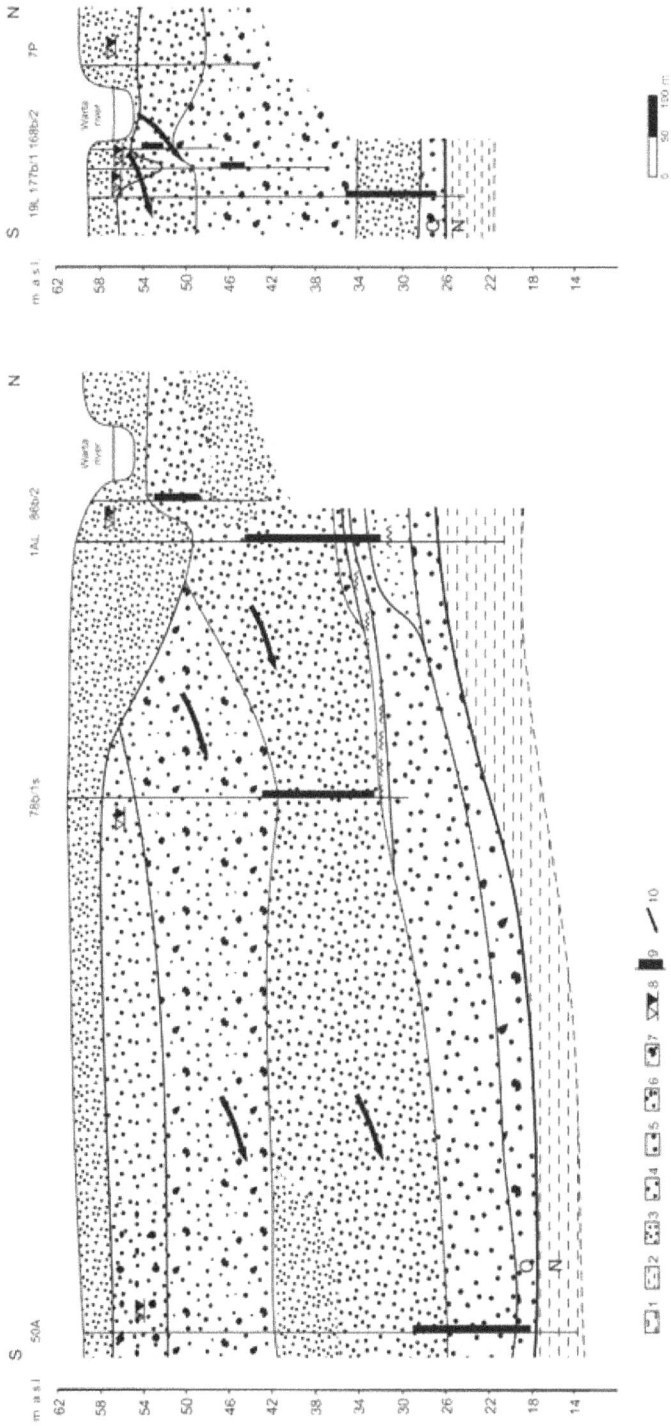

Figure 2. Hydrogeological cross-sections (lines of cross-sections are marked in Figure 1).

1 - silt, 2 - clay, 3 - fine-grained sand, 4 - medium-grained sand, 5 - coarse-grained sand, 6 - medium-grained sand and gravel, 7 - sand, gravel and pebble, 8 - groundwater level, 9 - well screen, 10 - groundwater flow directions. Q - Quaternary, N - Neogene

Two different well types are used for water extraction (Figure 1):

- a gallery of 29 vertical wells (RBF-c) on the left side of the Warta River located at a distance of 60–80 m from the river channel (Figure 3),
- a horizontal well (HW) with drains placed 5 m below the river bottom (Figure 3). The drains were installed by excavation (dredging) of the riverbed sediments.

At longer distances from the river (between 400 and 1000 m), the second well group is located on a higher terrace. This group includes 56 vertical wells. This part of the well field is not continuously exploited. For this study, only the portion of the well group shown in Figure 1 (RBF-f) was continuously pumped for a period of two years.

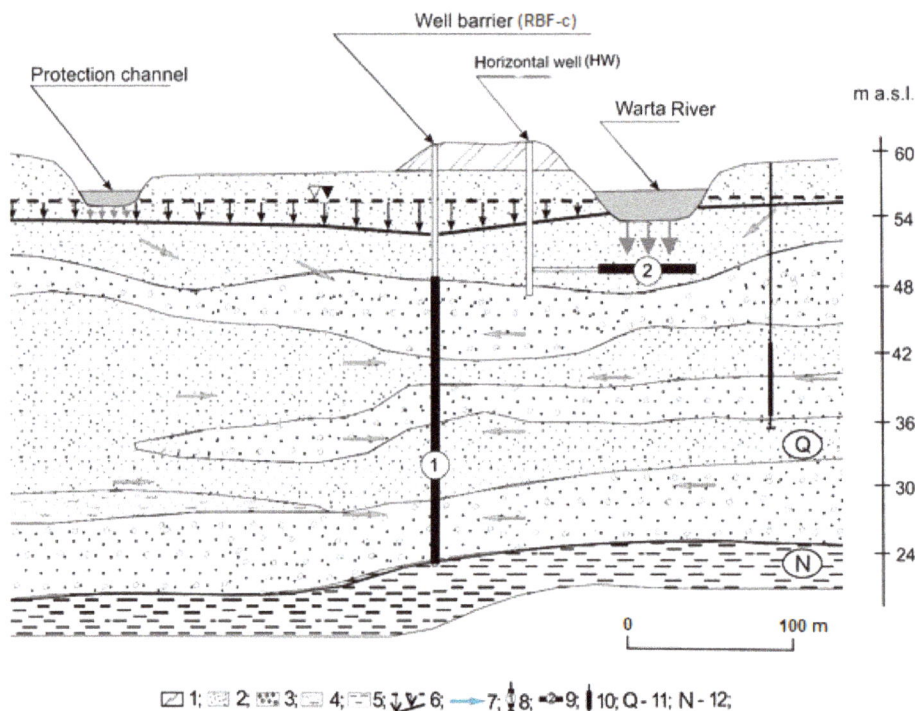

Figure 3. A scheme presenting the location of the horizontal drains of the collector well and positions of the RBF-c wells [20]. Legend: 1—the embankment; 2—sands; 3—gravels; 4—silts; 5—clays; 6—the static and dynamic water level; 7—groundwater flow directions; 8—the position of the RBF well screen; 9—the position of the HW drains; 10—other observation wells; 11—Quaternary; 12—Neogene.

2.2. Methods

To investigate groundwater chemistry changes in the RBF system, wells along two transects were selected for sampling. Transect I (shorter) was located between the river and the RBF-c production well and transect II (longer) was located between the river and the RBF-f well (Figure 1). The sampling points along transects were located along the flow paths, permitting the investigation of hydrochemical transformations associated with bank filtration at different distances from the source (river) water.

The monitoring programme included a one-year sampling campaign in two selected wells located on the transects (1AL and 19L) and an 18-month sampling campaign at sampling point H, which received mixed water from 15 wells located on the eastern side of the well gallery (Figure 1). Sampling was performed monthly between October 2016 and May 2018. Warta River water was also sampled during the investigation period. For pesticides and pharmaceuticals, 6 sampling series were planned at all sampling points located on the transects. In this article, the preliminary results from the first three series are presented along with the results of the first pilot sampling series for pharmaceutical residues. Additionally, Aquanet (waterworks operator) operational monitoring data were used, including the analyses of HW from January 2015 to May 2018.

The production wells were continuously pumped during sampling, while the observation wells were pumped using a portable pump (MP-1, Grundfos, Bjerringbro, Denmark). The water was stored in polyethylene bottles that were flushed three times before sampling. On the same day, watersamples were transported to the laboratory in a refrigerated container. Chemical analyses (Table 1) were performed at the Aquanet Laboratory (Poznań, Poland) with use of a Dionex ionic chromatograph (Thermo Fisher Scientific, Waltham, MA, USA) (NO_3^- and NO_2^-), a Varian Cary 50 spectrometer (Varian, Inc, Palo Alto, CA, USA) (NH_4^+), and a Shimadzu TOC-L-CSN IR spectrometer (Shimadzu Corporation, Kyoto, Japan), and filtered through a 0.45-μm membrane filter (DOC). Coliform bacteria were analysed with use of Quantitray Model 2X (IDEXX Laboratories, Westbrook, ME, USA). Pharmaceutical residues were measured in the laboratory of the Institute for Water Chemistry, TU Dresden (Germany), with an HPLC system (Agilent 1100, Agilent Technologies, Waldbronn, Germany) coupled with MS/MS detection (Sciex Q3200, AB Sciex Pte. Ltd, Woodlands, Singapore) after enrichment via solid-phase extraction. Pesticide measurements were performed at the laboratory of Plant Protection Institute, National Research Institute in Poznań (Department of Pesticide Residue Research) with use of liquid chromatograph (ACQUITY® UPLC, Waters, Milford, MA, USA).

Table 1. Statistical characteristics of the data set.

Parameters	Colour (mg Pt/L)	EC (μS/cm)	NO_3 (mg/L)	NO_2 (mg/L)	NH_4 (mg/L)	COD (mg O_2/L)	DOC (mg/L)	Coliform Bacteria MPN/100 mL
Warta river (*n* = 37)								
Average	25	624	18.6	0.09	0.10	28.2	8.4	6154
Median	25	619	14.0	0.09	0.06	28.0	8.0	5475
Minimum	20	542	0.5	0.03	0.02	17.0	5.0	308
Maximum	40	703	48.0	0.18	0.58	44.0	13.0	24,200
Standard deviation	4	49	13.0	0.04	0.13	6.6	1.8	5432
Horizontal well (HW) (*n* = 32)								
Average	17	626	18.4	0.02	0.02	16.5	6.2	1
Median	17	614	16.0	0.01	0.02	17.5	5.9	1
Minimum	10	539	3.6	0.00	0.00	4.0	3.8	0
Maximum	30	695	44.0	0.11	0.06	29.0	9.0	2
Standard deviation	5	56	12.0	0.03	0.02	6.2	1.5	1
Reduction/Increase (average)	32.5%	−0.4%	0.9%	74.8%	80.5%	41.5%	26.01%	99.98%
RBF barrier (Point H) (*n* = 21)								
Average	13	650	7.8	0.09	0.19	13.8	5.0	0
Median	15	662	6.4	0.11	0.18	13.0	5.0	0
Minimum	7.5	581	0.0	0.02	0.12	3.0	3.9	0
Maximum	15	695	18.0	0.16	0.25	29.0	6.6	0
Standard deviation	2	33	6.8	0.04	0.04	5.2	0.7	0
Reduction/Increase (average)	49.8%	−4.3%	58.1%	−6.9%	−91.5%	51.1%	40.3%	100%
Well 19L (*n* = 10)								
Average	15	614	0.58	0.03	0.19	9.8	5.0	0
Median	15	622	0.23	0.02	0.21	9.0	5.1	0
Minimum	10	580	0.00	0.01	0.10	3.0	4.3	0
Maximum	20	652	1.91	0.09	0.27	20.0	5.8	0
Standard deviation	4	29	0.68	0.03	0.07	6.83	0.6	0
Reduction/Increase (average)	42.1%	1.5%	96.9%	64.2%	−99.9%	65.3%	40.4%	100%

Table 1. *Cont.*

Parameters	Colour (mg Pt/L)	EC (μS/cm)	NO$_3$ (mg/L)	NO$_2$ (mg/L)	NH$_4$ (mg/L)	COD (mg O$_2$/L)	DOC (mg/L)	Coliform Bacteria MPN/100 mL
				Well 1AL (*n* = 12)				
Average	13.75	598	1.55	0.04	0.61	8.6	4.9	0
Median	10	612	1.23	0.03	0.42	8.5	4.9	0
Minimum	10	563	0.91	0.01	0.14	3.7	4.1	0
Maximum	25	618	2.83	0.12	1.18	15.0	5.4	0
Standard deviation	5	24	0.80	0.04	0.45	4.7	0.6	0
Reduction/Increase (average)	46.9%	4.1%	91.7%	54.0%	−529%	69.6%	42.4%	100%

TDS—total dissolved solids; EC—electrical conductivity; COD—chemical oxygen demand; DOC—dissolved organic carbon; *n*—number of analyses; (−)—increase.

3. Results

The statistical characteristics of the water samples are presented in Table 1. Figure 4 presents fluctuations in some parameter concentrations of RBF water relative to the source water in the Warta River. The most apparent difference is seen in the case of coliform bacteria. Despite the high concentration of bacteria in river water, almost no bacteria were found in bank filtrate. This is a common effect observed at RBF sites and a result of filtration and adsorption and inactivation or die-off with time. A high removal efficiency was also observed for parameters reflecting the occurrence of NOM in water. The chemical oxygen demand (COD) reflected good removal of NOM from source water. In the Warta River, the maximum concentration occasionally reached levels higher than 50 mg O$_2$/L (median 24.5 mg O$_2$/L) whereas in the bank filtrate the level of COD was much lower (maximum 27.0 mg O$_2$/L, median 13.0 mg O$_2$/L). The median DOC concentration was 8.2 mg/L and was quite high compared to other rivers. The DOC concentration showed large fluctuation in source water from 5.0 to 10 mg/L, while the concentration of DOC in bank filtrate was relatively stable and much lower (maximum concentration of 6.0 mg/L, median 5.0 mg/L). The relatively stable level of DOC achieved by RBF is important for post-treatment. In contrast to COD, the DOC concentration did not follow seasonal fluctuations in source water. The reduction of NOM caused a significant decrease in water colour. A 30–40 mg Pt/L decrease in colour to less than 15 mg Pt/L was observed in RBF wells.

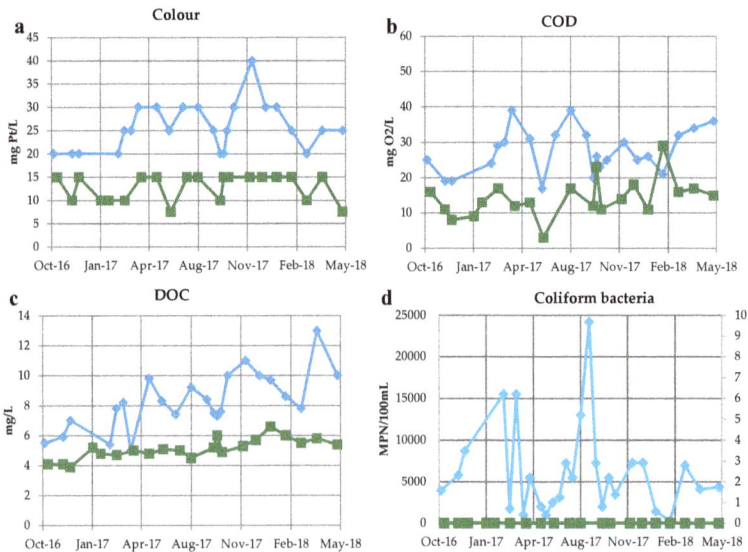

Figure 4. *Cont.*

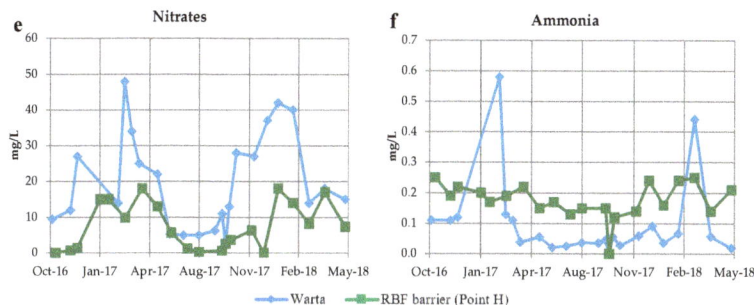

Figure 4. Temporal changes of selected parameters in bank filtrate and Warta River water. (**a**) Colour, (**b**) COD, (**c**) DOC, (**d**) coliform bacteria, (**e**) nitrates, (**f**) ammonia.

A high level of nitrogen reduction was observed during bank filtration. There were very high fluctuations of nitrate, nitrite, and ammonia in river water (Figure 4e,f). The seasonal variations in nitrogen concentrations are related to the growth periods of flora and fauna in the river, which result from seasonal temperature changes and are a major factor in regulating the biological processes that determine N-cycling [21]. During seasonal fluctuations, the changes related to extreme weather conditions overlap (mainly long-term drought and the influence of the wet season after droughts). It was observed [11] that high concentrations of nitrate up to 80 mg/L occurred after long-term drought as a result of flushing the accumulated contaminants in the environment. Bank filtrate displayed significantly lower nitrogen concentrations. The variability is related to nitrate, which was reduced from the maximum level of 50 mg/L (median 17.5 mg/L) in source water to a maximum level of 18.0 mg/L (median 6.4 mg/L) in bank filtrate during winter. In summer months, denitrification causes a strong decrease in nitrate concentration in bank filtrate. The concentration peaks of ammonia observed in river water (maximum concentration 0.58 mg/L) were buffered by RBF (maximum concentration 0.25 mg/L). However, the average ammonia level was higher in bank filtrate than in river water (median in bank filtrate 0.18 mg/L compared to 0.09 mg/L in source water), indicating a portion of ammonia coming from mixing with ambient groundwater.

Figure 5 presents the fluctuations of some parameter concentrations from the HW in relation to the source water in the Warta River. In the case of coliform bacteria, water treatment is usually effective, but during some periods, coliform bacteria were present in HW water. A distinct decrease in COD was observed in HW water. In the Warta River, periodic peaks were observed, mainly in summer because of biological activity in the river (maximum 60 mg O_2/L). The COD in the HW showed low fluctuation, usually significantly less than 20 mg O_2/L, with an increase to 30 mg O_2/L in spring 2018. The DOC behaviour in the HW followed the concentration peaks observed in the river, but the concentration level was significantly lower (maxima significantly lower than 6 mg/L with an increase in spring 2018 to a value of 9 mg/L). The decrease in water colour was evident in the HW, but in some periods, the high colour peak followed the colour of water in the river.

Low ammonia concentrations were found in the HW. In general, the high concentration peaks followed the behaviour of ammonia in the source river water, but the concentration of ammonia in the HW was significantly lower (maximum of 0.5–0.6 mg/L in the river compared to less than 0.2 mg/L in the HW). There was no removal of nitrate between the river and the HW. The behaviour of nitrate in HW strictly follows fluctuations observed in surface water. The minima and maxima observed in river water and the HW were almost identical with respect to time and range of concentration.

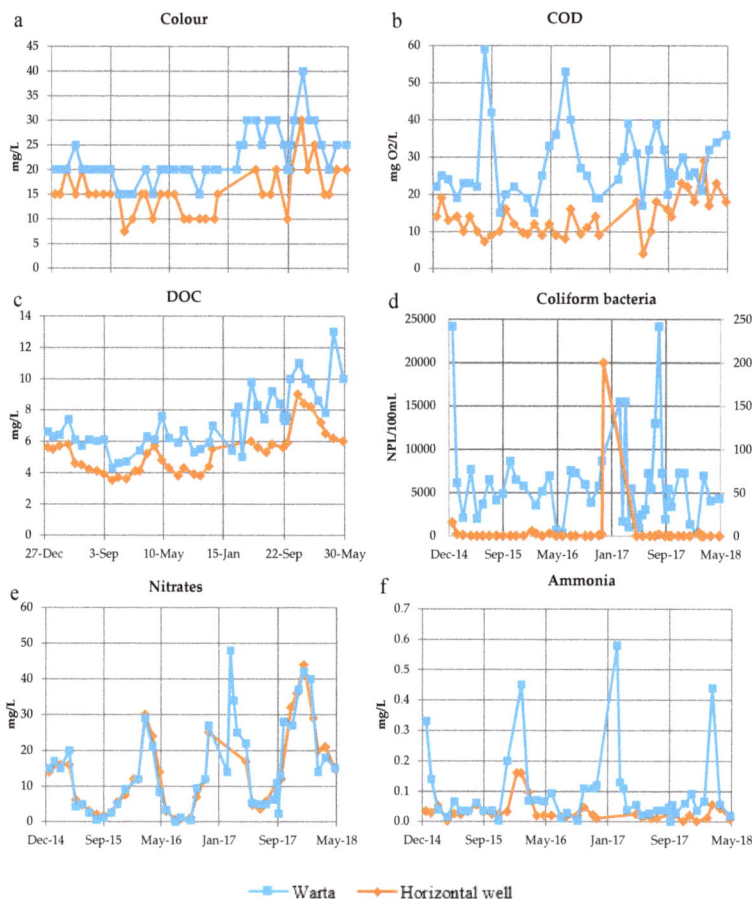

Figure 5. Temporal changes of selected parameters in the horizontal well (HW) and the Warta River. (**a**) Colour, (**b**) COD, (**c**) DOC, (**d**) coliform bacteria, (**e**) nitrates, (**f**) ammonia.

Preliminary results show the presence of some pharmaceutical compounds and other micropollutants in both source water and bank filtrate (Table 2). In total, 30 micropollutants were analysed. The following pharmaceutical residues were detected in the Warta river, but not in bank filtrate: diclofenac (15 ng/L), iohexol (20 ng/L), iomeprol (20 ng/L), iopamidol (20 ng/L), metoprolol (10 ng/L), and theophylline (40 ng/L). The pharmaceutical residues that were not found in river water nor in bank filtrate were as follows: 4-DMA-antipyrin, 4-IP-antipyrine, atenolol, bezafibrate, diazepam, loratidin, naproxen, paracetamol, phenazone, primidone, sulfadiazine, theophylline, aspartam, chloramphenicol, gemfibrozil, and phenobarbital.

Table 2. Concentration of relevant pharmaceuticals and other micropollutants in ng/L (Limit of Quantification (LOQ) < 5).

Sampling Point	1H-Benzotriazole	Carbamazepine	Caffeine	Sulfamethoxazole	Tolytriazole	Chlorothiazide	Ibuprofen	Sucralose	Sum
Warta River	120	40	60	15	30	<LOQ	20	40	450
Horizontal well	180	50	5	<LOQ	40	<LOQ	<LOQ	45	320
168b/2	180	60	15	<LOQ	30	<LOQ	<LOQ	55	340
177b/1	110	70	5	15	25	5	40	55	325
19L	50	60	<LOQ	<LOQ	15	10	<LOQ	40	175
1AL	80	60	5	<LOQ	15	5	<LOQ	50	215
78b/1s	<LOQ	30	<LOQ	<LOQ	<LOQ	<LOQ	<LOQ	10	40
50A	<LOQ	<LOQ	<LOQ	<LOQ	<LOQ	<LOQ	<LOQ	<LOQ	<LOQ

Figure 6 shows the total concentration of all 30 micropollutants analysed. In the Warta River, higher concentrations of pharmaceuticals and other micropollutants were detected. To get a rough estimate of the removal efficacy along flow paths, all results have been summed up, knowing that a single-compound assessment is more reliable but here not feasible due to an insufficient number of samples. Furthermore, it is of note that the concentration in the river water could have been lower or higher when the river water infiltrated which was abstracted as bank filtrate at the HW and other sampling points. The resulting total concentration from all 30 micropollutants in river water was 450 ng/L. Lower concentrations were documented in the HW and observation wells located close to the river and wells 168b/1 and 177b/1 (320, 340, and 325 ng/L, respectively). Much lower concentrations were documented in RBF-c wells 19L and 1AL (175 and 215 ng/L, respectively). Furthermore, from the river to well 78b/1s, the concentration decreased to 40 ng/L, while in the RBF-f well (50A), micropollutants were not detected.

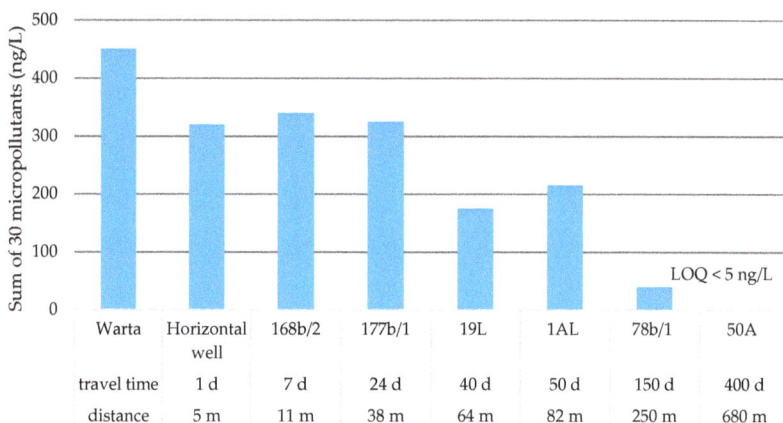

Figure 6. Changes in total micropollutant concentrations along the flow path. Limit of Quantification (LOQ).

The most common micropollutants found were the corrosion inhibitor benzotriazole and the pharmaceutical carbamazepine (Figure 7). Their concentrations in bank filtrate were higher than in river water. This finding may reflect the travel time influence on micropollutant behaviour and a higher concentration of these micropollutants in river water prior to the sampling period.

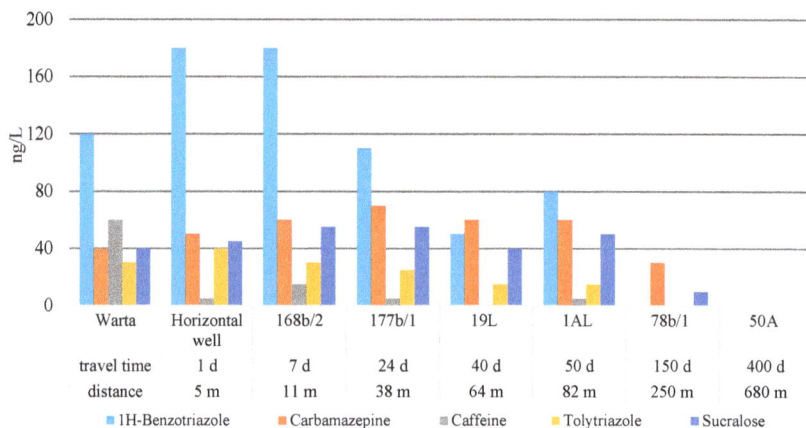

Figure 7. Changes in the detected micropollutant concentrations along the flow path.

The spatial distribution of pesticides was very similar to that of the other micropollutants in the region. The highest concentrations were found in Warta River water and in the HW (Figure 8).

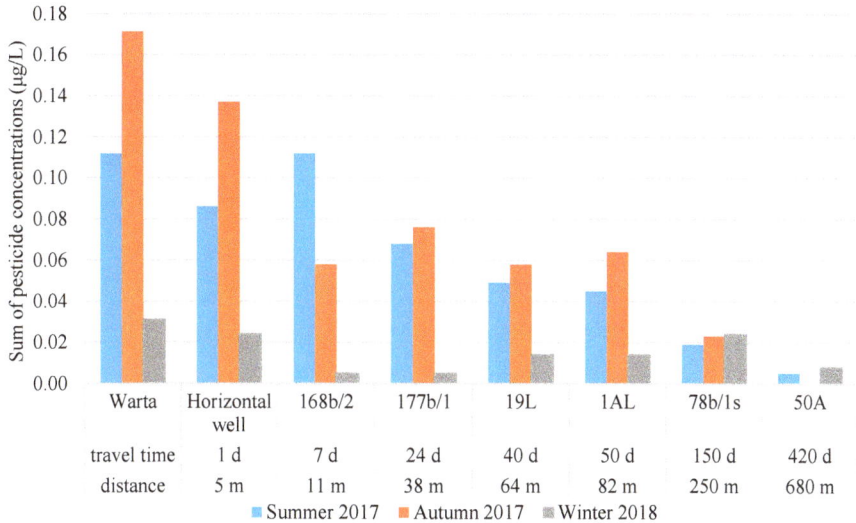

Figure 8. The sum of pesticide concentrations along the flow path.

Based on literature search on relevant pesticides found elsewhere at RBF sites and on indications for application in the study area, selected pesticides were analysed (Table 3). The concentrations of pesticides decreased with increasing distance from the river, and in the RBF-f well no pesticides were detected. During the sampling campaigns performed in 2017, pesticides were detected at a total concentration of 0.112 µg/L in summer, and a total concentration of 0.171 µg/L was detected in autumn (Table 3). In the winter sampling campaign, lower concentrations of pesticides were observed (0.031 µg/L). In the HW similar pesticide concentrations as in river water were found (0.086, 0.137, and 0.024 µg/L, respectively). Much lower concentrations were detected in the vertical wells. In the sampling campaign of summer 2017, three pesticides were detected, with total concentrations of 0.045 and 0.049 µg/L in wells 1AL and 19L, respectively. In the autumn campaign, the total pesticide levels were 0.064 and 0.058 µg/L in wells 1AL and 19L, respectively. Seven pesticide constituents were detected in well 1AL, and five were observed in well 19L. The concentrations of pesticides in piezometers located between the river and RBF-c wells were intermediate concentrations that reflected the successive concentration reduction with distance during RBF. Pesticides were not detected in the RBF-f well, and only isoproturon was detected in well 78b/ls at concentrations of 0.019, 0.023, and 0.024 µg/L during the three sampling campaigns.

Table 3. Concentration of pesticides in µg/L (LOQ < 0.005).

Sampling Period	Sampling Point	Imidacloprid	Isoproturon	M-Metalaxyl	Metazachlor	Nicosulfuron	Terbuthylazine	Chlortoluron	S-Metolachlor	Prometryn	Terbutryn	Sum
August 2017	Warta	0.007	0.008	0.005	0.036	0.012	0.008	<LOQ	<LOQ	<LOQ	<LOQ	0.112
	Horizontal well	0.005	<LOQ	0.007	0.015	0.021	0.008	<LOQ	<LOQ	<LOQ	<LOQ	0.086
	168b/2	0.009	0.006	0.015	<LOQ	0.03	0.012	0.012	0.006	0.006	<LOQ	0.112
	177b/1	0.011	<LOQ	0.006	<LOQ	0.012	0.008	0.017	0.006	0.008	<LOQ	0.068
	19L	0.007	<LOQ	<LOQ	<LOQ	0.017	<LOQ	0.018	<LOQ	0.007	<LOQ	0.049
	1AL	0.005	<LOQ	<LOQ	<LOQ	0.016	<LOQ	0.016	<LOQ	0.008	<LOQ	0.045
	78b/1s	<LOQ	0.012	<LOQ	<LOQ	<LOQ	<LOQ	0.007	<LOQ	<LOQ	<LOQ	0.019
	50A	<LOQ	<LOQ	<LOQ	<LOQ	<LOQ	<LOQ	<LOQ	<LOQ	<LOQ	<LOQ	
November 2017	Warta	<LOQ	0.009	0.005	0.03	0.023	0.006	0.047	0.009	<LOQ	0.007	0.171
	Horizontal well	0.007	0.007	0.005	0.021	0.019	0.007	0.032	0.008	<LOQ	0.007	0.137
	168b/2	0.008	<LOQ	<LOQ	0.007	0.018	0.012	0.005	0.005	<LOQ	0.006	0.058
	177b/1	0.007	<LOQ	<LOQ	0.007	0.017	0.007	0.008	0.006	0.006	0.008	0.076
	19L	0.007	<LOQ	<LOQ	<LOQ	0.024	<LOQ	0.01	0.005	0.006	0.006	0.058
	1AL	0.007	0.006	<LOQ	<LOQ	0.02	<LOQ	0.012	0.006	0.007	0.006	0.064
	78b/1s	<LOQ	0.014	<LOQ	<LOQ	<LOQ	<LOQ	0.009	<LOQ	<LOQ	<LOQ	0.023
	50A	<LOQ	<LOQ	<LOQ	<LOQ	<LOQ	<LOQ	<LOQ	<LOQ	<LOQ	<LOQ	
February 2018	Warta	0.005	0.008	<LOQ	<LOQ	<LOQ	<LOQ	0.012	<LOQ	0.006	<LOQ	0.031
	Horizontal well	<LOQ	0.006	<LOQ	<LOQ	<LOQ	<LOQ	0.013	<LOQ	0.005	<LOQ	0.024
	168b/2	<LOQ	<LOQ	<LOQ	<LOQ	<LOQ	<LOQ	0.005	<LOQ	<LOQ	<LOQ	0.005
	177b/1	<LOQ	<LOQ	<LOQ	<LOQ	<LOQ	<LOQ	<LOQ	<LOQ	<LOQ	<LOQ	
	19L	<LOQ	0.005	<LOQ	<LOQ	<LOQ	<LOQ	0.009	<LOQ	<LOQ	<LOQ	0.014
	1AL	<LOQ	0.005	<LOQ	<LOQ	<LOQ	<LOQ	0.009	<LOQ	<LOQ	<LOQ	0.014
	78b/1s	<LOQ	0.015	<LOQ	<LOQ	<LOQ	<LOQ	0.009	<LOQ	<LOQ	<LOQ	0.024
	50A	<LOQ	0.008	<LOQ	<LOQ	<LOQ	<LOQ	<LOQ	<LOQ	<LOQ	<LOQ	0.008

4. Discussion

The results presented here show a high efficacy of RBF for the removal of organic compounds, micropollutants, and coliform bacteria when (vertical) wells are located at least 60–80 m from the river bank. For all vertical wells a complete removal of coliform bacteria was observed. The reduction of DOC was about 40–42% (Table 1), and the reduction of COD more than 50% at point H and almost 70% at wells 1AL and 19L. These results are in accordance with others previously documented in the literature for other RBF sites in Europe [4]. The reduction of nitrates occurs at a high level during summer months with higher water temperature. Nitrates in RBF wells are reduced by 58%, in some wells up to 97% (1AL and 19L). The decrease in nitrate concentration is caused by denitrification and mixing with ambient groundwater. Denitrification was previously documented at RBF sites [22]. The mixing rate at the Krajkowo site shows a value of 65–86% of bank filtrate relative to the total water balance, but it should be emphasized that the amount of river water in the total water balance is a changeable factor and depends strongly on well exploitation. Lower well yield can lead to a decrease in river water portion in the total water balance in wells. This factor affects the nitrate concentration too. The seasonal peaks of high ammonia concentrations are strongly buffered and decreased (mainly by sorption in riverbed sediments and aquifer sediments), but the average concentration of ammonia is higher in bank filtrate.

In the HW, removal of NOM is also visible, but the removal rate is much lower than in RBF wells located at further distance from the river. The decrease in coliform bacteria is evident, but bacteria appear in the HW periodically. The reduction of COD was found to be 42% and that of DOC 26%, but seasonal changes of these parameters follow the fluctuations observed in river water (Figure 5). This makes the HW very vulnerable to extreme weather conditions, especially floods. During and after floods, the NOM content increases in river water and causes breakthroughs to the HW [6]. The distinct removal of ammonia is visible (especially in the peaks of high concentrations, which are damped), while the nitrates in the HW follow fluctuations observed in river water. This also proves limited attenuation of pollutants if the flow paths between the river bed and the screen of a well are short.

A high removal rate of organic micropollutants was determined at the Krajkowo site. Concentrations of pharmaceuticals in Warta River water were found similar to levels detected in other European rivers [19,23]. Among the 30 analysed micropollutants, 14 were detected in the Warta River. Non-steroidal anti-inflammatory drugs (diclofenac and ibuprofen) previously measured in the Warta River showed lower concentrations in current research than in 2007 [24], and the naproxen detected in 2007 was not detected in this study. This can be related to the high discharge rate of the Warta River during the wet period in September 2017. Out of the 14 substances detected in Warta River, 8 substances were detected in the bank filtrate (Table 2). The high attenuation potential is visible during water passage through the aquifer and depending on flow path length. The pharmaceutical concentrations in the HW and observation wells located close to the river are at levels observed in the source water, while after further aquifer passage, the concentrations decrease considerably. In wells located 60–80 m from the river (travel time 40–50 days), the concentrations are significantly lower (Figures 6 and 7), while at a distance of 250 m from the river (point 78b/1s), only three substances were detected. Further away from the river, no pharmaceutical residues were detected. Along the flow path, a low increase in carbamazepine and sucralose was visible (Figure 7), indicating that these compounds were present in the Warta River at higher concentrations before the sampling period. This finding shows the importance of regular sampling of source water and RBF water to assess the removal efficacy.

Pesticide levels were also reduced significantly during RBF (Figure 8). The similar constituents and concentrations detected in the Warta River and the HW indicate that the well is vulnerable to pollution from the river. Water passage through 5-m-thick sediments is not sufficient to remove micropollutants from the drained water. In vertical wells located 60–80 m from the river (RBF-c wells) pesticide concentrations were much lower than those in the river and HW, but some pesticides were still present (Figure 8).

A factor that influences concentrations in RBF wells is the mixing of bank filtrate with ambient groundwater. In RBF-c wells, 65–85% of water is derived from bank filtration, and in RBF-f wells, this percentage is ~40%. This mixing leads to the dilution of pollutants in bank filtrate, but it should be emphasized that the portion of river water in total water balance is changeable throughout the year. It is also changeable according to the wells' exploitation rate, which causes the mixing rate to change.

The European Union (EC 1998; EC 2006) and Polish regulations (Rozporządzenie 2017) have established a maximum acceptable concentration of 0.1 μg/L for individual pesticides and their degradation products and of 0.5 μg/L for the total pesticide concentration [25–27]. Pharmaceutical residue concentrations are currently not regulated in the European Union or Polish guidelines but are proposed to be lower than 0.1 μg/L. In the study area the maximum admissible limit of pesticides was not exceeded. Only benzotriazole was found to be present at levels higher than 0.1 μg/L in some wells (bank filtrate) and needs to be removed during post-treatment.

5. Conclusions

Investigations at the Krajkowo site show effective removal of NOM in the vertical RBF wells located at distances of 60–80 m from the river. The removal of DOC, COD, and colour was found in the ranges of 40–42%, 51–70%, and 42–50%, respectively. A much lower reduction of DOC (26%), COD (42%), and colour (33%) in horizontal well was observed. Furthermore, the horizontal well is more sensitive than vertical wells to changes in the NOM content, which is expressed by similar seasonal fluctuations in NOM content compared to river water.

Results of micropollutant investigations (mainly of pharmaceutical residues) demonstrate a gradual lowering of concentrations along the flow path. In the RBF wells the reduction rate of the sum of micropollutant concentrations is greater than 50%. Lower reduction rates (approximately 30%) were found for the HW and observation wells located 11 m and 38 m from the river. At a distance of 250 m from the river (travel time ~150 days) only carbamazepine and sucralose were detected. At a distance of 680 m (travel time ~420 days) pharmaceutical residues were not detected. The most persistent pharmaceutical is carbamazepine. The decrease of its concentration was observed at a distance of 250 m.

Results of pesticides investigation show also gradual decrease of concentrations along the flow path. High reduction rates are visible in RBF wells (about 80% for the sum of pesticide concentrations). In the RBF-f well pesticides were practically not detected, but in the 78b/s well (250 m from the river) isoproturon was detected at low concentrations.

The presented results prove a high efficacy of contaminant removal by the riverbank filtration system. Significantly lower contaminant removal was documented in the horizontal well, which received river water after a very short travel time. For RBF sites with similar conditions, the distance from the river should be at least 60 m. However, higher removal rates can be achieved for wells located at a distance of 250 m from the river.

The preliminary results of the organic micropollutant investigation show the need for further monitoring of emerging compounds in both source (river) water and extracted bank filtrate. In the case of increased concentrations in river water, operation of vertical wells at a longer distance from the wells should be favoured against operation of the HW. Regular monitoring of relevant micropollutants is important for water management purposes as well as for adjusting post-treatment technologies.

Author Contributions: J.G. and K.D. were responsible for the overall coordination of the research team; T.G. and D.D. took part in conceptualisation of water monitoring; R.K. took part in field work and performed graphical and statistical interpretations; K.D., J.G., and R.K. interpreted the data and were involved in discussing the study; K.D. prepared the manuscript; and all authors read and approved the manuscript.

Funding: This research was completed with support from the AquaNES project. This project has received funding from the European Union's Horizon 2020 Research and Innovation Program under grant agreement no. 689450.

Acknowledgments: The authors would like to thank Aquanet SA (Poznań Waterworks operator) for their contribution, Rico Bartak and Sebastian Paufler for water sampling for micropollutants, and Hilmar Börnick and Oliver Faber for analysis of pharmaceuticals.

Conflicts of Interest: The authors declare no conflicts of interest.

References

1. Hiscock, K.M.; Grischek, T. Attenuation of groundwater pollution by bank filtration. *J. Hydrol.* **2002**, *266*, 139–144. [CrossRef]
2. Foriczs, T.; Berecz, Z.; Molnar, Z.; Suveges, M. Origin of shallow groundwater of Csepel Island (south of Budapest. Hungary. River Danube): Isotopic and chemical approach. *Hydrol. Process.* **2005**, *19*, 3299–3312. [CrossRef]
3. Lasagna, M.; De Luca, D.A.; Franchino, E. Nitrates contamination of groundwater in the western Po Plain (Italy): The effects of groundwater and surface water interactions. *Environ. Earth Sci.* **2016**, *75*, 240. [CrossRef]
4. Grunheid, S.; Amy, G.; Jekel, M. Removal of bulk dissolved organic carbon (DOC) and trace organic compounds by bank filtration and artificial recharge. *Water Res.* **2005**, *39*, 3219–3228. [CrossRef] [PubMed]
5. Hoppe-Jones, C.; Oldham, G.; Drewes, J.E. Attenuation of total organic carbon and unregulated trace organic chemicals in U.S. riverbank filtration systems. *Water Res.* **2010**, *44*, 4643–4659. [CrossRef] [PubMed]
6. Górski, J.; Dragon, K.; Kruć, R. A comparison of river water treatment efficiency in different types of wells. *Geologos* **2018**, in press. [CrossRef]
7. Miettinen, I.T.; Martikainen, P.J.; Vartiainen, T. Humus transformation at the bank filtration water plant. *Water Sci. Techol.* **1994**, *30*, 179–187. [CrossRef]
8. Ray, C.; Soong, T.W.; Lian, Y.Q.; Roadcap, G.S. Effect of flood-induced chemical load on filtrate quality at bank filtration sites. *J. Hydrol.* **2002**, *266*, 235–258. [CrossRef]
9. Sandhu, C.; Grischek, T.; Kumar, P.; Ray, C. Potential for riverbank filtration in India. *Clean Technol. Environ. Policy* **2011**, *13*, 295–316. [CrossRef]
10. Hu, L.; Xu, Z.; Huang, W. Development of a river-groundwater interaction model and its application to a catchment in Northwestern China. *J. Hydrol.* **2016**, *543*, 483–500. [CrossRef]
11. Górski, J.; Dragon, K.; Kaczmarek, P. Nitrate pollution in the Warta River (Poland) between 1958 and 2016: Trend and causes. *Environ. Sci. Pollut. Res.* **2017**. [CrossRef] [PubMed]
12. Sui, Q.; Cao, X.; Lu, S.; Zhao, W.; Qiu, Z.; Yu, G. Occurrence, sources and fate of pharmaceuticals and personal care products in the groundwater: A review. *Emerg. Contam.* **2015**, *1*, 14–24. [CrossRef]
13. Bohlke, J.K. Groundwater recharge and agricultural contamination. *Hydrogeol. J.* **2002**, *10*, 153–179. [CrossRef]
14. Dragon, K. Groundwater nitrates pollution in the recharge zone of a regional Quaternary flow system (Wielkopolska region, Poland). *Environ. Earth Sci.* **2013**, *68*, 2099–2109. [CrossRef]
15. Guzzella, L.; Pozzoni, F.; Giuliano, G. Herbicide contamination of surficial groundwater in Northern Italy. *Environ. Pollut.* **2006**, *142*, 344–353. [CrossRef] [PubMed]
16. Loos, R.; Locoro, G.; Comero, S.; Contini, S.; Schwesig, D.; Werres, F.; Balsaa, P.; Gans, O.; Weiss, S.; Blaha, L.; et al. Pan-European survey on the occurrence of selected polar organic persistent pollutants in ground water. *Water Research* **2010**, *44*, 4115–4126. [CrossRef] [PubMed]
17. Köck-Schulmeyer, M.; Ginebreda, A.; Postigo, C.; Garrido, T.; Fraile, J.; López de Alda, M.; Barceló, D. Four-year advanced monitoring program of polar pesticides in groundwater of Catalonia (NE-Spain). *Sci. Total Environ.* **2014**, *470–471*, 1087–1098. [CrossRef] [PubMed]
18. Li, W.C. Occurrence, sources, and fate of pharmaceuticals in aquatic environment and soil. *Environ. Pollut.* **2014**, *187*, 193–201. [CrossRef] [PubMed]
19. Kovačević, S.; Radišić, M.; Laušević, M.; Dimkić, M. Occurrence and behavior of selected pharmaceuticals during riverbank filtration in The Republic of Serbia. *Environ. Sci. Pollut. Res.* **2017**, *24*, 2075–2088. [CrossRef] [PubMed]
20. Przybyłek, J.; Dragon, K.; Kaczmarek, P. Hydrogeological investigations of river bed clogging at a river bank filtration site along the River Warta, Poland. *Geologos* **2017**, *23*, 201–214. [CrossRef]
21. Howden, N.J.K.; Burt, T.P. Temporal and spatial analysis of nitrates concentrations from the Frome and Piddle catchments in Dorset (UK) for water years 1978 to 2007: Evidence for nitrates break through? *Sci. Total Environ.* **2008**, *407*, 507–526. [CrossRef] [PubMed]
22. Grischek, T.; Hiscock, K.M.; Metschies, T.; Dennis, P.F.; Nestler, W. Factors affecting denitrification during infiltration of river water into a sand and gravel aquifer in Saxony, Germany. *Water Research* **1998**, *32*, 450–460. [CrossRef]

23. Szymonik, A.; Lach, J.; Malińska, K. Fate and removal of pharmaceuticals and illegal drugs present in drinking water and wastewater. *Ecol. Chem. Eng. S* **2017**, *24*, 65–85. [CrossRef]
24. Kasprzyk-Hordern, B.; Dąbrowska, A.; Vieno, N.; Kronberg, L.; Nawrocki, J. Occurrence of Acidic Pharmaceuticals in the Warta River in Poland. *Chem. Anal.* **2008**, *52*, 289–303.
25. European Commission (EC). *Council Directive 98/83/EC of 3 November 1998 on the Quality of Water Intended for Human Consumption (L327/1. 22/12/2000)*; Official Journal of the European Union: Aberdeen, UK, 1998.
26. European Commission (EC). *Directive 2006/118/EC of the European Parliament and the Council of 12th December 2006 on the Protection of Ground Water Against Pollution and Degradation (L372/19. 27/12/2006)*; Official Journal of the European Union: Aberdeen, UK, 2006.
27. Internet System of Legal Acts (ISAP). *Regulation of the Ministry of Health on the Quality of Water Intended for Human Consumption (Dz.U. 2017 nr 2294)*; ISAP: Warsaw, Poland, 2017.

water

Article

The Impact of River Discharge and Water Temperature on Manganese Release from the Riverbed during Riverbank Filtration: A Case Study from Dresden, Germany

Sebastian Paufler [1,*], Thomas Grischek [1], Marcos Roberto Benso [1], Nadine Seidel [1] and Thomas Fischer [2]

[1]   Dresden University of Applied Sciences, Friedrich-List-Platz 1, 01069 Dresden, Germany;
      thomas.grischek@htw-dresden.de (T.G.); marcosbenso@hotmail.com (M.R.B.); nadine_85@msn.com (N.S.)
[2]   DREWAG Netz GmbH Dresden, Kohlenstraße 23, 01189 Dresden, Germany;
      Thomas_Fischer@drewag-netz.de
*   Correspondence: sebastian.paufler@htw-dresden.de; Tel.: +4-935-1463-2631

Received: 19 July 2018; Accepted: 15 October 2018; Published: 19 October 2018

Abstract: The climate-related variables, river discharge, and water temperature, are the main factors controlling the quality of the bank filtrate by affecting infiltration rates, travel times, and redox conditions. The impact of temperature and discharge on manganese release from a riverbed were assessed by water quality data from a monitoring transect at a riverbank filtration site in Dresden-Tolkewitz. Column experiments with riverbed material were used to assess the Mn release for four temperature and three discharge conditions, represented by varying infiltration rates. The observed Mn release was modeled as kinetic reactions via Monod-type rate formulations in PHREEQC. The temperature had a bigger impact than the infiltration rates on the Mn release. Infiltration rates of <0.3 $m^3/(m^2{\cdot}d)$ required temperatures >20 °C to trigger the Mn release. With increasing temperatures, the infiltration rates became less important. The modeled consumption rates of dissolved oxygen are in agreement with results from other bank filtration sites and are potentially suited for the further application of the given conditions. The determined Mn reduction rate constants were appropriate to simulate Mn release from the riverbed sediments but seemed not to be suited for simulations in which Mn reduction is likely to occur within the aquifer. Sequential extractions revealed a decrease of easily reducible Mn up to 25%, which was found to reflect the natural stratification within the riverbed, rather than a depletion of the Mn reservoir.

Keywords: riverbank filtration; organic matter degradation; manganese; riverbed; climate change; floods; droughts; column experiments; PHREEQC

---

1. Introduction

Riverbank filtration (RBF) has been successfully used as a natural and cost-efficient water treatment method in many countries in Europe [1,2], the USA [3,4], Africa [5,6], and Asia [7,8]. RBF can naturally occur or can be induced by pumping, whereby wells are placed adjacent to the river that creates a hydraulic potential gradient from the river towards the wells. RBF triggers a variety of natural attenuation processes that can largely improve the water quality of the bank filtrate (BF) and lower the post-treatment effort [9,10]. For example, some organic micropollutants are effectively removed within the first meter of infiltration under oxic conditions [11]. A series of redox processes along the flow path of the infiltrate can adversely affect the BF quality. Depletion of dissolved oxygen (DO), followed by denitrification and the reduction of manganese (Mn) minerals in the riverbed and the aquifer can cause elevated Mn concentrations, which require subsequent treatment [12–14].

However, the resulting Mn concentration depends on various factors, including residence time and water temperature [15]. Assuming a steady well operation, the residence time can change due to river water level fluctuations, which are usually connected to river discharge variations (floods/droughts). For example, decreasing water levels usually lower the hydraulic gradient and prolong the distance between the river and the RBF wells. The lowered hydraulic gradient also affects the infiltration rate. Additionally, the infiltration rate is impacted by a usually smaller infiltration area at lower water levels. Water temperature variations affect the viscosity of the water and, therefore, also affect the residence times and infiltration rates. Additionally, according to the rule of Van 't Hoff [16], increasing temperatures lead to increased biological degradation rates, whereby Mn release is considered to be largely biologically mediated.

Current climate forecasts are thought-provoking at many RBF sites [17–19]. Droughts potentially lower the river discharge, extend travel times, and promote anaerobic conditions along the flow path, while floods can shorten travel times or cause, for example, breakthroughs of pathogens and organic micropollutants. The current increase of climate extremes in Europe is expected to continue, with a higher frequency of heat waves, long-lasting droughts in some regions, heavy precipitation events and river floods [20]. For the German state of Saxony, the air temperature is projected to increase by 3–3.5 °C and the mean summer rainfall is to decrease by 20–25% by the year 2100 [21].

The Waterworks (WW) Dresden-Tolkewitz (Saxony, Germany) was built in 1898 and is one of the oldest RBF schemes in Europe. This study investigated the impact of the climate-related variables, temperature and discharge, on the BF quality. Water quality data of a 10-year time span from a monitoring transect were examined to identify discharge- and temperature-related patterns. To assess the potential Mn release from the riverbed in Dresden-Tolkewitz, three columns filled with riverbed sediment from the Elbe river emulated three infiltration rates and four temperature conditions. In order to use the results from the column experiment for a planned modeling of the transect, the observed Mn release was reproduced by hydrogeochemical modeling with PHREEQC. Additionally, a sequential extraction procedure was applied to the riverbed sediment from the columns after the experiment. Based on that data, the implications of the results to the redox-related BF quality in a potentially changing climate are discussed with a focus on Mn.

2. Materials and Methods

2.1. Description of the RBF Waterworks Dresden-Tolkewitz

The WW Dresden-Tolkewitz is located at the upper Elbe river in Germany (Figure 1). Three siphon well galleries with 72 vertical wells abstract up to 1500 m^3/h (36,000 m^3/d). The portion of riverbank filtrate is around 83% during mean flow and 70–75% during low flow conditions [22]. The focus of this study was a 95 m wide transect between the Elbe river and a production well (PW), which already was the focus of previous riverbed clogging studies [22]. The PW fully penetrates the aquifer and the 4 m long filter screen is located directly above the aquitard. The transect has three observation wells (OW 1, 2 and 3). Each OW has sampling points at three depths (upper, middle, lower = OW i-1, i-2, i-3). During the mean flow, the nearest OW (OW 1) is around 21–30 m apart from the riverbank. OW 2 and OW 3 are around 40 and 80 m apart from the bank during mean flow. The average travel time along the transect is between 24 and 30 days [23].

The mean discharge of the Elbe river in Dresden is 332 m^3/s (at 1.84 m river stage). The discharge varies during the mean low and high flow periods between 110 m^3/s (0.75 m) and 1700 m^3/s at a water level of 5.47 m [22]. The climate in Dresden is humid continental with warm summers. The alluvial aquifer is unconfined, composed of gravel and coarse sand with a saturated thickness of 11–14 m and has a hydraulic conductivity of 1–2 × 10^{-3} m/s [22].

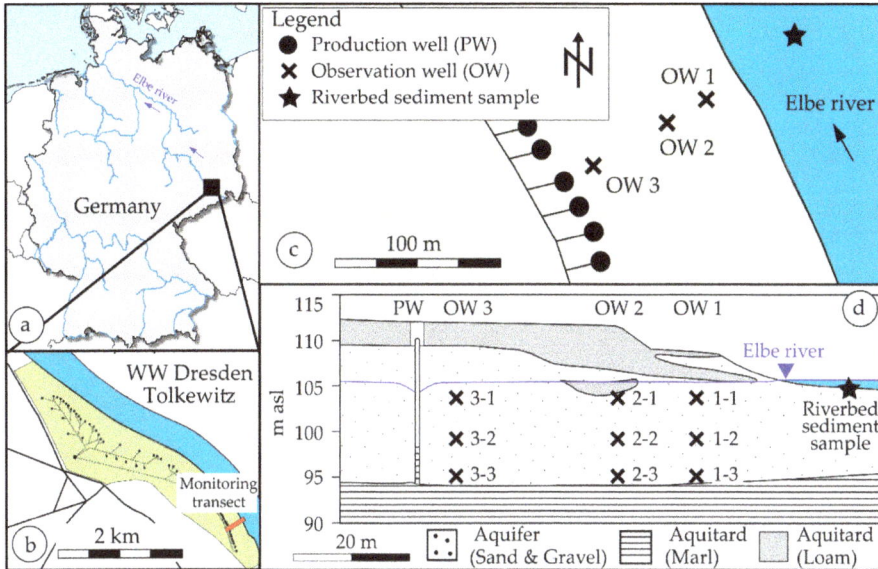

Figure 1. (**a**) The location of the RBF Waterworks Dresden-Tolkewitz, (**b**) Observed transect between the Elbe river and the pumping well (PW), (**c**) Location of the observation wells within the transect and sampling point of the riverbed sediment for the column experiment, (**d**) cross-section of the transect and locations of the observation points for each observation well (OW).

2.2. Regular Monitoring in Dresden-Tolkewitz

In this study, the evaluation period of the water quality data from the WW Dresden-Tolkewitz was a 10-year time span from 1 January 2006 to 31 December 2016. Regular samples were taken twice a year from all sampling points of each OW by the waterworks staff. Additional event-based samples were taken during low flow periods and, if possible, during high flow periods. Sampling was carried out according to DVGW [24] and corresponding to earlier guidelines. Water quality data for the Elbe river were taken from the database of the Saxon state ministry [25]. The relevant sampling point is located at river km 43.5, around 3 km upstream of the WW. Water levels in this study refer to the federal water level "Dresden Augustusbruecke" at river km 55.63 [26].

2.3. Set-Up of the Column Experiments

To understand the behavior of Mn at the RBF site in Dresden-Tolkewitz, three columns with riverbed sediments were set up in the laboratory of the University of Applied Sciences Dresden (Figure S1). Each column was 1 m long, had an inner diameter of 0.08 m, and was made up of galvanized steel. The filling material was riverbed sediment from the Elbe River. The riverbed material was collected in front of the investigated monitoring cross-section of the RBF Waterworks Dresden-Tolkewitz (Figure 1). The riverbed sediment was recovered around 20 m apart from the riverbank during a low discharge period. Due to the very low gradient of the riverbed towards the riverbank, the area around the sampling point was already flooded at slightly higher water levels, which still would occur during mean low flow conditions. Hence, the area around the sampling point can be considered as a potential infiltration area. Because of the relatively coarse riverbed, undisturbed sampling was not possible. Thus, the upper 5 cm of the riverbed where scratched first to represent the clogging layer. Subsequently, the deeper riverbed material was dug out layer-wise and sieved in place to a grain size <4 mm. Immediately after transporting it to the lab, the wet riverbed material was filled into columns in ≈0.1 m thick, separately compacted layers with the clogging layer on top.

During the filling of the columns, the sediment mass was measured using a balance to calculate the bulk density and assess the compaction of the material in the columns. The mean travel time (t_a) and effective porosity (n_e) for both columns were determined from electrical conductivity (NaCl) breakthrough curves from tracer experiments performed before start-up.

To adjust the temperature regiment, all three columns with riverbed material were stored in a thermostatic cabinet. A second thermostatic cabinet contained three storage containers with Elbe river water, collected in Dresden once per week. The outflow of the columns flowed into three additional containers within the second cabinet. The investigated temperatures were 10, 20, 30, and 35 °C (Table 1).

Table 1. The experimental design of the column experiment.

Column					1, 2, and 3							
Temperature in °C		10			20			30			35	
Flow in mL/min	1	2	4	1	2	4	1	2	4	1	2	4
n Samples/event	3	3	3	3	3	3	3	3	3	3	3	3

Temperature (T), dissolved oxygen (DO), the pH-value, and electrical conductivity (EC) were determined using WTW Multi 3430 and appropriate electrodes (WTW, Weilheim, Germany) before the columns in the storage containers and after the columns in a flow-through cell. A series of valves allowed sending the outflow of each column separately through the cell.

At most RBF sites, the infiltration rates depend for example on abstraction rates of the wells, clogging of the riverbed, distance between the river and the wells and the infiltrating area, and are therefore very site specific. To represent the low, mean, and high infiltration rates of 0.3, 0.6 and 1.1 $m^3/(m^2 \cdot d)$, the flow through the columns was adjusted to 1, 2, and 4 mL/min. The flow rate was adjusted individually for each column using ProMinent Beta diaphragm pumps (ProMinent, Heidelberg, Germany). Each of the 12 possible flow and temperature conditions run until 15 to 20 pore volumes (PV) of every column were exchanged. Sampling started after around 5 PV and continued until at least 10 and 15 PV. Up to three intermediate samples were taken if possible (e.g., weekends were skipped). At one sampling event, the water samples from all three columns were taken at separate sampling taps after the columns before the outflow container. Thus, the presented results for each temperature and flow rate represent the mean value of three similarly prepared, and independently operated columns. Samples from the storage containers (=inflow) were taken once per week. Alkalinity was determined at every sampling event by alkalimetric titrations with 0.1 M hydrochloric acid (HCl).

2.4. Water Analysis

Water samples of the regular monitoring at the OW's in Dresden-Tolkewwitz were analyzed for >100 parameters in the lab of DREWAG Netz GmbH (DIN EN ISO/IEC 17025 certified). Water samples from the column experiment were filtered immediately after sampling through 0.45 μm membrane filters (VWR International GmbH, Darmstadt, Germany). The samples for cation analysis were preserved with 0.1 M nitric acid (HNO_3). Major cations K^+, Na^+, Ca^{2+}, Mg^{2+} and dissolved metals As, Fe, Mn, Si, and Sr were measured with ICP-OES (Optima 4300 DV, PerkinElmer, Waltham, MA, USA). Br^-, Cl^-, F^-, NO_2^-, NO_3^-, PO_4^{3-}, and SO_4^{2-} were determined with ion-chromatography (autosampler AS50, eluent generator EG50, gradient pump GP50, electrochemical detector ED50, separation column AS19, all from Dionex) at the Institute for Water Chemistry, TU Dresden, Germany.

2.5. Sequential Extraction of the Riverbed Sediment

To estimate the mobilization behavior of Mn and to assess the mineralogical composition, a 4-step sequential extraction procedure was applied to the filling material of the columns after the experiment

(Table S1). Rauret et al. [27] and Sutherland and Tack [28] described the applied procedure in detail. The total extractable Mn was determined by microwave acid digestion with HNO_3 for separate samples from the same sampling points. Samples were taken after 0.05 m (below the clogging layer), and at 0.3, 0.6, and 0.9 m along the columns, before being immediately filled into airtight sample containers and stored at 4 °C before analysis.

2.6. Estimation of Reduction Constants for the Elbe Riverbed with PHREEQC

Elevated Mn concentrations at many bank filtration sites are linked to the microbiological reduction of Mn minerals within the riverbed and the aquifer [1,29]. The degradation (oxidation) of organic matter (OM, simplified CH_2O) is the driving force for the associated redox reactions (Equations (1)–(3)).

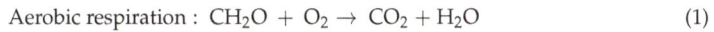

$$\text{Aerobic respiration}: CH_2O + O_2 \rightarrow CO_2 + H_2O \tag{1}$$

$$\text{Denitrification}: 5CH_2O + 4NO_3^- + 4H^+ \rightarrow 5CO_2 + 2N_2 + 7H_2O \tag{2}$$

$$\text{Mn(IV) reduction}: CH_2O + 2MnO_2(s) + 4H^+ \rightarrow 2Mn^{2+} + 3H_2O + CO_2 \tag{3}$$

The results from the column experiments are considered to represent the potential Mn release from the riverbed. In order to use the results for a planned modeling of the transect, the observed Mn release was reproduced by chemical modeling with PHREEQC [30]. By applying the approach of Henzler et al. [14], the relevant redox reactions were modeled as kinetic reactions using Monod-type rate formulations (Equations (4)–(7)). Because neither increasing Fe concentrations nor decreasing sulfate concentrations were observed along the transect, additional redox reactions accounting for iron and sulfate reduction were excluded.

$$r_{ox} = -f_{reac} \times Y_{ox}^{-1} \times k_{ox} \times \left(\frac{C_{ox}}{C_{ox} + K_{ox}} \right) \times f_T \tag{4}$$

$$r_{nit} = -f_{reac} \times Y_{nit}^{-1} \times k_{nit} \times \left(\frac{C_{nit}}{C_{nit} + K_{nit}} \right) \times \left(\frac{K_{inhb_{nit}}^{ox}}{C_{ox} + K_{inhb_{nit}}^{ox}} \right) \times f_T \tag{5}$$

$$r_{mn} = f_{reac} \times Y_{mn}^{-1} \times k_{mn} \times \left(\frac{K_{inhb_{mn}}^{ox}}{C_{ox} + K_{inhb_{mn}}^{ox}} \right) \times \left(\frac{K_{inhb_{mn}}^{nit}}{C_{ox} + K_{inhb_{mn}}^{nit}} \right) \times f_T \tag{6}$$

$$r_{OM} = Y_{ox} \times r_{ox} + Y_{nit} \times r_{nit} - Y_{mn} \times r_{mn} \tag{7}$$

The parameters r_{ox}, r_{nit} and r_{mn} denote the production and consumption rates (positive and negative) of dissolved O_2, NO_3^-, and Mn^{2+}. Rate constants for OM degradation under oxic, nitrate and manganese reducing condition are represented by k_{ox}, k_{nit} and k_{mn}. C_{ox} and C_{nit} are the dissolved oxygen and nitrate concentrations. K_{ox} and K_{nit} denote Monod-half saturation constants. The inhibition of nitrate and manganese reduction under oxic conditions was implemented by the inhibition constants $K_{inhb_{nit}}^{ox}$ and $K_{inhb_{mn}}^{nit}$. Accordingly, $K_{inhb_{mn}}^{nit}$ represents the inhibition constant for manganese reduction under nitrate-reducing conditions. The overall turnover rate of OM r_{OM} (Equation (7)) is the sum of the reaction rates r_{ox}, r_{nit}, and r_{mn} that are multiplied with the stoichiometric coefficients Y_{ox}, Y_{nit}, and Y_{mn} corresponding to the redox reactions (Equations (1)–(3)). Following a similar modeling approach of Greskowiak et al. [31], the parameter f_{reac} was included to simulate a zone of increased reactivity at the first section of the infiltration path [14]. The application of Equations (4)–(7) implied two assumptions. First, MnO_2 was present in excess. Hence, MnO_2 was not rate limiting and the implementation of a Monod-half saturation constant for Mn was not necessary in Equation 6. Second, the OM content was assumed to be infinitely available (or redelivered by the river) and would not be exhausted during the simulation period.

Similar to Diem et al. [32], Greskowiak et al. [31], and Sharma et al. [33], an additional temperature factor f_T was implemented that accounted for the impact of temperature changes on the degradation rates (Equation (8)).

$$f_T = exp\left[\alpha + \beta \times T \times \left(1 - 0.5 \times \frac{T}{T_{opt}}\right)\right]$$

(8)

T_{opt} denotes the optimal temperature for a maximal degradation rate and α as well as β are fitting parameters. Applying the results from Diem et al. [32], Henzler et al. [14], Greskowiak et al. [33] and Sharma et al. [33], none of the three parameters were to be changed from the initial data set.

In PHREEQC, a 1 m long column was represented by 50 cells with a length of each of them being 0.02 m (Table S2). Porosity and pore velocity were known from the tracer test. The dispersion and diffusion coefficients for the model were calibrated for the NaCl breakthrough curves from the tracer tests and non-reactive transport. Subsequently, the data from Henzler et al. [14] for the rate constants k_{ox}, k_{nit}, and k_{mn}, as well as for the inhibition constants $K^{ox}_{inhb_{nit}}$, $K^{ox}_{inhb_{mn}}$, and $K^{nit}_{inhb_{mn}}$, were used as the initial parameter set for reactive modeling. Calibration was initially carried out with PEST [34]. Since the inhibition constants $K^{ox}_{inhb_{nit}}$, $K^{ox}_{inhb_{mn}}$ and $K^{nit}_{inhb_{mn}}$ did not change during the initial calibration runs, and in order to speed up the calibration, the inhibition constants were held fixed at the initial values during further calibration. The following calibration of the rate constants k_{ox}, k_{nit} and k_{mn} was first carried out by adjusting the parameters for best fit by hand (trial-and-error). Afterward, the trial-and-error results were checked with PEST.

The calibration targets were the determined median values of pH, DO, NO_3^- and Mn^{2+} in the outflow water of the columns for each of the three flow and four temperature conditions. Hence, the calibration each resulted in 12 values for k_{ox}, k_{nit}, and k_{mn}.

3. Results

3.1. Seasonal Fluctuation of Redox-Sensitive Parameters Close to the Riverbank

Most relevant redox parameters of the Elbe river undergo strong seasonal fluctuations (Table S3). The median value for water temperature was 10.9 °C (3.0–21.3 °C, 10–90%ile, $n = 267$) during the entire 10-year observation period. Median values for DO, NO_3^-, DOC, TOC, and Mn were 10.8 mg/L (8.4–13.8 mg/L, $n = 269$), 15.0 mg/L (12.0–20.0 mg/L, $n = 279$), 5.2 mg/L (4.6–6.0 mg/L, $n = 325$), 6.3 mg/L (5.2–8.2 mg/L, $n = 292$), and 0.01 mg/L (0.01–0.03 mg/L, $n = 278$). During the cold winter season (December–March), the water temperature decreased down to 3 °C and the TOC concentration to <6 mg/L, whereas DO and NO_3^- increased to around 13 and 19 mg/L. During hot summer months (June–September), the water temperature increased to >21 °C and the TOC concentration to >7 mg/L, whereas DO and NO_3^- usually decreased to ≈8.5 mg/L and ≈13 mg/L. Mn in the Elbe river showed no noticeable seasonal fluctuations.

Along the transect from the Elbe river towards the PW seasonal fluctuations were also noticeable (Table S4). As expected, OW 1-1 and 2-1 in the upper aquifer showed the strongest variations. The temperatures at both observation points varied from winter to summer from 6.1 to 20.0 °C and 8.3 to 19.9 °C (Figures 2 and 3). Temperature fluctuations at the deeper observation points OW 1-2 and 2-2 were in the same order of magnitude. At the lowest observation points OW 1-3 and 2-3, as well as further along the flow path at OW 3, the temperature variations were ±3 °C compared to the median values.

DO showed similar patterns during the year. The higher DO concentration in the Elbe river in winter resulted in >6.5 mg/L at OW 1-1, >2.5 mg/L at OW 1-2 and >1.0 mg/L at OW 2-1. Further, along the flow path, DO was mostly found <0.5 mg/L. During summer, the DO was almost depleted at OW 1-1. Nitrate showed a similar behavior. During winter, NO_3^- concentrations were >20 mg/L at OW 1-1, >7 mg/L at OW 1-2 and >13 mg/L at OW 2-1, whereas in summer, the nitrate was with <1.0, <0.5 and <3.0 mg/L almost neglectable.

Mn concentration varied primary at OW 1-1 during the year. In winter, the Mn concentration was very low at OW 1-1, but increased up to 0.42 mg/L during summer time. At all other observation points, variations were found but without distinct patterns. Along the entire flow path after OW 1-1, the Mn concentration was always >0.1 mg/L and mostly below 0.35 mg/L (Table S4).

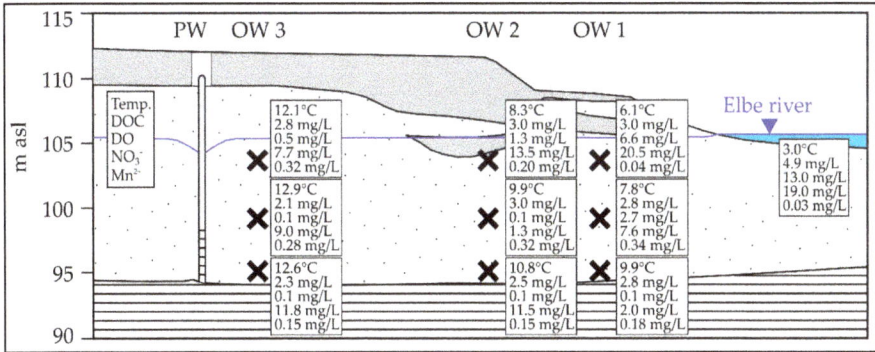

Figure 2. The mean values (n = 2) of the relevant redox parameters along the transect in February (Table S4).

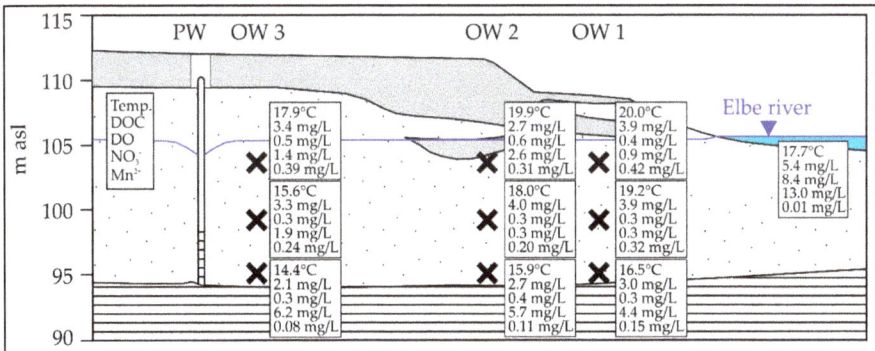

Figure 3. The mean values (n = 2) of the relevant redox parameters along the transect in September (Table S4).

3.2. Mn Release During Low Discharge Periods of the Elbe River

The water level of the Elbe river shows strong annual fluctuations. Long-lasting mean low discharge conditions are represented by a water level of ≤0.75 m and are rare. From 1998 to 2006, the Elbe water level was between 0.7 and 0.8 m only in 2003 (for almost 3 months). During the observation period from 2006 onwards, the Elbe river decreased to mean low discharge conditions in 2008, 2009, and 2016, which lasted for two weeks at maximum. In 2015, the latest low discharge period was observed that lasted for more than 3 months and the water level dropped to as low as 0.5 m [26].

During this 153 day long low discharge period in 2015, the median water level was 0.74 m (0.62–1.05 m, 10–90%ile, n = 153) and the mean water temperature 20.1 °C (11.0–24.6 °C, n = 153, Table S5). The Mn concentration at OW 1-1 increased up to 0.69 mg/L (median 0.19 mg/L, n = 6, Table S5) after the water temperature already fell below 20 °C (Figure 4). At the two deeper OW 1-2 and OW 1-3, the Mn concentration did not change noticeably. Further, along the flow path at OW 2-1, Mn increased up to 0.42 mg/L (median 0.23 mg/L, n = 6, Figure S2) and at OW 3-1 up to 0.39 mg/L (median 0.26 mg/L, $n = 6$, Figure S3). At OW 2-2, 2-3, 3-2, and 3-3, the Mn concentration remained almost constant.

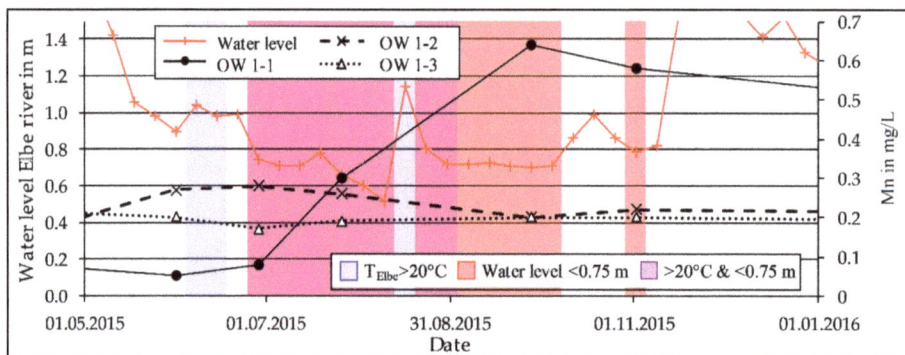

Figure 4. The Mn concentration at OW 1 during a low discharge period in 2015.

3.3. Mn Release Depending on the Temperature and Infiltration Rate During the Column Experiments

To investigate the effect of varying temperature to the Mn release from the riverbed, three columns filled with riverbed sediments run at 10, 20, 30, and 35 °C. Varying flow rates of 1, 2, and 4 mL/min represented low, mean and high infiltration rates for all four temperature regimes.

After changing temperature and/or flow, the Mn concentration changed within 5 pore volumes (PV) and was stable after 8–10 PV (e.g., Figure S4). At high infiltration rates (4 mL/min, pore velocity $v_a = 3.44 \cdot 10^{-5}$ m/s, residence time $t_R = 8.1$ h) and water temperatures of 10 and 20 °C, the median Mn release was <<0.01 mg/L ($n = 9$ and 12) and almost neglectable (Figure 5, Table S6). At 30 °C, the Mn concentration increased in the outflow slightly to around 0.03 mg/L (median, $n = 21$). Only at 35 °C was Mn released (median of 0.51 mg/L).

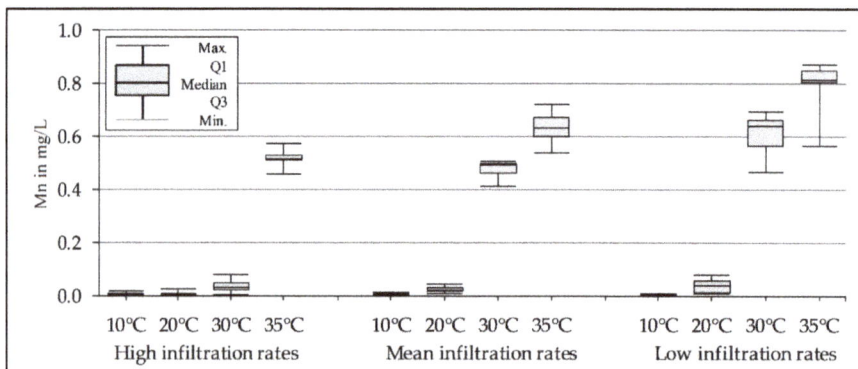

Figure 5. The mean Mn concentration in the outflow of the columns with riverbed sediment for high, mean, and low infiltration rates (4, 2 and 1 mL/min) at different temperatures (see Table S6 for no. of samples), Q_1 and Q_3 correspond to the 1st and 3rd quartile.

At mean infiltration rates (2 mL/min, $v_a = 1.72 \cdot 10^{-5}$ m/s, $t_R = 16.2$ h), a minor Mn release was observed at 10 and 20 °C with median values of 0.01 and 0.02 mg/L, respectively. With increasing temperature, the Mn release increased to 0.49 mg/L at 30 °C (median, $n = 9$) and 0.63 mg/L at 35 °C (median, $n = 15$).

For low infiltration rates (1 mL/min, $v_a = 8.60 \cdot 10^{-6}$ m/s, $t_R = 32.4$ h) at 10 °C, no Mn release was observed. The Mn concentration in the outflow increased already at 20 °C and stabilized around 0.04 mg/L (median, $n = 21$). After the temperature rose to 30 °C, Mn increased sharply to ≈0.64 mg/L (median, $n = 18$, Figure S4). The subsequent temperature increase to 35 °C led to around 0.81 mg/L Mn (median, $n = 15$).

The Mn concentration in the feed water maximally was 0.02 mg/L (median <LOD, limit of detection 0.005 mg/L Mn, $n = 23$).

3.4. Reduction Constants of O_2, NO_3^- and $Mn(IV)$ as Electron Acceptors

The observed Mn release was reproduced by chemical modeling with PHREEQC using Monod-type rate formulations and the reduction rate constants k_{ox}, k_{nit}, and k_{mn} as calibration parameters. The reduction rate constant k_{ox} showed the highest value at high infiltration rates and at a temperature of 10 °C (1.8×10^{-9} mol/(L·s)). With increasing temperature, k_{ox} decreased down to 4.3×10^{-10} mol/(L·s) (Table 2, Figure S5). For mean and low infiltration rates, a similar behavior was found. The lowest k_{ox} of 9.7×10^{-11} mol/(L·s) was determined at low infiltration rates and a temperature of 35 °C.

Table 2. The calibrated reduction constants of this study compared to the literature data.

This Study					
Temperature	v_a	k_{ox}	k_{nit}	k_{mn}	Notes
	m/s	mol/(L·s)	mol/(L·s)	mol/(L·s)	
10 °C	8.60×10^{-6}	5.18×10^{-10}	2.00×10^{-12}	2.50×10^{-10}	Low infiltration rate (1 mL/min)
	1.72×10^{-5}	8.65×10^{-10}	7.00×10^{-12}	2.00×10^{-10}	Mean infiltration rate (2 mL/min)
	3.44×10^{-5}	1.80×10^{-9}	5.00×10^{-12}	1.00×10^{-10}	High infiltration rate (4 mL/min)
20 °C	8.60×10^{-6}	2.17×10^{-10}	1.53×10^{-10}	1.10×10^{-10}	Low infiltration rate (1 mL/min)
	1.72×10^{-5}	4.50×10^{-10}	5.50×10^{-11}	1.50×10^{-10}	Mean infiltration rate (2 mL/min)
	3.44×10^{-5}	8.90×10^{-10}	5.00×10^{-12}	1.00×10^{-10}	High infiltration rate (4 mL/min)
30 °C	8.60×10^{-6}	1.02×10^{-10}	1.88×10^{-10}	1.90×10^{-9}	Low infiltration rate (1 mL/min)
	1.72×10^{-5}	2.14×10^{-10}	2.80×10^{-10}	2.20×10^{-9}	Mean infiltration rate (2 mL/min)
	3.44×10^{-5}	4.70×10^{-10}	5.00×10^{-12}	1.00×10^{-10}	High infiltration rate (4 mL/min)
35 °C	8.60×10^{-6}	9.70×10^{-11}	1.35×10^{-10}	1.62×10^{-9}	Low infiltration rate (1 mL/min)
	1.72×10^{-5}	2.14×10^{-10}	3.50×10^{-11}	1.40×10^{-9}	Mean infiltration rate (2 mL/min)
	3.44×10^{-5}	4.28×10^{-10}	5.00×10^{-12}	1.95×10^{-9}	High infiltration rate (4 mL/min)
10%ile		1.13×10^{-10}	5.00×10^{-12}	1.00×10^{-10}	
Median		$\mathbf{4.39 \times 10^{-10}}$	$\mathbf{2.10 \times 10^{-11}}$	$\mathbf{2.25 \times 10^{-10}}$	
90%ile		8.88×10^{-10}	1.85×10^{-10}	1.95×10^{-9}	
n		12	12	12	
Literature Data					
Temperature	v_a	k_{ox}	k_{nit}	k_{mn}	Source
	m/s	mol/(L·s)	mol/(L·s)	mol/(L·s)	
Variable	Variable	1.52×10^{-10}	3.81×10^{-11}	8.91×10^{-13}	[31]
Variable	Variable	2.00×10^{-10}	1.00×10^{-10}	1.70×10^{-12}	[14]
22 °C	7.60×10^{-6}	3.50×10^{-8}	3.40×10^{-8}	3.00×10^{-13}	[35]
n.a.	n.a.	3.98×10^{-10}	3.98×10^{-11}	6.31×10^{-14}	[36]
Variable	Variable	1.57×10^{-9}	1.00×10^{-11}	n.a.	[37]
Variable	Variable	1.30×10^{-9}	8.00×10^{-10}	n.a.	[33] for DOC
Variable	Variable	1.90×10^{-11}	1.20×10^{-11}	n.a.	[33] for SOM

n.a.—not available, v_a—pore velocity.

For the reduction rate constant k_{nit}, no distinct pattern was found. At high infiltration rates, k_{nit} remained constant at 5.0×10^{-12} mol/(L·s). For mean and low infiltration rates at 10 °C, k_{nit} was in the same order of magnitude (7.0 and 2.0×10^{-12} mol/(L·s)). With increasing temperatures of 20 °C and 30 °C, k_{nit} increased up to 2.8×10^{-10} mol/(L·s). At 35 °C, k_{nit} decreased again at the mean and low infiltration rates.

The reduction rate constant k_{mn} at 10 °C and 20 °C for high, mean and low infiltration rates were in the order of magnitude of 1.0 to 2.5×10^{-10} mol/(L·s). At 30 °C and high infiltration rates, k_{mn} remained in this range. For mean and low infiltration rates at 30 °C and 35 °C, k_{mn} increased to $1.4–2.0 \times 10^{-9}$ mol/(L·s).

With a percental error of −2.4% compared to the measured values (median, −6.7–0.2%, 10–90%ile, $n = 12$), the simulated DO concentrations showed the largest errors of relevant redox parameters (Table S7). The error for NO_3^- with a median of 0.4% and a span of −0.6 to 1.1% (10 to 90%ile, $n = 12$) was lower. With an error of 0.2%, the simulated Mn^{2+} concentrations showed the smallest median deviation but a comparable large span of −15.7 to 8.5% ($n = 12$). The median error for the simulated pH was −1.0% with a span of −8.7 to 1.7% ($n = 12$).

3.5. The Decrease of Easily Reducible Mn Along the Flow Path

To estimate the mobilization behavior of Mn, a 4-step sequential extraction procedure was applied to the filling material of the three columns after the experiment. All three columns showed qualitatively similar results and all the following values represent median values with $n = 3$ (Table S8). The total Mn mass (Mn_{tot}, as the sum of all 4 extracted fractions) was around 270 mg/kg at the inlet after 0.05 m (Figure 6). Further along the flow path, the total Mn mass decreased to ≈150 mg/kg down to a minimum of ≈125 mg/kg the minimum at the outlet after 0.9 m. The total extractable Mn (microwave acid digestion) was around 112 mg/kg at the inlet and increased to 133, 250, and 473 mg/kg at the outlet (Table S8).

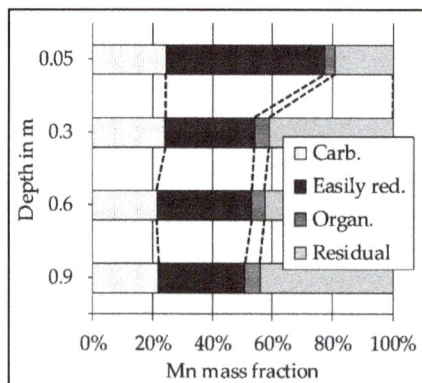

Figure 6. The median values ($n = 3$) of the absolute Mn mass fractions along the column.

The mass fraction of soluble and carbonate bound Mn ("Carb.") decreased from around 65 mg/kg to 30 mg/kg along the flow path. On a percentage base, the soluble and carbonate bound Mn always remained between 21–24% (Figure 7). Easily reducible Mn dropped from around 140 mg/kg (53% of the total mass) at the inlet to 36 mg/kg (<30%) after 0.9 m. Organically bound Mn ("Organ.") remained stable between 6–9 mg/kg (3–5%). The residual Mn fraction was found almost constant at 51–64 mg/kg, but the percentage increased from around 20 to >40%.

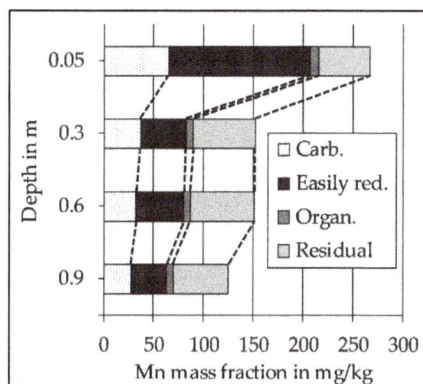

Figure 7. The median values (*n* = 3) of percentage Mn mass fractions along the columns.

4. Discussion

4.1. The Significance of the Calibrated Degradation Rate Constants

In order to use the results from the column experiment for a planned modeling of the transect, the observed Mn release was simulated using PHREEQC, focusing on relevant redox reactions. During an initial automated calibration with PEST, the inhibition constants $K^{ox}_{inhb\_nit}$, $K^{ox}_{inhb\_mn}$, and $K^{nit}_{inhb\_mn}$ did not change and were held fixed at the initial values during further calibration. Hence, only the reduction rate constants k_{ox}, k_{nit} and k_{mn} were used for calibration.

The calibrated k_{ox} had an overall median value of 4.4×10^{-10} mol/(L·s) (*n* = 12, Table 2). These results are in fair agreement with results of Henzler et al. [14], who determined a k_{ox} of 2.0×10^{-10} mol/(L·s) for a bank filtration site at Lake Tegel in Berlin, Germany. Greskowiak et al. [31] used a k_{ox} of 1.52×10^{-10} mol/(L s) to simulate the consumption of DO at an infiltration pond in Berlin. Sharma et al. [33] used a more than ten times higher degradation rate with $k_{ox} = 1.30 \times 10^{-9}$ mol/(L·s) to simulate a transect at the RBF Waterworks Flehe in Düsseldorf, Germany. The average water temperature in their study was 13.5 °C and the BF traversed the ≈60 m wide stretch from the riverbank to the well within 60 days [33]. This corresponds to a pore velocity (v_a) of ≈1.2×10^{-5} m/s. Comparing this with the results of this study of $k_{ox} = 8.65 \times 10^{-10}$ mol/(L·s) at 10 °C and $v_a \approx 1.72 \times 10^{-5}$ m/s, shows a good agreement, too.

The overall median value for the calibrated k_{nit} was 2.1×10^{-11} mol/(L·s) (*n* = 12, Table 2). This is about five times lower than a reported k_{nit} of 1.0×10^{-10} mol/(L·s) by Henzler et al. [14] and 40-times lower than what Sharma et al. [33] reported with a $k_{nit} = 8.0 \times 10^{-10}$ mol/(L·s). Prommer and Stuyfzand [37] and Greskowiak et al. [31] reported a k_{nit} of 1.0 and 3.8×10^{-11} mol/(L·s), which correspond to the values of this study.

For k_{mn}, the overall median value was 2.3×10^{-10} mol/(L·s) (*n* = 12, Table 2). This is ≈100- to 1000-times higher than the reported data of $k_{mn} = 1.7 \times 10^{-12}$ mol/(L·s) [14], 8.9×10^{-13} mol/(L·s) [31] or 3.0×10^{-13} mol/(L·s) [35]. Matsunga et al. [35] modeled a column experiment and included the precipitation of Rhodochrosite ($MnCO_3$) as a sink for Mn^{2+}. The precipitation of $MnCO_3$ as a sink for Mn^{2+} has also been shown at other bank filtration sites [38]. Taking the average water quality of the outflow from the columns at 20 °C (pH 8, 50 mg/L Ca^{2+}, 120 mg/L HCO_3^-) and mean infiltration rates for equilibrium in PHREEQC, the water would be supersaturated with respect to $MnCO_3$ with around 0.1 mg/L Mn^{2+} in equilibrium (saturation index 0.77). Thus, the precipitation of$MnCO_3$ would probably control the Mn concentration for longer residence times but did not along the columns due to slow reaction kinetics [39]. As a consequence, the given k_{mn} seem to be representative in simulating an Mn release from a (highly reactive) riverbed, but are not suited for simulations of longer transects, in which an Mn reduction is likely to occur within the aquifer and can be controlled by $MnCO_3$ precipitation, for example.

4.2. Impact of the Discharge on the Observed Redox Patterns

River discharge indirectly affects the quality of the BF in multiple ways. Low river discharge and related low water levels are often associated with less dilution of wastewater effluent, higher loads of dissolved organic matter (DOM), prolonged travel times, and they are expected to promote DO consumption [19]. Low discharge periods further result in reduced shear stress at the riverbed and more intense clogging at the bottom of the water body, which can also promote anoxic conditions [40,41]. High discharge conditions/floods can cause a partial removal of the clogging layer, resulting in a better hydraulic conductivity of the riverbed and higher water levels lead to higher infiltration rates [18]. Increased flow velocities combined with shorter flow paths cause shorter travel times. Furthermore, high flow conditions and shorter travel times were linked to an increased input of TOC (total organic carbon) [42,43].

To the authors' knowledge, very few researchers have addressed the impact of river discharge to Mn release from the riverbed or the Mn concentration in the BF. Previous work has focused on the impact of discharge on the degradation of natural organic matter (NOM) and DO consumption. In a column experiment, von Rohr et al. [44] evaluated the role of discharge on NOM degradation during RBF. The experiment was set up at 20 °C and with four flow rates, resulting in residence times (t_R) of 40, 20, 12, and 4 h. At residence times of 4 h, oxic conditions persisted through the entire column. With a residence time of 20 h, anoxic conditions were observed within 20 cm along the flow path. Results from the column experiment in this study confirm the flow/infiltration rate dependency of oxygen depletion. At 20 °C and at high infiltration rates (t_R = 8.1 h), the columns remained oxic, but became anoxic at residence times of 16–32 h (Table S9). Diem et al. [32] found river discharge to be correlated with an enhanced POM input and higher DO consumption during flood events. Due to a generally higher DO consumption at temperatures >15 °C, the correlation was only found to be true for temperatures below 15 °C [17]. Diem et al. [32] successfully modeled these observations using higher DO consumption for higher discharges. In this study, rates for DO consumption (k_{ox}) in PHREEQC were highest at high infiltration rates and decreased at mean and low infiltration rates, which is in agreement with the findings of Diem et al. [32].

Groffman and Crossey [45] found slightly increasing Mn concentrations within the upper aquifer section during periods with lower discharge at Rio Calaveras (New Mexico, USA). The observations from the transect in Dresden-Tolkewitz did not reveal a clear discharge/water level dependency. Figure 4 indicates an increasing Mn concentration at OW 1-1 during a long-lasting low discharge period of the Elbe River in 2015. In addition, Mn at OW 2-1 and 3-1 increased during this period (Figures S2 and S3). Contrary to this, at OW 1-1 Mn, already decreased when the water level was still low, whereas Mn further increased at OW 2-1 and 3-1. Hence, the field observations did not indicate a mere discharge-dependency of Mn along the transect.

Manganese release during the column experiment with riverbed material was low at 10 °C and 20 °C for the high, mean, and low infiltration rates (Figure 5). At 30 and 35 °C, Mn increased from high to mean and again to low infiltration rates. Correlation coefficients of −0.99 for 30 °C and −0.87 for 35 °C were very high and significant ($p < 0.05$, data not shown).

The relation between river discharge and the release of Mn from the riverbed is ambiguous. On the one hand, the literature data [44] and the results from the column experiments in this study suggest that periods of low discharge lead to elevated Mn concentrations due to lower infiltration rates and longer residence times. Contrary to that, e.g., Diem et al. [17] found higher river discharge to be correlated with higher DO consumption, which was also reinforced by modeling [32]. Since higher DO consumption is often related to increasing Mn concentrations during RBF [1], periods of higher discharge may cause Mn release. Shorter travel times during periods of high river discharge would interfere with this. In addition, the results of the column experiment in this study showed that the temperature effect distorts this discharge dependency (see the following paragraph).

4.3. Manganese Release Controlled by Temperature

A temperature-dependent fluctuation of the Mn concentration has been shown at many bank filtration sites [1,15,38,46–50].

Manganese along the transect in Dresden-Tolkewitz varied the most at OW 1-1, 2-1, and 2-2 (Figures 4 and S2), which are closest to the Elbe river (≈20 m). Furthermore, those observation points showed the largest temperature fluctuations (Figures 2 and 3). Especially the Mn concentration at OW 1-1, which is closest to the riverbank, responded to higher water temperatures during low and mean flow conditions (Figures S6–S8). However, at the more distant observation points (OW 1-3, 2-3 and OW 3), Mn was comparably stable during the year. This can partially be attributed to extended travel times (compared to the average travel times), especially towards the deeper observation points. Similar observations were made at RBF sites at the Lot river in France [51] and at the Glatt river in Switzerland [50]. Both found a seasonal trend for Mn, with elevated concentrations during the warm summer months and lower concentrations during the winter season. For the RBF site in Dresden-Tolkewitz, no seasonal trend for Mn was observed at any OW, which is probably due to the long sampling interval. Temperature seems to be a very important factor controlling initial Mn release from the riverbed in Dresden-Tolkewitz, but more research is needed in order to assess this parameter quantitatively. Since the assessment is only possible during temporary constant, low water levels [23], future research must focus on the rare long-lasting low discharge periods.

Results from the column experiments indicate a high-temperature dependency of Mn release. At constant infiltration rates (high, mean, or low), the Mn concentration increased with increasing temperature (Figure 5). At high infiltration rates (residence time t_R = 8.1 h), the Mn concentration increased in the outflow sharply from 0.03 to 0.51 mg/L at higher temperatures of 30 and 35 °C. For mean and low infiltration rates (t_R = 16.2 and 32.4 h), a similar Mn increase happened after water temperature increased from 20 to 30 °C. The calibrated k_{mn} in this study reflects the sharp increases, showing a 20-fold increase from 30 to 35 °C at high infiltration rates as well as a 15- and 17-times higher k_{mn} after temperature raised from 20 to 30 °C at mean and low infiltration rates (Table 2).

Bourg and Bertin [51] reported a threshold water temperature of 10 °C to trigger a microbiologically mediated Mn reduction. Hoehn et al. [52] observed extensive denitrification, stronger reducing conditions, and elevated Mn concentrations at water temperatures above 14 °C. The findings of the column experiment support these observations. At water temperatures of 10 °C, the Mn release was neglectable for all infiltration rates. In addition, the results from the column experiment and the determined reduction constants can expand these statements for Mn containing riverbeds. For high infiltration rates of 1.1 $m^3/(m^2 \cdot d)$ and above, Mn release from the riverbed is unlikely for surface water temperatures that are typical in temperate climate zones (≤30 °C). If the BF infiltrates at infiltration rates in the order of magnitude of 0.6 $m^3/(m^2 \cdot d)$ and below, water temperatures >20 °C are sufficient to trigger extensive Mn release from the riverbed.

4.4. Depletion of the Mn Reservoir Within the Riverbank

After the column experiments were finished, a sequential extraction procedure was applied to the riverbed sediment (Figures 6 and 7).

Manganese contents in the riverbed sediments are very site-specific and cannot be limited to regional or climatic differences [53]. Moreover, the Mn content depends highly on the grain size. Jain and Ram [54] and Jain and Sharma [55] found 230–650 mg/kg Mn in the grain size fraction with the highest mass fraction (36%, d = 0.21–0.25 mm), whereas the grain size fraction with the lowest mass fraction of only 2% (d < 0.075 mm) contained up to 2800 mg/kg Mn.

However, the order of magnitude of ≈150–300 mg/kg Mn was low compared to many other riverbed sediments, e.g., 960 mg/kg Mn in the Rhine sediment [56] or 1700 mg/kg Mn in the

Garonne [53]. Nonetheless, the values are in agreement with the former analysis of the riverbed sediment from the Elbe river. Grischek et al. [57] determined ≈50 mg/kg Mn_{tot} (n = 3) in riverbed sediment at Torgau around 100 km downstream of Dresden. In Dresden-Tolkewitz at the same transect, Paufler [58] determined around 250 mg/kg Mn_{tot} (n = 3) for the upper 5 cm of the riverbed and only ≈80 mg/kg Mn_{tot} (n = 6) for the depths 5–30 cm.

At the Glatt river in Switzerland, von Gunten et al. [13] hypothesized that the repetitive exhaustion of Mn deposits in the river sediments towards the fall season could be one reason for seasonal Mn variations.

Integrating the released Mn from one column over the entire experimental time in this study (all four infiltration rates and three temperature regimes), this results in an overall mass output of approx. 140 mg Mn (data not shown). After the experiment, the total Mn mass was around 1600 mg within a column (Figure 7, ≈10 kg of sediment, calculated dry weight). Considering the large deviations from the total extractable Mn (Table S8), and the potential error of sequential extraction procedures [59], conclusions about a possible exhaustion of Mn during the column experiment would not be reasonable. Taking the observations of Paufler [58] into account, the observed steps along the columns in this study may reflect the natural stratification within the riverbed, rather than a depletion of the Mn reservoir.

With around 5 mg/L of filterable substances and a Mn content of 3500 mg/kg of the suspended matter (both median, n > 100 [25]), the Elbe river water contains approximately 0.04 mg/L Mn, which is bound to suspended solids. Considering the mean abstraction of 22,000 m^3/d in Dresden-Tolkewitz and a bank filtrate portion of 70%, about 15,500 m^3 BF are abstracted per day. Assuming an even distribution of suspended solids in the infiltrating river water, the Elbe river delivers around 6×10^5 mg Mn per day into the riverbed, which is probably not entirely reducible. With the same BF portion and the observed Mn release of 0.1 to 0.8 mg/L, the Mn output from the riverbed into the BF would be 1.5×10^6 to 1.2×10^7 mg/d. Thus, a depletion of the Mn reservoir seems to be possible but more research is needed to evaluate this conclusively.

4.5. Implications for (River-)Bank Filtration Sites

Current climate forecasts show an increase in river discharge seasonality, with increasing high discharges and decreasing low discharges. Furthermore, global mean river water temperatures are projected to increase by 0.8–1.6 °C by 2100 with Europe, the United States, parts of southern Africa, Australia and eastern China facing the largest changes [60].

The results of this study showed stronger Mn release for low infiltration rates compared to high infiltration rates when the temperature was equal. RBF sites in regions with a colder climate probably will not notice different Mn concentrations in the future, since Mn release was neglectable at water temperatures of 10 °C for all infiltration rates. In regions with a temperate climate, Mn release from the riverbed is unlikely during high discharge events with infiltration rates ≥1.1 m^3/(m^2·d). At infiltration rates of around 0.6 m^3/(m^2·d) and below, water temperatures >20 °C are sufficient to trigger extensive Mn release from the riverbed. Thus, RBF sites in temperate climate zones at rivers with large seasonal discharge fluctuations may have to deal with increasing Mn concentrations in the future. Additionally, the effect of seasonal rivers showing the highest water temperatures during low discharge periods [60] may intensify this effect. Surface water temperatures of >30 °C are rare in the temperate climate zone.

Typically, rivers in dry/arid or tropical climates like the Nile river [61] or the Mekong river [62] show temperatures in this order of magnitude. Thus, the results of this study suggest that RBF sites in arid or tropical climate zones should be aware of an intensified Mn release from the riverbed in the future, which can distort the quality of the BF even at high infiltration velocities. In such cases, biological post-treatment could be a viable option to remove Mn from the BF [63].

Nonetheless, elevated temperatures are not necessarily causing elevated Mn concentrations during RBF. During the summer in 2003, maximum river water temperatures of 25 °C were observed at the RBF site at the Rhine river in Germany. No elevated Mn concentrations were detected, although anaerobic conditions developed within the aquifer [64].

The field observations in Dresden-Tolkewitz and at other RBF sites prove that an actual high Mn release from the riverbed must not necessarily lead to elevated Mn concentrations in the pumped raw water [1,47]. Sorption, (re-)oxidation, and precipitation along the flow path are potential sinks for Mn released from a riverbed. Especially if the aquifer material contains Mn(hydr)oxides, Mn concentrations of the BF can change largely along the flow path due to the high affinity of Mn^{2+} for Mn(hydr)oxides [63]. Thus, a long distance between the riverbank and the pumping wells could be of advantage to buffer elevated Mn concentrations and temperature. However, such an advantage may have to be checked against a lower portion of abstracted bank filtrate and a higher portion of potentially Fe- and Mn-rich landside groundwater [65].

For the RBF site in Dresden-Tolkewitz, the temperature was found to be the driving force for Mn release from the riverbed and high discharge and infiltration rates limit the release at lower temperatures. In Dresden-Tolkewitz, the mean infiltration rate is around 0.2 $m^3/(m^2 \cdot d)$, which corresponds to the infiltration rate that was found to be sustainable for RBF sites along the River Rhine, Elbe, and other European Rivers [66]. After Soares [67], an infiltration rate of 0.32 $m^3/(m^2 \cdot d)$ would be still sustainable for this RBF site. Ahrns [23] observed infiltration rates up to 0.95 $m^3/(m^2 \cdot d)$ in Dresden-Tolkewitz. Thus, the investigated low infiltration rate of 0.3 $m^3/(m^2 \cdot d)$ in this study represents the mean infiltration rate in Dresden-Tolkewitz. Such low infiltration rates must go in hand with temperatures above 20 °C to trigger a Mn release. With increasing temperatures, the infiltration rate becomes less important and at water temperatures around 30 °C, an extensive Mn release from the riverbed can be expected even at mean infiltration rates.

5. Conclusions

Current climate forecasts project increasing river discharge seasonality and water temperatures and thus, are thought-provoking at many (river)bank filtration sites. Water quality data of a 10-year time span from a monitoring transect at the RBF site Dresden-Tolkewitz and accompanying column experiments were used to assess the potential Mn release from the riverbed with respect to the climate-related variables, temperature and discharge. Temperature was found to be more important than discharge. Low infiltration rates 0.3 $m^3/(m^2 \cdot d)$ required temperatures above 20 °C to trigger Mn release. With an increasing temperature, the discharge becomes less important and at 30 °C, the infiltration rates of $\approx$0.6 $m^3/(m^2 \cdot d)$ can already cause an extensive Mn release from the riverbed. The subsequent modeling of the column experiment with PHREEQC resulted in degradation rates for DO that are applicable at other RBF sites for several water temperatures and flow velocities. The determined Mn reduction rate constants are appropriate to simulate an Mn release from riverbed sediments but are not suited for simulations in which Mn reduction is likely to occur within the aquifer and the Mn concentrations can be limited by precipitation, for example.

Supplementary Materials: The following are available online at http://www.mdpi.com/2073-4441/10/10/1476/s1,

- Figure S1: The flow scheme of the column experiments inside the thermostatic cabinets;
- Figure S2: The Mn concentration at OW 2 during a low discharge period in 2015;
- Figure S3: The Mn concentration at OW 3 during a low discharge period in 2015;
- Figure S4: The increase of the Mn concentration within the first 5 PV at 30 °C and 1 mL/min after lowering from 2 mL/min, Median, 10- and 90-%ile (each $n = 3$);
- Figure S5: The calibrated reduction rate constants for the column experiment;
- Figure S6: The Mn concentration at OW 1 during a low discharge period in 2007;
- Figure S7: The Mn concentration at OW 2 during a low discharge period in 2007;
- Figure S8: The Mn concentration at OW 3 during a low discharge period in 2007;
- Table S1: The chemical reagents and analytical conditions for the optimized BCR sequential extraction procedure [27,28];
- Table S2: The input data of the column in PHREEQC;
- Table S3: The statistical data of the Elbe river at Dresden for the entire observation period 2006–2016 [25];
- Table S4: The monthly and 10-year median values of selected parameters along all three OW's of the transect in the WW Dresden-Tolkewitz for the entire observation period 2006–2016;
- Table S5: The statistical data for the low flow period in 2015;

- Table S6: The statistical data for Mn release during the column experiment with riverbed sediment from the Elbe river in Dresden-Tolkewitz, Q_1 and Q_3 correspond to the 1st and 3rd quartile;
- Table S7: The comparison of the measured parameter values during the column experiment and the modeled parameters from PHREEQC;
- Table S8: The results and statistical data of the sequential extraction of the riverbed sediment after the column experiment ($n = 3$ for each statistical data);
- Table S9: The median concentration of DO in the outflow during the column experiment.

Author Contributions: Data curation, M.B., N.S. and T.F.; Investigation, S.P. and T.G.; Methodology, S.P. and T.G.; Project administration, T.G.; Modeling, S.P.; Supervision, S.P. and T.G.; Writing—original draft, S.P.; Writing—review & editing, S.P., T.G., M.B., N.S. and T.F.

Funding: This research received no external funding.

Acknowledgments: This work was performed as cooperation between DREWAG Netz GmbH and the Division of Water Sciences at the University of Applied Sciences Dresden. The authors are grateful to the ESF for the financial support to S. Paufler (grant no. 200031585). The paper was completed by further analysis supported by the AquaNES project, which has received funding from the European Union's Horizon 2020 research and innovation programme under grant agreement No. 689450.

Conflicts of Interest: The authors declare no conflict of interest. The founding sponsors had no role in the design of the study; in the collection, analyses, or interpretation of data; in the writing of the manuscript, and in the decision to publish the results.

References

1. Bourg, A.C.M.; Bertin, C. Biogeochemical processes during the infiltration of river water into an alluvial aquifer. *Environ. Sci. Technol.* **1993**, *27*, 661–666. [CrossRef]
2. Grischek, T.; Schoenheinz, D.; Worch, E.; Hiscock, K. Bank filtration in Europe – An overview of aquifer conditions and hydraulic controls. In *Management of Aquifer Recharge for Sustainability*; Dillon, P., Ed.; Balkema Publ.: Lisse, The Netherlands, 2002; pp. 485–488.
3. Ray, C.; Melin, G.; Linsky, R.B. *Riverbank Filtration—Improving Source Water Quality*; Kluwer: Dordrecht, The Netherlands, 2003; 364p.
4. Regnery, J.; Barringer, J.; Wing, A.D.; Hoppe-Jones, C.; Teerlink, J.; Drewes, J.E. Start-up performance of a full-scale riverbank filtration site regarding removal of DOC, nutrients, and trace organic chemicals. *Chemosphere* **2015**, *127*, 136–142. [CrossRef] [PubMed]
5. Blanford, W.; Boving, T.; Al-Ghazawi, Z.; Shawaqfah, M.; Al-Rashdan, J.; Saadoun, I.; Schijven, J.; Ababneh, Q. River bank filtration for protection of Jordanian surface and groundwater. In Proceedings of the World Environmental and Water Resources Congress 2010: Challenges of Change; ASCE: Providence, RI, USA, 2010; pp. 776–781.
6. Ghodeif, K.; Paufler, S.; Grischek, T.; Wahaab, R.; Souaya, E.; Bakr, M.; Abogabal, A. Riverbank filtration in Cairo, Egypt—Part I: Installation of a new riverbank filtration site and first monitoring results. *Environ. Earth. Sci.* **2018**, *77*, 270. [CrossRef]
7. Hu, B.; Teng, Y.; Zhai, Y.; Zuo, R.; Li, J.; Chen, H. Riverbank filtration in China: A review and perspective. *J. Hydrol.* **2016**, *541*, 914–927. [CrossRef]
8. Sandhu, C.; Grischek, T.; Kumar, P.; Ray, C. Potential for riverbank filtration in India. *Clean Technol. Environ. Policy* **2011**, *13*, 295–316. [CrossRef]
9. Hiscock, K.M.; Grischek, T. Attenuation of groundwater pollution by bank filtration. *J. Hydrol.* **2002**, *266*, 139–144. [CrossRef]
10. Kuehn, W.; Mueller, U. Riverbank filtration—An overview. *J. Am. Water Works Assoc.* **2000**, *92*, 60–69. [CrossRef]
11. Bertelkamp, C.; Reungoat, J.; Cornelissen, E.R.; Singhal, N.; Reynisson, J.; Cabo, A.J.; van der Hoek, J.P.; Verliefde, A.R.D. Sorption and biodegradation of organic micropollutants during river bank filtration: A laboratory column study. *Water Res.* **2014**, *52*, 231–241. [CrossRef] [PubMed]
12. Bourg, A.C.M.; Richard-Raymond, F. Spatial and temporal variability in the water redox chemistry of the M27 experimental site in the Drac river calcareous alluvial aquifer (Grenoble, France). *J. Contam. Hydrol.* **1994**, *15*, 93–105. [CrossRef]

13. Von Gunten, H.R.; Karametaxas, G.; Krähenbühl, U.; Kuslys, M.; Giovanoli, R.; Hoehn, E.; Keil, R. Seasonal biogeochemical cycles in riverborne groundwater. *Geochim. Cosmochim. Acta* **1991**, *55*, 3597–3609. [CrossRef]
14. Henzler, A.F.; Greskowiak, J.; Massmann, G. Seasonality of temperatures and redox zonations during bank filtration—A modeling approach. *J. Hydrol.* **2016**, *535*, 282–292. [CrossRef]
15. Gross-Wittke, A.; Gunkel, G.; Hoffmann, A. Temperature effects on bank filtration: Redox conditions and physical–chemical parameters of pore water at Lake Tegel, Berlin, Germany. *J. Water Clim. Chang.* **2010**, *1*, 55–66. [CrossRef]
16. Van't Hoff, J.H.; Cohen, E. *Studies on Chemical Dynamics*; Studien zur chemischen Dynamik, Etudes de dynamique chemique; Frederik Muller & Co.: Amsterdam, The Netherlands, 1896; p. 128.
17. Diem, S.; von Rohr, M.R.; Hering, J.G.; Kohler, H.-P.E.; Schirmer, M.; von Gunten, U. NOM degradation during river infiltration: Effects of the climate variables temperature and discharge. *Water Res.* **2013**, *47*, 6585–6595. [CrossRef] [PubMed]
18. Schoenheinz, D.; Grischek, T. Behavior of dissolved organic carbon during bank filtration under extreme climate conditions. In *Riverbank Filtration for Water Security in Desert Countries*; Shamrukh, M., Ed.; NATO Science for Peace and Security Series C: Environmental Security; Springer: Dordrecht, The Netherlands, 2011; pp. 151–168, ISBN 978-94-007-0025-3.
19. Sprenger, C.; Lorenzen, G.; Hulshoff, I.; Gruetzmacher, G.; Ronghang, M.; Pekdeger, A. Vulnerability of bank filtration systems to climate change. *Sci. Total Environ.* **2011**, *409*, 655–663. [CrossRef] [PubMed]
20. EEA European Environment Agency. *Climate Change, Impacts and Vulnerability in Europe 2016*; An Indicator-Based Report; Publications Office of the European Union: Luxembourg, 2017; ISBN 978-92-9213-835-6.
21. Kreienkamp, F.; Spekat, A.; Enke, W. *WEREX V: Provision of an Ensemble of Regional Climate Projections for Saxony*, 1st ed.; Saxon State Office for the Environment, Agriculture and Geology: Dresden, Germany, 2011; 71p. (In German)
22. Grischek, T.; Bartak, R. Riverbed clogging and sustainability of riverbank filtration. *Water* **2016**, *8*, 604. [CrossRef]
23. Ahrns, J. Modeling the Exchange between Surface Water and Groundwater Using Temperature as Tracer. Diploma Thesis, Dresden University of Applied Sciences, Dresden, Germany, 2008. (In German)
24. DVGW German Technical and Scientific Association for Gas and Water. *Guideline DVGW W 112 (A), Principles of Groundwater Sampling from Groundwater Monitoring Wells*; WVGW: Bonn, Germany, 2011; ISSN 0176-3504. (In German)
25. SMUL Saxon State Ministry for Environment and Agriculture. Water Quality Data for the Elbe River. 2018. Available online: www.umwelt.sachsen.de/umwelt/wasser/ (accessed on 15 June 2018).
26. WSV Federal Waterways and Shipping Administration. Hydrological Information System. 2018. Available online: www.wsv.de (accessed on 14 June 2018).
27. Rauret, G.; López-Sánchez, J.F.; Sahuquillo, A.; Rubio, R.; Davidson, C.; Ure, A.; Quevauviller, P. Improvement of the BCR three step sequential extraction procedure prior to the certification of new sediment and soil reference materials. *J. Environ. Monit.* **1999**, *1*, 57–61. [CrossRef] [PubMed]
28. Sutherland, R.A.; Tack, F.M.G. Determination of Al, Cu, Fe, Mn, Pb and Zn in certified reference materials using the optimized BCR sequential extraction procedure. *Anal. Chim. Acta* **2002**, *454*, 249–257. [CrossRef]
29. Farnsworth, C.E.; Hering, J.G. Inorganic geochemistry and redox dynamics in bank filtration settings. *Environ. Sci. Technol.* **2011**, *45*, 5079–5087. [CrossRef] [PubMed]
30. Parkhurst, D.L.; Appelo, C.A.J. Description of Input and Examples for Phreeqc Version 3: A Computer Program for Speciation, Batch-Reaction, One-Dimensional Transport, and Inverse Geochemical Calculations. U.S. Geological Survey Techniques and Methods, book 6, chap. A43; 2013. Available online: http://pubs.usgs.gov/tm/06/a43 (accessed on 14 June 2018).
31. Greskowiak, J.; Prommer, H.; Massmann, G.; Nutzmann, G. Modeling seasonal redox dynamics and the corresponding fate of the pharmaceutical residue phenazone during artificial recharge of groundwater. *Environ. Sci. Technol.* **2006**, *40*, 6615–6621. [CrossRef] [PubMed]
32. Diem, S.; Cirpka, O.A.; Schirmer, M. Modeling the dynamics of oxygen consumption upon riverbank filtration by a stochastic–convective approach. *J. Hydrol.* **2013**, *505*, 352–363. [CrossRef]
33. Sharma, L.; Greskowiak, J.; Ray, C.; Eckert, P.; Prommer, H. Elucidating the effects on seasonal variations of biogeochemical turnover rates during riverbank filtration. *J. Contam. Hydrol.* **2012**, *428–429*, 104–115. [CrossRef]

34. Doherty, J. *PEST. Model-Independent Parameter Estimation*, 7th ed.; Watermark Numerical Computing: Brisbane, Australia, 2018.

35. Matsunaga, T.; Karametaxas, G.; von Gunten, H.R.; Lichtner, P.C. Redox chemistry of iron and manganese minerals in river-recharged aquifers: A model interpretation of a column experiment. *Geochim. Cosmochim. Acta* **1993**, *57*, 1691–1704. [CrossRef]

36. Mayer, K.U.; Benner, S.G.; Frind, E.O.; Thornton, S.F.; Lerner, D.N. Reactive transport modeling of processes controlling the distribution and natural attenuation of phenolic compounds in a deep sandstone aquifer. *J. Contam. Hydrol.* **2001**, *53*, 341–368. [CrossRef]

37. Prommer, H.; Stuyfzand, P.J. Identification of temperature-dependent water quality changes during a deep well injection experiment in a pyritic aquifer. *Environ. Sci. Technol.* **2005**, *39*, 2200–2209. [CrossRef] [PubMed]

38. Massmann, G.; Pekdeger, A.; Merz, C. Redox processes in the Oderbruch polder groundwater flow system in Germany. *Appl. Geochem.* **2004**, *19*, 863–886. [CrossRef]

39. Jensen, D.L.; Boddum, J.K.; Tjell, J.C.; Christensen, T.H. The solubility of rhodochrosite ($MnCO_3$) and siderite ($FeCO_3$) in anaerobic aquatic environments. *Appl. Geochem.* **2002**, *17*, 503–511. [CrossRef]

40. Heeger, D. Investigation of Clogging in Rivers. Ph.D. Thesis, Dresden University of Technology, Dresden, Germany, 1987. (In German)

41. Grischek, T. Management of bank filtration sites along the Elbe River. Ph.D. Thesis, Faculty of Forestry, Geo and Hydro Sciences, Dresden University of Technology, Dresden, Germany, 2003. (In German)

42. Eckert, P.; Rohns, H.P.; Irmscher, R. Dynamic processes during bank filtration and their impact on raw water quality. In Proceedings of the Aquifer Recharge, Recharge Systems for Protecting and Enhancing Groundwater Resources, Berlin, Germany, 11–16 June 2005; pp. 17–22.

43. Eckert, P.; Irmscher, R. Over 130 years of experience with riverbank filtration in Düsseldorf, Germany. *Aqua* **2006**, *55*, 283–291. [CrossRef]

44. Von Rohr, R.M.; Hering, J.G.; Kohler, H.P.; von Gunten, U. Column studies to assess the effects of climate variables on redox processes during riverbank filtration. *Water Res.* **2014**, *61*, 263–275. [CrossRef] [PubMed]

45. Groffman, A.R.; Crossey, L.J. Transient redox regimes in a shallow alluvial aquifer. *Chem. Geol.* **1999**, *161*, 415–442. [CrossRef]

46. Bourg, A.C.M.; Darmendrail, D.; Ricour, J. Geochemical filtration of riverbank and migration of heavy metals between the Deûle River and the Ansereuilles alluvion-chalk aquifer (Nord, France). *Geoderma* **1989**, *44*, 229–244. [CrossRef]

47. Jacobs, L.A.; von Gunten, H.R.; Keil, R.; Kuslys, M. Geochemical changes along a river-groundwater infiltration flow path: Glattfelden, Switzerland. *Geochim. Cosmochim. Acta* **1988**, *52*, 2693–2706. [CrossRef]

48. Kedziorek, M.; Geoffriau, S.; Bourg, A.C.M. Organic matter and modeling redox reactions during river bank filtration in an alluvial aquifer of the Lot River, France. *Environ. Sci. Technol.* **2008**, *42*, 2793–2798. [CrossRef] [PubMed]

49. Maeng, S.K.; Ameda, E.; Sharma, S.K.; Grützmacher, G.; Amy, G.L. Organic micro-pollutant removal from wastewater effluent-impacted drinking water sources during bank filtration and artificial recharge. *Water Res.* **2010**, *44*, 4003–4014. [CrossRef] [PubMed]

50. Von Gunten, H.R.; Karametaxas, G.; Keil, R. Chemical processes in infiltrated riverbed sediments. *Environ. Sci. Technol.* **1994**, *28*, 2087–2093. [CrossRef] [PubMed]

51. Bourg, A.C.M.; Bertin, C. Seasonal and spatial trends in manganese solubility in an alluvial aquifer. *Environ. Sci. Technol.* **1994**, *28*, 868–876. [CrossRef] [PubMed]

52. Hoehn, E.; Zobrist, J.; Schwarzenbach, R.P. Infiltration of river water into groundwater—Hydrogeological and hydrochemical investigations in the Glattal. *Gas Wasser Abwasser* **1983**, *63*, 401–410. (In German)

53. Martin, J.M.; Meybeck, M. Elemental mass-balance of material carried by major world rivers. *Mar. Chem.* **1979**, *7*, 173–206. [CrossRef]

54. Jain, C.K.; Ram, D. Adsorption of lead and zinc on bed sediments of the river Kali. *Water Res.* **1997**, *31*, 154–162. [CrossRef]

55. Jain, C.K.; Sharma, M.K. Adsorption of Cadmium on bed sediments of River Hindon: Adsorption models and kinetics. *Water Air Soil Poll.* **2001**, *137*, 1–19. [CrossRef]

56. Förstner, U.; Salomons, W. Trace metal analysis on polluted sediments, II. Evaluation of environmental impact. *Environ. Technol. Lett.* **1980**, *1*, 506–517. [CrossRef]

57. Grischek, T.; Dehnert, J.; Nestler, W.; Treutler, H.C.; Freyer, K. Description of system conditions during bank filtration through accompanying drilling core investigations. In Proceedings of the 2nd Dresdner Grundwasserforschungstage, Dresden, Germany, 25 March 1993; pp. 207–220. (In German)

58. Paufler, S. Management of Riverbank Filtration Sites to Reduce Manganese Concentrations in Raw Water. Diploma Thesis, Dresden University of Applied Sciences, Dresden, Germany, 2015. (In German)

59. Nirel, P.M.V.; Morel, F.M.M. Pitfalls of sequential extractions. *Water Res.* **1990**, *24*, 1055–1056. [CrossRef]

60. Van Vliet, M.T.H.; Franssen, W.H.P.; Yearsley, J.R.; Ludwig, F.; Haddeland, I.; Lettenmaier, D.P.; Kabat, P. Global river discharge and water temperature under climate change. *Glob. Environ. Chang.* **2013**, *23*, 450–464. [CrossRef]

61. Abdalla, F.A.; Shamrukh, M. Quantification of river Nile/Quaternary aquifer exchanges via riverbank filtration by hydrochemical and biological indicators, Assiut, Egypt. *J. Earth Syst. Sci.* **2016**, *125*, 1697–1711. [CrossRef]

62. Buschmann, J.; Berg, M.; Stengel, C.; Sampson, M.L. Arsenic and manganese contamination of drinking water resources in Cambodia: Coincidence of risk areas with low relief topography. *Environ. Sci. Technol.* **2007**, *41*, 2146–2152. [CrossRef] [PubMed]

63. Brandhuber, P.; Clark, S.; Knocke, W.; Tobiason, J. (Eds.) *Guidance for the Treatment of Manganese*; Water Research Foundation: Denver, Colorado, 2013; 148p, ISBN 978-1-60573-187-2.

64. Rohns, H.P.; Forner, C.; Eckert, P.; Irmscher, R. Efficiency of riverbank filtration considering the removal of pathogenic microorganisms of the River Rhine. In *Recent Progress in Slow Sand and Alternative Biofiltration Processes*; Gimbel, R., Graham, N.J.D., Collins, M.R., Eds.; IWA Publishing: London, UK, 2006; pp. 539–546, ISBN 9781843391203.

65. Grischek, T.; Paufler, S. Prediction of iron release during riverbank filtration. *Water* **2017**, *9*, 317. [CrossRef]

66. Grischek, T.; Schoenheinz, D.; Eckert, P.; Ray, C. Sustainability of river bank filtration: Examples from Germany. In *Groundwater Quality Sustainability*; Maloszewski, P., Witczak, S., Malina, G., Eds.; Taylor & Francis Group: London, UK, 2012; pp. 213–227.

67. Soares, M. The influence of high infiltration rates, suspended sediment concentration and sediment grain size on river and lake bed clogging. Ph.D. Thesis, Technical University of Berlin, Berlin, Germany, 2015.

Article

A Water Quality Appraisal of Some Existing and Potential Riverbank Filtration Sites in India

Cornelius Sandhu [1,*], Thomas Grischek [1], Hilmar Börnick [2], Jörg Feller [3] and Saroj Kumar Sharma [4]

1   Division of Water Sciences, University of Applied Sciences Dresden, D-01069 Dresden, Germany; thomas.grischek@htw-dresden.de
2   Institute for Water Chemistry, TU Dresden, D-01069 Dresden, Germany; hilmar.boernick@tu-dresden.de
3   Faculty of Agriculture, Environment & Chemistry, University of Applied Sciences Dresden, D-01069 Dresden, Germany; joerg.feller@htw-dresden.de
4   Environmental Engineering and Water Technology Department, UNESCO-IHE Institute for Water Education, P.O. Box 3015, 2601 DA Delft, The Netherlands; s.sharma@un-ihe.org
*   Correspondence: cornelius.sandhu@htw-dresden.de; Tel.: +49-351-462-2661

Received: 14 November 2018; Accepted: 25 January 2019; Published: 28 January 2019

Abstract: There is a nationwide need among policy and decision makers and drinking water supply engineers in India to obtain an initial assessment of water quality parameters for the selection and subsequent development of new riverbank filtration (RBF) sites. Consequently, a snapshot screening of organic and inorganic water quality parameters, including major ions, inorganic trace elements, dissolved organic carbon (DOC), and 49 mainly polar organic micropollutants (OMPs) was conducted at 21 different locations across India during the monsoon in June–July 2013 and the dry non-monsoon period in May–June 2014. At most existing RBF sites in Uttarakhand, Jammu, Jharkhand, Andhra Pradesh, and Bihar, surface and RBF water quality was generally good with respect to most inorganic parameters and organic parameters when compared to Indian and World Health Organization drinking water standards. Although the surface water quality of the Yamuna River in and downstream of Delhi was poor, removals of DOC and OMPs of 50% and 13%–99%, respectively, were observed by RBF, thereby rendering it a vital pre-treatment step for drinking water production. The data provided a forecast of the water quality for subsequent investigations, expected environmental and human health risks, and the planning of new RBF systems in India.

Keywords: bank filtration; drinking water treatment; inorganic chemicals; organic micropollutants; Ganga; Yamuna; Damodar

1. Introduction

The substantial discharge of untreated to partially treated industrial and domestic wastewater into surface water (SW) in India, accompanied by the very high turbidity during monsoon, frequently interrupt the production of drinking water by conventional plants. These plants directly abstract surface water and treat it by flocculation, sedimentation, rapid sand filtration, and disinfection. By using wells installed in the banks of flowing rivers, riverbank filtration (RBF) combines the advantage of easy access to large volumes of SW with the benefit of natural filtration during aquifer passage. Field investigations conducted mainly on urban drinking water production systems at various locations across India have confirmed that there is a large potential to use RBF as an alternative or a supplement to directly abstracted SW for drinking water production [1]. The main advantage of using RBF is that it provides an ecosystem service by effectively removing pathogens and turbidity, especially during the monsoon [2]. A significant removal of total coliforms, *Escherichia coli* (*E. coli*), turbidity, adenoviruses, and noroviruses by up to 90%–99.99% ($\geq$4 Log_{10} removal) is attained at RBF sites in northern India [2–7]. This is due to the superior surface water quality in Uttarakhand in contrast to the extremely polluted (with domestic sewage and industrial wastewater) stretch of the Yamuna river in the central part of Delhi (downstream of Uttarakhand). Other key water quality benefits of RBF are the removal of dissolved organic carbon (DOC) and organic contaminants, which are often responsible for the color of water. High concentrations of DOC and organic contaminants require high doses of chlorine and thereby create a greater risk for formation of carcinogenic disinfection byproducts, as reported for an RBF site in Mathura by the Yamuna, 150 km downstream of Delhi [8,9]. Furthermore, the use of RBF for rural water supply in the southwest Indian state of Karnataka has demonstrated a removal of total coliforms and *E. coli* of 1–3 Log_{10} and 2–4 Log_{10}, respectively [10,11].

However, RBF does not present an absolute barrier to other substances of concern (e.g., ammonium), and some inorganic elements (e.g., arsenic) may even be mobilized, as has been observed in central Delhi. There, infiltrating sewage-contaminated river water is the primary source of the ammonium contamination in the aquifer (35 mg/L), leading to reducing conditions that probably trigger the release of geogenic arsenic (0.146 mg/L) [12]. In light of the growing concern of emerging pollutants in the environment, recent studies on the occurrence of organic micropollutants (OMPs) in SW and their removal by RBF in Delhi [13–16] and Mathura [15,16], and potentially by RBF in Agra [16], have confirmed that the compounds with the highest relevance at these sites are diuron (37%–91% removal by RBF with respect to source river water concentration), 1H-benzotriazole (77%–98%), acesulfame, theophylline (56%–99%), diclofenac (37%–80%), gabapentin (91%–100%), and paracetamol (46%–50%) [16]. Overall, most studies have concluded that RBF is advantageous as a pretreatment step that improves water quality compared to directly abstracted and conventionally treated SW.

The water quality of only a few RBF sites (e.g., Uttarakhand, Delhi, Mathura, Agra, and rural Karnataka) have been monitored for periods long enough to include seasonal effects (≥ 1 year). Other than these sites, very limited holistic and systematic water quality information, especially for concentrations of inorganic and organic substances (including OMPs), is available for existing and potential RBF sites in other locations in India. Field visits to some conventional drinking water treatment plants, which directly abstract surface water, have shown that the main quality parameters that are usually and routinely determined are physical field parameters, total hardness and total alkalinity, major anions, and often only the presence or absence of bacteriological coliform indicators. Moreover, bacteriological indicators are only occasionally quantified as counts/100 mL of sample. These limited number of parameters do not cover the entire list of parameters in the Indian drinking water standards [17]. Moreover, despite the advantages of being a sustainable natural process, an element of integrated water resources management, and a component of managed aquifer recharge [18,19], RBF is intentionally used for pretreatment only at some places in India, resulting in a low portion (<0.1%) of drinking water produced therefrom [20].

In order to effectively implement RBF by starting with a suitable site for an exploratory well and to subsequently make an informed decision to expand it into a full-scale RBF system, knowledge of site-specific geohydraulic and water quality parameters is essential. Therefore, the objective of this article was to obtain an initial assessment of water quality parameters for the selection and subsequent development of new riverbank filtration sites in India. Post-treatment options have been discussed for sites where inorganic parameters exceed the Indian Standard or WHO guideline value for drinking water [17,21]. Mostly urban, but also some rural sites located in countrywide diverse hydroclimatic conditions, were investigated. The design parameters of these RBF systems and a summary of their hydrogeological settings were presented in a previous publication [22] and are thus not repeated here. Present post-treatment conducted and future post-treatment requirements at these sites supplement this information.

2. Materials and Methods

2.1. General Sampling Strategy

A random snapshot screening of water quality parameters (as specified subsequently), including instant physical field parameters, major ions, inorganic elements, DOC, and 49 mainly polar OMPs (those of environmental relevance in Europe) was conducted at a total of 21 different locations across India during the monsoon in June–July 2013 and the dry non-monsoon period in May–June 2014 (Figure 1). In 2013, 49 samples from 17 locations were collected (Figure 1, green circles/squares). In 2014, 75 samples from 11 locations were taken (Figure 1, red circles/squares). Out of the 21 locations sampled, 7 locations were sampled both during wet (monsoon, 2013) and dry (non-monsoon, 2014) seasons (Figure 1, orange circles).

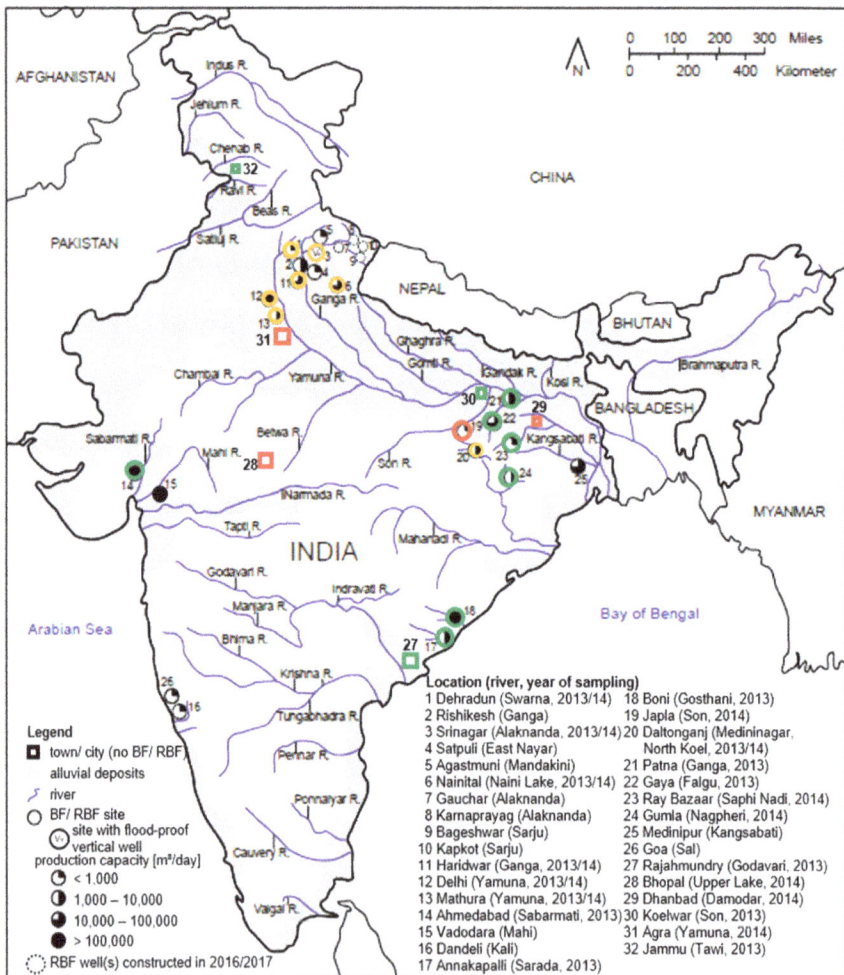

Figure 1. Locations having riverbank filtration (RBF) wells in India and those sampled in 2013 (monsoon, green circles), 2014 (non-monsoon, red circles), and both in 2013 and 2014 (orange circles), including six locations where it was not conclusively established that the existing wells abstract bank filtrate (squares) (modified from [22]).

A "random snapshot screening" means that in each case a single (random) sample was taken, but not a composite sample over a certain time period, from different sites or water depths, which were screened for a set of different parameters and different single compounds (targeted screening for higher numbers of OMPs). The sampling locations were selected using a four-stage methodology derived for the investigation of potential and existing RBF sites, for which no or only limited data exist in the public domain [20]. Accordingly, the sampling locations were selected within the first stage "initial site assessment" and with the support of the National Institute of Hydrology in Roorkee and its regional centers across India (see "Acknowledgements"). This included site visits and a visual assessment, interaction with the local water supply and research organizations, the subsequent documentation of verbal and onsite archived information, and finally the random snapshot water quality sampling. Sampling from rural areas was conducted to take into account potential non-point pollution of SW [23], especially as a result of monsoon runoff.

In earlier field investigations [20], no health-relevant organic trace compounds were detected using non-target screening analysis with gas chromatography–mass spectrometry (GC/MS) of Ganga River water in Haridwar in 2005. Consequently, a mobile solid-phase extraction (SPE) unit was developed to enrich the sample onsite for subsequent analyses in the laboratory. The mobile SPE unit was developed in order to mitigate the effects of long transport times of the samples to the laboratory, which usually occur between sampling and subsequent laboratory enrichment and analysis in India. Furthermore, enrichment of the sample is very important because of the low concentrations of OMPs found in the environment [24]. Thus, while volumes of up to 500 mL per sample suffice for the enrichment of OMPs from moderately to highly polluted waters, nearly 1 L is required per sample for waters expected to have a low pollution. Additionally, the (air) transport of such large-volume samples to distant laboratories is extremely limited. Depending on the group of parameters sampled for (inorganic chemical parameters, DOC, and OMPs), around 130 samples in total were collected from SW bodies, RBF wells, and in some case ambient groundwater (GW) from sites in the states of Uttarakhand, Uttar Pradesh, Bihar, Jharkhand, Andhra Pradesh, Madhya Pradesh, Gujarat, and the cities of Jammu and Delhi during both sampling campaigns in 2013 and 2014.

2.2. Sampling and Analysis of Water for Inorganic Chemical Parameters and DOC

The temperature, pH, dissolved oxygen, and electrical conductivity (EC) of the water samples were determined onsite using a WTW multi 3430 instrument (Wissenschaftlich-Technische Werkstätten GmbH (WTW), Weilheim, Germany). Two 100-mL water samples were collected from each source of water at the sampling location, for DOC and for ions (including inorganic trace elements). All samples were filtered with a 0.45-µm Whatman syringe filter. Subsequently, the samples for the determination of DOC were conserved with nitric acid. The analyses for anions and trace metals were conducted using inductively coupled plasma optical emission spectrometry (ICP-OES, Spektrometer Optima 4300 DV, PerkinElmer) in a radial viewing configuration in the Division of General and Inorganic Chemistry at the Faculty of Agriculture, Environment & Chemistry in the University of Applied Sciences Dresden. DOC analyses were conducted by the Institute for Water Chemistry (IWC) at the TU Dresden.

A spectrum of 18 inorganic (trace) elements, including trace metals and radionuclides, were determined, which comprised iron (Fe), manganese (Mn), strontium (Sr), barium (Ba), zinc (Zn), silicon (Si), chromium (Cr), cobalt (Co), nickel (Ni), copper (Cu), aluminum (Al), selenium (Se), lead (Pb), cadmium (Cd), mercury (Hg), arsenic (As), silver (Ag), and nickel (Ni). The objective was to determine if the concentrations of these elements exceeded the guideline value [17,21] or if the concentrations were unusually high, thereby indicating a possible contamination.

2.3. Water Sampling and Analyses of Organic Micropollutants

The enrichment of OMPs was conducted onsite by the mobile SPE unit from 0.5 L–1 L filtered water samples with an enrichment factor of 1000 [25]. One-hundred and twenty-four water samples were collected and enriched in total (in 2013 and 2014). Cartridges (OASIS, Waters, Milford, MA, USA) were used for the enrichment. Subsequently, a target screening analysis using RP-HPLC (Agilent, 1100, Santa Clara, CA, USA) and ESI-MS/MS (QTRAP®, Q3200, Sciex, Framingham, MA, USA) was conducted for 49 polar organic compounds (pharmaceuticals, pesticides and transformation products, antibiotics, medical contrast media, corrosion inhibitors, and stimulants such as caffeine that are not micropollutants). The analyses for the OMPs were conducted by the IWC [26].

3. Results and Discussion

3.1. Inorganic Chemical Parameters

The concentrations of major ions and 18 other inorganic elements in surface water analyzed at most RBF locations in Uttarakhand, Jharkhand, Andhra Pradesh, and Bihar did not exceed the Indian drinking water guideline values [17] and thus the surface water quality is generally suitable for RBF in terms of these parameters (Table 1). However, the extremely high salinity of Yamuna river (R.) water (EC 1665–1700 µS/cm) and adjacent GW (EC 1455–3400 µS/cm) between Delhi and Agra gives the drinking water derived from the waterworks a brackish taste.

The concentrations of mainly Fe, Mn, and As exceeded the Indian drinking water guideline requirement (acceptable limit, [17]) only in some source waters and occasionally in drinking water, and in some cases the permissible limit in the absence of an alternate source (Table 2). Mn is naturally occurring in many surface water and groundwater sources, particularly in anaerobic or low oxidation conditions [21]. This explains its comparatively high concentration, especially in the Yamuna R. in Delhi, because of the very high input of wastewater (industrial and domestic) and correspondingly very low dissolved oxygen concentrations of 0.1–0.3 mg/L in river and groundwater (hand pumps). The Indian standard for drinking water [17] requires an acceptable limit of 0.1 mg/L and a permissible limit in the absence of an alternate source of 0.3 mg/L for Mn. Mn can be removed by chlorination followed by filtration [21], as is practiced in Jharkhand, where the bank filtrate subsequently undergoes post-treatment comprising aeration, flocculation, rapid sand filtration, and disinfection. Similarly, the water from a radial collector well (RCW) supplying raw water for drinking to the township of the oil refinery in Mathura also undergoes similar conventional post-treatment [8,9].

Arsenic exceeded the required acceptable limit of 10 µg/L for drinking water [17,21] in groundwater (hand pumps) near the Yamuna riverbank in central Delhi and Mathura (Table 2). The As concentrations of 44–66 µg/L found in the water from hand pumps in central Delhi (Table 2) were consistent in magnitude to concentrations of 27–56 µg/L that were determined by Lorenzen et al. [27] for shallow depths (6–13 m below ground level) in the same area.

In the RBF RCW constructed within the riverbed in Mathura, the acceptable limit was exceeded only during the non-monsoon 2014 (32 µg/L), but not in monsoon 2013. However, the concentration was a <50 µg/L limit in the absence of an alternate source [17]. This indicates a decrease in concentration by mixing with a greater portion of bank filtrate abstracted on account of a higher hydraulic head in monsoon. This is a positive effect of RBF, as was also observed for the Palla RBF site in Delhi where As was found to be <10 µg/L due to a high portion of bank filtrate abstracted from the high-capacity vertical well field [27]. In contrast, ambient groundwater in different areas in Delhi was found to have high As concentrations (range 17–100 µg/L, mean 40 µg/L), as determined by Lalwani et al. [28]. For all other sites, the arsenic concentration was a <10 µg/L detectable limit.

Table 1. Field parameters and major ions of source water, RBF well water, and drinking water at various locations sampled in 2013 and 2014 (n = 1 for each source at each location).

Location (State) Season	Source of Water Sample	Name of SW Body / Treatment	T_W (°C)	pH	EC (μS/cm)	DO	Ca²⁺	Mg²⁺	Na⁺ (mg/L)	K⁺	Cl⁻	NO₃⁻	SO₄²⁻
Dehradun (UK) non-mon. 2014	SW	Song & Asan	25.0	7.8	n.d.	n.d.	92	38	3.0	1.4	3.4	4.9	301
	DW	SWA	26.0	7.7	n.d.	n.d.	96	40	3.4	1.5	5.2	2.7	343
Jammu (JK) mon. 2013	SW	Tawi	27.0	8.5	539	n.d.	26	5.2	3.9	1.9	<5	<1	<10
	Potentially RBF from Tawi		21.9	8.3	591	4.0	25	6.7	2.7	0.8	n.d.	<5	<10
Mathura (UP) non-mon. 2014	SW	Yamuna	30.8	7.9	1700	6.3	76	34	199	21	280	14	97
	RBF	Yamuna	34.7	7.3	1455	0.1	69	29	161	17	215	4.4	69
	DW (RBF, small RO)		38.2	6.4	150	1.1	2	0.6	17	1.7	15	<1	<5
Agra (UP) non-mon. 2014	SW	Yamuna	29.6	8.8	1665	18.5	77	35	206	21	290	22	92
	DW	SWA	30	7.5	1645	0.1	73	33	199	20	279	13	97
	GW (private well)		28.5	6.8	3400	3.3	121	116	372	5.1	661	74	353
Bhopal (MP) non-mon. 2014	SW	Bhopal L.	29.6	8.1	337	7.1	19	7.6	11	2.8	0.2	n.d.	<5
	DW	SWA	30.4	8.1	257	7.2	22	8.2	12	2.5	12	<1	7.2
Daltonganj (JH) non-mon. 2014	SW	N. Koel	27.7	7.6	251	5.0	22	6.7	17	2.2	5.3	<1	6.8
	RBF	N. Koel	31.2	7.6	272	5.2	22	5.9	15	2.0	5.9	<1	8.0
	DW	RBF-CT	30.3	7.9	255	6.9	23	6.9	18	2.5	7.1	<1	9.1
Japla (JH) non-mon. 2014	SW	Son	32.8	8.6	172	8.1	16	5.1	9.0	2.1	5.9	<1	11
	RBF	Son	32.0	8.5	177	7.5	17	5.2	9.7	7.2	9.5	<1	13
	DW	RBF-CT	30.6	8.3	193	7.0	17	4.9	9.3	2.1	5.3	<1	10
Ray Bazaar (JH) non-mon. 2014	SW	Saphi	37.6	8.6	253	8.8	23	6.9	17	2.4	8.8	<1	14
	RBF	Saphi	26.8	7.2	313	4.4	29	8.1	19	3.4	11	1.2	14
	DW	RBF-CT	29.2	7.9	325	7.1	30	7.8	19	3.8	12	1.1	17
Gumla (JH) non-mon. 2014	SW	Nagpheri	31.0	8.2	131	8.1	12	3.0	9.5	1.8	5.5	<1	<5
	RBF	Nagpheri	30.5	7.5	135	5.6	12	3.0	9.6	2.0	5.5	<1	<5
	DW	RBF-CT	30.3	7.3	141	6.9	14	2.7	8.5	2.3	5.2	1.3	8.4

Table 1. Cont.

Location (State) Season	Source of Water Sample	Name of SW Body / Treatment	Tw °C	pH -	EC µS/cm	DO	Ca²⁺	Mg²⁺	Na⁺ (mg/L)	K⁺	Cl⁻	NO₃⁻	SO₄²⁻
Dhanbad/Jamadoba (JH) non-mon. 2014	SW	Damodar	34.6	**8.9**	347	11.8	27	12	20	6.3	15	2.1	60
	DW	SWA	34.6	8.1	371	7.9	26	13	20	6.3	15	3.0	61
Gaya (B) mon. 2013	SW	Falgu	36.0	**9.2**	185	8.1	18	5.3	7.8	2.9	6.2	<1	<10
	RBF	Falgu	32.0	7.9	155	n.d.	62	17	11	1.9	<5	<1	<10
Anakapalli (AP) mon. 2013	SW	Sarada	31.5	**9.0**	398	12.5	20	18	35	4.5	35	<1	17
	RBF	Sarada	30.7	8.1	540	7.1	25	17	29	5.6	33	<1	17
Ahmedabad (GJ) mon. 2013	SW	Sabarmati	30.1	**9.2**	558	9.6	20	8	6.5	1	6.1	<1	<10
	DW	CT-RBF/DIS	29.8	**9.1**	611	12.7	21	7.6	6.2	1	6.4	<1	<10

Values highlighted in bold font: Values exceeding the permissible limit in the absence of an alternate source [17] or exceptionally high values for electrical conductivity (EC) resulting in a noticeable saline taste. Abbreviations: RBF: Bank filtration well water; DW: Drinking water (after post-treatment/disinfection); GW: Groundwater; SW: Surface water; SWA: Surface water abstraction followed by conventional treatment; RBF-CT: Bank filtration followed by conventional treatment; DIS: Only disinfection by chlorination as post-treatment; RO: Reverse osmosis; states: UP: Uttar Pradesh; UK: Uttarakhand; JH: Jharkhand; B: Bihar; AP: Andhra Pradesh; GJ: Gujarat; seasons: non-mon. 2014: Non-monsoon (May–June 2014); mon. 2013: Monsoon (June–July 2013); n.d.: Not determined.

142

Table 2. Summary of inorganic parameters exceeding acceptable limits [17] determined in 2013 and 2014.

Element	Source Water	Dissolved Concentration (mg/L) *	Location and Description
Fe	GW	0.7–11	Delhi (central), handpumps on Yamuna River east bank
	RBF	2	Mathura (only in monsoon 2013)
	SW	0.3–1.4	Agra, Keetham Lake water intake for Mathura and Yamuna River
	SW, RBF, GW	0.2–0.6	Delhi (central), Yamuna River, handpumps on river east bank and 5 RBF RCW
	RBF	0.6–1	Mathura, in monsoon 2013 and non-monsoon 2014
Mn	GW	0.3	Mathura, hand pump near RBF well, in monsoon 2013
	GW	1.3	Gaya, mixed sample from various vertical wells within Falgu riverbed downstream of city possibly also receiving wastewater, in monsoon 2013
	RBF and DW	0.4–0.9	Ray Bazaar, RBF RCW (0.9 mg/L) within riverbed and subsequently conventionally treated DW (0.4 mg/L), in non-monsoon 2014
	SW	0.2	Chas (Bokaro–Dhanbad area, Jharkhand), Garga River at confluence with Damodar River, receiving substantial amount of wastewater from Bokaro Steel City
	RBF	0.2–0.3	Nainital, vertical wells numbers 2 and 4
As	SW	30 µg/L	Koelwar, Son River water near sand mining site within riverbed
	GW	44-66 µg/L	Delhi, hand pumps on Yamuna riverbank
	GW	20 µg/L	Mathura, hand pumps on Yamuna riverbank near RBF well, in monsoon 2013
	RBF	32 µg/L	Mathura, RBF RCW within Yamuna riverbed, in non-monsoon 2014 only
Al	SW	30-40 µg/L	Coastal Andhra Pradesh, Godavari, Sarada and Thatpudi rivers
	DW	278 µg/L	Bhopal, conventionally treated drinking water from PHED DW plant

* Except for As and Al; RBF: Bank filtration well water; DW: Drinking water; GW: Groundwater; SW: Surface water; RBF RCW: Riverbank filtration radial collector well; PHED: Public Health and Engineering Department.

Aluminum was found above the detectable limit of 10 µg/L mainly in surface waters, and substantially exceeded (up to 278 µg/L) the acceptable required drinking water guideline value of 30 µg/L [17] only in one drinking water sample from the water treatment plant in Bhopal that conventionally treats surface water (from Bhopal Upper Lake) using aluminum-based coagulants (Table 2). While the Indian Drinking Water Standard permissible limit [17] in the absence of an alternate source is 200 µg/L, the WHO guideline [21] advocates a practicable concentration of ≤100 µg/L for large water treatment facilities using aluminum-based coagulation processes. Otherwise, Al concentrations ≥30 µg/L were mainly found in the sampled rivers of Andhra Pradesh in monsoon 2013 (Table 2). The detectable naturally occurring Al in surface water in Andhra Pradesh was attributed to the surface runoff from the substantial bauxite deposits found in the Eastern Ghats (hills) through which these rivers flow. Al concentrations were below the detectable limit of 10 µg/L in most other samples from RBF and groundwater production wells and in drinking water.

At all sites sampled in Figure 1, barium (0.04–0.49 mg/L), cadmium (<1 µg/L), chromium (<2 µg/L), copper (<15 µg/L), nickel (<4 µg/L), and selenium (<40 µg/L) were below the WHO guideline limits [21] of 1.3 mg/L, 3 µg/L, 50 µg/L (provisional), 2 mg/L, 70 µg/L, and 40 µg/L (provisional), respectively, in all samples. In all samples, cobalt was <2 µg/L (detection limit), and zinc ranged from <4 µg/L up to 0.71 mg/L. As there are no guideline limits for cobalt and zinc in drinking water, these two parameters, as well as Ba, Cd, Cr, Cu, Ni, and Se, are not of concern to human health at the sampled sites.

3.2. Dissolved Organic Carbon

Of all the surface waters sampled, it was found that the stretch of the Yamuna River starting in central Delhi (ITO Bridge) up to ~200 km downstream in Agra, had the highest DOC concentration of around 12 mg/L (Figure 2). In Figure 2, it can be observed that the DOC concentrations in the Yamuna River at Delhi and Mathura were significantly lower during the monsoon in 2013 compared to the non-monsoon in 2014. The annual monsoon thus had a positive effect on highly polluted surface waters (e.g., in Delhi and Mathura) in terms of lowering the DOC concentration by dilution. On the other hand, for surface waters already having a relatively low ambient or background DOC concentration, such as that observed in Uttarakhand (Haridwar and Nainital in Figure 2) and Jharkhand (Daltonganj in Figure 2 and Dhanbad in Figure 3), it may even increase during monsoon, probably due to surface runoff.

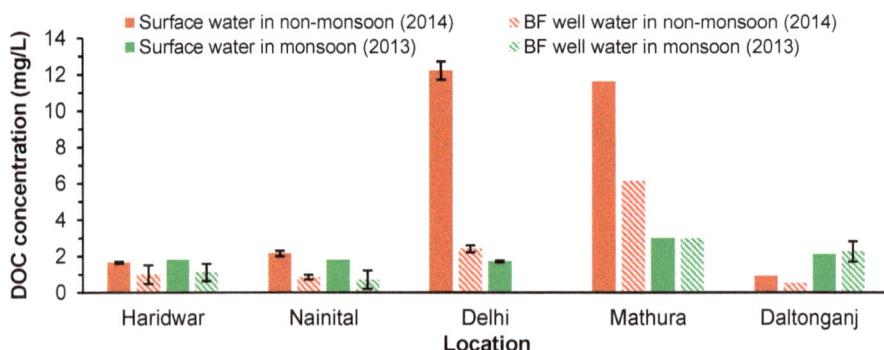

Figure 2. Median dissolved organic carbon (DOC) concentrations in surface water and bank filtration wells at selected sites, except for the RBF wells in Delhi, where no samples were collected in monsoon (2013). Error bars indicate the standard deviation for numbers of samples (n_m: Monsoon; n_{nm}: Non-monsoon) ≥ 2. Haridwar: Ganga, $n_m = 1$, $n_{nm} = 2$; RBF, $n_m = 9$, $n_{nm} = 13$; Nainital: Naini Lake, $n_m = 1$, $n_{nm} = 2$; BF, $n_m = 3$, $n_{nm} = 6$; Delhi: Yamuna, $n_m = 2$, $n_{nm} = 2$; RBF, $n_{nm} = 5$; Mathura: Yamuna, $n_m = 1$, $n_{nm} = 1$; RBF, $n_m = 1$, $n_{nm} = 1$; Daltonganj: North Koel, $n_m = 1$, $n_{nm} = 1$; RBF, $n_m = 2$, $n_{nm} = 1$.

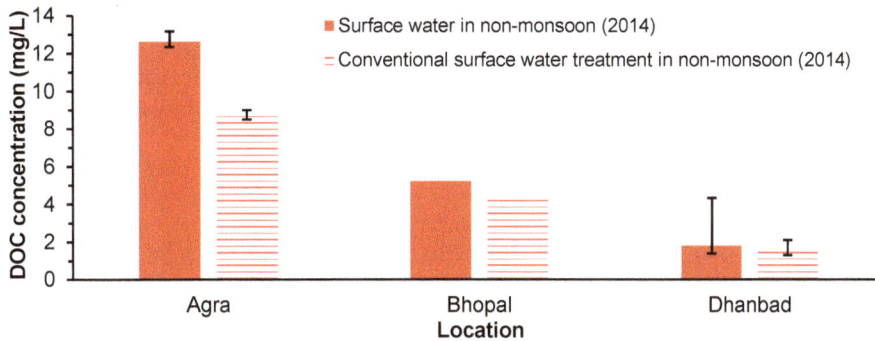

Figure 3. Median DOC concentrations in surface water and conventionally treated drinking water derived from direct surface water abstraction at selected sites in non-monsoon 2014. Error bars indicate the standard deviation for number of samples (n) $\geq$ 2. Agra: Yamuna, n = 3, n_{CT} = 2; Bhopal: Upper Lake, n = 1, n_{CT} = 1; Dhanbad: Damodar, n = 3, n_{CT} = 2.

In Figure 2, the DOC concentration was observed to be slightly lower in the bank filtration wells in Haridwar and Nainital, which have a caisson and vertical well design, respectively, when compared to surface water concentrations. At these locations, the travel time of the bank filtrate was longer (weeks to months) compared to Daltonganj (Figure 2). In Daltonganj, the travel time was only in the range of minutes to hours on account of the radial collector wells being installed within the riverbed at a shallow depth (1–6 m, [22]), and consequently no significant removal of DOC was observed. As the RBF well in Mathura also has a radial collector design, albeit with longer travel time (1.5–3 days, [8]) compared to Daltonganj, the DOC concentration in well water was similar to the river water during monsoon. Nevertheless, the advantage of RBF in Mathura is visible in Figure 2 during non-monsoon conditions, when nearly 50% DOC removal occurred.

The DOC concentration in directly abstracted surface water and drinking water derived thereof by conventional treatment in non-monsoon (2014) was compared for the cities of Agra, Bhopal, and Dhanbad (Figure 3). It was observed that at least for highly polluted surface waters with a high DOC concentration (e.g., Yamuna River in Agra), the removal of DOC by conventional treatment systems (which do not use activated carbon) was lower than that observed for RBF systems with similar source water quality, such as by the RBF wells in Delhi and Mathura upstream of Agra (in Figure 2).

The DOC concentration in surface water at the RBF sites of Haridwar (Ganga River and Upper Ganga Canal), Srinagar (Alaknanda River), and Nainital (Nainital Lake) in Uttarakhand, and in the Asan River that flows past the industrial area in Dehradun city, was relatively low at 1.1–2.4 mg/L, with only a minor difference between the non-monsoon and monsoon seasons. In the corresponding RBF wells, the DOC concentrations were 0.4–2.3 mg/L and generally lower than the respective surface water, except for one RBF well in Haridwar and Srinagar that had slightly higher DOC concentrations (compared to their surface water sources) of 2.5 and 2.7 mg/L, respectively. The higher DOC concentration in the RBF well in Haridwar could be attributed to human activities, such as washing and bathing, that take place at the well.

At the RBF sites of Daltonganj (North Koel R.), Gumla (Nagpheri R.), Ray Bazaar (Saphi Nadi R.), and Japla (Son R.) in Jharkhand, the surface water and water from the RBF wells contained DOC in the range of 0.9–3 mg/L and mostly 0.5–1.7 mg/L, respectively, with higher DOC observed in monsoon (Figure 2, location Daltonganj). The water from two wells showed exceptionally higher DOC of 2.7 and 3 mg/L. However, while these RBF sites are generally affected by low surface water pollution on account of them being located mostly in the upstream areas of the towns, and the impact of anthropogenic activities (agricultural and industrial activities) is low, the observed DOC removal was generally low due to the very short travel time of the bank filtrate to the wells [22].

In the river water sampled in coastal Andhra Pradesh (Figure 1, locations 17, 18, and 27), DOC in monsoon (June 2013) was 3–4.6 mg/L. In the RBF well water in Annakapalli (Figure 1, location 17), DOC was 3 mg/L. The RBF well in Annakapalli is similar in design to that in Daltonganj (Figure 1, location 20, [22]), and thus was expected to have short travel times of bank filtrate in the range of hours. Therefore, a large removal of DOC was not expected.

In monsoon 2013 at other locations in India, the DOC in surface water in Gaya (Falgu R.), Patna (Ganga R.), Koelwar (Son R.), and Ahmedabad (Sabarmati R.) was 1.9–2.3 mg/L and 3.2–3.8 mg/L upstream and downstream, respectively, of the Tawi R. in Jammu, with lower concentrations of 0.9–1.5 mg/L in the RBF and groundwater abstraction wells. In these towns, the RBF systems are also located in the upstream areas, and consequently a lower anthropogenic impact is noticeable.

However, two different RBF site examples to those discussed previously are in Daltonganj and Gaya, where the RBF wells have been inappropriately sited within the riverbed and downstream of the towns such that they are directly impacted by wastewater discharged locally or upstream. Consequently, not only was the DOC concentration in the surface water higher, but due to the discharge and accumulation of domestic wastewater directly at and around the wells, which can also be potentially contaminated by flood water, the DOC concentration in the abstracted well water was nearly 4 mg/L.

3.3. Organic Micropollutants

Out of 49 mainly polar OMPs screened, only 22 could be detected (Table 3). Although not regarded an OMP, caffeine was detected in nearly all surface water samples and in many groundwater, RBF well water, and treated water samples in concentrations ranging from <10 ng/L up to 400 ng/L (ubiquitous presence, hence not in Table 3). The occurrence of these 22 OMPs and caffeine is summarized in Table A1. Overall, 26 OMPs were not detected (Table A2).

Similar to the highest DOC concentrations found in the Yamuna R. water between Delhi and Agra, the highest occurrence and also nearly the highest concentrations of OMPs comprising pharmaceutical, medical contrast media, personal care product, corrosion inhibitor, insecticide, and herbicide compounds were also found along this stretch in the river water and also partly in the RBF RCW in Mathura and handpumps located near the river (groundwater, Table 3). The removal efficiency of OMPs by RBF can be demonstrated by taking the Mathura RBF site as an example, where the Yamuna R. had comparably (to Delhi and Agra) high concentrations of OMPs (Figure 4).

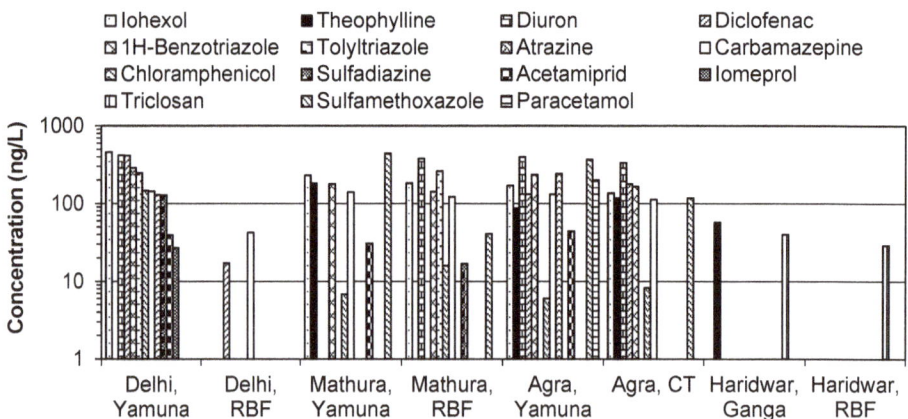

Figure 4. Median concentrations (except for Mathura) of organic micropollutants (OMPs) in surface water, RBF well water (RBF), and conventionally treated (CT) drinking water at selected sites sampled in non-monsoon 2014. Number of samples (*n*): Delhi: Yamuna, *n* = 2, n_{RBF} = 5; Mathura: Yamuna, *n* = 1, n_{RBF} = 1; Agra: Yamuna, *n* = 3, n_{CT} = 2.

Table 3. Concentrations of OMPs at selected sites where their presence was detected in non-monsoon 2014, given as median and maximum (separated by /) values in ng/L. Values in parentheses () indicate the number of samples these values are based on (if ≠ n in column 2), or the number of samples wherein the OMP was not detected (n.d.).

Location, River	Source of Sample/n	1H-Benzotriazole	Acetamiprid	Atrazine	Carbamazepine	Chloramphenicol	Cotinine	Diclofenac	Diuron	Ibuprofen	Iohexol	Iomeprol	Iopromide	Naproxen	Paracetamol	Phenazone	Phenobarbital	Simazine	Sulfadiazine	Sulfamethoxazole	Theophylline	Tolyltriazole	Triclosan
Limit of detection		30	10	10	30	30	10	10	10	10	10	10	10	10	10	10	10	10	10	10	10	10	30
Delhi (central), Yamuna	SW/2	285/320	39.0/49.9	145/153	143/145	129 (1)	d.	413 (1)	418 (1)	d.	457/465	26.7/30.8	52.3/54.9	n.d.	d.	d. & 12.2	220/371	10.5/10.8	127/179	d.	d.	245/253	d.
	GW/2	n.d.	n.d.	n.d.	64.7/78.8	n.d.	n.d.	307/382	n.d.	25.9 (1)	n.d.	n.d.	n.d.	161 (1)	n.d.	d.	133/177	n.d.	28.5/44.1	n.d.	34.8 (1)	n.d.	n.d.
	RBF/5	n.d.	n.d.	n.d.	42.3 (2)	n.d.	n.d.	17.1/24.8 (3)	n.d.	n.d.	n.d.	n.d.	n.d.	n.d.	n.d.	n.d.	n.d.	n.d.	n.d.	d.	n.d.	n.d.	n.d.
Mathura, Yamuna	SW/1	177	30.7	d.	140	n.d.	46.2	d.	d.	n.d.	229	d.	12.4	n.d.	n.d.	158	39.0	n.d.	d.	440	182	d.	n.d.
	RBF/1	142	n.d.	15.9	122	n.d.	14.4	d.	379	n.d.	182	d.	15.5	n.d.	n.d.	40.0	n.d.	n.d.	16.8	40.9	n.d.	259	n.d.
	RO/1	36.8	n.d.	n.d.	n.d.	n.d.	n.d.	43.9	23.7	n.d.	n.d.	n.d.	n.d.	n.d.	n.d.	n.d.	n.d.	n.d.	n.d.	d.	n.d.	43.2	n.d.
Agra, Yamuna	SW/3	226/240	43.9/72.5	d. (2) & 23.3	131/186	240 & n.d.(2)	44.9/69.8	131/246	398/463	d.	169/256	d.	d. (2) & 16.8	94.2 (1)	198 (1)	235/446	67.5/69.8	d. & n.d.(2)	n.d.	368/400	86.7/448	d.	d.
	CT/2	164/190	n.d.	d.	113/124	n.d.	20.6/23.2	177 (1)	333/418	n.d.	136/162	d.	d.	n.d.	n.d.	54.3 (1)	33.8/56.2	n.d.	n.d.	117/196	d. & 33.5	d.	n.d.
Japla (Jh.), Sone	SW/1	n.d.	n.d.	13.2	n.d.	n.d.	n.d.	n.d.	n.d.	n.d.	n.d.	n.d.	n.d.	n.d.	d.	n.d.	n.d.	n.d.	n.d.	n.d.	n.d.	n.d.	42.7
	RBF/1	n.d.	n.d.	13.6	n.d.	n.d.	18.8	n.d.	n.d.	108	n.d.	n.d.	n.d.	n.d.	76.2	n.d.	n.d.	n.d.	n.d.	n.d.	140	n.d.	n.d.
	CT/1	n.d.	n.d.	13.6	n.d.	n.d.	d.	n.d.	n.d.	n.d.	n.d.	n.d.	n.d.	n.d.	d.	n.d.	n.d.	n.d.	n.d.	d.	n.d.	n.d.	n.d.
DTJ (Jh.), N. Koel	SW/1	n.d.	n.d.	n.d.	n.d.	n.d.	13.5	n.d.	n.d.	n.d.	n.d.	n.d.	n.d.	n.d.	n.d.	n.d.	n.d.	n.d.	n.d.	d.	n.d.	n.d.	41.4
	RBF/1	n.d.	n.d.	n.d.	n.d.	n.d.	n.d.	n.d.	n.d.	n.d.	n.d.	n.d.	n.d.	n.d.	n.d.	n.d.	n.d.	n.d.	n.d.	d.	n.d.	n.d.	n.d.
	CT/1	n.d.	n.d.	n.d.	n.d.	n.d.	21.4	n.d.	n.d.	n.d.	10/100	n.d.	n.d.	n.d.	n.d.	n.d.	341	n.d.	n.d.	d.	17.8	n.d.	30.1
Dhanbad, Damodar	SW/4	n.d.	n.d.	n.d.	58.2 & n.d.(3)	n.d.	126 & n.d.(3)	46.9 & n.d.(3)	72.5 & n.d.(3)	n.d.	n.d.	n.d.	n.d.	n.d.	50.5/146	59.7 & n.d.(3)	25.8 & n.d.(3)	n.d.	n.d.	39.6 & n.d.(3)	94.3/n.d.(3)	88.9/102	n.d.
	CT/1	n.d.	n.d.	n.d.	n.d.	n.d.	n.d.	n.d.	n.d.	n.d.	d.	n.d.	n.d.	n.d.	n.d.	n.d.	n.d.	n.d.	n.d.	d.	n.d.	67.0	n.d.
Nainital	SW/2	n.d.	n.d.	n.d.	n.d.	n.d.	d. & 13.0	n.d.	89.5/112	n.d.	n.d.	n.d.	n.d.	n.d.	d. & 105	n.d.	n.d.	n.d.	n.d.	d.	d. & 35.1	n.d.	n.d.
	BF/6	n.d.	n.d.	n.d.	n.d.	n.d.	n.d.	n.d.	n.d.	n.d.	n.d.	n.d.	n.d.	n.d.	n.d.	n.d.	n.d.	n.d.	n.d.	n.d.	n.d.	n.d.	n.d.

Abbreviations: n.d.: Not detected; d.: Detected but not quantifiable (qualitative result only); SW: Surface water; GW: Groundwater; BF/RBF: bank filtration (lake)/riverbank filtration; RO: Household reverse osmosis unit; CT: Conventional treatment; DTJ: Daltonganj (Medininagar, Jharkhand); Dhanbad (Jharkhand).

In Figure 4, it is observed that the concentrations of some OMPs in the RBF well water were 13%–99% lower than river water (RCW design, fast travel time), whereas others were not present in well water or else detectable in very low concentrations that were not quantifiable. Although this observation was made from the interpretation of a limited number of samples (Figure 4) in this study (collected once during non-monsoon 2014), similar ranges for removal of OMPs by RBF were observed for the same locations in a subsequent study (September 2015 to June 2018) [16]. The higher concentrations of atrazine, diuron, iopromide, sulfadiazine, and tolyltriazole in RBF water (and diuron after reverse osmosis) in Mathura compared to river water could be attributed to the possible effect of high concentrations of these substances in landside groundwater [16]. Thus, in general, dilution effects (groundwater, monsoon) cannot yet be ruled out. Among the herbicides, atrazine was found in the Yamuna R. water at all the sampled locations, with the highest concentration of 153 ng/L being found in Delhi (Table 3). Although atrazine was found in Mathura and Agra, its concentration in the Yamuna R. water was very low and could not be quantified.

Cotinine, diuron, paracetamol, triclosan, theophylline, and sulfamethoxazole were present or only detectable in very low concentrations (but not quantifiable) in the surface water in Haridwar (Figure 4), Srinagar, and Nainital (Table 3) in Uttarakhand. Of these OMPs, only sulfamethoxazole was detectable in nearly all RBF water samples, albeit in unquantifiable amounts (<10 ng/L), and triclosan was present in Haridwar (29 ng/L) and in Srinagar (38 ng/L, only one out of three samples). No other OMPs were detectable in the RBF wells at these sites.

In the state of Jharkhand at the RBF site of the town of Japla, surrounded by predominantly rural and agricultural areas, atrazine was found in surface water, bank filtrate (RCW), and subsequently conventionally treated water with a concentration of 13–14 ng/L, indicating no removal, most likely also on account of the very short travel time of the bank filtrate due to the shallow RCW design of the wells. However, in the more industrialized and densely populated coal mining city of Dhanbad, a wide range of pharmaceutical compounds (paracetamol, sulfamethoxazole, phenazone, diclofenac, theophylline, tolyltriazole, and carbamazepine), contrast media (iohexol), and the herbicide diuron were found in concentrations ranging from 10 to 126 ng/L in the adjacent Damodar R. (Table 3). The Damodar R. receives wastewater from the industrial cities of Dhanbad and Bokaro. In the conventionally treated drinking water derived by direct abstraction from the Damodar, sulfamethoxazole and iohexol were found in unquantifiable amounts (<10 ng/L). Tolyltriazole was found in drinking water (67 ng/L) slightly below the concentration in surface water (88–102 ng/L, Table 3). This indicates on one hand not much removal by conventional treatment, but on the other hand also indicates a natural removal of the other compounds within the river water by degradation and/or dilution and low effect of mixing with landside groundwater.

4. Summary and Conclusions

The Yamuna R. water quality between Delhi and Agra was observed to have the highest organic pollution (concentration of DOC and OMPs), and the surface as well as the groundwater was characterized by a high salinity, giving the drinking water derived therefrom a brackish taste. However, the concentrations of some OMPs in the RBF well water were 13%–99%, and DOC was 50% lower than in river water. The removal of DOC and some OMPs by RBF was considerably greater compared to direct surface water abstraction and subsequent conventional treatment (e.g., in Agra).

At most RBF locations in Jharkhand, Andhra Pradesh, Bihar, and Jammu, surface water quality was generally good with respect to all inorganic parameters. The design and location of the radial collector wells (RCWs) in Jharkhand and Andhra Pradesh within the riverbed ensures the year-round abstraction of water, even during the non-monsoon, when very low to negligible surface water flow is observed. However, the travel time of bank filtrate for such riverbed RCW systems is too short, and thus the removal of organics is lower, and breakthroughs of pathogens and turbidity are likely to occur. One advantage at these locations is that the surface water itself has relatively low concentrations of DOC and OMPs. Iron and manganese can also occur in the abstracted water from such systems.

Thus, the bank filtrate subsequently undergoes post-treatment comprising aeration, flocculation, rapid sand filtration, and disinfection. In some coastal and peninsular (hard rock) areas of India (Jharkhand, Odisha, Andhra Pradesh, and Tamil Nadu), RBF is the only viable means of obtaining water compared to direct surface water or even groundwater, and in this context, RBF buffers the quantity of water required through bank or bed storage and can thus be considered to be an element of managed aquifer recharge and integrated water resources management.

This is the first country-wide overview of critical water quality parameters for some existing and potential RBF sites in India. Although this study was based on limited data for some locations, the data were well supported by results from recent investigations [16,29]. The data were insufficient to describe the hydrogeochemical and attenuation processes of inorganic parameters and organic substances in detail, but it nevertheless reiterated the fact that where in use, RBF at least serves as an effective pre-treatment step. For RBF sites at extremely polluted surface waters (such as in Delhi, Mathura, and potentially Agra by the Yamuna River), post-treatment should be made mandatory, e.g., by using activated carbon or advanced oxidation, which is less costly and easier to maintain if RBF is used for pre-treatment [16]. Moreover, the study provides a forecast within an initial site assessment of the water quality to be expected for subsequent investigations ranging from long-term monitoring and hydrogeological process description to environmental and human health risk assessments, and eventually the planning of new full-scale RBF systems in India. The water quality knowledge base generated in this study and the spatial distribution of the investigated sites, both with existing RBF systems and those with potential to develop new systems, forms the basis to develop a master plan for RBF water supply in India [30].

Author Contributions: Conceptualization, C.S., T.G., and S.K.S; data curation, H.B. and J.F.; formal analysis, C.S.; investigation, C.S., H.B., and J.F.; methodology, C.S., T.G., H.B., J.F., and S.K.S; project administration, T.G.; resources, H.B., J.F., and S.K.S; supervision, T.G.; validation, T.G. and H.B.; visualization, C.S.; writing—original draft, C.S.; writing—review and editing, T.G., H.B., and S.K.S.

Funding: Primary data collection was funded by the European Commission, within the project "Saph Pani—Enhancement of natural water systems and treatment methods for safe and sustainable water supply in India" (grant no. 282911). Data analyses and interpretation were funded by the European Commission, within the project "AquaNES—Demonstrating Synergies in Combined Natural and Engineered Processes for Water Treatment Systems" (grant no. 689450) and by the Federal Ministry of Education and Research (BMBF) of Germany, within the project "NIRWINDU—Safe and sustainable drinking water production in India by coupling natural and innovative techniques" (grant no. 02WCL1356A). A part of the mobility of the first author to conduct sampling in India was funded by the German Academic Exchange Service (DAAD), within the project "Nachhaltige Trinkwasserversorgung in Uttarakhand (Sustainable Drinking Water Supply in Uttarakhand)" (grant no. 56040107) as part of the DAAD's program "A New Passage to India" (2013–2014).

Acknowledgments: The support for sampling and analyses by the students M. Ullmann (2013) and T. Jacob (2014) is greatly appreciated. The authors also gratefully acknowledge logistical support from the Indian Saph Pani and AquaNES project partners, namely N.C. Ghosh, B. Chakraborty, M. Nema, Y.R.S. Rao, T. Thomas, Pradeep Kumar, M. Ronghang, S. Kumari, A. Gupta, S.K. Sharma, P.C. Kimothi, D.K. Singh, and R.K. Rohela. Furthermore, Prashant Singh, Manindra Mohan, Anupam K. Singh, S. Prasad, P K. Singh, and S.K. Dawn are thanked for their support.

Conflicts of Interest: The authors declare no conflicts of interest.

Appendix A

Table A1. Summary of occurrence of organic micropollutants.

OMP	Number of Samples Wherein Detected	Source Water
Iomeprol	2	Yamuna R.
Naproxen	2	Delhi groundwater, Keetham Lake Agra
Simazine	2	Yamuna R.
Chloramphenicol	3	Yamuna R.
Iopromid	5	Yamuna R. and RBF wells along Yamuna
Sulfadiazine	5	Yamuna R. and RBF wells along Yamuna
Ibuprofen	6	Scattered locations
Acetamiprid	7	Mainly Yamuna region
Atrazine	7	Yamuna region and Japla (Jharkhand)
Phenazone	8	Yamuna and Damodar river regions
1H-Benzotriazole	11	Mainly Yamuna region
Paracetamol	11	Scattered locations
Phenobarbital	11	Mainly Yamuna region
Iohexol	12	Mainly Yamuna region
Diuron	13	Mainly Yamuna region and Nainital Lake
Diclofenac	14	Yamuna and Damodar river regions
Sulfamethoxazole	14	Scattered locations
Cotinine	15	Mainly Yamuna region, Jharkhand and Nainital
Tolyltriazole	15	Mainly Yamuna region and Jharkhand
Carbamazepine	16	Yamuna and Damodar river regions
Theophylline	17	Scattered locations
Triclosan	23	Scattered locations
Caffeine	41	Ubiquitously present nearly everywhere

Table A2. List of organic micropollutants not detected in any sample.

Ametryne	Clofibric acid	Linuron	Atenolol
Atrazine-desethyl	Clothianidine	Loratadine	Ciprofloxacin Ditrizoate
Atrazine-desisopropyl	Diazepam	Primidone	Gabapentin
Bentazone	Gemfibrozil	Propanil	Metformin
Carbofuran	Iopamidol	Terbutryn	Metoprolol
Chlorothiazide	Isoproturon	Trimethoprim	N,N-Dimethylsulfamide Ranitidin

References

1. Sandhu, C.; Grischek, T.; Kumar, P.; Ray, C. Potential for Riverbank filtration in India. *Clean Technol. Environ. Policy* **2011**, *13*, 295–316. [CrossRef]
2. Sandhu, C.; Grischek, T. Riverbank filtration in India—Using ecosystem services to safeguard human health. *Water Sci. Technol. Water Supply* **2012**, *12*, 783–790. [CrossRef]
3. Dash, R.R.; Mehrotra, I.; Kumar, P.; Grischek, T. Lake bank filtration at Nainital, India: Water-quality evaluation. *Hydrogeol. J.* **2008**, *16*, 1089–1099. [CrossRef]
4. Dash, R.R.; Bhanu Prakash, E.V.P.; Kumar, P.; Mehrotra, I.; Sandhu, C.; Grischek, T. River bank filtration in Haridwar, India: Removal of turbidity, organics and bacteria. *Hydrogeol. J.* **2010**, *18*, 973–983. [CrossRef]
5. Gupta, A.; Singh, H.; Ahmed, F.; Mehrotra, I.; Kumar, P.; Kumar, S.; Grischek, T.; Sandhu, C. Lake bank filtration in landslide debris: Irregular hydrology with effective filtration. *Sustain. Water Resour. Manag.* **2015**, *1*, 15–26. [CrossRef]
6. Ronghang, M.; Gupta, A.; Mehrotra, I.; Kumar, P.; Patwal, P.; Kumar, S.; Grischek, T.; Sandhu, C. Riverbank filtration: A case study of four sites in the hilly regions of Uttarakhand, India. *Sustain. Water Resour. Manag.* **2018**. [CrossRef]
7. Sprenger, C.; Lorenzen, G.; Grunert, A.; Ronghang, M.; Dizer, H.; Selinka, H.-C.; Girones, R.; Lopez-Pila, J.M.; Mittal, A.K.; Szewzyk, R. Removal of indigenous coliphages and enteric viruses during riverbank filtration from highly polluted river water in Delhi (India). *J. Water Health* **2014**, *12*, 332–342. [CrossRef] [PubMed]

8.   Singh, P.; Kumar, P.; Mehrotra, I.; Grischek, T. Impact of riverbank filtration on treatment of polluted river water. *Environ. Manag.* **2010**, *91*, 1055–1062. [CrossRef] [PubMed]
9.   Kumar, P.; Mehrotra, I.; Boernick, H.; Schmalz, V.; Worch, E.; Schmidt, W.; Grischek, T. Riverbank Filtration: An Alternative to Pre-chlorination. *J. Indian Water Works Assoc.* **2012**, 50–58.
10. Cady, P.; Boving, T.B.; Choudri, B.S.; Cording, A.; Patil, K.; Reddy, V. Attenuation of Bacteria at a Riverbank Filtration Site in Rural India. *Water Environ. Res.* **2013**, *85*, 2164–2175. [CrossRef]
11. Boving, T.B.; Choudri, B.S.; Cady, P.; Cording, A.; Patil, K.; Reddy, V. Hydraulic and Hydrogeochemical Characteristics of a Riverbank Filtration Site in Rural India. *Water Environ. Res.* **2014**, *86*, 636–649. [CrossRef] [PubMed]
12. Groeschke, M.; Frommen, T.; Taute, T.; Schneider, M. The impact of sewage-contaminated river water on groundwater ammonium and arsenic concentrations at a riverbank filtration site in central Delhi, India. *Hydrogeol. J.* **2017**, *25*, 2185–2198. [CrossRef]
13. Mutiyar, P.K.; Mittal, A.K.; Pekdeger, A. Status of organochlorine pesticides in the drinking water well-field located in the Delhi region of the flood plains of river Yamuna. *Drink. Water Eng. Sci.* **2011**, *4*, 51–60. [CrossRef]
14. Kumar, P.; Mehrotra, I.; Gupta, A.; Kumari, S. Riverbank Filtration: A sustainable process to attenuate contaminants during drinking water production. *J. Sustain. Dev. Energy Water Environ. Syst.* **2018**, *6*, 150–161. [CrossRef]
15. Krishan, G.; Singh, S.; Sharma, A.; Sandhu, C.; Grischek, T.; Gosh, N.C.; Gurjar, S.; Kumar, S.; Singh, R.P.; Glorian, H.; et al. Assessment of water quality for river bank filtration along Yamuna river in Agra and Mathura. *Int. J. Environ. Sci.* **2016**, *7*, 56–67.
16. Glorian, H.; Börnick, H.; Sandhu, C.; Grischek, T. Water quality monitoring in northern India for an evaluation of the efficiency of bank filtration sites. *Water* **2018**, *10*, 1804. [CrossRef]
17. IS 10500. *Drinking Water—Specification*; Indian Standard (Second Revision); Bureau of Indian Standards: New Delhi, India, 2012.
18. Grischek, T.; Ray, C. Bank filtration as managed surface—Groundwater interaction. *Int. J. Water* **2009**, *5*, 125–139. [CrossRef]
19. Grischek, T.; Schoenheinz, D.; Sandhu, C. Water Resources Management and Riverbank Filtration. In *Drinking Water—Source, Treatment and Distribution*; Dobhal, R., Grischek, T., Uniyal, H.P., Uniyal, D.P., Sandhu, C., Eds.; Uttarakhand State Council for Science and Technology (UCOST): Dehradun, India, 2011; pp. 1–7.
20. Sandhu, C. A Concept for the Investigation of Riverbank Filtration Sites for Potable Water Supply in India. Ph.D. Thesis, TU Dresden, Faculty of Environmental Sciences, HTW Dresden, Division of Water Sciences, Dresden, Germany, 2015.
21. WHO. *Guidelines for Drinking-Water Quality*, 4th ed.; Incorporating 1st Addendum; World Health Organization: Geneva, Switzerland, 2017; ISBN 978-9241548151.
22. Sandhu, C.; Grischek, T.; Ronghang, M.; Mehrotra, I.; Kumar, P.; Ghosh, N.C.; Rao, Y.R.S.; Chakraborty, B.; Patwal, P.S.; Kimothi, P.C. Overview of bank filtration in India and the need for flood-proof RBF systems. In *Natural Water Treatment Systems for Safe and Sustainable Water Supply in the Indian Context—Saph Pani*; Wintgens, T., Nättorp, A., Lakshmanan, E., Asolekar, S., Eds.; IWA Publishing: London, UK, 2016; pp. 17–38, ISBN 9781780407104.
23. Mutiyar, P.K.; Mittal, A.K. Status of organochlorine pesticides in Ganga river basin: Anthropogenic or glacial? *Drink. Water Eng. Sci. Discuss.* **2012**, *5*, 1–30. [CrossRef]
24. Huntscha, S.; Singer, H.P.; McArdell, C.S.; Frank, C.E.; Hollender, J. Multiresidue analysis of 88 polar organic micropollutants in ground, surface and wastewater using online mixed-bed multilayer solid-phase. *J. Chromatogr. A* **2012**, *1268*, 74–83. [CrossRef]
25. Ullmann, M. Untersuchung zu Arzneimittelwirkstoffen bei der Uferfiltration in Indien (Investigation of Pharmaceutical Compounds during Riverbank Filtration in India). Master's Thesis, HTW Dresden, Dresden, Germany, 2013.
26. Schnitzler, H.; Börnick, H.; Ullmann, M.; Sandhu, C.; Grischek, T. Bestimmung von polaren organischen Spurenstoffen in Gewässern in Indien mittels vor-Ort-Anreicherung und LC-MS/MS (Determination of polar organic micropollutants in waters in India by on-site enrichment and LCMS/MS). In Proceedings of the Jahrestagung der Wasserchemischen Gesellschaft, Wasser, Haltern am See, Germany, 26–28 May 2014; pp. 479–483, ISBN 978-3-936028-83-6.

27. Lorenzen, G.; Sprenger, C.; Taute, T.; Pekdeger, A.; Mittal, A.; Massmann, G. Assessment of the potential for bank filtration in a water-stressed megacity (Delhi, India). *Environ. Earth Sci.* **2010**, *61*, 1419–1434. [CrossRef]
28. Lalwani, S.; Dogra, T.D.; Bhardwaj, D.N.; Sharma, R.K.; Murty, O.P.; Vij, A. Study on arsenic level in ground water of Delhi using hydride generator accessory coupled with atomic absorption spectrophotometer. *Indian J. Clin. Biochem.* **2004**, *19*, 135–140. [CrossRef] [PubMed]
29. Lapworth, D.J.; Das, P.; Shaw, A.; Mukherjee, A.; Civil, W.; Petersen, J.O.; Gooddy, D.C.; Wakefield, O.; Finlayson, A.; Krishan, G.; et al. Deep urban groundwater vulnerability in India revealed through the use of emerging organic contaminants and residence time tracers. *Environ. Pollut.* **2018**, *240*, 938–949. [CrossRef] [PubMed]
30. Grischek, T.; Sandhu, C.; Ghosh, N.C.; Kimothi, P.C. A Conceptual Master Plan for RBF Water Supply in India—Science, Policy & Implementation Aspects. In Proceedings of the Indo-German Conference on Sustainability, Chennai, India, 27–28 February 2016; pp. 118–123.

water MDPI

Article

Water Quality Monitoring in Northern India for an Evaluation of the Efficiency of Bank Filtration Sites

Heinrich Glorian [1],*, Hilmar Börnick [1], Cornelius Sandhu [2] and Thomas Grischek [2]

[1] Institute for Water Chemistry, TU Dresden, 01062 Dresden, Germany; hilmar.boernick@tu-dresden.de
[2] Division of Water Sciences, University of Applied Sciences Dresden, 01069 Dresden, Germany;
    cornelius.sandhu@htw-dresden.de (C.S.); thomas.grischek@htw-dresden.de (T.G.)
* Correspondence: heinrich.glorian@tu-dresden.de; Tel.: +49-351-463-36140

Received: 30 October 2018; Accepted: 6 December 2018; Published: 8 December 2018

Abstract: The study presents results of five sampling campaigns at riverbank filtration sites at the Yamuna and Ganges Rivers in Uttarakhand, Uttar Pradesh and New Delhi 2015–2018. Samples were analyzed for organic micropollutants and general water quality parameters. In New Delhi and Uttar Pradesh, 17 micropollutants were detected frequently at relevant concentrations. Out of the detected micropollutants, 1H-benzotriazole, caffeine, cotinine, diclofenac, diuron, gabapentin and paracetamol were frequently detected with concentrations exceeding 1000 ng/L. Sites in Uttarakhand showed only infrequent occurrence of organic micropollutants. The mean concentration of micropollutants in the well water was lower compared to the river water. For all sites, removal rates for all micropollutants were calculated from the obtained data. Thereby, the capacity of riverbank filtration for the removal of organic micropollutants is highlighted, even for extremely polluted rivers such as the Yamuna.

Keywords: riverbank filtration; organic micropollutants; water quality; environmental monitoring

1. Introduction

The appearance of organic micropollutants (OMPs) in surface water bodies on a global scale is an unwelcome reality [1,2]. Micropollutants may find their way into water bodies either from point or diffuse sources [3,4]. Conventional wastewater treatment plants are unable to remove all micropollutants and discharge contaminated effluents directly into surface water bodies [1]. This discharge of contaminated effluents is called a point source. Diffuse sources like agriculture are even more difficult to control and pose a further risk for the water quality. Due to this issue, it is of major interest for water companies to know about the water quality of the source water to produce safe drinking water without risks for human health. The number of known trace compounds in the environment is ever-increasing due to further development of more sensitive analytical and sample preparation methods. With these modern methods it is also possible to measure so called emerging micropollutants. This group includes polar and persistent or pseudo-persistent (degradable, but always occurring due to continuous input) compounds, such as pharmaceutical, personal care and industrial compounds as well as pesticides and their transformations products [5,6]. All these compounds pose a potential risk for human health, if present in drinking water. So far, there are no limits or threshold concentrations defined for many emerging pollutants in water quality guidelines. Therefore, a comprehensive water quality monitoring is necessary to determine potential problems and risks concerning drinking water quality and to adapt the water treatment technology.

The available data regarding the occurrence of emerging pollutants in northern India are insufficient. The presence of emerging pollutants in source water at riverbank filtration (RBF) sites in northern India and their removal by RBF has only been investigated for an extremely polluted stretch of the Yamuna River in central Delhi [7] and their removal by RBF has only been monitored sporadically for some sites. Apart from several studies on general water quality parameters and heavy metals [8–11], only a few studies have shown the occurrence of OMPs. In this context, monitoring data for organochlorine pesticides have mainly been published [12–18]. Studies on other micropollutants like pharmaceuticals were carried out rarely [19]. The existing drinking water treatment technologies are not sufficient to remove all micropollutants from source water [15,20,21]. Consequently, and in light of the imminent risk from OMPs, the selection of source water has to be done very carefully, especially if surface water is directly abstracted for drinking water production. In case of polluted river water, RBF provides a pre-treatment for the removal of, among others, OMPs [10,22]. The removal rate depends on compound-specific properties (biodegradability and adsorption behavior) as well as on water quality, geochemical composition of aquifer material and hydraulic boundary conditions [23].

The aim of the presented study is to expand the knowledge of organic and inorganic water quality at RBF sites over four years in the upper part of the Ganges and Yamuna Rivers. This includes the occurrence and removal of organic micropollutants as well as general inorganic water quality parameters. Furthermore, the efficiency of RBF at these sites to remove OMPs is characterized for the first time (other than Delhi [7]). For this purpose, different general parameter (e.g., main anions and cations, DOC) and selected typical anthropogenic organic micropollutants were analyzed.

2. Materials and Methods

2.1. Study Sites

The monitoring was done at nine selected sites in Uttarakhand, Uttar Pradesh and in New Delhi, India (Figure 1, Table 1). At every site one surface water sample and one well water sample was taken. The sites are located along the rivers Yamuna, Ganges and their tributaries. Depending on the distance from the riverbank, the production rate and duration of operation (h/day), the wells abstract a low (Agastmuni) to high (Haridwar) portion of bank filtrate. Due to the lower density of population and industry in Uttarakhand, the six sites there (Haridwar, Srinagar, Agastmuni, Gauchar, Karnaprayag and Satpuli) should show a lower level of contamination compared to the three sites in New Delhi, Mathura and Agra. Overall five sampling campaigns were conducted in September 2015, May and September 2016, September 2017 and June 2018. The wells in Agra and Mathura were constructed in 2017. Because of the relatively low number of repeated measurements the obtained results are giving a first general overview of the water quality in this region but give little information about the interim periods between the sampling campaigns. Furthermore, in this part of India the monsoon can have an influence on the concentration of water constituents due to dilution effects. The obtained data do not allow deeper statements regarding this topic.

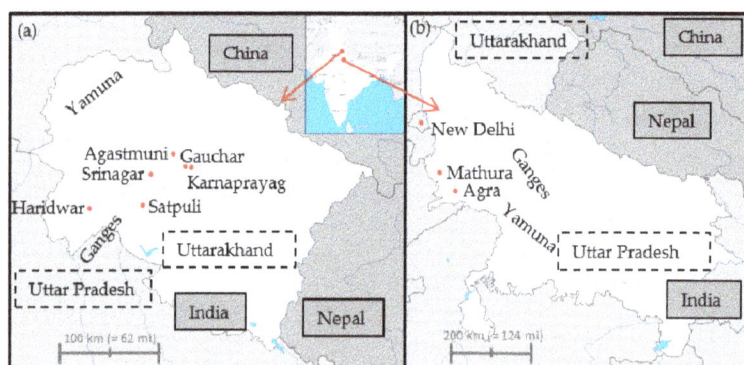

Figure 1. Sampling locations in (**a**) Uttarakhand and (**b**) Uttar Pradesh and New Delhi, India (adapted from [24]).

Table 1. Summary of RBF sites in this study, and wells that were monitored (adapted from [25]).

Location	River	Well Type (# of Wells)	Total Production of all Wells at Site (m³/d)	Depth (m)	Distance: River–Nearest Well(s) (m)	Portion of Bank Filtrate (%)
Agastmuni [26]	Mandakini	V (1)	>280	30	33	25–35
Karnaprayag (Kaleshwar)	Alaknanda	C (1) and V (1) [M]	5760	14.7 (C) 20 (V)	≤1[**]–25[*]	n.d.; assumed > 50
Gauchar	Alaknanda	C (1) and V (1) [M]	4320	14.7	61	n.d.
Srinagar [26]	Alaknanda	V (7)	1300–8000	18–44	10[**]–102[*]	36–72
Satpuli [26]	East Nayar	V (1)	756	26	43–45	95–100
Haridwar [10]	Ganga and UGC	C (22) (well #18 [M])	59,000–67,000	7–10	4–110	40–90
New Delhi [7]	Yamuna	R (8) (well #P4 [M])	n.d.	19–31	~1300 (well #P4)	n.d.
Mathura	Yamuna	V (1)	n.d.	(under construction)		n.d.
Agra [27]	Yamuna	V (1)	n.d.	20	140	n.d.

[M] well(s) that were monitored; [*] during non-monsoon; [**] during monsoon; C: caisson well; V: vertical well; R: radial collector well; UGC: Upper Ganga Canal; n.d.: not determined yet as these wells were constructed in 2016–2017 and became operational in 2017–2018.

2.2. Analytical Methods

Sampling was done wearing gloves to prevent any contamination of the sample with trace compounds for example from hand cream. For the micropollutants samples, a glass vial was rinsed with the sampling water two times and emptied. A second glass vial was rinsed two times too and filled half. From the second vial a volume of 5 mL was taken and transferred to the first vial. 250 μL of an internal standard was added using a Hamilton syringe. The vial was closed and shaken. The spiked sample was taken with a one-way syringe and filtered through a 0.45 μm membrane filter (Chromafil® Xtra RC-20/25, Macherey-Nagel, Düren, Germany) and filled into a HPLC analysis vial after the first ml was wasted. Samples for main anions and cations were filtered as well (Chromafil® GF/PET-45/25, Macherey-Nagel, Düren, Germany). Samples for cations were preserved using nitric acid. All samples were cooled until analysis.

The analysis of 32 micropollutants (Table 2) was carried out with a LC-MS/MS (6500+ QTRAP®, Sciex, Framingham, MA, USA) using a Luna® Omega 1.6 μm Polar C18 column (Phenomenex, Torrance, CA, USA) and a H₂O/ACN eluent (0.02% CH₃COOH). The samples of the first two sampling campaigns were measured using a 3200 QTRAP® (Sciex, Framingham, MA, USA) after samples were enriched using SPE cartridges (SiliaPrepX HLB, SiliCycle Inc., Quebec City, QC, Canada). Acesulfame could not be enriched and was not measured here. Main anions (Table 3) were analyzed by IC (DX-100, Dionex/Thermo Fisher Scientific, Waltham, MA, USA) and cations (Table 3) by ICP-MS (4500, Agilent, Santa Clara, CA, USA). Dissolved organic carbon (DOC) was determined as non-purgeable organic carbon with a TOC-5000 (Shimadzu, Kyōto, Japan) according to the standard DIN EN 1484 H3 [28].

Table 2. List of analyzed micropollutants and range of quantification.

Compound	Limit of Determination in ng/L	Compound	Limit of Determination in ng/L
Pharmaceuticals		**Pesticides**	
Azithromycin	800	Acetamiprid	2
Bezafibrate	2	Atrazine	10
Carbamazepine	1	Dimethoate	1
Clarithromycin	60	Diuron	4
Diclofenac	1	Imidacloprid	2
Erythromycin	60	Irgarol	1
Fluoxetine	1	Isoproturon	1
Gabapentin	8	Nicosulfuron	2
Ibuprofen	1	Terbutryn	6
Iomeprol	1	**Industrial chemicals**	
Metoprolol	1	1H-Benzotriazole	6
Naproxen	2	Bisphenol A	6
Paracetamol	2	Tolyltriazole	4
Roxithromycin	1	**Other micropollutants**	
Sulfamethoxazole	1	Acesulfame	2
Triclosan	60	Caffeine	2
		Cotinine	2
		Theophylline	6

Table 3. List of main anions and cations analyzed and range of quantification.

Cation	Limit of Determination in mg/L	Anion	Limit of Determination in mg/L
Arsenic (As^{3+})	0.0005	Chloride (Cl^-)	0.127
Calcium (Ca^{2+})	0.1	Fluoride (F^-)	0.028
Iron (Fe^{2+})	0.02	Nitrate (NO_3^-)	0.022
Magnesium (Mg^{2+})	0.1	Nitrite (NO_2^-)	0.004
Manganese (Mn^{2+})	0.002	Phosphate (PO_4^{3-})	0.004
Sodium (Na^+)	0.1	Sulfate (SO_4^{2-})	0.020

3. Results

3.1. Main Ions and DOC

As expected, water quality at the study sites varied widely because of the strong differences in demographic density and industrial settlements. The sites in Uttarakhand rarely showed a critical contamination from OMP (Appendix B, Table A3) whereas data for New Delhi, Mathura and Agra pointed out much higher contamination. This difference in water quality is confirmed by the DOC concentrations in Appendix Tables A1 and A4 for all sites. DOC concentrations measured in river water samples from New Delhi, Mathura and Agra are six time higher than in river water samples from Uttarakhand. The DOC is the sum of all dissolved organic carbon in the water, including the organic carbon from the targeted OMP. It can be assumed, that input of waste water is connected with increased organic content as well as occurrence of higher concentrations of OMP [29]. Therefore, DOC can be used as an indicator for the organic pollution of water bodies. Nevertheless, the natural organic background always has to be considered. The monitoring of main anions and cations is largely unobtrusive and indicates a good water quality (Appendix A Table A1). Known issues like a high nitrate concentration in Srinagar (c_{max} = 78.9 mg/L, Appendix B Table A4) [9] and relatively high arsenic concentrations for example in New Delhi, Mathura and Agra (c_{max} = 0.01–0.10 mg/L) can be confirmed [30]. Furthermore, high concentrations of nitrite (c_{max} = 1.95–4.34 mg/L) and manganese (c_{max} = 0.07–2.25 mg/L) were detected at Mathura and Agra (Appendix A Table A1). These concentrations for nitrate, nitrite and manganese exceed the threshold concentrations given by the German and Indian Drinking Water Ordinance. The threshold concentrations are 45–50 mg/L, 0.01 mg/L and 0.05–0.1 mg/L, respectively.

3.2. Micropollutants

In the well sample from Haridwar in September 2017 unusual high concentrations of diclofenac (2000 ng/L) and gabapentin (4090 ng/L) were measured. At the same time a relatively high chloride concentration of 133 mg/L was measured. Normally, well water in Haridwar shows chloride concentrations around 12.5 mg/L. This would indicate a temporal infiltration of urban waste water into the well. Here, the consistently polluted sites are described and discussed in detail with a focus on removal rates during RBF. In New Delhi, Mathura and Agra, 17 micropollutants were detected nearly in every sample. Mean and maximum concentrations for each compound and site are shown in Appendix A Table A2.

3.3. Pharmaceuticals

Out of 16 analyzed pharmaceuticals eight compounds were found to be regularly present in both river and well water at all three sites. Gabapentin was found with the highest mean concentrations from 832 to 5380 ng/L followed by paracetamol (114–1550 ng/L), sulfamethoxazole (733–1260 ng/L) and diclofenac (199–994 ng/L). Whereby, paracetamol was never detected in samples from Mathura (Figure 2). The mean concentration of gabapentin in river samples from Agra (5380 ng/L) exceeds the scale of the bar chart of Figure 2. Four additional pharmaceuticals were detected frequently but at lower concentrations (Figure 3). Naproxen, metoprolol, ibuprofen and carbamazepine showed mean concentrations of 102–423 ng/L, 171–395 ng/L, 9–333 ng/L and 96–112 ng/L, respectively. In all cases the concentration of the detected pharmaceuticals was substantially lower in the RBF well samples compared to the river samples. The well water showed a decreased mean concentration by 91–100% for gabapentin, 46–50% for paracetamol, 41–95% for sulfamethoxazole, 37–80% for diclofenac, 39–100% for naproxen, 78–100% for metoprolol, 19–74% for ibuprofen and 15–73% for carbamazepine.

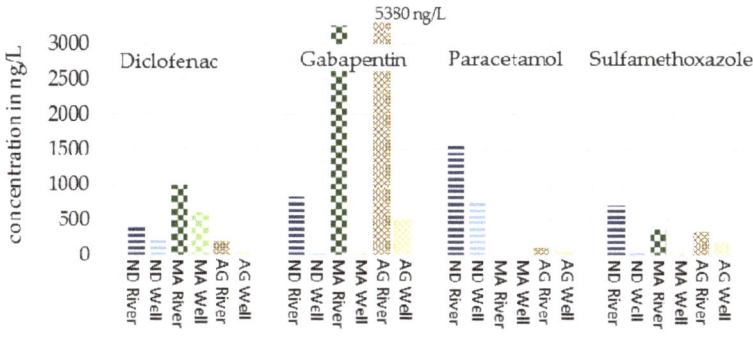

Figure 2. Pharmaceuticals with highest mean concentrations in New Delhi (ND), Mathura (MA) and Agra (AG) in river and RBF well water.

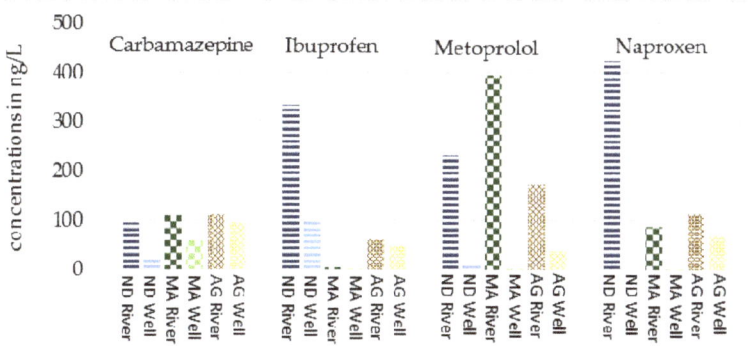

Figure 3. Mean concentrations of four pharmaceuticals with lower concentration in New Delhi (ND), Mathura (MA) and Agra (AG) in river and RBF well water.

3.4. Herbicides and Pesticides

Out of nine analyzed herbicides and pesticides only acetamiprid, diuron and imidacloprid were detected frequently in river and well water. The highest concentrations by far were determined for diuron (Figure 4). In surface water, diuron reached mean concentrations from 2710 ng/L in Agra to >3450 ng/L in New Delhi and to 4810 ng/L in Mathura. Acetamiprid and imidacloprid were detected in river water with a 100 times lower concentration of 7–65 ng/L and 9–18 ng/L, respectively. RBF well water samples showed a decreased mean concentration by 37–91% for diuron, 64–88% for acetamiprid and 22–89% for imidacloprid. Pesticide concentrations at all sites in Uttarakhand were found to be much lower, except in Srinagar, where bank filtrate has a very long flow path and is affected by inputs from the urban area and agriculture [10].

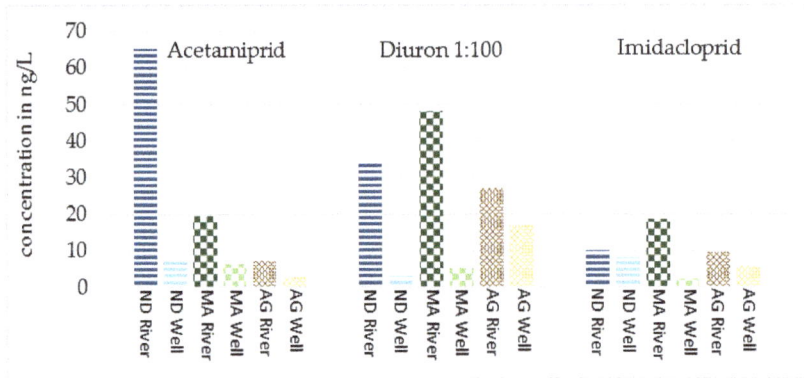

Figure 4. Mean concentrations of herbicides and pesticides in New Delhi (ND), Mathura (MA) and Agra (AG) in river and RBF well water; Diuron is shown in a scale 1:100, the sample was measured undiluted.

3.5. Industrial Products

1H-benzotriazole and tolyltriazole were detected in all surface and well water samples from New Delhi, Mathura and Agra. Bisphenol A only was detected sporadically. 1H-benzotriazole showed higher mean concentrations than tolyltriazole (Figure 5). The mean concentrations of 1H-benzotriazole in river water were 271–1050 ng/L and of tolyltriazole 98–418 ng/L. The concentration levels in RBF well water were in all cases lower than in surface water. 1H-benzotriazole showed a decrease in concentration by 77–98% and tolyltriazole by 33–100%.

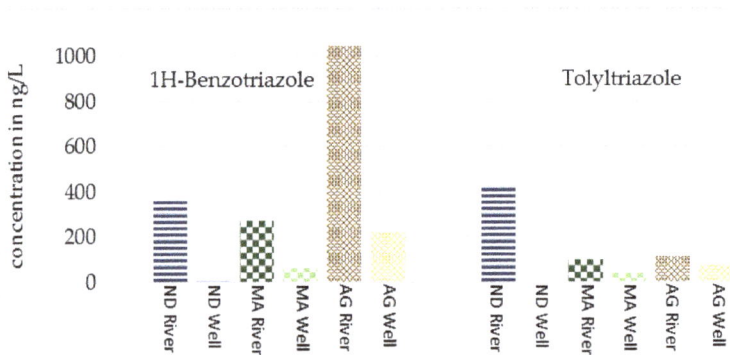

Figure 5. Mean concentrations of industrial products in New Delhi (ND), Mathura (MA) and Agra (AG) in river and RBF well water.

3.6. Other Micropollutants

Acesulfame, caffeine, cotinine and theophylline were detected in all river water samples in New Delhi, Mathura and Agra with mean concentrations of 376–988 ng/L, 279–3360 ng/L, 38–1180 ng/L and 431–1350 ng/L, respectively (Figure 6). The mean concentration of caffeine in the river sample from New Delhi (3360 ng/L) exceeds the scale of the bar chart of Figure 6. The mean concentrations for the compounds caffeine, cotinine and theophylline decreased in the RBF well water samples by 71–99%, 5–95% and 56–99%, respectively. The mean concentrations of the artificial sweetener acesulfame in samples of RBF wells are in the same or lower range in comparison to the river water samples in Agra, New Delhi and Mathura (Figure 6). The findings from Mathura, with a short travel time and >75% bank filtrate, underline the very low removal of acesulfame.

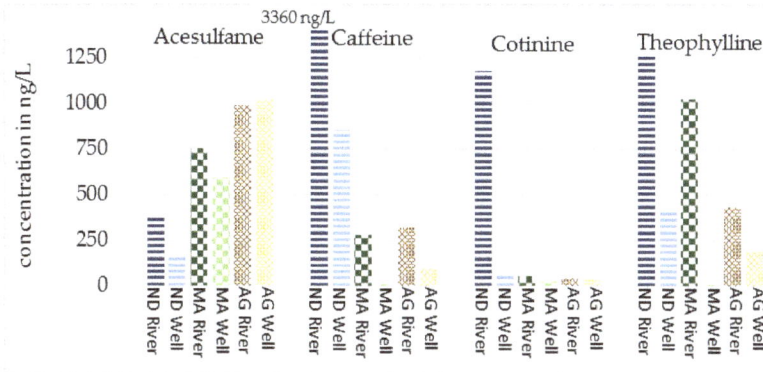

Figure 6. Mean concentrations of other micropollutants in New Delhi (ND), Mathura (MA) and Agra (AG) in river and RBF well water.

4. Discussion

4.1. Acesulfame

Acesulfame is persistent and shows low degradation during wastewater treatment [29,31]. Therefore, acesulfame is found in waters influenced by anthropogenic activities and became a favorable indicator for domestic wastewater. It can be used to estimate the portion of bank filtrate in the wells at RBF sites, assuming no occurrence of acesulfame in the groundwater [32]. Mean concentrations of acesulfame in the Yamuna River and RBF wells at the three sites would indicate a mean portion of bank filtrate of 46% in New Delhi, 79% in Mathura and around 100% in Agra. Apparently, the calculated removal rates in Mathura and Agra are influenced not only by adsorption and biodegradation in the aquifer but also by dilution of the bank filtrate with land-side groundwater.

4.2. Pharmaceuticals

Most of the regarded pharmaceuticals possess a wide variety of functional groups and are therefore medium to highly polar and very mobile in water. Adsorption on suspended particles and sediment is limited. Furthermore, many pharmaceuticals are resistant to degradation in the human body as well as in wastewater treatment plants to a great extent [3,5]. Consequently, 50% of the targeted pharmaceuticals were found in river water and bank filtrate at a high frequency. Only two of them, azithromycin and roxithromycin, could not be detected at all at the sampling points in northern India.

The calculated mean removal rate for pharmaceuticals in New Delhi and Mathura is nearly 80% and significantly higher than in Agra (51%) comparing only river water and RBF well water. One reason for this will be the different portion of bank filtrate in the well water. Based on data for acesulfame and supported by data for chloride and other ions (Appendix A Table A1), the portion of bank filtrate is near to 100% in Agra, thus there is no effect of mixing with less polluted groundwater. Higher removal rates can be expected also for sites with longer flow paths and travel times.

Diclofenac showed with around 39% the lowest removal rate in New Delhi and Mathura. In Agra, a significantly higher removal rate of 80% was observed. This shows that apart from flow path length and travel time of bank filtrate, other parameters can have an influence of the removal of micropollutants as well. Additionally, the attenuation of organic micropollutants depends on redox conditions during RBF. For example, diclofenac shows significantly higher removal rates during aerobic and denitrifying conditions [33] whereas carbamazepine is better removed under anaerobic conditions [32]. Carbamazepine shows removal rates in New Delhi and Mathura of 73% and 46%, in Agra only 15%. Thus, removal rates for diclofenac and carbamazepine indicate anaerobic conditions during RBF in New Delhi and Mathura and aerobic/denitrifying conditions in Agra. Strong differences in occurrence of paracetamol in river and well samples at the sites New Delhi and Mathura were observed. Analyzed concentrations of this analgesic in New Delhi were frequently in µg/L-range. In contrast, this compound was never determined in samples from Mathura (detection limit: 2 ng/L). Paracetamol shows a relatively good biodegradability particularly at aerobic conditions. Possibly due to longer residence times and the availability of sufficient oxygen in Mathura, a complete degradation already in the river can occur. In Delhi, other boundary conditions may limit degradation processes and/or higher input from industry is possible.

The concentrations of X-ray contrast media (e.g., iomeprol) was found to be low in comparison to many European rivers [34]. This may be associated with the different level of medical care or with the use of other contrast agents in India.

4.3. Herbicides and Pesticides

Out of all analyzed herbicides and pesticides, only three compounds were detected frequently. Two of them, acetamiprid and imidacloprid, were found with very low concentrations. But diuron was detected with a maximum concentration of >10,000 ng/L in the Yamuna River at New Delhi. Even with a high removal rate of 91% in New Delhi, 89% in Mathura and 37% in Agra, mean concentrations between 299 and 1700 ng/L were determined in the RBF well water samples. Applications in sugar cane production (common in the region) are assumed to be the major source of diuron. If river water having such high concentration has to be used as raw water for drinking water production, an extensive post-treatment is needed, since diuron is potentially carcinogenic [35]. The German Drinking Water Ordinance gives threshold concentrations for a single herbicide or pesticide of 100 ng/L and for the sum of all compounds out of this group of 500 ng/L. The diuron concentrations are exceeding this threshold concentration in river and well water.

4.4. Industrial Products

Especially in New Delhi, the occurrence of 1H-benzotriazole and tolyltriazole in the river water should not be a problem for the drinking water production if RBF is used. As a result of a removal rate for 1H-benzotriazole of 100% and for tolyltriazole of 98% the bank filtrate is nearly free of these complexing agents used in the industry. In Mathura und Agra the removal rates do not exceed 80%. Therefore, despite a relatively high removal rate, these industrial products were found in the RBF well water in Mathura and Agra. Studies show that both compounds should be relatively stable only with a certain potential for biodegradation under aerobic conditions [36]. However, through cometabolic processes at higher DOC, the biodegradation can be enhanced [29]. Since the DOC at the sites is relatively high, this could be an explanation for the high removal rates.

4.5. Other Micropollutants

Caffeine, its degradation product theophylline, and cotinine that is a degradation product of nicotine, can be regarded as qualitative indicators for untreated wastewater. In New Delhi, Mathura and Agra those three compounds were found in nearly every sample. This is clear evidence for inadequate wastewater management in these areas. Caffeine, theophylline and cotinine are usually degradable to a high degree during RBF [37–40]. This fact was confirmed within this study for Delhi, Mathura and Agra (Appendix A Table A2), Haridwar, Srinagar and Karnaprayag (Appendix B Table A4). In Agastmuni, higher concentrations of caffeine were observed in RBF wells than in river water, indicating potential contamination via local wastewater input.

5. Conclusions

Five sampling campaigns from 2015 to 2018 provided an overview of the organic and inorganic water quality of the surface water and RBF well water at selected sites in Uttarakhand, Uttar Pradesh and New Delhi along the Ganges and Yamuna rivers. While pollution by the selected organic micropollutants was found to be very low in Uttarakhand, it was significant higher in New Delhi and Uttar Pradesh. The potential of RBF to remove organic trace pollutants was demonstrated particularly for the more polluted sites. The results support findings from RBF sites worldwide proving the pre-treatment efficiency of RBF to significantly improve raw water quality for drinking water production. Nevertheless, micropollutants were frequently detected in the RBF well water samples in New Delhi, Mathura and Agra due to a very high pollution of the river water. Diuron, 1H-benzatriazole, acesulfame, theophylline, diclofenac, gabapentin and paracetamol are the compounds with highest relevance at these sites. Thus, post-treatment such as activated carbon or advanced oxidation should be mandatory, which are less costly and easier to maintain if RBF is used for pre-treatment. The results provide a good basis for further sophisticated and comprehensive investigations, whereby a higher frequency of sampling, an extension of micropollutants spectrum, and the stringent consideration of meteorological data and geological material properties are scheduled.

Author Contributions: Conceptualization and project administration, H.B. and T.G.; Literature review, H.G. and H.B.; Selection of the study sites, C.S. and H.G.; Sampling, H.G. and C.S., Analyses, H.G., Writing Original Draft H.G.; Editing, H.B., C.S., T.G.

Funding: This research was funded by the Federal Ministry of Education and Research Germany (BMBF) within the project NIRWINDU (02WCL1356C). The financing of the LC-MS/MS system was supported by European Fund for Regional Development and by the Free State of Saxony.

Acknowledgments: The authors are very grateful to Dr. Gopal Krishan of the National Institute of Hydrology Roorkee for providing logistical support to the sampling in Delhi, Mathura and Agra.

Conflicts of Interest: The authors declare no conflict of interest.

Appendix A

Table A1. Mean (Ø) and maximum concentrations in mg/L of main anions and cations and DOC in the river and RBF well water at sampling points in Uttar Pradesh and New Delhi, 2015–2018 (n.d.—not determined), κ = conductivity.

Parameter in mg/L		New Delhi (n = 5)		Mathura (n = 2)		Agra (n = 2)	
		River	Well	River	Well	River	Well
As^{3+}	Ø	0.02	0.02	0.01	0.02	0.01	0.03
	Max	0.04	0.03	0.01	0.03	0.02	0.10
Ca^{2+}	Ø	97.2	76.2	48.3	69.7	59.0	109
	Max	124	107	74.9	106	78.4	136
Fe^{2+}	Ø	0.07	0.05	<0.02	0.18	<0.02	0.75
	Max	0.21	0.19	0.03	0.63	<0.02	2.97
Mg^{2+}	Ø	30.4	29.0	22.6	27.0	28.8	46.3
	Max	34.3	42.4	43.9	40.3	47.1	59.5
Mn^{2+}	Ø	0.52	0.33	0.04	0.42	0.07	0.58
	Max	0.60	0.59	0.07	1.12	0.23	2.25
Na^+	Ø	79.3	101	92.9	86.7	97.1	122
	Max	82.5	104	95.5	88.7	103	132
Cl^-	Ø	170	120	234	215	212	252
	Max	237	209	352	228	354	376
F^-	Ø	0.29	0.31	0.36	0.41	0.37	0.33
	Max	0.39	0.60	0.53	0.77	0.61	0.63
NO_3^-	Ø	3.73	13.0	3.50	0.55	9.25	12.5
	Max	10.5	23.5	3.9	1.3	15.4	30.5
NO_2^-	Ø	0.17	0.03	2.87	0.01	2.68	0.49
	Max	0.50	0.07	4.34	0.04	3.53	1.95
PO_4^{3-} *	Ø	1.24	<0.004	2.40	0.33	0.92	0.66
	Max	3.03	<0.004	4.58	1.33	2.94	2.62
SO_4^{2-}	Ø	84.1	46.8	81.2	58.4	82.9	79.7
	Max	118	63.4	93.2	125	107	123
DOC	Ø	6.48	1.78	8.41	1.85	8.59	5.18
	Max	8.80	2.38	10.6	2.38	9.80	8.30
κ in	Ø	1410	1190	1250	1390	1340	1370
μS/cm	Max	1560	1270	1370	1860	1730	1500

* No data in May 2016.

Table A2. Mean (∅) and maximum concentrations in ng/L of the most prominent organic micropollutants in the river and RBF well water at sampling points in Uttar Pradesh and New Delhi, 2015–2018 (n.d.—not determined).

Parameter in ng/L		New Delhi (n = 5)		Mathura (n = 2)		Agra (n = 2)	
		River	Well	River	Well	River	Well
1H-Benzotriazole	∅	376	<6	271	61.9	1050	224
	Max	378	<6	427	161	1850	266
Tolyltriazole	∅	418	n.d.	98.6	38.1	113	75.2
	Max	733	n.d.	157	38.2	179	88.5
Acetamiprid	∅	65.3	7.52	19.5	6.4	7.20	2.56
	Max	125	13.9	30.9	6.8	7.33	4.11
Diuron	∅	3450	299	4810	512	2710	1700
	Max	>10,000	599	5860	855	4670	4200
Imidacloprid	∅	10.2	7.97	18.6	<2	9.59	5.93
	Max	19.4	14.9	19.9	<2	18.2	7.57
Acesulfame	∅	377	172	749	592	989	1020
	Max	584	255	873	709	1060	1160
Caffeine	∅	3360	852	279	<2	322	93.8
	Max	>10,000	1700	308	<2	674	187
Cotinine	∅	1180	64.2	49.9	22.6	38.7	36.8
	Max	3260	128	74.1	39.1	91.5	54.4
Theophylline	∅	1350	402	1020	<6	431	188
	Max	3810	800	1700	<6	977	239
Carbamazepine	∅	96.0	25.7	109	59.4	112	95.1
	Max	185	56.0	122	106	114	124
Diclofenac	∅	410	248	995	623	199	40.5
	Max	1220	267	2040	1990	232	92.4
Gabapentin	∅	833	<8	3260	<8	5380	508
	Max	1200	<8	5340	<8	10,000	527
Ibuprofen	∅	334	101	3.83	<1	59.9	48.8
	Max	346	202	9.50	<1	164	117
Metoprolol	∅	233	9.0	395	<1	172	37.3
	Max	420	27.1	609	<1	436	62.2
Naproxen	∅	424	<2	87.4	<2	113	68.7
	Max	424	<2	102	<2	128	85.7
Paracetamol	∅	1550	771	n.d.	n.d.	115	61.7
	Max	2990	771	n.d.	n.d.	199	61.7
Sulfamethoxazole	∅	719	37.1	381	18.4	347	205
	Max	1260	73.2	742	35.8	733	205

Appendix B

Table A3. Mean (Ø) and maximum concentrations in ng/L of most prominent organic micropollutants in the river and RBF well water at sampling points in Uttarakhand, 2015–2018 (n.d.—not determined).

Parameter in ng/L		Haridwar (n = 5)		Srinagar (n = 5)		Karnaprayag (n = 5)		Gauchar (n = 5)		Agastmuni (n = 5)		Satpuli (n = 5)	
		River	Well	River	Well	River	Well	River	Well	River	Well	River	Well
1H-Benzotriazole	Ø	<6	n.d.	<6	55.1	<6	<6	<6	<6	<6	<6	<6	<6
	Max	<6	n.d.	<6	61.7	<6	<6	<6	<6	<6	<6	<6	<6
Tolyltriazole	Ø	n.d.	31.7	32.1	13.8	12.5	n.d.	n.d.	n.d.	42.6	<4	n.d.	n.d.
	Max	n.d.	60.3	56.6	27.6	15.6	n.d.	n.d.	n.d.	90.6	<4	n.d.	n.d.
Diuron	Ø	<4	72.8	<4	759	<4	<4	n.d.	n.d.	95.5	273	<4	n.d.
	Max	<4	143	<4	1510	<4	<4	n.d.	n.d.	280	407	<4	n.d.
Imidacloprid	Ø	n.d.	n.d.	n.d.	2.60	7.14	<2	<2	n.d.	<2	<2	<2	<2
	Max	n.d.	n.d.	n.d.	4.19	10.2	<2	<2	n.d.	<2	<2	<2	<2
Acesulfame	Ø	22.4	11.6	30.5	29.2	45.5	27.8	<2	97.2	20.5	270	56.2	51.5
	Max	43.9	22.1	60.1	57.5	50.1	54.5	<2	144	40.0	429	58.6	54.6
Caffeine	Ø	436	<2	674	<2	158	<2	<2	<2	6.40	11.1	<2	<2
	Max	1310	<2	2020	<2	393	<2	<2	<2	17.2	21.1	<2	<2
Cotinine	Ø	n.d.	n.d.	n.d.	n.d.	6.0	n.d.	n.d.	n.d.	n.d.	11.2	n.d.	n.d.
	Max	n.d.	n.d.	n.d.	n.d.	11.0	n.d.	n.d.	n.d.	n.d.	11.2	n.d.	n.d.
Theophylline	Ø	<6	<6	53.3	<6	164	<6	<6	<6	66.5	108	<6	<6
	Max	<6	<6	102	<6	204	<6	<6	<6	128	210	<6	<6
Diclofenac	Ø	<1	1000	<1	<1	<1	<1	<1	<1	106	<1	<1	<1
	Max	<1	2000	<1	<1	<1	<1	<1	<1	210	<1	<1	<1
Gabapentin	Ø	<8	2050	<8	294	<8	<8	<8	<8	<8	42.0	<8	<8
	Max	<8	4090	<8	581	<8	<8	<8	<8	<8	77.0	<8	<8
Ibuprofen	Ø	53.9	<1	<1	<1	n.d.	<1	<1	<1	50.1	52.9	<1	<1
	Max	107	<1	<1	<1	n.d.	<1	<1	<1	50.1	105	<1	<1
Paracetamol	Ø	107	<2	14.2	<2	10.4	n.d.	<2	17.5	18.5	<2	<2	n.d.
	Max	147	<2	27.5	<2	19.7	n.d.	<2	17.5	36.0	<2	<2	n.d.
Sulfamethoxazole	Ø	<1	<1	<1	31.6	15.1	<1	<1	n.d.	n.d.	<1	<1	<1
	Max	<1	<1	<1	59.1	29.2	<1	<1	n.d.	n.d.	<1	<1	<1

Table A4. Mean (Ø) and maximum concentrations in mg/L of main cations and anions and DOC in the river and RBF well water at sampling points in Uttarakhand, 2015–2018 (n.d.—not determined; * no data in May 2016). κ = conductivity.

Parameter in mg/L		Haridwar (n = 5)		Srinagar (n = 5)		Karnaprayag (n = 5)		Gauchar (n = 5)		Agastmuni (n = 5)		Satpuli (n = 5)	
		River	Well	River	Well	River	Well	River	Well	River	Well	River	Well
As^{3+}	Ø	0.00	n.d.	0.01	0.02	0.00	0.03	0.00	0.00	n.d.	n.d.	n.d.	n.d.
	Max	0.01	n.d.	0.05	0.03	0.01	0.13	0.01	0.01	n.d.	n.d.	n.d.	n.d.
Ca^{2+}	Ø	16.9	36.2	60.8	33.9	15.5	23.1	26.5	33.6	15.6	24.6	13.1	13.1
	Max	23.0	64.0	236	42.7	29.7	33.4	53.7	51.3	36.2	33.9	16.3	22.1
Fe^{2+}	Ø	0.02	<0.02	<0.02	<0.02	<0.02	0.06	<0.02	0.05	0.03	0.02	<0.02	0.06
	Max	0.05	0.03	0.05	0.04	0.03	0.28	0.03	0.18	0.11	0.05	<0.02	0.23
Mg^{2+}	Ø	4.50	8.14	17.4	10.2	3.82	6.70	7.34	8.72	2.54	6.56	4.28	3.97
	Max	6.39	17.1	72.6	11.8	7.70	10.8	16.9	14.5	8.05	11.9	6.09	8.48
Mn^{2+}	Ø	<0.002	0.02	0.01	0.02	0.01	0.04	0.00	0.01	0.01	0.02	<0.002	0.00
	Max	0.01	0.04	0.03	0.04	0.02	0.21	0.01	0.01	0.01	0.10	<0.002	0.01
Na^{+}	Ø	2.33	4.11	1.44	6.96	1.34	5.62	2.48	3.80	4.09	9.09	5.83	7.49
	Max	2.86	7.12	2.10	7.60	2.13	18.2	5.42	5.17	12.3	16.0	8.47	12.9
Cl^{-}	Ø	3.08	51.3	8.25	2.85	1.64	10.4	2.85	11.0	8.53	12.6	3.34	6.53
	Max	8.35	133	36.3	2.95	4.64	45.3	7.13	35.6	39.5	17.4	5.69	14.5
F^{-}	Ø	0.60	0.45	0.43	0.34	0.40	0.43	0.44	0.19	0.26	0.24	0.49	0.47
	Max	1.40	0.83	1.31	0.38	1.35	1.39	1.44	0.55	0.73	0.51	1.23	1.23
NO_3^{-}	Ø	6.5	11.9	17.0	4.6	3.06	15.9	6.95	27.0	1.08	23.4	0.84	7.54
	Max	21.5	26.1	78.9	6.0	11.5	66.8	19.0	88.0	2.48	36.9	1.65	26.7
NO_2^{-}	Ø	0.01	0.04	0.01	<0.004	<0.004	0.04	0.01	<0.004	0.02	0.01	<0.004	<0.004
	Max	0.02	0.15	0.03	<0.004	0.01	0.16	0.01	0.01	0.06	0.05	<0.004	0.01
PO_4^{3-} *	Ø	<0.004	<0.004	<0.004	n.d.	<0.004	<0.004	<0.004	<0.004	<0.004	<0.004	<0.004	<0.004
	Max	<0.004	<0.004	<0.004	n.d.	<0.004	<0.004	<0.004	<0.004	<0.004	<0.004	<0.004	<0.004
SO_4^{2-}	Ø	16.5	24.9	13.2	36.7	15.2	16.4	11.6	8.7	9.07	10.8	7.48	9.62
	Max	23.5	35.0	18.3	37.9	23.1	27.5	20.2	15.2	13.5	15.6	11.8	14.2
DOC	Ø	1.36	0.87	1.36	1.43	1.03	1.31	0.87	0.73	0.97	0.93	1.61	1.10
	Max	3.54	1.64	2.36	1.60	1.40	1.90	1.20	0.95	1.47	1.24	2.48	1.28
κ in µS/cm	Ø	159	581	156	911	144	289	136	517	88	334	82	136
	Max	194	940	229	1290	155	312	148	705	95	372	107	160

References

1. Kim, M.-K.; Zoh, K.-D. Occurrence and removals of micropollutants in water environment. *Environ. Eng. Res.* **2016**, *21*, 319–332. [CrossRef]
2. Duan, W.; He, B.; Takara, K.; Luo, P.; Nover, D.; Sahu, N.; Yamashiki, Y. Spatiotemporal evaluation of water quality incidents in Japan between 1996 and 2007. *Chemosphere* **2013**, *93*, 946–953. [CrossRef] [PubMed]
3. Schwarzenbach, R.P.; Escher, B.I.; Fenner, K.; Hofstetter, T.B.; Johnson, C.A.; von Gunten, U.; Wehrli, B. The challenge of micropollutants in aquatic systems. *Science* **2006**, *313*, 1072–1077. [CrossRef] [PubMed]
4. Duan, W.; He, B.; Nover, D.; Yang, G.; Chen, W.; Meng, H.; Zou, S.; Liu, C. Water Quality Assessment and Pollution Source Identification of the Eastern Poyang Lake Basin Using Multivariate Statistical Methods. *Sustainability* **2016**, *8*, 133. [CrossRef]
5. Lapworth, D.J.; Baran, N.; Stuart, M.E.; Ward, R.S. Emerging organic contaminants in groundwater: A review of sources, fate and occurrence. *Environ. Pollut.* **2012**, *163*, 287–303. [CrossRef] [PubMed]
6. Deblonde, T.; Cossu-Leguille, C.; Hartemann, P. Emerging pollutants in wastewater: A review of the literature. *Int. J. Hyg. Environ. Health* **2011**, *214*, 442–448. [CrossRef] [PubMed]
7. Kumar, P.; Mehrotra, I.; Gupta, A.; Kumari, S. Riverbank Filtration: A Sustainable Process to Attenuate Contaminants during Drinking Water Production. *J. Sustain. Dev. Energy Water Environ. Syst.* **2017**, *6*, 150–161. [CrossRef]
8. Rana Rajender, S.; Singh, P.; Singh, R.; Sanjay, G. Assessment of physico-chemical pollutants in pharmaceutical industrial wastewater of pharma city, Selaqui, Dehradun. *Int. J. Res. Chem. Environ.* **2014**, *4*, 136–142.
9. Gupta, A.; Ronghang, M.; Kumar, P.; Mehrotra, I.; Kumar, S.; Grischek, T.; Sandhu, C.; Knoeller, K. Nitrate contamination of riverbank filtrate at Srinagar, Uttarakhand, India: A case of geogenic mineralization. *J. Hydrol.* **2015**, *531*, 626–637. [CrossRef]
10. Bartak, R.; Page, D.; Sandhu, C.; Grischek, T.; Saini, B.; Mehrotra, I.; Jain, C.K.; Ghosh, N.C. Application of risk-based assessment and management to riverbank filtration sites in India. *J. Water Health* **2015**, *13*, 174–189. [CrossRef]
11. Krishan, G.; Singh, S.; Sharma, A.; Sandhu, C.; Grischek, T.; Gosh, N.C.; Gurjar, S.; Kumar, S.; Singh, R.P.; Glorian, H.; et al. Assessment of water quality for river bank filtration along Yamuna river in Agra and Mathura. *Int. J. Environ. Sci.* **2016**, *7*, 56–67.
12. Aleem, A.; Malik, A. Genotoxicity of the Yamuna River water at Okhla (Delhi), India. *Ecotoxicol. Environ. Saf.* **2005**, *61*, 404–412. [CrossRef] [PubMed]
13. Kaushik, C.P.; Sharma, H.R.; Jain, S.; Dawra, J.; Kaushik, A. Pesticide residues in river Yamuna and its canals in Haryana and Delhi, India. *Environ. Monit. Assess.* **2008**, *144*, 329–340. [CrossRef] [PubMed]
14. Kumar, B.; Singh, S.K.; Mishra, M.; Kumar, S.; Sharma, C.S. Assessment of polychlorinated biphenyls and organochlorine pesticides in water samples from the Yamuna River. *J. Xenobiotics* **2012**, *2*, 6. [CrossRef]
15. Mukherjee, I.; Gopal, M. Organochlorine insecticide residues in drinking and ground water in and around Delhi. *Environ. Monit. Assess.* **2002**, *76*, 185–193. [CrossRef]
16. Mutiyar, P.K.; Mittal, A.K.; Pekdeger, A. Status of organochlorine pesticides in the drinking water well-field located in the Delhi region of the flood plains of river Yamuna. *Drink. Water Eng. Sci.* **2011**, *4*, 51–60. [CrossRef]
17. Mutiyar, P.K.; Mittal, A.K. Status of organochlorine pesticides in Ganga river basin: Anthropogenic or glacial? *Drink. Water Eng. Sci.* **2013**, *6*, 69–80. [CrossRef]
18. Singh, R.P. Comparison of organochlorine pesticide levels in soil and groundwater of Agra, India. *Bull. Environ. Contam. Toxicol.* **2001**, *67*, 0126–0132. [CrossRef]
19. Mutiyar, P.K.; Gupta, S.K.; Mittal, A.K. Fate of pharmaceutical active compounds (PhACs) from River Yamuna, India: An ecotoxicological risk assessment approach. *Ecotoxicol. Environ. Saf.* **2018**, *150*, 297–304. [CrossRef]
20. Thacker, N.; Bassin, J.; Deshpande, V.; Devotta, S. Trends of organochlorine pesticides in drinking water supplies. *Environ. Monit. Assess.* **2008**, *137*, 295–299. [CrossRef]

21. Sharma, H.R.; Trivedi, R.C.; Akolkar, P.; Gupta, A. Micropollutants levels in macroinvertebrates collected from drinking water sources of Delhi, India. *Int. J. Environ. Stud.* **2003**, *60*, 99–110. [CrossRef]

22. Ghosh, N.C.; Mishra, G.C.; Sandhu, C.S.S.; Grischek, T.; Singh, V.V. Interaction of Aquifer and River-Canal Network near Well Field. *Groundwater* **2015**, *53*, 794–805. [CrossRef] [PubMed]

23. Gutiérrez, J.P.; van Halem, D.; Rietveld, L. Riverbank filtration for the treatment of highly turbid Colombian rivers. *Drink. Water Eng. Sci.* **2017**, *10*, 13–26. [CrossRef]

24. Glorian, H.; Schmalz, V.; Lochyński, P.; Fremdling, P.; Börnick, H.; Worch, E.; Dittmar, T. Portable Analyzer for On-Site Determination of Dissolved Organic Carbon—Development and Field Testing. *Int. J. Environ. Res. Public. Health* **2018**, *15*, 2335. [CrossRef] [PubMed]

25. Wintgens, T.; Nättorp, A.; Elango, L.; Asolekar, S.R. (Eds.) *Natural Water Treatment Systems for Safe and Sustainable Water Supply in the Indian Context, Saph Pani*; IWA Publishing: London, UK, 2016; ISBN 978-1-78040-710-4.

26. Ronghang, M.; Gupta, A.; Mehrotra, I.; Kumar, P.; Patwal, P.; Kumar, S.; Grischek, T.; Sandhu, C. Riverbank filtration: A case study of four sites in the hilly regions of Uttarakhand, India. *Sustain. Water Resour. Manag.* **2018**. [CrossRef]

27. Krishan, G.; Singh, S.; Sharma, A.; Sandhu, C.; Kumar, S.; Kumar, C.P.; Gurjar, S. Assessment of river Yamuna and groundwater interaction using isotopes in Agra and Mathura area of Uttar Pradesh, India. *Int. J. Hydrol.* **2017**, *1*, 86–89. [CrossRef]

28. DIN Deutsches Institut für Normung e.V., Arbeitsausschuss Chemische Terminologie (AChT). *D. I. für Normung, DIN EN 1484–1997: Anleitungen zur Bestimmung des Gesamten Organischen Kohlenstoffs (TOC) und des Gelösten Organischen Kohlenstoffs (DOC)*; Beuth Verlag GmbH: Berlin, Germany, 1997.

29. Reemtsma, T.; Miehe, U.; Duennbier, U.; Jekel, M. Polar pollutants in municipal wastewater and the water cycle: Occurrence and removal of benzotriazoles. *Water Res.* **2010**, *44*, 596–604. [CrossRef] [PubMed]

30. Groeschke, M.; Frommen, T.; Taute, T.; Schneider, M. The impact of sewage-contaminated river water on groundwater ammonium and arsenic concentrations at a riverbank filtration site in central Delhi, India. *Hydrogeol. J.* **2017**, *25*, 2185–2197. [CrossRef]

31. Scheurer, M.; Storck, F.R.; Brauch, H.-J.; Lange, F.T. Performance of conventional multi-barrier drinking water treatment plants for the removal of four artificial sweeteners. *Water Res.* **2010**, *44*, 3573–3584. [CrossRef]

32. Jekel, M.; Dott, W. *RISKWA Leitfaden: Polare Organische Spurenstoffe als Indikatoren im Anthropogen Beeinflussten Wasserkreislauf*; DECHEMA e.V.: Frankfurt, Germany, 2013.

33. Schmidt, C.K.; Lange, F.T.; Brauch, H.-J. Assessing the impact of different redox conditions and residence times on the fate of organic micropollutants during riverbank filtration. In Proceedings of the 4th International Conference on Pharmaceuticals and Endocrine Disrupting Chemicals in Water, Minneapolis, MN, USA, 13–15 October 2004; Volume 13, p. 2004.

34. Ens, W.; Senner, F.; Gygax, B.; Schlotterbeck, G. Development, validation, and application of a novel LC-MS/MS trace analysis method for the simultaneous quantification of seven iodinated X-ray contrast media and three artificial sweeteners in surface, ground, and drinking water. *Anal. Bioanal. Chem.* **2014**, *406*, 2789–2798. [CrossRef]

35. ECHA Classifications-CL Inventory-Diuron. Available online: https://echa.europa.eu/de/information-on-chemicals/cl-inventory-database/-/discli/details/446 (accessed on 18 October 2018).

36. Liu, Y.-S.; Ying, G.-G.; Shareef, A.; Kookana, R.S. Biodegradation of three selected benzotriazoles under aerobic and anaerobic conditions. *Water Res.* **2011**, *45*, 5005–5014. [CrossRef] [PubMed]

37. Regnery, J.; Wing, A.D.; Alidina, M.; Drewes, J.E. Biotransformation of trace organic chemicals during groundwater recharge: How useful are first-order rate constants? *J. Contam. Hydrol.* **2015**, *179*, 65–75. [CrossRef] [PubMed]

38. Bertelkamp, C.; Reungoat, J.; Cornelissen, E.R.; Singhal, N.; Reynisson, J.; Cabo, A.J.; van der Hoek, J.P.; Verliefde, A.R.D. Sorption and biodegradation of organic micropollutants during river bank filtration: A laboratory column study. *Water Res.* **2014**, *52*, 231–241. [CrossRef] [PubMed]

39. Bertelkamp, C.; Schoutteten, K.; Vanhaecke, L.; Vanden Bussche, J.; Callewaert, C.; Boon, N.; Singhal, N.; van der Hoek, J.P.; Verliefde, A.R.D. A laboratory-scale column study comparing organic micropollutant removal and microbial diversity for two soil types. *Sci. Total Environ.* **2015**, *536*, 632–638. [CrossRef] [PubMed]
40. Nivala, J.; Kahl, S.; Boog, J.; van Afferden, M.; Reemtsma, T.; Müller, R.A. Dynamics of emerging organic contaminant removal in conventional and intensified subsurface flow treatment wetlands. *Sci. Total Environ.* **2019**, *649*, 1144–1156. [CrossRef] [PubMed]

water MDPI

Article

Performance of Riverbank Filtration under Hydrogeologic Conditions along the Upper Krishna River in Southern India

T.B. Boving [1,*], K. Patil [2], F. D'Souza [2], S.F. Barker [3], S.L. McGuinness [3], J. O'Toole [3], M. Sinclair [3], A.B. Forbes [3] and K. Leder [3]

[1] Department of Geosciences, Department of Civil and Environmental Engineering, University of Rhode Island, Kingston, RI 02881, USA

[2] The Energy and Resources Institute (TERI), Goa 403202, India; kavitah@teri.res.in (K.P.); fraddry.dsouza@teri.res.in (F.D.)

[3] School of Public Health and Preventive Medicine, Monash University, Melbourne, Victoria 3004, Australia; fiona.barker@monash.edu (S.F.B.); sarah.mcguinness@monash.edu (S.L.M.); joanne.otoole@monash.edu (J.O.); martha.sinclair@monash.edu (M.S.); andrew.forbes@monash.edu (A.B.F.); karin.leder@monash.edu (K.L.)

* Correspondence: boving@uri.edu; Tel.: +1-401-874-7053; Fax: +1-401-874-2195

Received: 10 November 2018; Accepted: 13 December 2018; Published: 21 December 2018

Abstract: Riverbank filtration (RBF) systems were installed in four rural villages along a 64 km stretch of the upper Krishna River in southern India; with each one designed to supply approximately 2500 people. Site selection criteria included hydrogeological suitability, land availability and access, proximity to villages and their population sizes, and electric power supply. Water samples were collected from the river and the RBF wells over more than one year (November 2015 to December 2017) and were analyzed for *Escherichia coli* bacteria, major ions, and a range of other physicochemical and chemical parameters. The shallow groundwater at the study sites was also sampled, but less frequently. The hydrogeology of the four RBF systems was described in terms of bore-log data, mixing of river and groundwater, pumping test data, and vertical water column profiling. *E. coli* removal percentages of >99.9% were observed immediately before and during the monsoon, when *E. coli* concentrations in the river were the highest. The results provide evidence that RBF installations are challenging but possible under the climate and hydrogeologic conditions prevailing in this part of southern India. Specifically, when installing RBF wells in the study, area one needs to balance the well depth and set-back distance from the river against the limited extent of alluvial deposits. The viability of RBF systems as a domestic water source is also influenced by other factors that are not limited to southern India, including surface water and groundwater salinity, agricultural practices surrounding RBF wells, and the reliability of the power grid.

Keywords: riverbank filtration (RBF); Krishna River; southern India; water treatment; water quality; salinity

1. Introduction

Access to safe drinking water is essential to human health, but affordable and sustainable solutions remain out of the reach of many communities, particularly in rural areas of developing countries, such as India. Despite India being ranked among the top ten water rich countries with 4% of the world's fresh water resources, access to fresh water in the world's second most populous nation is problematic [1,2]. Water availability in India is highly variable spatially and temporally, and it is strongly influenced by the southwest monsoon. Coupled with widespread pollution of both surface- and groundwater resources, the people of India increasingly face water shortages and water borne disease outbreaks [3]. Herein, an affordable and sustainable water treatment approach to produce water suitable for domestic use is presented.

While India's urban citizens typically have access to improved water sources and sanitation facilities, several hundred million people in rural areas still cannot access adequate water supplies for sanitation and consumption [4,5]. Rural villages typically receive a mixture of public and/or private water supplies, mainly from groundwater wells, piped, or, if available, truck-delivered. For those living in close proximity to a surface water body, river and/or lake water may provide a significant proportion of water for domestic uses, including for drinking. Groundwater supplies are vulnerable to chemical contamination due to geological formations and leaching, as well as microbiological contamination that is associated with subsurface infiltration and/or surface runoff entering the well, notably occurring with shallow aquifers. River water supplies are prone to contamination with industrial wastewater and human-derived effluent, resulting in both chemical and microbiological contamination.

While effective drinking water treatment options exist, such as reverse osmosis systems, these are typically out of the reach of the rural poor, and there remains a need for affordable and sustainable water treatment solutions. One such technology is Riverbank Filtration (RBF), which has been used in Europe for over 100 years [6,7]. In RBF systems, water is withdrawn from one or more wells near a river. Wells may either be vertical or horizontal and are typically installed at least 50 m away from the river [8]. Pumping a RBF well lowers the water table and river water, together with some groundwater from the land-side, is induced to flow through porous riverbed (alluvial) sediments [9]. As raw surface water travels towards the RBF well, pathogens and suspended contaminants are removed or significantly reduced via a combination of physical, chemical, and biological processes [6]. Bacterial pathogen removal efficiencies of >99.9% can be achieved and heavy metal concentrations are reduced [10–12]. When compared to direct surface water abstraction, a disadvantage of RBF could be an increase in salinity or hardness due to the dissolution of minerals in the aquifer or mixing with brackish groundwater.

RBF technology relies on natural, auto-regenerative treatment processes, so properly engineered RBF systems can essentially remain indefinitely effective. Also, the depth to groundwater in the vicinity of rivers is relatively shallow in most areas, which generally makes RBF wells located near rivers less costly to drill and highly productive [6,13]. For these reasons, RBF technology is well suited for use in both developing and industrial countries [10,13,14].

When compared to the north of the country, there are limited data available about the performance of RBF in southern India [10,15], where the climate and hydrogeological conditions along most major rivers are generally less favorable for RBF. This is because the lower amounts of annual precipitation east of the Western Ghats Mountains and the absence of snow melt [16] cause some rivers to flow intermittently or exhibit large stage fluctuations. Also, groundwater salinity levels tend to increase toward south India and the country's arid western states [17]. Therefore, installing RBF systems under the conditions prevailing in southern India poses unique, but little studied challenges. This paper, one of the few reports documenting RBF performance under conditions that are frequently encountered in southern India, highlights some of these challenges. The water quality and hydrogeologic data presented herein and a description of experiences with operating RBF under conditions frequently encountered in southern India contribute to an enhanced understanding of the performance of RBF outside the well-researched locations in northern India. The results of this study might be of interest to water supply authorities and regulators seeking inexpensive water treatment solutions for villages that are in close proximity to surface water bodies without current access to public/private water supply system.

2. Materials and Methods

2.1. Study Area

The study was carried out in the Athani Taluka, Belagavi district, in northwestern Karnataka, India (Figure 1). According to the 2011 census [18], the total population of the Taluka is 525,832 and the population density is 120/km^2. Agriculture is the main occupation, with sugar cane being the main commercial crop. Approximately 94% of the geographical area of the Athani Taluka (199,500 ha) is under irrigation [19]. During several field visits in 2015, four villages were identified as suitable from a hydrogeological perspective for RBF well installation. The maximum distance between these four sites is 64 km of river length. The four study locations (Village 1 through 4) along the Krishna River are illustrated in Figure 1.

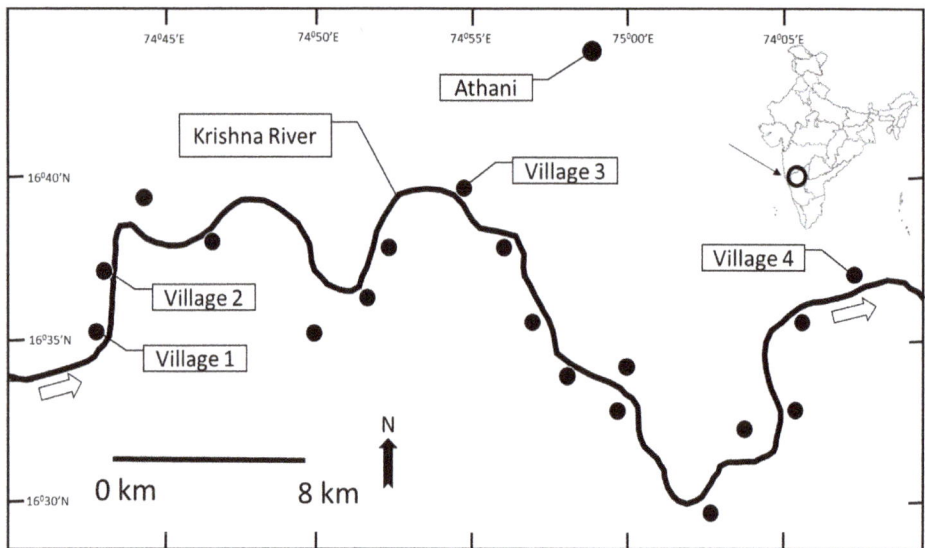

Figure 1. Location of the four villages served by Riverbank Filtration (RBF) water. All villages are located along the Krishna River near Athani in the northwestern part of the state of Karnataka, India. River flow direction indicated by arrows. ● Villages.

The study area is part of the Krishna river basin, which is India's fourth largest, spreading across 7.9% of its surface area. The river Krishna, along with its tributaries Ghataprabha and Malaprabha, is perennial and it flows in an easterly direction. Flow in the river is dominated by the southwest monsoon climate, which induces large inter-seasonal variations in the river stage. In addition, river water flow outside of the monsoon season is dependent upon water releases from dams located upstream of the study area. The climate in the study area is semi-arid. Most of the mean annual rainfall (582 mm) is received during the period June to October. May is typically the hottest month, with temperatures exceeding 39 °C.

Geologically, the area is dominated by the Basalt of the Tertiary Deccan Trap formation. At depth, leached alumina clay is found on top of the weathered massive trap. At the surface, the basaltic bedrock is predominantly weathered to Vertisol, known in India as Black Cotton Soil. The soil cover in the area ranges in thickness from zero to 25 m and it is generally fine grained and has a low porosity but permits the percolation of rainwater to the deep bedrock [19]. Alluvial deposits along the Krishna River are limited in extent and thickness. Composed of coarse sand, sandy-loam, and loams, these sediments have good infiltration characteristics [20].

Deccan basalt is the primary, multilayer aquifer having low to medium permeability [21]. Intra-trap red bole beds act locally as aquicludes. Shallow unconfined aquifers comprised of unconsolidated deposits above the bedrock are exploited locally. Groundwater occurrence in the Deccan basalt is generally controlled by secondary porosity and it occurs under mostly confined and semi-confined conditions. There are about 2750 borewells and 14,676 dugwells reported in the Athani Taluka [20]. Borewells are usually 40 m to 175 m deep, with yields ranging from 40 to 1440 m^3/day. Water level fluctuations of 5 m to 10 m are common as river stages fluctuate between seasons. The main source of recharge to all aquifers is precipitation and water applied for irrigation [19]. In the Athani Taluka, the annual recharge is in the range of 100 mm to 150 mm [20].

2.2. Site Selection

The upper Krishna River watershed was selected for this study because it shares many characteristics of southern India's many polluted rivers [22,23]. That is, local sewage treatment plants, where present, are typically working beyond capacity, so most wastewater is discharged into the river untreated. Also, open defecation is widely practiced in the predominantly rural watershed, resulting in the entry of high levels of human faecal runoff into the river [24]. In addition, most rural villages in the study area lack public water treatment facilities, resulting in villagers abstracting untreated water for domestic use directly from the river or relying on shallow dugwells or handpumps. The study sites were selected based on several criteria, including (1) similar village sizes (~2500 people), (2) community receptiveness to installation of RBF systems, (3) use of untreated river water as a primary source of drinking water, (4) availability of suitable land close to the river (<100 m) for RBF well drilling, (5) proximity of RBF wells to villages to limit pipeline construction, and (6) access to electric power lines.

Prior to drilling, a geophysical survey (Schlumberger geoelectric method) was conducted at potential sites with permission for drilling being granted by the District Groundwater Office, Belagavi, India. In addition, data on water quality and quantity were reviewed [19–21,25].

2.3. Well Construction

Besides a number of exploration wells, six RBF wells were installed at the four villages (Table 1). All RBF wells were drilled by the rotary air drilling method to a diameter of 10 inches (25 cm) and cased with 8 inch (20 cm) PVC pipe, inserted at least 1 m into the bedrock. Below that level (well sump), the well diameter was 7.5 inches (19 cm). Pipes were slotted in the field, using 3 mm slot cutters. The slotted sections, ranging from 7.7 m to 9.2 m in length, were set at the contact of the bedrock and the unconsolidated sediments above (Table 1). Solid PVC pipe was used for the remaining length to the surface. Drill cuttings filled the angular space between the pipe and the borehole wall, except for the uppermost 1 m where either bentonite or concrete was used to minimize infiltration from the surface. The top of each well was set at about 0.6 m above surface, capped, and protected with a 1 m by 1 m concrete foundation and a metal cage (Figure S1). Depending on the well yield, 5 HP or 7.5 HP electric pumps were installed, together with flow meters, designated sampling ports, and water level loggers (Solinst, Georgetown, ON, Canada). All wells were disinfected immediately after drilling following the American Water Works Association standard procedures [26] amended with information from the Washington Department of Health [27]. As needed, disinfection procedures were repeated either after flooding of the well field during monsoon and/or after major pump maintenance.

2.4. Water Sampling and Analysis

Between November 2015 to January 2018, water samples were collected from each RBF well field and the adjacent Krishna River at each of the four villages. Measured weekly, the field parameters pH, electrical conductivity (EC), dissolved oxygen (DO), and temperature were determined with calibrated hand-held digital meters (Hanna pH-HI98128, EC-HI983003; DO-HI9146-04, Hanna Instruments, Woonsocket, RI, USA). Turbidity was measured with a Hanna HI98703 instrument. Samples for *E. coli* bacteria were collected in sterile plastic bottles, stored in coolers during transport, and analyzed within 12 to 24 h in a field laboratory that was set-up in Athani for the duration of this study. In the field lab, the U.S. EPA approved IDEXX Colilert-18 method [28] was used to quantify *E. coli* bacteria. All bacterial data were reported as Most Probable Number (MPN per 100 mL). Duplicate and negative control samples were analyzed for quality assurance. For *E. coli* data, non-detects were set equal to 0.5 MPN/100 mL to permit graphing on a logarithmic scale [29].

On a monthly basis, a commercial laboratory was used to analyze river and RBF water for major cations (Na, K, Ca, Mg) and anions (Cl, SO_4, NO_3, F, HCO_3, CO_3), the parameters NO_2, PO_4, B, and dissolved silica SiO_2, as well as Total Organic Carbon (TOC), Total Dissolved Solids (TDS), Total Suspended Solids (TSS), Biological Oxygen Demand (BOD) and Chemical Oxygen Demand (COD), total alkalinity, and hardness. Groundwater data was collected on two occasions in December 2017 and February 2018 and was analyzed for the same parameters as river and RBF water. The laboratory followed the Bureau of Indian Standards (BIS) IS 3025 (Part 45) Method of Sampling and Test (Physical and Chemical) for Water and Wastewater (First Revision) or the American Public Health Association (APHA) Standard Methods for the Examination of Wastewater (see Table S1 for details). Where appropriate, the results were related to the drinking water limits of the Bureau of Indian Standards [30] (Table S2). The heavy metal and pesticide data that were collected as part of this study are not presented herein.

3. Results and Discussion

3.1. RBF Settings

The geology encountered during drilling of the RBF wells consisted of bedrock between 15.4 m and 20.8 m (Table 1) and unconsolidated sediment above it. The sediment was weathered, silty to sandy silt Black Cotton soil (Vertisol). In villages 3 and 4, alluvial sediments consisting of sand and fine gravel were encountered in discontinuous layers that were no more than 2.5 m thick. In villages 1 and 2, water-bearing, more silty sediments dominated at the contact with the bedrock. A geologic profile, including well construction information, is shown in Figure S2.

Table 1. RBF well characteristics.

Location	Water Table (mbgl)	Depth Bedrock (mbgl)	RBF Well 1			RBF Well 2			Total Yield (m³/h)	Lpcd
			Depth (m)	L (m)	Yield (m³/h)	Depth (m)	L (m)	Yield (m³/h)		
Village 1	11.1	19.7	30.8	22	3	30.8	24	4	7	8 to 16
Village 2	8.3	20.8	30.8	30	5	NA	NA	NA	5	14 to 28
Village 3	8.3	18.5	24.6	40	12	NA	NA	NA	12	112
Village 4	4.6	15.4	18.5	50	5	24.6	25	7	12	15 to 30

RBF well field yields and supply of liters per capita per day (Lpcd) based on duration of daily power supply available (24 h at Village 3. Everywhere else: 6 h, except 3 h during April to June). Also provided: depth to water table (post monsoon) and bedrock, depth of well, and distance to river (L). All depths are reported in meters below ground level (mbgl). NA: Not applicable.

At all locations, the wells were drilled approximately 5 m to 11 m above the post-monsoon river stage. Six RBF wells were installed across the four villages (Table 1). The two wells at Village 1 are discussed together, because the site constraints made it necessary to install them in close proximity (2 m apart). The set-back distance (L) between the river and the RBF wells ranged from 22 m to 50 m. The depth to the water table ranged from 4.6 m to 11.0 m, depending on the site (Table 1). During monsoon, water table elevations rose by several meters, mirroring the river stage. For the first time in recent history, the Krishna River dried up in April and May 2016. During that time, the RBF wells still yielded water, which was apparently fed by baseflow. Flooding of the Village 4 well field occurred during the heavier than usual 2016 Monsoon, but not during the following year. In response to the flooding, an additional well (RBF2) was drilled at a more protected location about 1 km downstream. All well locations were surrounded by irrigated agricultural land, with sugar cane being the dominant crop.

Based on the village size (2358 ± 261 people) and around-the-clock power supply, a RBF well field yield of 5.4 ± 0.4 m^3/h would have been sufficient to meet the 55 liters per capita per day (Lpcd) target that was established by the Government of India for rural villages [31]. The measured yields of the four RBF well field locations ranged from 3 m^3/h to 12 m^3/h with an average of 6 m^3/h (Table 1). However, only one well field (Village 3) received continuous power from the electric grid serving the village. In all other villages, access to power was limited to six hours daily during most of the year and three hours per day during the driest months (April through June). Due to the scheduled power outages, the 55 Lpcd target could not be met at these locations.

3.2. Water Quality

The major ion concentration data for the RBF wells (n_{RBF} = 46), the river (n_{River} = 32) and local groundwater (n_{GW} = 12) are presented in Table S3. The average ion balance error was 8.6%. Because of the elevated ion balance error, the major ion data were regarded as estimates. Data for the field parameters, including *E. coli* bacteria, are presented in Table S4 (n_{River} = 442 and n_{RBF} = 320).

3.2.1. Groundwater

Shallow groundwater samples from wells <36.5 m deep were collected from local handpumps and borewells (n = 12), except in Village 3, where no wells were accessible (Supplementary Tables S2 and S3). All of the sampled wells were located 180 m or less land-ward from the RBF well fields. *E. coli* concentrations were low (non-detects to 2 MPN per 100 mL) in Villages 1 and 4, but they ranged from 5 to 17 MPN per 100 mL in Village 2. The average pH was 7.8 ± 0.5. The measured EC was high in all groundwater wells, with an average of 4906 µS/cm. That value was much higher than the 2420 µS/cm average reported by the Government of India [20] for the study area (Athani Taluka) and also outside the range (1500 to 3000 µS/cm) reported by the Central Ground Water Board of Karnataka [21]. However, EC readings were within the range that was reported by Purandara [25]. The highest recoded EC reading (8692 µS/cm) was measured in a shallow (18.3 m deep) exploration well that was drilled in Village 1. The comparatively high standard deviation (±1593 µS/cm) reflects high spatial and temporal variability in groundwater EC. Turbidity was up to 11.9 NTU (average: 4.4 NTU). Since the tested groundwater wells were not used on regular basis, the turbidity readings were considered to be artificially elevated because of the disturbance that is caused by the sampling. BOD ranged from 0.1 to 2.8 mg/L (average: 0.7 mg/L) and COD from 21.8 to 174.2 mg/L (average: 58.0 mg/L). The highest COD values were measured in Village 4.

High major ion concentrations were recorded in groundwater samples, particularly Ca, Mg, Cl, SO_4, and NO_3 (Table S3). Most concentrations exceeded the BIS permissible drinking water limits (Table S2) and were detected primarily in water samples that were collected from handpumps. High major ion concentrations were also reflected in the recorded total dissolved solids (3412 ± 1136 mg/L) and total hardness data (1801 ± 842 mg/L). Hardness at this level impacts the ability of water to form lather with soap and increases the boiling point of water [32]. Fluoride concentrations above the BIS limits (1 mg/L) were detected in Village 1 only (1.2 to 1.4 mg/L; Table S3). Overall, the high degree and extent of groundwater salinity in this part of the Krishna River watershed was not well documented prior to this study. The liberal application of fertilizers (particularly $MgSO_4$) on sugar cane fields was repeatedly observed, together with a reliance on flood irrigation. Together, these practices likely contributed to high ion concentrations in the local groundwater [33]. High groundwater salinity is a problem that is not unique to this study area, but it is widespread in this part of southern India [21].

3.2.2. River Water

Across the four study sites, the average pH was 8.3 ± 0.4 and the water temperature was 25 ± 0.4 °C (Table S4). The ranges and averages of the EC, turbidity, and *E. coli* parameters are summarized in Table 2.

Table 2. Electric conductivity (EC), turbidity and *E. coli* data for RBF and river water samples.

		River Water				
Parameter		Village 1 (n = 110)	Village 2 (n = 111)	Village 3 (n = 109)	Village 4 (n = 112)	Average
EC (μS/cm)	Average	965	855	1000	1440	
	Range	196–3728	196–3139	196–2943	177–4905	1065
	SD	624	499	650	939	
Turbidity(NTU)	Average	11.5	9.2	11.5	13.4	
	Range	0.5–60	0.7–75	0.4–73	0.3–97	11.5
	SD	10	10	12	14	
E.coli (MPN/100 mL)	Average	57.6	29.0	24.9	22.4	
	Range	0–2851	1–2420	0–1011	0–2420	34.2
	SD	533.6	441.4	180.4	420.6	
		RBF Well Water				
Parameter		Village 1 (n = 60)	Village 2 (n = 86)	Village 3 (n = 76)	Village 4 RBF 1 (n = 60)	Village 4 RBF 2 (n = 36)	Average
EC (μS/cm)	Average	2638	4506	2936	2993	3357	
	Range	1903–5337	2236–7710	1687–5297	1765–3924	2590–3885	3286
	SD	688	994	555	395	306	
Turbidity (NTU)	Average	3.4	2.7	3.1	9.7	7.1	
	Range	0.4–42	0.4–16	0.3–14	0.4–88	0.6–31	5.2
	SD	0.7	3.0	3.0	13.6	7.0	
E.coli (MPN/100 mL)	Average	0.8	0.8	0.6	0.8	0.8	
	Range	0–22	0–30	0–16	0–30	0–73	1.5
	SD	3.3	4.6	2.6	4.6	13.5	

EC and turbidity data are reported as averages (arithmetic means) with the ranges and standard deviations (SD). *E.coli* averages are reported as geometric means. n = number of samples. Note: One outlier (867 MPN/100 mL) was removed from the Village 3 RBF *E. coli* data set.

As a representative example, the river water data for Village 3 show that *E. coli* concentrations and turbidity spiked during the monsoon season, whereas EC declined to <250 μS/cm (Figure 2A through Figure 2C). The inflow of raw sewage and sediments, particularly at the start of the rainy season, explains the elevated bacteria and turbidity levels in the river, while dilution by rainwater reduced the river's EC values. Similar observations were reported during a RBF study along the Kali River, Karnataka [10]. Immediately after the monsoon, when entering the dry period, EC increased steadily to values exceeding 2000 μS/cm. The highest measured EC (2943 μS/cm) was observed in April and May 2016, when an exceptional drought caused the Krishna river to almost dry out. The BIS limit for turbidity (1 NTU) was consistently exceeded in the river over the observation period. The average BOD and COD were 1.6 mg/L and 27.6 mg/L, respectively, with the highest concentrations being recorded at Village 4. The average total hardness was 277 mg/L.

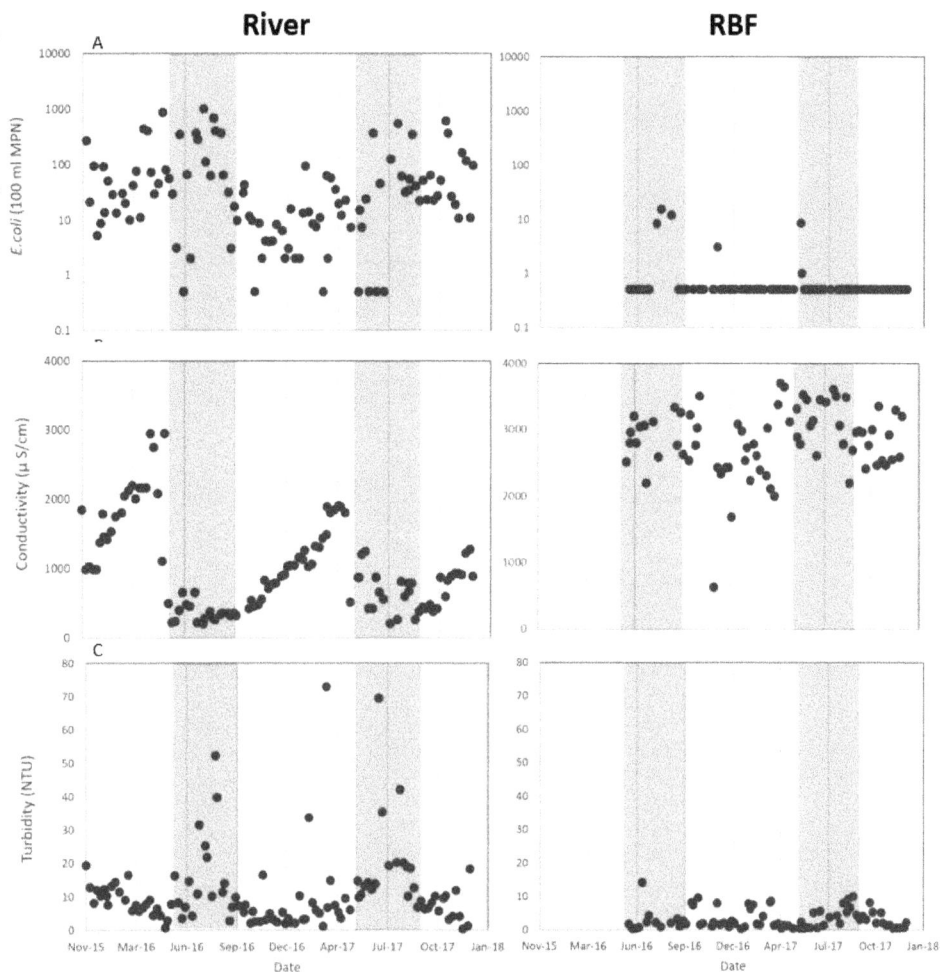

Figure 2. Measurements of (**A**) *E. coli*, (**B**) electrical conductivity, and (**C**) turbidity in the river (**left**) and RBF (**right**) water at Village 3. Shaded areas signify the monsoon season.

In Village 3, the *E. coli* concentrations in the river water in excess of 1000 MPN per 100 mL were measured during and immediately after the monsoon. High *E. coli* concentrations (>100 MPN per 100 mL) were also recorded during the pre-monsoon drought (April/May 2016) when only stagnant pools remained in the riverbed. Outside the pre-monsoon and monsoon season, large variations in *E.coli* counts were noted between sampling events, ranging from non-detects to greater than 500 MPN per 100 mL. Even larger fluctuations, up to 2851 MPN per 100 mL (Village 1), were recorded in the other three villages. Overall, the annual average of *E. coli* counts from Krishna River samples that were collected in the study area were comparatively lower (average: 34 MPN per 100 mL; Table 3) than those reported for many other rivers in India. This observation was attributed to the low population density in the watershed (120 people/km^2) relative to northern India, where five-times higher densities are common [10].

Table 3. Ranges, averages (arithmetic means), and means of major cation and anion analysis, including dissolved silicate and ion balance in RBF and river water samples. All concentrations are in mg/L.

RBF Wells

Parameter	Village 1 (n = 9)			Village 2 (n = 7)			Village 3 (n = 15)			Village 4 (n = 15)		
	Range	Average	SD	Range	Average	SD	Range	Average	SD	Range	Average	SD
Ca	28.9–60.1	40.3	9.7	20.8–128.3	67.6	35.5	24.1–100.2	64.7	20.3	24.1–216.4	101	49.2
Mg	36.0–113.8	88.3	21.4	161.5–367.2	270.3	77.2	75.4–178.0	117.6	26.8	26.8–256.8	174.8	54.8
Na	283.0–363.0	325.8	27.5	405.5–700.0	535.1	92.1	179.0–630.0	369	101.1	182.0–297.0	244.7	36.6
K	0.8–7.0	3.2	1.9	0.1–8.0	3.7	2.6	0.4–6.8	2.9	1.6	0.3–4.0	1.9	0.9
B	0	0	0	0	0	0	0	0	0	0	0	0
HCO_3	424.0–492.0	460	23.4	252.2–434.0	377.3	53.4	154.0–447.0	291.1	85.1	48.0–332.0	218.1	87.7
CO_3	30.0–118.0	70.2	23.2	10.0–102.0	68	31.5	10.0–52.0	27.2	12.5	10–224.0	45	72.1
Cl	92.3–564.5	220.1	128.02	450.9–752.6	569.8	105.2	230.8–568.0	397.8	86.5	404.7–699.4	504.1	66.8
SO_4	75.7–343.4	181	72.1	255.2–996.9	664.9	255.2	140.8–1209.7	351.4	268.3	130.0–576.6	240.8	113.5
F	0.4–0.6	0.5	0.1	0.9–1.0	1	0	0.2–0.7	0.5	0.1	0.4–0.6	0.5	0.1
NO_3	13.5–47.8	31.1	11.4	11.5–35.2	27.3	8.1	24.5–61.5	44.9	10.3	33.5–237.5	81.1	46.5
NO_2	0–0.1	0	0	0–0.1	0	0	0–0.0	0	0	0–0.3	0.1	0.1
PO_4	0.1–0.9	9	0.2	0.7–1.1	0.9	0.2	0–0.3	0.1	0.1	0–0.2	0.1	0.3
SIO_2	20.7–55.6	38.1	13.3	28.7–67.3	54.1	15.8	28.0–94.8	67.2	15.1	11.7–53.7	36.9	11.3
Ion Balance	9.0%			9.0%			7.9%			9.3%		

River Water

Parameter	Village 1 (n = 8)			Village 2 (n = 8)			Village 3 (n = 8)			Village 4 (n = 8)		
	Range	Average	SD	Range	Average	SD	Range	Average	SD	Range	Average	SD
Ca	14.4–50.3	28.3	12.7	9.6–48.1	26.8	14.3	16.0–76.2	35.1	20.2	19.2–124.3	46.8	36.6
Mg	2.9–29.2	18.2	8.8	7.3–46.7	19.6	13.2	2.3–75.4	23.4	21.5	3.9–107.0	34.9	32.5
Na	11.2–114.0	44.2	30.8	13.0–100.0	44.4	28.5	11.1–208.0	56.9	60.1	11.3–326.0	93.5	107.0
K	1.7–3.6	2.6	0.7	0.1–12.0	3.2	3.5	0.8–6.0	2.7	1.5	0.9–8.0	2.9	2.1
B	0	0	0	0	0	0	0	0	0	0	0	0
HCO_3	40.0–146.0	97.9	33.7	38.8–134.0	92	35.3	36.0–148.0	89.5	37.4	42.0–136.0	86.1	31.2
CO_3	10–20.0	11	5.5	0.1–30.0	12.5	9.2	0.1–26.0	13	10.5	0.1–22.0	10.1	8.0
Cl	17.8–88.8	49.7	24.9	14.2–106.5	50.1	29.7	10.7–234.3	66	68.4	10.7–447.3	126.9	147.9
SO_4	12.1–98.9	38.3	27.2	5.6–148.6	39.1	43.2	7.9–215.0	46.3	64.4	7.9–343.4	93.1	114.4
F	0.1–0.3	0.3	0.2	0.1–0.8	0.3	0.2	0.1–0.7	0.3	0.2	0–0.9	0.3	0.3
NO_3	1.4–33.4	11.8	9.5	2.0–15.9	8	4.4	0.1–14.2	8	4.9	2.5–14.2	7.2	4.6
NO_2	0–0.1	0	0	0–0.1	0	0	0–0.1	0	0	0–0.1	0	0
PO_4	0–0.3	0.1	0.1	0–0.1	0	0	0–0.3	0.1	0.1	0	0	0
SIO_2	0.2–28.7	14	7.9	4.0–28.9	16.1	7.1	10.7–27.5	18.7	5.8	11.5–25.1	17.3	4.0
Ion Balance	7.7%			6.9%			9.3%			8.3%		

The major ion concentration ranges, averages (arithmetic mean), standard deviation, including ion balance and including dissolved silica, at the four villages are summarized in Table 2. At all four locations, concentrations for Ca and Mg exceeded the BIS limits in 17.3% and 50% of samples, respectively (n = 52). Chloride and SO_4 exceedances were less frequent, being 11.5% and 9.6%, respectively. The NO_3 limit was exceeded only once and there were no exceedances for fluoride. Phosphate (as PO_4) and nitrite (as NO_2) were <0.4 mg/L and <0.1 mg/L, respectively. Boron was found to be below the detection limit of 0.1 mg/L in all samples.

3.2.3. RBF Well Water

Besides chemical and bacteriological measurements, each RBF well that contributed to the water supply of a village was tested for EC, pH, and temperature. The average *E. coli* concentration ranged from undetectable to 3.8 MPN per 100mL across all four villages (Table 2), with the highest measurement being recorded in Village 4 in the monsoon season (72 MPN per 100 mL). A similar spike of *E. coli* during monsoon was observed in all villages and is also noticeable in Figure 2B. The RBF well reduced peak *E. coli* concentrations by approximately two orders of magnitude. Beyond these spikes, *E. coli* concentrations were at or below the detection limit. In general, *E. coli* removal percentages of >99.9% were observed, equivalent to three $\log_{10}$ units immediately before and during the monsoon, when *E. coli* concentrations in the river where also highest (Figure 2A). Outside the monsoon season, when *E. coli* concentrations in the river water were low (<100 MPN per 100 mL), the minimum bacteria removal capacities that could be quantified with our method ranged from 90% to 99%.

The average EC across all sites was 3821 µS/cm (Table 3); 0.67 times lower than the groundwater (average: 4906 µS/cm) but 3.1 times higher than the river (Table 2). As the data in Figure 2B indicate, the EC in the RBF water remained elevated during the monsoon and afterwards. The average turbidity was 5.3 NTU, which was comparable to the groundwater (4.4 NTU) and approximately half as low as the river (Table 2). Also, as expected, Figure 2C clearly shows that the turbidity spikes observed in the river during monsoon as well as during the dry season were attenuated in the RBF water. The average pH was 7.7 (Table S4) and the water temperature was 26.4 °C on average. The BOD (average: 0.8 mg/L) of RBF treated water was half that of the river water and was similar to the groundwater. The COD (average: 32.1 mg/L) was similar to that of the river and 0.55 times lower than the groundwater. Total hardness (average: 852 mg/L) was three times higher than in the river, but was 2.1 times lower than the groundwater. All major ion BIS limits were frequently exceeded at all four villages (n = 78; Table S3). Higher then acceptable fluoride concentrations (>1.0 mg/L) were recorded in a small fraction (7.7%) of all samples (up to 1.4 mg/L; Village 2).

In terms of major ion concentrations (Table 3), the hydrogeochemical characteristics of the RBF water were closer to that of the shallow groundwater than the river (Figure 3A). However, differences existed between sites (Figure 3B–E). For instance, the composition of the RBF water at Village 2 was very similar to the GW (Figure 3D), except for Ca and NO_3 concentrations, which were both lower in the RBF well water. At Village 4 (Figure 3E), the RBF water was closer in composition to the river water, which suggests that a greater fraction of river water contributed to this RBF well field.

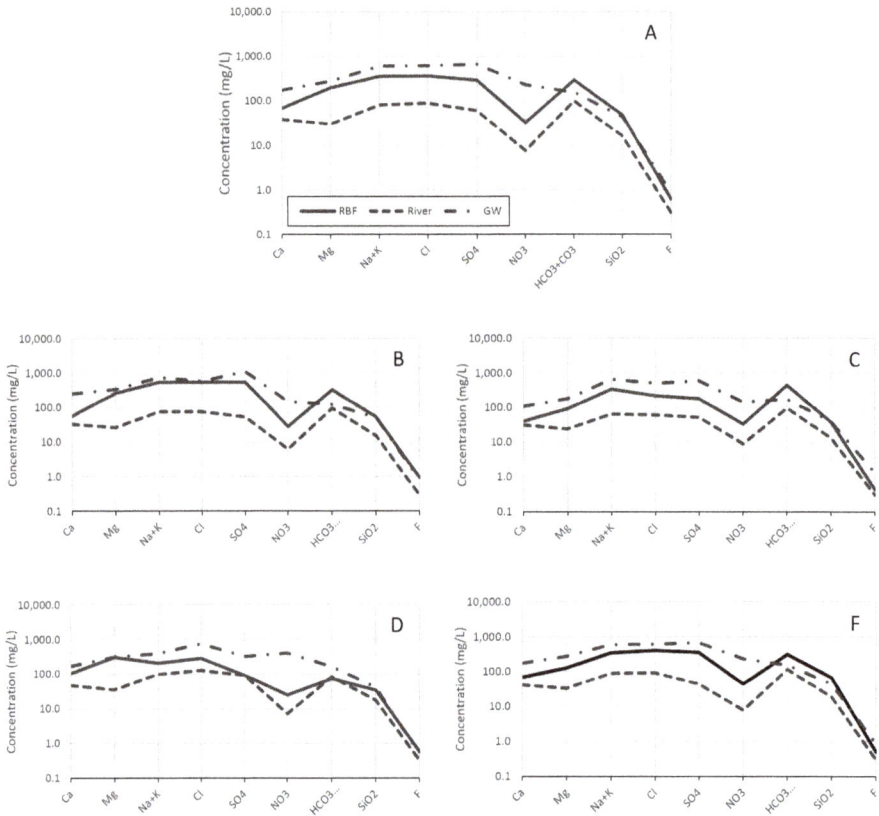

Figure 3. Major ion concentrations (mg/L) in the RBF wells (RBF, n = 52), the Krishna River (River, n = 43), and the shallow groundwater (GW, n = 10). (**A**) Averages across all villages, (**B**) Village 1, (**C**) Village 2, (**D**) Village 3, and (**E**) Village 4 RBF 2. No GW was analyzed in Village 3; the average GW composition was graphed instead.

A closer examination of the cation chemistry reveals that Na was generally the dominant cation in the RBF water, followed by Mg and Ca, with K only a minor constituent ($\leq$3.7 mg/L). In the RBF water at Village 4, Mg rather than Na was the dominant cation. In river water, Na was also the dominant cation, but Ca was of greater abundance relative to Mg. Cation concentrations in shallow groundwater followed a similar pattern as RBF and river water, but they were between 1.7 and 2.6 times higher when compared to RBF water (Table S3).

Regarding anions, Cl generally was the dominant anion in RBF water, followed by SO_4, and then HCO_3 + CO_3. The observed application of magnesium sulfate fertilizer to sugar cane fields surrounding RBF well fields likely contributed to the high SO_4 (and Mg) concentrations in RBF water. Average NO_3 concentrations ranged from 22.6 to 44.6 mg/L (as NO_3), while F was <1 mg/L. High NO_3 concentrations can result from geogenic and anthropogenic sources [34]; but with NO_3 leaching phyllitic and quartzite bedrock being absent, anthropogenic sources, including animal and human waste, were the likely origins of nitrate in the study area. In the Molwad RBF well, HCO_3 + CO_3 surpassed Cl as the dominant anion. When compared to Villages 3 and 4 (<0.3 mg/L), the phosphate concentrations were higher in Villages 1 and 2 (up to 1.2 mg/L), suggesting that more fertilizer was applied in these villages. As in the river, boron was found to be below the detection limit of 0.1 mg/L in all samples.

The dissolved major ionic species milliequivalents were plotted on a Piper diagram [35] to further investigate the hydrogeochemical character of RBF water (n = 45) in relation to river (n = 32) and shallow groundwater (n = 12) (Figure 4). The diagram indicates that 28.1% and 15.6% of the river water samples plot in the Ca-Mg-HCO$_3$ (I) and Na-Cl (III) segments, respectively, while the majority of river water samples (56.3%) are of mixed type (Ca-Mg-Na-Cl-HCO$_3$; V). In comparison, 46.1% of RBF samples are of mixed type (Ca-Mg-Na-Cl-HCO$_3$), 28.9% Na-Cl (III), and 25.0% Ca-Mg-Cl (II). At Village 1, the RBF water had a strong (62.5%) Na-Ca-HCO$_3$-Cl (VI) signature, setting this water apart from all other RBF sites. Shallow groundwater samples are a mixture of Ca-Mg-Cl (II), Na-Cl (III), and Cal-Mg-Cl-HCO$_3$ (V) types. Overall, the hydrogeochemical signatures of the RBF water at all sites, except Village 1, have the characteristics of predominantly mixed-type river and Na-Cl/Mg-SO$_4$ type groundwater. The unique hydrogeochemistry of the RBF water at Village 1 suggests that local conditions, such as inflow of Na-bicarbonate rich water typically found in deeper portions of aquifers, must be considered.

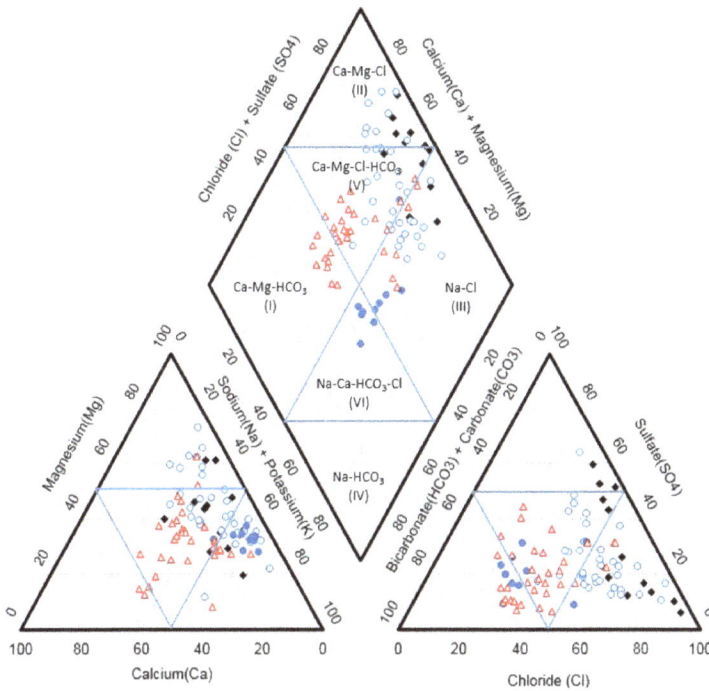

Figure 4. Piper diagram showing the hydrogeochemical character of the RBF water in comparison to the river and groundwater. Red triangles: River, Open blue circles: RBF, Bold blue circles: RBF in Village 1, Black diamonds: Groundwater.

In summary, the water quality results show that *E. coli* and turbidity values were lower in the RBF wells as compared with the Krishna River. While RBF water was highly mineralized relative to the river year-round, it was not as saline as the local shallow groundwater. Still, most major ion concentrations (magnesium, chloride, and sulfate, in particular) in RBF water exceeded the recommended BIS drinking water limits.

3.3. Well Field Hydraulics

A continuously pumped RBF well is expected to achieve the strongest hydraulic connection to the nearby river. Therefore, it must be expected that an intermittent power supply negatively impacts RBF water quality, because flow from the river to the well is interrupted during power outages, leading to a higher fraction of higher mineralized groundwater entering the well when pumping resumes. To investigate the impact of intermittent versus continuous pumping on RBF water quality, the pump in the RBF2 well in Village 4 was temporarily connected to a continuous electricity supply (diesel electric generator) and was pumped continuously for 11 days in September/October 2017. During the test, chloride and nitrate concentrations decreased by 6.2% and 16.2%, respectively (Figure 5). Additionally, the concentrations were lower (9% to 15% in the case of chloride, and 33% to 43% for nitrate) when compared to samples that were collected immediately before and after the test when pumping was intermittent. However, no corresponding changes were observed for other major ions, TDS, or EC. Hence, the pumping test did not conclusively answer whether a continuous power supply and a longer pumping duration or higher pumping rate could have increased the portion of bank filtrate in the pumped water, thereby enhancing the water quality in the RBF well.

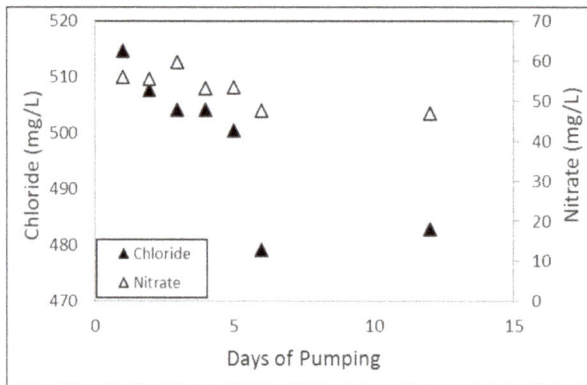

Figure 5. Chloride and nitrate concentrations decreased during continuous pumping of RBF2 well at Village 4.

In addition to continuous pumping, it was investigated if the well water quality, measured as EC, changed with depth. The assumption was that lower EC values that are associated with inflow of river water dominates at shallower depth, whereas higher EC values, which is water containing higher TDS loads, indicate the inflow of water from deeper parts of the aquifer. This kind of vertical stratification was previously reported for a well field located in southern India (Goa) [36]. The data show that the EC did indeed increase with depth, i.e. form approximately 3000 µS/cm to 3800 µS/cm (Figure S2). The EC in the upper part of the water column was 0.77 times lower than in the deeper part. That is, comparatively lower EC values coincided with the alluvial sediments (sand and medium gravel) that were encountered during well drilling 5.5 to 11 m below the water table, while a distinct jump in EC was observed between 11 m and 14 m below the water table. This depth coincided with the fractured bedrock depth. The vertical EC profile data suggests that river water with comparatively low EC (699 µS/cm at time of test) contributed to the RBF well.

The magnitude of the contribution of river water versus groundwater to a RBF well can be estimated from mixing calculations (Equation (1)), where the percentage of river water contribution to the RBF well is equal to:

$$\%River = \frac{(C_{RBF} - C_{GW})}{(C_R - C_{GW})} \tag{1}$$

where C_{RBF}, C_R, and C_{GW} are the concentrations of a compound in the RBF well water, the river, and the groundwater, respectively. Equation (1) was solved for the dominating ions (Na, Mg, and Cl) and EC, using field data collected in mid-February 2018. On that date, concurrent samples were collected from all three water sources (groundwater, RBF, and river), except at Village 3, where no GW wells were available for sampling in the vicinity of the RBF well. It is noted that all wells were installed along fairly straight, gaining sections of the river, as opposed to areas inside a meander where the groundwater flow towards a RBF well would be less important (See Figure 1).

The RBF wells in Villages 1 and 4 received between 40% and 74% (estimated) river water at the time of testing. At Village 2, the mixing percentage was 15% and 0% in terms EC and ion concentrations, respectively. That data set in Table 4 suggests a prevalence of groundwater flow to the RBF well and minimal river water mixing.

Table 4. Contribution (%) of river water to RBF well water based on mixing ratios from ion, dissolved silica, and electric conductivity (EC) measurements. Note: Both RBF wells at Village 1 were lumped together because of their proximity.

Village	Mg	Na	Cl	EC	Average
Village 1	84%	62%	80%	68%	74%
Village 2	0%	0%	0%	15%	4%
Village 4 RBF1	61%	47%	68%	69%	61%
Village 4 RBF2	36%	34%	48%	42%	40%

There was no apparent correlation between the degree of mixing and the local geology, i.e., RBF wells installed in sand-rich alluvial sediments at Village 4 RBF 2 did not draw-in more river water than wells that there were installed in less permeable, more silty deposits, such as Village 4 RBF 1. However, the well with the highest river water contribution (Village 1) was also the RBF well that was installed closest to the river (22 m; Table 1). This observation suggests that the setback distance from the river was of greater significance than the presence of more permeable deposits along this part of the Krishna River. This assessment is further supported by data from Village 2, where the geology was similar to Village 1, but where the set-back distance was greater, resulting in the lowest mixing ratio of all sites. Further, the comparative close vicinity of the groundwater wells (less than 180 m from RBF wells) and their similar depth (36.5 m or less) suggest that possible influences of location, well depth, or other boundary conditions on the mixing calculations were minimal.

Since the mixing calculations were based on only one sampling campaign (February 2018) and a limited number of samples from groundwater wells (handpumps/boreholes; n = 12), it was not possible to evaluate whether these ratios changed throughout the season. However, higher river water contributions should be expected during and after the monsoon when the Krishna River stage is high. Conversely, lower contributions are likely towards the height of the dry season (April to June), when the flow of water in river at its lowest.

4. Conclusions

The results of this study provide evidence that RBF installations are challenging but possible under the climate and hydrogeologic conditions prevailing in this part of southern India, where there is generally a lack of the typically thick, highly conductive alluvial sediments that are more common along the major rivers of northern India, such as the Ganga or Yamuna [14].

In this study, RBF well fields were installed at four sites along the upper Krishna River watershed and water quality and quantity studies were performed to characterize the RBF systems operation. RBF water treatment resulted in expected reductions in *E. coli* and turbidity, even though the RBF well setback distances from the river were less than what is considered to be typical for RBF sites in northern India, i.e., 50 m or more [14]. However, RBF treatment performance was hindered by the overall suboptimal hydraulic and hydrogeochemical conditions in combination with highly saline shallow groundwater in the study area.

Although the sediments that were encountered at the Krishna River study sites were less permeable than those typically observed in northern India, the data confirm that RBF well yields in this setting are sufficient to supply populations of 2000 to 3000. However, appropriate site selection is critical, and when installing RBF wells one needs to balance the well depth and set-back distance from the river against the presence and composition of alluvial deposits. This can be difficult because geologic and hydrogeologic conditions along major rivers in this part of southern India appear to be more variable and often less supportive for RBF. The viability of RBF systems as a domestic water source is also influenced by other factors not limited to southern India, including surface water and groundwater salinity, agricultural practices in the vicinity of RBF wells, and the reliability of the power grid.

Identifying appropriate RBF sites in this part of southern India will remain challenging until more detailed hydrogeological data become available. To aid in the installation of future RBF systems, a better characterization of local groundwater quality, more detailed mapping of alluvial sediments along the river, enhancements of electrical power supply (e.g., introduction of solar or other renewable energy sources), and access to more suitable sites is required. Also, additional studies along other rivers in south India, ideally in combination with simulations of groundwater flow conditions, would be required to obtain a better understanding of RBF performance in this part of India.

Supplementary Materials: The following are available online at http://www.mdpi.com/2073-4441/11/1/12/s1, Figure S1: RBF well in Village 3 during logger data retrieval by co-author Mrs. K. Patil (Source: T. Boving), Figure S2: Vertical EC profile measured in RBF2 well at Village 4 in relation to the geologic profile and the well construction diagram. Reference level: Water table. At this location, the depth from the surface to the water table was 4.6 m. The brown loam above the alluvial sediment continued to the surface, Table S1: Methods used for analysis of water samples. IS 3025—Bureau of Indian Standards (BIS) (Part 45) Method of Sampling and Test (Physical and Chemical) for Water and Wastewater (First Revision). APHA: Standard Methods for the Examination of Wastewater, America Public Health Association. Washington, DC, USA, Table S2: Water quality parameters analyzed and BIS (2012) drinking water quality standards, where available. ND = Not defined, Table S3: Sampling results for major ion concentrations in RBF wells, the Krishna River, and groundwater (mg/L). HP: Handpump, BW: Borewell. The listed data only include samples that passed quality control, Table S4: Field parameter and *E. coli* concentrations in RBF wells, the Krishna River, and groundwater (mg/L). EC: electrical conductivity, DO: Dissolved Oxygen, T: Temperature, HP: Handpump, BW: Borewell.

Author Contributions: Conceptualization, T.B.B., K.P., F.D., J.O., A.B.F. and K.L.; Data curation, K.P., S.F.B., S.L.M., J.O. and A.B.F.; Formal analysis, T.B.B., K.P., S.F.B., S.L.M., J.O. and K.L.; Funding acquisition, T.B.B., K.P., M.S. and K.L.; Investigation, T.B.B., K.P., F.D., S.L.M., J.O. and K.L.; Methodology, T.B.B., K.P., F.D., S.L.M., J.O., M.S., A.B.F. and K.L.; Project administration, K.P., F.D., M.S. and K.L.; Resources, K.P., F.D., J.O. and K.L.; Software, S.F.B.; Supervision, T.B.B., K.P., F.D. and K.L.; Validation, K.P., S.F.B., S.L.M. and J.O.; Visualization, T.B.B., S.L.M. and J.O.; Writing—original draft, T.B.B., S.L.M., J.O. and K.L.; Writing—review & editing, K.P., F.D., M.S. and A.B.F.

Funding: This work was supported by an Australian National Health and Medical Research Council (NHMRC) project grant (APP1083408).

Acknowledgments: We gratefully acknowledged the help of our field assistant Mr. R.K. Vhaval. Also, we thank the villagers and their leaders for supporting our study.

Conflicts of Interest: The authors declare no conflict of interest.

References

1. CIA (Central Intelligence Agency). CIA World Fact Book—India. 2018. Available online: https://www.cia.gov/library/publications/the-world-factbook/geos/in.html (accessed on 12 December 2018).
2. FAO (Food and Agriculture Organization of the United Nations). Review of World Water Resources by Country. 2003. Available online: http://www.fao.org/docrep/005/Y4473E/Y4473E00.HTM (accessed on 12 December 2018).
3. WHO (World Health Organization). Global Health Observatory Country Views. 2018. Available online: http://apps.who.int/gho/data/node.country.country-IND?lang=en (accessed on 12 December 2018).
4. Black, R.E. Where and why are 10 million children dying every year? *Child Care Health Dev.* **2003**, *361*, 2226–2234. [CrossRef]
5. Fan, V.Y.-M.; Mahal, A. What prevents child diarrhoea? The impacts of water supply, toilets, and hand-washing in rural India. *J. Dev. Eff.* **2011**, *3*, 340–370. [CrossRef]
6. Ray, C.; Melin, G.; Linsky, R.B. *Riverbank Filtration. Improving Source Water Quality*; Kluwer Academic Publishers: Dordrecht, The Netherlands, 2003.

7.   Schmidt, C.K.; Lange, F.T.; Kuehn, H.J.B. Experiences with riverbank filtration and infiltration in Germany. *ResearchGate* **2003**. Available online: https://www.researchgate.net/publication/267779083_Experiences_with_riverbank_filtration_and_infiltration_in_Germany (accessed on 12 December 2018).

8.   Grischek, T.; Schoenheinz, D.; Ray, C. Siting and Design Issues for Riverbank Filtration Schemes. In *Riverbank Filtration*; Ray, C., Melin, G., Linsky, R.B., Eds.; Kluwer Academic Publishers: Dordrecht, the Netherlands, 2003; Volume 43, pp. 291–302.

9.   Hiscock, K.M.; Grischek, T. Attenuation of groundwater pollution by bank filtration. *J. Hydrol.* **2002**, *266*, 139–144. [CrossRef]

10.   Cady, P.; Boving, T.B.; Choudri, B.S.; Cording, A.; Patil, K.; Reddy, V. Attenuation of Bacteria at a Riverbank Filtration Site in Rural India. *Water Environ. Res.* **2013**, *85*, 2164–2174. [CrossRef]

11.   Kelly, B.P.; Rydlund, J.; Paul, H. Water-Quality Changes Caused by Riverbank Filtration between the Missouri River and Three Pumping Wells of the Independence, Missouri, Well Field 2003-05. 2006; 48p. Available online: http://purl.access.gpo.gov/GPO/LPS75695 (accessed on 12 December 2018).

12.   Kumar, P.; Mehrotra, I. Riverbank Filtration for Water Supply: Indian Experience. In *World Environmental and Water Resources Congress*; American Society of Civil Engineers: Reston, VA, USA, 2009.

13.   Schubert, J. Hydraulic aspects of riverbank filtration—Field studies. *J. Hydrol.* **2002**, *266*, 145–161. [CrossRef]

14.   Sandhu, C.; Grischek, T.; Kumar, P.; Ray, C. Potential for Riverbank filtration in India. *Clean Technol. Environ. Policy* **2011**, *13*, 295–316. [CrossRef]

15.   Boving, T.B.; Choudri, B.S.; Cady, P.; Cording, A.; Patil, K.; Reddy, V. Hydraulic and Hydrogeochemical Characteristics of a Riverbank Filtration Site in Rural India. *Water Environ. Res.* **2014**, *86*, 636–648. [CrossRef] [PubMed]

16.   Mukherjee, A.; Saha, D.; Harvey, C.F.; Taylor, R.G.; Ahmed, K.M.; Bhanja, S.N. Groundwater systems of the Indian Sub-Continent. *J. Hydrol. Reg. Stud.* **2015**, *4*, 1–14. [CrossRef]

17.   MIT. Electrodialysis, powered by solar panels, could provide drinking water for villages in India. *Membr. Technol.* **2014**, *2014*, 9–10. [CrossRef]

18.   Census. *Census of 2011*; Office of the Registrar General & Census Commissioner: New Delhi, India, 2011.

19.   GoK (Government of Karnataka). *District Geological Survey Report of Belagavi District*; Government of Karnataka: Bangalore, India, 2017.

20.   GoI (Government of India). *Groundwater Information Booklet. Belgaum District*; Ministry of Water Resources, Central Ground Water Board, Government of India: Bangalore, India, 2012.

21.   CGWB (Central Ground Water Board). *Aquifer Systems of Karnataka*; Central Ground Water Board, Ministry of Water Resources, Government of India: Bangalore, India, 2012.

22.   ITT (Institute for Transformative Technologies). *Technology Breakthroughs for Global Water Security: A Deep Dive into South Asia*; Institute for Transformative Technologies: Mumbai, India, 2018.

23.   The Hans India.   River Krishna among the Highly Polluted Rivers.   2016.   Available online: http://www.thehansindia.com/posts/index/Andhra-Pradesh/2016-07-19/River-Krishna-among-the-highly-polluted-rivers/243080 (accessed on 12 December 2018).

24.   MITRA (Maharashtra Pollution Control Board). *Comprehensive Study Report on Krishna River Stretch*; Maharashtra Pollution Control Board: Mumbai, India, 2014.

25.   Purandara, B.K. Groundwater Quality Studies in the Belgaum District. CS/AR-2999-2000. 2000. Available online: http://www.indiawaterportal.org/sites/indiawaterportal.org/files/Groundwater%20Qulaity%20Studies%20in%20Belgaum%20District%20_NIH_1999-2000.pdf (accessed on 12 December 2018).

26.   AWWA (American Water Works Association). *Disinfection of Water Storage Facilities*; AWWA C652-11; American Water Works Association: Denver, CO, USA, 2011.

27.   DOH (Washington Department of Health). *Emergency Disinfection of Small Water Systems*; DOH 331-242; Washington Department of Health: Washington, DC, USA, 2018.

28.   Dichter, G. *IDEXX Colilert*-18 and Quanti-Tray* Test Method for the Detection of Fecal Coliforms in Wastewater*; DEXX Laboratories, Inc.: Westbrook, TX, USA, 2011.

29.   Luby, S.P.; Halder, A.K.; Huda, T.M.; Unicomb, L.; Islam, M.S.; Arnold, B.F.; Johnston, R.B. Microbiological Contamination of Drinking Water Associated with Subsequent Child Diarrhea. *Am. J. Trop. Med. Hyg.* **2015**, *93*, 904–911. [CrossRef] [PubMed]

30.   BIS (Bureau of Indian Standards). *Indian Standard Drinking Water IS 10500:2012—Specifications*; 2nd Rev.; Bureau of Indian Standards: New Delhi, India, 2012.

31. NRDWP. National Rural Drinking Water Programme Guidelines—2013. Available online: http://www. indiaenvironmentportal.org.in/content/265047/national-rural-drinking-water-programme-guidelines/ (accessed on 12 December 2018).

32. Krishan, G.; Singh, S.; Sharma, A.; Sandhu, C.; Grischek, T.; Gosh, N.C.; Gurjar, S.; Kumar, S.; Singh, R.P.; Glorian, H.; et al. Assessment of Water Quality for River Bank Filtration along Yamuna River in Agra and Mathura. *Int. J. Environ. Sci.* **2016**, *7*, 56–67.

33. Brouwer, C.; Goffeau, A.; Heibleom, M. *Introduction to Irrigation. Irrigation Water Management: Training Manual No. 1*; FAO: Rome, Italy, 1985.

34. Gupta, A.; Ronghang, M.; Kumar, P.; Mehrotra, I.; Kumar, S.; Grischek, T.; Sandhu, C.; Knoeller, K. Nitrate contamination of riverbank filtrate at Srinagar, Uttarakhand, India: A case of geogenic mineralization. *J. Hydrol.* **2015**, *531*, 626–637. [CrossRef]

35. Piper, A.M. A graphic procedure in the geochemical interpretation of water-analyses. *Trans. Am. Geophys. Union* **1944**, *25*, 914. [CrossRef]

36. Boving, T.B.; Patil, K. Riverbank Filtration at the Nexus of Water-Energy-Food. In *Water-Energy-Food Nexus: Principles and Practices (Geophysical Monograph Series)*; Chp. 18; AGU: Washington, DC, USA; Wiley: Hoboken, NJ, USA, 2017.

water **MDPI**

Article

Lithologic Control of the Hydrochemistry of a Point-Bar Alluvial Aquifer at the Low Reach of the Nakdong River, South Korea: Implications for the Evaluation of Riverbank Filtration Potential

Md Moniruzzaman [1], Jeong-Ho Lee [1], Kyung Moon Jung [1,2], Jang Soon Kwon [3], Kyoung-Ho Kim [4] and Seong-Taek Yun [1,*]

[1] Department of Earth and Environmental Sciences, Korea University, Seoul 02841, Korea; monir@korea.ac.kr (M.M.); earth1977@hanmail.net (J.-H.L.); jkmoon16@naver.com (K.M.J.)
[2] Sunjin Engineering and Architecture Co., Anyang 14057, Korea
[3] Korea Atomic Energy Research Institute, Daejeon 34057, Korea; jskwon@kaeri.re.kr
[4] Korea Environment Institute, Sejong 30147, Korea; khkim@kei.re.kr
* Correspondence: styun@korea.ac.kr; Tel.: +82-02-3290-3176

Received: 31 October 2018; Accepted: 28 November 2018; Published: 1 December 2018

Abstract: To assess the groundwater−river water interaction in a point-bar alluvial aquifer as a crucial step in site assessment for riverbank filtration, hydrochemical and hydrogeologic investigations were performed on a riverine island at the low reach of the Nakdong River, South Korea. The site was evaluated for the application of large-scale bank filtration. Unconsolidated sediments (~40 m thick) of the island comprise fine- to medium-grained sand (upper aquifer), silty sand with clay intercalations, and sandy gravel (lower aquifer) in descending order. The intermediate layer represents an impermeable aquitard and extends below the river bottom. A total of 66 water samples were collected for this study; groundwater (n = 57) was sampled from both preexisting irrigation wells, and three multi-level monitoring wells (each 35 m deep). Groundwater chemistry is highly variable, but it shows a distinct hydrochemical change with depth: shallow groundwater (<25 m deep) from the upper aquifer is characteristically enriched in NO_3^- and SO_4^{2-}, due to agricultural contamination from the land surface, while deeper groundwater (>25 m deep) from the lower aquifer is generally free of NO_3^- and relatively rich in F. The lower aquifer groundwater is also higher in pH, and concentrations of K^+, Mg^{2+}, and HCO_3^-, indicating that the aquifer is likely fed by regional groundwater flow. Such separation of groundwater into two water bodies is the result of the existence of an impermeable layer at intermediate depth. In addition, the hyporheic flow of river water is locally recognized at the upstream part of the upper aquifer as the zone of low TDS (Total Dissolved Solids) values (<200 mg/L). This study shows that the study site does not seem to be promising for large-scale riverbank filtration because 1) the productive, lower aquifer is not directly connected to the bottom of the river channel, and 2) the upper aquifer is severely influenced by agricultural contamination. This study implies that the subsurface hydrogeologic environment should be carefully investigated for site assessment for riverbank filtration, which can be aided by a detailed survey of groundwater chemistry.

Keywords: point-bar alluvial setting; riverbank filtration; site investigation; hydrochemistry; subsurface geology

1. Introduction

Riverside alluvial aquifers are a target of bank filtration in many countries [1–6]. The river bank filtration (RBF) has been frequently used as a pre-treatment process for domestic water supply, to overcome various surface water quality problems [7–13]. In many European countries, a significant amount of drinking water supply is obtained by RBF: for example, 80% in Switzerland, 50% in France, 48% in Finland, 40% in Hungary, 16% in Germany, and 7% in the Netherlands [14,15]. Recently, many Asian countries such as India [16–18] and China [4,5,19,20] have also become interested in the use of RBF. RBF has also been used in a few localities in South Korea (e.g., Changwon City, 80,000 m^3/day; Gimhae City, 180,000 m^3/day) [21–26].

In RBF, a large portion of water from a nearby stream or river is filtered through alluvial sediments during induced flow toward pumping wells. During filtration, significant changes in water chemistry occur via physical and biogeochemical processes [27–29]. Thus, the successful implementation of RBF technology is fully dependent on the site-specific conditions of hydrogeology (esp., the flow path and thickness of the aquifer) and water quality [1,30–34]. The zone of hyporheic exchange also plays an important role in governing contaminant exchange and transformation [35–40]. More specifically, the performance of RBF is controlled by many factors such as well type, pumping rate, flow paths, and travel time of water to wells, thickness and hydrogeological properties of alluvial sediments (and soil), quality of surface water and background groundwater, and biogeochemical reactions occurring in aquifer sediments [4,5,9,13,41–44].

However, riverside alluvial deposits are often used for agricultural activities in many countries, including South Korea, causing significant contamination of alluvial aquifers [45,46]. Therefore, in addition to the site-specific hydrogeology, the groundwater quality of an alluvial aquifer should be carefully examined for the sustainable use of RBF.

The current study on hydrogeology and water chemistry on a riverine island was initiated to evaluate the potential use of large-scale RBF technology in the study area. The main aims of this study are to: (1) examine groundwater chemistry in relation to the geologic section of an alluvial aquifer; (2) understand the groundwater−river water interaction in a point-bar sedimentary environment; (3) evaluate the applicability of large-scale bank filtration. In particular, the sequence of alluvial sedimentary strata was carefully examined to explain the observed hydrochemical features.

2. Study Area

The study area is a small riverine island in Gimhae City at the low reach of the Nakdong River, South Korea (Figure 1a) and it has been evaluated for the application of large-scale (180,000 m^3/day) bank filtration using a number of horizontal collector wells in gravel aquifer [47]. The Nakdong River is the longest river in South Korea (about 506 km long with a watershed of about 23,000 km^2), and flows from the Taebaek Mountains to the South Sea or the Korean Strait. The climate of Gimhae City is temperate monsoon with an annual average precipitation of about 1300 mm, of which more than 70% occurs during the summer months (June through September). The air temperature is highest in August (24 °C) and lowest in January (−2 °C), with an annual average of 12 °C. The study area is surrounded by mountains with elevations up to about 280−380 m above sea level (asl) at the northeast, southeast, and northwest. The topography of the island is relatively flat, and it ranges from 2.6 m asl at the northeastern part, to 6.8 m asl at the central part (Figure 1b). The flat island has been used for year-round agricultural activities (especially for strawberry production); thus, large amounts of synthetic fertilizers are applied on the fields during the growing season (spring through fall) and even during winter, using greenhouses.

In the study area, the Nakdong River meanders strongly from south to north, and then from north to south. The Miryang River joins the Nakdong River from the north to the south as a tributary, and a small tributary also joins the Nakdong River from the southeast and the northwest (Figure 1c). The deepest of the Nakdong river channel near the island in this study is −17.8 m asl [47]. The island in this study was recently formed by the point-bar accretion of alluvial sediments at the direct upstream of the confluence. A comparison of topographic maps printed after 1950 shows that there was a dynamic change of sedimentation and subsequent erosion over approximately the past 70 years (Figure 1c); the island was formed by extensive sedimentation before 1991, and then it was separated into two islands before 2006 by erosion along the confluence with a small northwest-trending tributary. Our study was initiated in 2006 with the installation of multilevel monitoring wells. The island has been frequently flooded in recent years.

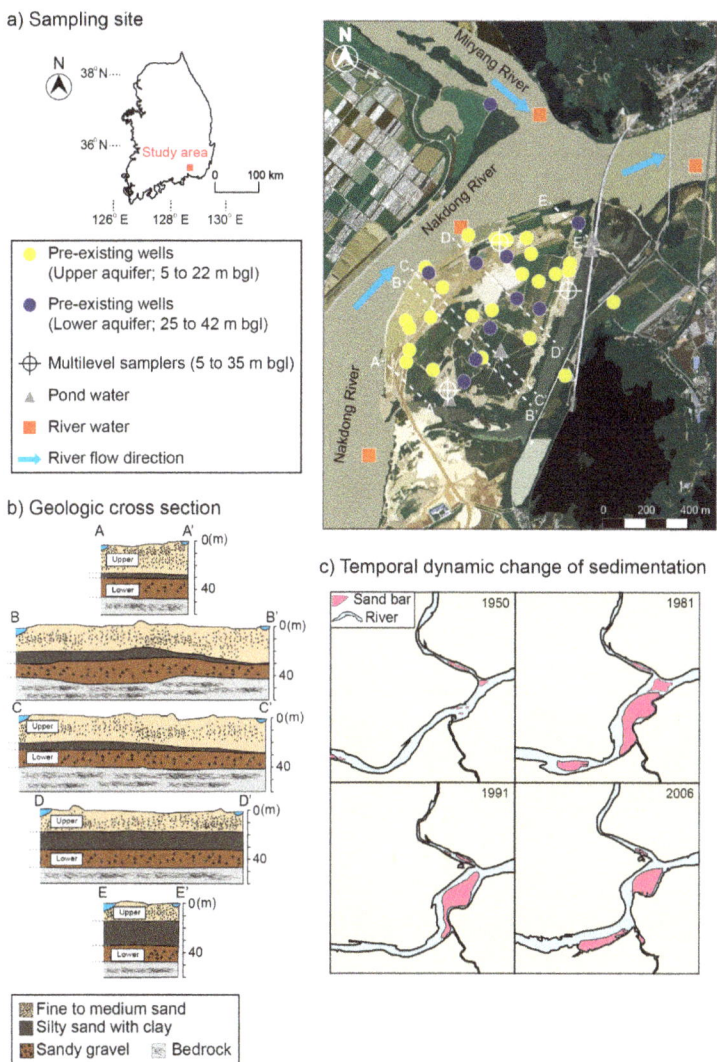

Figure 1. (a) Location of sampling sites in the study area at the lower reach of the Nakdong River, (b) The geological cross sections (A–A′ to E–E′ in (a)), and (c) Temporal dynamic change of sedimentation (modified after Daewoo Construction Co. [47]). bgl = below ground level.

3. Materials and Methods

A total of 66 water samples, including river water (n = 4), pond water (n = 5), and groundwater (n = 57) were collected for this study (see Figure 1a for localities). Alluvial groundwater was sampled from preexisting irrigation wells (n = 36) with known depths (26 samples from depths between 5 and 22 m, and 10 samples from depths between 30 and 42 m) and from three multi-level monitoring wells (n = 21). Core drilling was conducted to a depth of 35 m below the ground surface to install the multi-level monitoring wells. The monitoring wells were constructed with polyethylene tubes (0.5 cm diameter) with variable lengths with 5 m intervals (i.e., seven tubes (samplers) in each well). The tips of the tubes were wrapped with a stainless steel screen to allow groundwater inflow.

Water samples were collected using peristaltic pumps after sufficient purging (at least more than two well volumes). Unstable parameters such as pH, redox potential (Eh), dissolved oxygen (DO), and temperature were measured in situ using a flow-through chamber to minimize contact with the air. Alkalinity was measured by using an acid titration method. Samples for chemical analysis were immediately filtered through 0.45 µm cellulose membranes. Sampling bottles were soaked in 1:1 diluted HCl solution for 24 h, washed three times with deionized water, and washed again prior to each sampling with the filtrates.

Samples for the analysis of major cations and dissolved forms of silica, Fe, and Mn were acidified to pH < 2 by adding several drops of ultra-pure nitric acid. The samples were kept at 4 °C. The cations, silica, Fe, and Mn were analyzed using ICP-AES (Inductively Coupled Plasma Atomic Emission Spectroscopy; Perkin Elmer Optima 3000), while major anions were analyzed using ion chromatography (Dionex 120). The quality of chemical analysis was carefully examined by taking and analyzing blanks and duplicate samples, and by checking ion balances. Statistical analysis of data such as Mann-Whitney U test was performed using Statistica software (version 10). Maps showing the spatial distribution of the potentiometric head levels and some hydrochemical parameters were drawn by using a kriging method in Surfer 12 software (version 12).

4. Results and Discussion

4.1. Subsurface Geology and Hydrogeologic Condition

The alluvial sediments in the study area overlie bedrocks consisting of Yucheon group volcanic rocks (mainly, andesitic rocks and tuffs) and granitoids of the Cretaceous age [25,47]. Geologic logging showed that unconsolidated alluvial deposits (approximately 40−42 m thick) of the island comprise fine- to medium-grained sand, silty sand with clay intercalations, and sandy gravel, in descending order (Figure 1b). Silty sand with clayey intercalations occurs at intermediate depths between approximately 23−30 m below the land surface, and they represent an impermeable layer. This clayey silt layer is considered to represent deposition during the stage of sea level rise (around 5.4−8.0 ka) before the transition toward a prodelta terrestrial environment [26,48]. The transverse cross sections (A–A' to E–E') show that (1) the intermediate layer as an aquitard seems to extend continuously below the river bottom and (2) toward the downstream direction (i.e., the confluence with the Miryang River), the thickness of an impermeable silty layer at intermediate depth tends to increase, while the upper sandy layer tends to generally decrease in thickness (Figure 1b). Except for the area adjacent to the river, the top of the upper sandy layer is generally covered with silt-rich sediment of thickness variations by recent frequent flooding, which forms the surface soil for agricultural activities [47,48].

Groundwater for agricultural use on the island is pumped from the upper sandy layer (upper unconfined aquifer) and the deeper sandy gravel layer (lower confined aquifer); among these aquifers, the lower aquifer is more productive [47]. Hydrogeologic pumping tests conducted in the study island showed that the lower sandy gravel aquifer (thickness = 9 to 19 m) has hydraulic conductivity values (K) ranging from 2.6×10^{-4} to 2.9×10^{-4} m/s [47], which agrees with the values (about 3×10^{-4} m/s) reported at an RBF site that is about 12 km upstream of the study area [24].

Figure 2 shows the distribution of potentiometric head levels (m asl) measured in the upper and lower aquifers during the period of this study. The potentiometric head levels are clearly different in the island between the upper sandy aquifer and the lower aquifer (Figure 2). In the upper sandy aquifer, the pattern of groundwater levels tends to correspond to topographic features; higher groundwater levels occur at central and eastern parts of the island, although pumping of groundwater for irrigation locally results in the drawdown of groundwater levels. It is also noteworthy that lower groundwater tables are observed at the southwestern part of the island (Figure 2), which may indicate the presence of the lateral flow of river water (i.e., hyporheic flow). Thus, the upper aquifer represents an unconfined condition, with groundwater recharge from the infiltration of rain and irrigation water. On the other hand, potentiometric head levels measured in the lower aquifer overall tends to direct from northeast toward the river, and tend to be slightly higher than those for the upper aquifer (Figure 2). We consider that groundwater in the lower aquifer represents a regional groundwater flow discharging to the river under a confined condition. Therefore, it is obvious that the two aquifers in the study area have different hydrologic conditions.

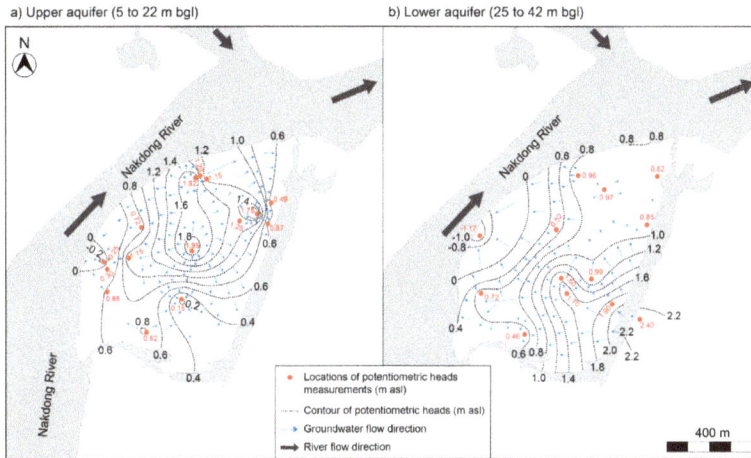

Figure 2. Locations of the measurements of potentiometric head levels, and the contours showing potential groundwater flow in the upper and lower aquifers in the study island. asl = above sea level, bgl = below ground level.

4.2. General Hydrochemistry

Hydrochemical data of four kinds of water samples (n = 66) are summarized in Table 1. The amount of total dissolved solid (TDS) in the water samples ranges widely from 87.4 to 901 mg/L. The dissolved oxygen (DO) is generally higher in river water and pond water, than groundwater.

The plots on a Piper diagram (Figure 3) show that the water samples are highly variable in hydrochemical characteristics, widely ranging from the Ca$-$Cl($-$NO$_3$) type to the Na$-$HCO$_3$ type. The chemical composition of river water is similar to that of the upper aquifer groundwater; river water and the upper aquifer groundwater are dominantly of the Ca$-$Na$-$HCO$_3$$-$Cl type. The lower aquifer groundwater is also variable in chemical features, ranging from the Na$-$HCO$_3$ type to the Ca$-$Mg$-$HCO$_3$$-$Cl type. Groundwater samples could not be collected from the intermediate silty-clayey layer.

However, careful examination of data shows that there is a noticeable difference in anionic composition between the upper aquifer groundwater and the lower aquifer groundwater: the lower aquifer groundwater tends to be relatively enriched in HCO$_3$$^-$, but it also tends to be more depleted in SO$_4$$^{2-}$ and NO$_3$$^-$. In cationic composition, the lower aquifer groundwater also tends to be relatively higher in Mg than the upper aquifer groundwater. The tendency of the enrichments of NO$_3$$^-$ and SO$_4$$^{2-}$ in many samples of the upper aquifer groundwater are attributed to the infiltration of contaminants from agricultural activities [49–53].

To compare the hydrochemical characteristics of pond water, river water, upper aquifer groundwater, and lower aquifer groundwater in the study area, we examined the data with box plots (Figure 4). The statistical differences in hydrochemical parameters between water groups were also tested by the Mann–Whitney U test. The main findings are as follows Figure 4: (1) most of the parameters examined are not statistically different (*p*-Values > 0.05) between the upper aquifer groundwater and the surface water, except DO, K, and F^-, and (2) the lower aquifer groundwater is statistically higher in pH, EC, TDS, K^+, Mg^{2+}, and F^- but is lower in SO_4^{2-} and NO_3^- (see also Table 2).The higher pH and TDS values, together with higher concentrations of K^+, Mg^{2+}, and F^- in the lower aquifer groundwater, indicate the influence of regional groundwater flow (i.e., base flow) whose chemistry is largely controlled by water–rock interactions. On the other hand, the enrichments of SO_4^{2-} and NO_3^- in the upper aquifer groundwater reflect the influence of anthropogenic contamination.

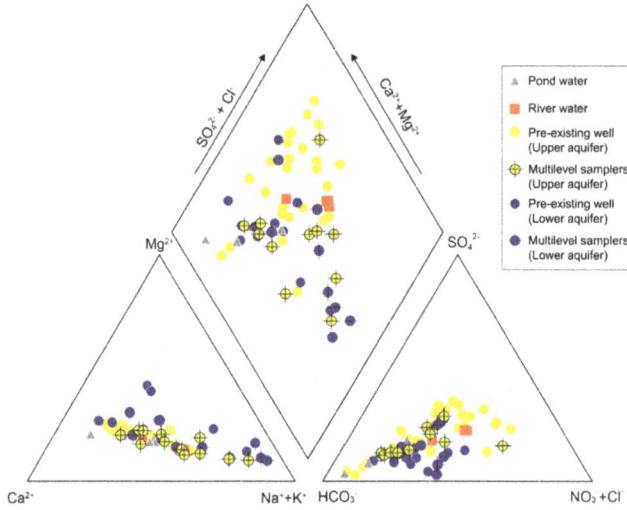

Figure 3. Piper's diagram of surface and ground water samples (n = 66) in the study area.

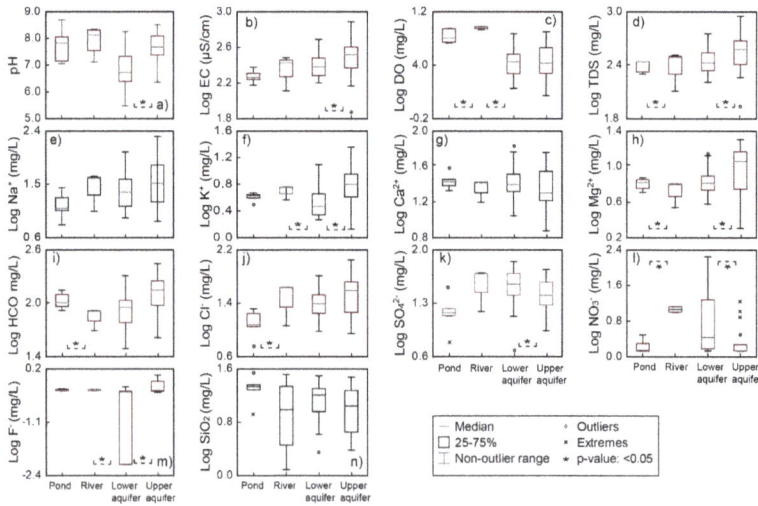

Figure 4. Box plots to compare the hydrochemical parameters of pond water, river water, upper aquifer groundwater, and lower aquifer groundwater samples in the study area. The notation with an asterisk shows that there is a significant difference (*p*-Value < 0.05) between neighboring groups, based on the Mann–Whitney U test.

Table 1. Statistical summary of hydrochemical data of water samples from the study area.

Title		pH	EC (μS/cm)	DO	Na⁺	K⁺	Ca²⁺	Mg²⁺	Cl⁻	NO₃⁻	SO₄²⁻	HCO₃⁻	F⁻	Fe	Mn	SiO₂	TDS
											Unit: mg/L						
Pond water (n = 5)	Min	6.8	150	5.4	6.6	3.1	21.3	5.1	5.9	1.3	6.2	101	0.5	0.0	0.1	9.0	200
	Median	7.5	183	6.4	12.4	4.2	26.7	6.6	13.7	1.4	15.0	128	0.5	0.0	0.1	21.8	277
	Max	8.2	239	8.9	27.5	4.7	38.2	7.5	25.6	3.1	31.3	186	0.5	0.1	0.7	34.2	277
	STD	0.6	33.6	1.7	8.1	0.6	6.4	1.0	7.7	0.7	9.1	35.3	0.0	0.0	0.3	9.0	39.1
River water (n = 4)	Min	6.9	130	8.5	11.0	3.7	15.7	3.5	13.1	10.0	15.5	54.9	0.5	0.0	0.0	1.2	129
	Median	7.7	269	9.3	39.5	5.6	26.1	6.3	57.3	11.8	48.7	98.4	0.5	0.0	0.0	10.6	310
	Max	7.9	305	9.5	43.2	5.8	26.6	6.4	58.3	13.3	50.1	101	0.5	0.0	0.0	33.0	328
	STD	0.5	77.5	0.4	14.9	1.0	5.3	1.4	22.3	1.8	16.8	22.2	0.0	0.0	0.0	13.9	93.7
Upper (5 to 22 m deep) aquifer groundwater (n = 38)	Min	5.4	159	1.1	8.6	1.8	11.1	3.8	10.7	1.4	4.8	32.0	0.0	0.0	0.0	2.2	163
	Median	6.5	240	2.8	23.2	2.9	24.9	6.5	30.9	2.7	35.3	111	0.4	2.0	2.7	16.2	268
	Max	7.8	488	7.4	110	12.4	67.3	13.7	92.6	180	70.7	291	0.6	27.7	7.0	31.2	568
	STD	0.6	89.2	1.4	19.7	2.2	11.6	2.6	17.4	32.1	15.2	59.6	0.2	6.4	2.0	7.5	96.4
Lower (25 to 42 m deep) aquifer groundwater (n = 19)	Min	6.2	69.8	0.9	7.6	1.3	7.5	2.0	9.8	1.4	8.7	44.2	0.4	0.0	0.0	2.4	87.4
	Median	7.3	331	2.7	32.8	6.3	20.0	11.4	51.5	1.4	25.3	186	0.5	0.5	0.2	11.1	381
	Max	8.1	776	7.9	205	22.7	57.0	20.3	171	17.6	55.7	418	1.1	19.4	5.4	30.0	901
	STD	0.5	160	1.9	51.9	5.4	14.0	5.3	41.6	4.3	13.5	92.9	0.2	4.7	1.6	8.1	186

Max = maximum, Min = minimum, STD = Standard deviation, Max = maximum, Min = minimum, STD = Standard deviation.

Table 2. Statistical summary of hydrochemical data obtained from multilevel monitoring wells (MLWs) in the study area.

Aquifer		pH	EC (μS/cm)	DO	Na⁺	K⁺	Ca²⁺	Mg²⁺	Cl⁻	NO₃⁻	SO₄²⁻	HCO₃⁻	F⁻	Fe	Mn	SiO₂	TDS
											mg/L						
Upper (5 to 22 m deep) aquifer groundwater (n = 12)	Min	6.4	164	1.8	11.1	2.9	11.1	3.8	12.7	1.4	15.5	87.0	0.4	0.0	0.1	5.9	191
	Median	7.3	198	3.2	24.5	4.2	19.7	5.3	25.3	2.0	24.1	124	0.5	0.2	0.7	9.2	236
	Max	7.8	462	7.4	110	12.4	67.3	13.2	66.9	180	51.1	291	0.6	3.8	3.4	19.6	568
	STD	0.4	105	1.7	28.0	2.8	14.5	2.5	15.8	51.3	15.4	53.9	0.0	1.1	0.9	4.1	129
Lower (25 to 42 m deep) aquifer groundwater (n = 9)	Min	7.3	147	1.9	10.7	3.2	9.3	4.4	14.0	1.4	12.1	94.6	0.4	0.1	0.0	2.4	184
	Median	7.6	299	4.0	38.7	6.3	16.6	7.0	51.5	1.4	34.1	133	0.5	0.2	0.1	10.6	325
	Max	8.1	776	7.9	205	22.7	35.1	13.8	171	10.5	55.7	418	0.8	1.6	0.4	21.3	901
	STD	0.3	188	2.0	63.7	6.8	7.9	3.8	50.0	3.0	15.8	109	0.2	0.5	0.1	6.9	227
p-Value *		0.049	0.169	0.602	0.310	0.069	0.310	0.111	0.049	0.049	0.862	0.193	0.023	0.508	0.000	0.917	0.129

Max = maximum, Min = minimum, STD = Standard deviation. *p*-Value *was obtained from the Mann–Whitney U-test to compare hydrochemical data between upper aquifer and lower aquifer (Bold characters denote the significant differences (*p*-Value < 0.05)).

4.3. Vertical Change of Hydrochemistry

Combined with the interpretation of groundwater flow using the spatial distributions of potentiometric head levels in the study area, the vertical change of hydrochemical parameters of water samples was further examined to elucidate potential pathways of water flow (Figure 5). A remarkable change of hydrochemistry with depth is observed for pH, TDS, Na^+, K^+, Mg^{2+}, HCO_3^-, Cl^-, SO_4^{2-}, NO_3^-, and F^-. The pH values of a few groundwater samples from the upper sandy aquifer are remarkably low (<6.5), compared to those of surface water (i.e., pond water and river water). Most groundwater samples from the upper unconfined aquifer are typically enriched in NO_3. These two observations indicate the acidification of groundwater due to nitrification [54–57].

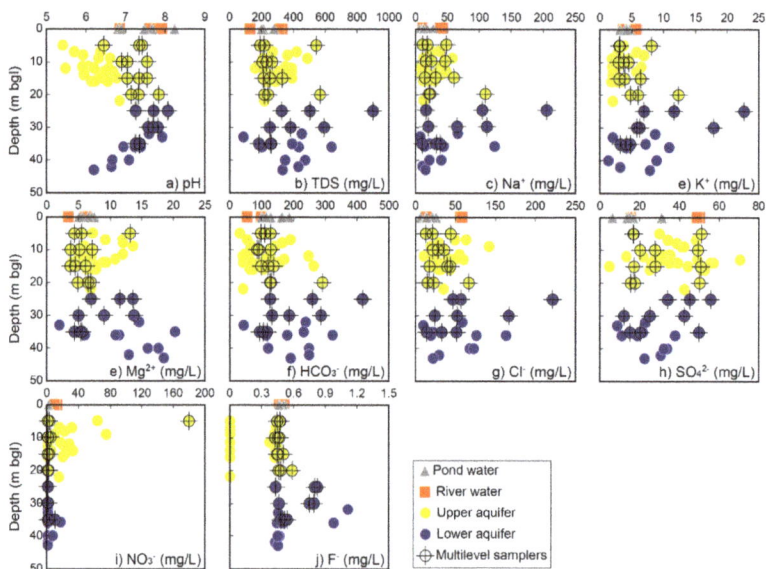

Figure 5. Vertical changes in some hydrochemical parameters with depth, showing notable differences in hydrochemistry between groundwater samples from the upper and lower aquifers. bgl = below ground level.

On the other hand, a few groundwater samples from the lower confined aquifer tend to be enriched in K^+, Mg^{2+}, HCO_3^-, and F (Figure 5). In particular, fluoride is relatively enriched (up to 1.1 mg/L) in the lower aquifer groundwater. In South Korean groundwater, such enrichment of fluoride was interpreted as the result of dissolution of F-bearing silicate minerals during water–rock interaction [56,58–60]. The enrichments of Na and Cl (and increasing TDS) in some samples from the lower gravel aquifer are possibly due to remnant seawater that was entrapped in the intermediate silty clay layer during the sea level rise, and that has been subsequently washed out during the deposition of recent point-bar sediment [26,61].

Interestingly, the enrichments of dissolved Fe and Mn are observed in many samples from the upper aquifer (Figure 6). It is noticeable because high concentrations of dissolved Fe or Mn frequently create a problem in the use of RBF technology [62,63]. Many groundwater samples from the upper aquifer are significantly lower in DO values than river water and pond water, approaching sub-oxic conditions (DO < 4 mg/L). These observations can be interpreted as the oxidation of organic matter preferentially beneath agricultural lands [27,64], which facilitates the sub-oxic condition to derive the reductive dissolution of Fe- and Mn-hydroxides.

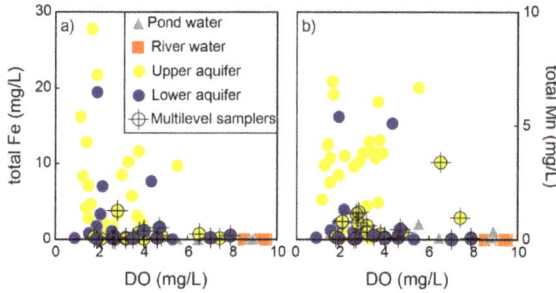

Figure 6. The relationship of dissolved iron (Fe$_{total}$); (**a**) and manganese (Mn$_{total}$) (**b**) to dissolved oxygen (DO) in groundwater.

4.4. Evaluation of Potential Flow Paths

Based on the distribution pattern of some hydrochemical parameters in the upper and lower aquifers (Figure 7), we attempted to evaluate potential flow paths in the studied riverine island. The parameters (pH, TDS, Na$^+$, K$^+$, Mg^{2+}, HCO$_3^-$, NO$_3^-$, and F) were selected based on their features, which show a meaningful difference between the two aquifers. The results were also compared with the groundwater flow suggested from the distribution of potentiometric head levels (see Figure 2).

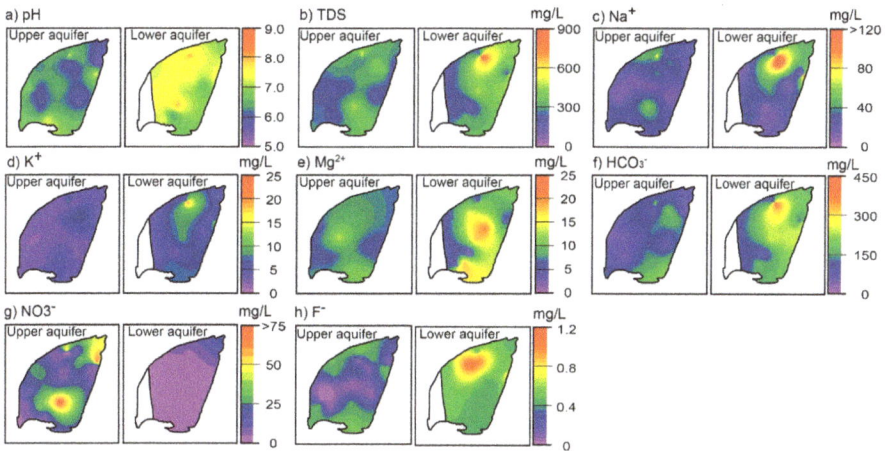

Figure 7. The spatial distribution of some hydrochemical parameters, showing distinct hydrochemical features between the groundwater from the upper aquifer and the lower aquifer in the study area.

4.4.1. Flow Path 1: Hyporheic Flow

The hyporheic zone is defined as the region beneath and alongside a stream bed, where the mixing of surface water with groundwater occurs via hyporheic flow or underflow [38,39,65]. At the southwestern and western part (i.e., upstream part) of the island, a zone of low TDS (< 200 mg/L) and Mg concentrations occurs locally in the upper unconfined aquifer. As suggested by the distribution of potentiometric head levels in the upper aquifer (see Figure 2), this zone is thought to represent the zone of horizontal hyporheic flow that results in the dilution of the upper aquifer groundwater. In this zone, other parameters such as Na$^+$, K$^+$, and HCO$_3^-$ tend to decrease.

4.4.2. Flow Path 2: Regional Groundwater Flow

Figure 7 shows that most parameters examined, except NO_3^-, tend to be higher in the lower aquifer than the upper aquifer. Such parameters (pH, TDS, Na^+, K^+, Mg^{2+}, HCO_3^-, and F^-) generally increase with water–rock (silicates) interaction [59]. Thus, together with the interpretation of the distribution of potentiometric head levels for the two aquifers in the island (see Figure 2), the hydrochemical distinction between the lower aquifer groundwater and the upper aquifer groundwater evidences that the two aquifers on the island are separated and not interconnected with each other, because of the presence of a silty clay layer at intermediate depth.

In the productive lower aquifer, the zone of higher concentrations of TDS, Na^+, K^+, Mg^{2+}, HCO_3^-, and F^- occurs toward the middle northeastern part of the island (Figure 7). This zone generally corresponds to the zone with higher potentiometric head levels (Figure 2) and possibly represents the initial discharge zone of regional groundwater flow (i.e., base flow). However, more in-depth surveys on the recharge and flow of groundwater in the study area are needed.

4.5. Potential Evaluation of the Use of River Bank Filtration

The successful implementation of RBF technology requires a detailed, site-specific investigation of the hydrogeological and biogeochemical characteristics of aquifer and water quality status. In recent years, the construction of an RBF facility has been considered in the study site, with a target of the lower aquifer for the water supply by pumping. However, the current study on the hydrogeological and hydrochemical characterization of a riverine alluvial aquifer shows that there are two separate aquifers that are different in hydrochemical characteristics. Furthermore, geologic loggings show that the lower confined aquifer (at depths between approximately 25 and 40 m) is not directly connected to the bottom of the river (see Figure 1b). Our data also indicate that the lower aquifer groundwater possibly recharges from regional groundwater flow (base flow). Thus, direct induced flow of a large proportion of river water cannot be expected to occur toward the lower productive aquifer. A recent hydrogeologic study in a nearby bank filtration site [26] also suggested that the lower gravel aquifer seems to be hydraulically isolated from the upper sandy aquifer. As an alternative to bank filtration, the upper aquifer (<25 m thick) can be considered; however, the aquifer is significantly contaminated by agricultural activities and is higher in dissolved Fe and Mn. In summary, we consider that the study site seems to be less promising for the application of large-scale riverbank filtration.

5. Summary and Conclusions

In this work, we characterized the geology, hydrochemistry, and hydrogeology of a riverine alluvial aquifer in the low reaches of the Nakdong River in South Korea. The use of bank filtration was designed to pump water from the lower productive aquifer. The major summary and implications of this study are given below.

1.   The island that was studied was recently formed by point-bar sedimentation and subsequent erosion near the confluence of the Miryang River to the Nakdong River. The alluvial sediments are about 40 m thick and consist of fine- to medium-grained sand (upper aquifer; depth to about <25 m), silty sand with clay intercalations, and sandy gravel (lower aquifer; at depths between approximately 25 and 40 m) in descending order. Due to the presence of the intermediate layer as an aquitard that extends below the bottom of the river, the upper unconfined aquifer and lower confined aquifer are not hydrologically interconnected. Measurements of potentiometric head levels in the two aquifers support the different hydrogeologic conditions of the two aquifers in the riverine island.

2.   Dissolved Mn is originated from agricultural activities on the surface, and likely, recharges from the direct infiltration. The separation of the two alluvial aquifers are also indicated by hydrochemical characteristics. Groundwater chemistry on the small island is highly variable, from $Ca-Cl(-NO_3)$ type to $Na-HCO_3$ type. The upper aquifer groundwater is

highly contaminated by nitrate and dissolved Fe in rainwater, and hyporheic flow of river water. The zone of horizontal hyporheic flow is recognized by the zone of low TDS and Mg^{2+} concentrations in the upper aquifer. On the other hand, the lower aquifer groundwater is enriched in TDS, Na^+, K^+, Mg^{2+}, HCO_3^-, and F^-, likely due to the water–rock interaction during regional groundwater flow. The quality of the lower aquifer is also influenced by remnant seawater under freshening.

3.   The results of this study indicate that the lower aquifer is not directly connected to the river channel. Therefore, sustainable large-scale bank filtration is not promising at the study site. This study implies that careful examination of groundwater chemistry can be very helpful to evaluate the potential of the use of RBF.

Author Contributions: The authors have contributed to this work as follows: conceptualization, S.-T.Y. and J.S.K.; methodology, M.M., J.-H.L., and K.M.J.; formal analysis, M.M. and K.-H.K.; investigation, M.M., K.M.J., and J.S.K.; writing—original draft preparation, M.M. and J.H.L.; writing—review and editing, S.-T.Y.; supervision, S.-T.Y.; project administration, K.-H.K and S.-T.Y.; funding acquisition, S.-T.Y.

Funding: This work was initiated with support from Dohwa Eng. Co., and was completed with support from Korea Environment Industry & Technology Institute (KEITI) through the Subsurface Environment Management (SEM) Project, funded by the Korea Ministry of Environment.

Acknowledgments: Comments and suggestions from five anonymous reviewers were very helpful to clarify and improve the manuscript.

Conflicts of Interest: The authors declare no conflict of interest.

References

1.   Doussan, C.; Ledoux, E.; Detay, M. River-groundwater exchanges, bank filtration, and groundwater quality: Ammonium behavior. *J. Environ. Qual.* **1998**, *27*, 1418–1427. [CrossRef]
2.   International Riverbank Filtration Conference. Proceedings of the International Riverbank Filtration Conference: 2–4 November 2000. Available online: https://d-nb.info/962327611/04 (accessed on 15 September 2018).
3.   Tufenkji, N.; Ryan, J.N.; Elimelech, M. The promise of bank filtration. *Environ. Sci. Technol.* **2002**, *36*, 422A–428A. [CrossRef] [PubMed]
4.   Ray, C.; Grischek, T.; Schubert, J.; Wang, Z.; Speth, T.F. A perspective of riverbank filtration. *J. Am. Water Works Assoc.* **2002**, *94*, 149–160. [CrossRef]
5.   Ray, C.; Melin, G.; Linsky, R.B. *Riverbank Filtration: Improving Source Water Quality*; Water Science and Technology Library: New York, NY, USA, 2002.
6.   Stuyfzand, P.J.; Juhasz-Holterman, M.H.A.; De Lange, W.J. Riverbed filtration in the Netherlands: Well fields, clogging and geochemical reactions. In *Riverbank Filtration Hydrology—Impacts on System Capacity and Water Quality*; Hubbs, S.A., Ed.; Springer: Dordrecht, The Netherlands, 2013; pp. 119–153.
7.   Von Gunten, H.R.; Kull, T.P. Infiltration of inorganic compounds from the Glatt river, Switzerland, into a groundwater aquifer. *Water Air Soil Pollut.* **1986**, *29*, 333–346. [CrossRef]
8.   Kühn, W.; Müller, U. Riverbank filtration. *J. Am. Water Works Assoc.* **2000**, *92*, 60–69. [CrossRef]
9.   Bouwer, H. Artificial recharge of groundwater: Hydrogeology and engineering. *Hydrogeol. J.* **2002**, *10*, 121–142. [CrossRef]
10. Dash, R.R.; Mehrotra, I.; Kumar, P.; Grischek, T. Lake bank filtration at Naintal, India: Water quality evaluation. *Hydrogeol. J.* **2008**, *16*, 1089–1099. [CrossRef]
11. Dash, R.R.; Bhanu Prakash, E.V.P.; Kumar, P.; Mehrotra, I.; Sandhu, C.; Grischek, T. Riverbank filtration in Haridwar, India: Removal or turbidity, organics, and bacteria. *Hydrogeol. J.* **2010**, *18*, 973–983. [CrossRef]
12. Sudhakaran, S.; Lattemann, S.; Amy, G.L. Appropriate drinking water treatment processes for organic micropollutants removal based on experimental and model studies-a multi-criteria analysis study. *Sci. Total Environ.* **2013**, *442*, 478–488. [CrossRef] [PubMed]
13. Gianni, G.; Richon, J.; Perrochet, P.; Vogel, A.; Brunner, P. Rapid identification of transience in streambed conductance by inversion of floodwave responses. *Water Resour. Res.* **2016**, *52*, 2647–2658. [CrossRef]

14. Grischek, T.; Schoenheinz, D.; Worch, E.; Hiscock, K.M. Bank filtration in Europe-an overview of aquifer conditions and hydraulic controls. In *Management of Aquifer Recharge for Sustainability*; Dillon, P., Ed.; Balkema: Rotterdam, The Netherlands, 2002; pp. 485–488.

15. Grischek, T.; Schoenheinz, D.; Ray, C. Siting and design issues for riverbank filtration schemes. In *Riverbank Filtration*; Ray, C., Melin, G., Linksy, R.B., Eds.; Kluwer Academic Publishers: Dordrecht, The Netherlands, 2002; pp. 291–302.

16. Sandhu, C.; Grischek, T.; Kumar, P. Potential for riverbank filtration in India. *Clean Technol. Environ. Policy* **2011**, *13*, 295–316. [CrossRef]

17. Ojha, C.S.P. Simulating turbidity removal at a river bank filtration site in India using SCS-CN approach. *J. Hydrol. Eng.* **2012**, *17*, 1240–1244. [CrossRef]

18. Singh, A.K.; Shah, G.; Sharma, V. Revival of defunct radial collector wells for urban water supply using river bank filtration technique in India. *J. Indian Water Works Assoc.* **2012**, *49*, 24–39.

19. Chang, L.C.; Ho, C.C.; Yeh, M.S.; Yang, C.C. An integrating approach for conjunctive-use planning of surface and subsurface water system. *Water Resour. Manag.* **2011**, *25*, 59–78. [CrossRef]

20. Wang, L.; Ye, X.; Du, X. Suitability evaluation of river bank filtration along the second Songhua River, China. *Water* **2016**, *8*, 176. [CrossRef]

21. Korea Water Corporation. *Unpublished Internal Report on Pilot Survey of Hydraulic Property of Fluvial Deposits for Water Resource Utilization*; Korea Water Corporation: Daejeon, Korea, 1995; p. 132.

22. Hamm, S.-Y.; Cheong, J.-Y.; Ryu, S.M.; Kim, M.J.; Kim, H.S. Hydrogeological characteristics of bank storage area in Daesan-Myeon, Changwon City, Korea. *J. Geol. Soc. Korea* **2002**, *38*, 595–610.

23. Hamm, S.-Y.; Cheong, J.-Y.; Kim, H.S.; Hahn, J.S.; Cha, Y.H. Groundwater flow modeling in a riverbank filtration area, Deasan-Myeon, Changwon City. *Econ. Environ. Geol.* **2005**, *38*, 67–78. (In Korean)

24. Cheong, J.-Y.; Hamm, S.-Y.; Kim, H.-S.; Ko, E.-J.; Yang, K.; Lee, J.-H. Estimating hydraulic conductivity using grain-size analyses, aquifer tests, and numerical modeling in a riverside alluvial system in South Korea. *Hydrogeol. J.* **2008**, *16*, 1129–1143. [CrossRef]

25. Seo, J.A.; Kim, Y.C.; Kim, J.S.; Kim, Y.J. Site prioritization for artificial recharge in Korea using GIS mapping. *J. Soil Groundwater Environ.* **2011**, *16*, 66–78. (In Korean) [CrossRef]

26. Lee, S.-H.; Hamm, S.-Y.; Ha, K.; Kim, Y.C.; Koh, D.-C.; Yoon, H.; Kim, S.-W. Hydrogeologic and paleo-geographic characteristics of riverside alluvium at an artificial recharge site in Korea. *Water* **2018**, *10*, 835. [CrossRef]

27. Hiscock, K.M.; Grischek, T. Attenuation of groundwater pollution by bank filtration. *J. Hydrol.* **2002**, *266*, 139–144. [CrossRef]

28. Diem, S.; Cirpka, O.A.; Schirmer, M. Modeling the dynamics of oxygen consumption upon river bank filtration by a stochastic-convective approach. *J. Hydrol.* **2013**, *505*, 352–363. [CrossRef]

29. Huntscha, S.; Rodriguez Velosa, D.M; Schroth, M.H; Hollender, J. Degradation of polar organic micropollutants during riverbank filtration: Complementary results from spatiotemporal sampling and push-pull tests. *Environ. Sci. Technol.* **2013**, *47*, 11512–11521. [CrossRef] [PubMed]

30. Kvitsand, H.M.L.; Myrmel, M.; Fiksdal, L.; Østerhus, S.W. Evaluation of bank filtration as a pretreatment method for the provision of hygienically safe drinking water in Norway: Results from monitoring at two full-scale sites. *Hydrogeol. J.* **2017**, *25*, 1257–1269. [CrossRef]

31. Sontheimer, H. Experience with river bank filtration along the Rhine River. *J. Am. Water Works Assoc.* **1980**, *72*, 386–390. [CrossRef]

32. Bourg, A.C.M.; Darmendrail, D.; Ricour, J. Geochemical filtration of riverbank and migration of heavy metals between the Deule River and the Ansereuilles alluvion-chalk aquifer (Nord, France). *Geoderma* **1989**, *4*, 229–244. [CrossRef]

33. Bertin, C.; Bourg, A.C.M. Radon-222 and chloride as natural tracers of the infiltration of river water into an alluvial aquifer in which there is significant river/groundwater mixing. *Environ. Sci. Technol.* **1994**, *28*, 794–798. [CrossRef] [PubMed]

34. Squillace, P.J. Observed and simulated movement of bank storage water. *Ground Water* **1996**, *34*, 121–134. [CrossRef]

35. Triska, F.J.; Duff, J.H.; Avanzino, R.J. The role of water exchange between a stream channel and its hyporheic zone in nitrogen cycling at the terrestrial-aquatic interface. *Hydrobiologia* **1993**, *251*, 167–184. [CrossRef]

36. Hoehn, E. Hydrogeological issues of riverbank filtration—A review. In *Riverbank Filtration: Understanding Contaminant Biogeochemistry and Pathogen Removal*; Ray, C., Ed.; Kluwer Academic Publishers: Dordrecht, The Netherlands, 2002; pp. 17–41.

37. Conant Jr., B.; Cherry, J.A.; Gillham, R.W. A PCE groundwater plume discharging to a river: Influence of the streambed and near-river zone on contaminant distributions. *J. Contam. Hydrol.* **2004**, *73*, 249–279. [CrossRef] [PubMed]

38. Bencala, K.E.; Gooseff, M.N.; Kimball, B.A. Rethinking hyporheic flow and transient storage to advance understanding of stream-catchment connections. *Water Resour. Res.* **2011**, *47*, W00H03. [CrossRef]

39. Boano, F.; Harvey, J.W.; Marion, A.; Packman, A.I.; Revelli, R.; Ridolfi, L.; Worman, A. Hyporheic flow and transport processes: Mechanisms, models, and biogeochemical implications. *Rev. Geophys.* **2015**, *52*, 603–679. [CrossRef]

40. Hunt, H.; Schubert, J.; Ray, C. Conceptual design of riverbank filtration systems. In *Riverbank Filtration: Improving Source-Water Quality*; Ray, C., Melin, G., Linksy, R.B., Eds.; Kluwer Academic Publishers: Dordrecht, The Netherlands, 2003; pp. 19–27.

41. Gollnitz, W.D.; Whitteberry, B.L.; Vogt, J.A. Riverbank filtration: Induced filtration and groundwater quality. *J. Am. Water Works Assoc.* **2004**, *96*, 98–110. [CrossRef]

42. Massmann, G.; Nogeitzig, A.; Taute, T.; Pekdeger, A. Seasonal and spatial distribution of redox zones during lake bank filtration in Berlin, Germany. *Environ. Geol.* **2008**, *54*, 53–65. [CrossRef]

43. Hubbs, S.A. Laboratory-simulated RBF particle removal processes. *J. Am. Water Works Assoc.* **2010**, *102*, 57–66. [CrossRef]

44. Su, X.; Lu, S.; Gao, R.; Su, D.; Yuan, W.; Dai, Z.; Papavasilopoulos, E.N. Groundwater flow path determination during riverbank filtration affected by groundwater exploitation: A case study of Liao River, Northeast China. *Hydrol. Sci. J.* **2017**, *62*, 2331–2347. [CrossRef]

45. Min, J.H.; Yun, S.T.; Kim, K.; Kim, H.S.; Kim, D.J. Geologic controls on the chemical behavior of nitrate in riverside alluvial aquifers, Korea. *Hydrol. Proc.* **2003**, *17*, 1197–1211. [CrossRef]

46. Choi, B.Y.; Yun, S.T.; Mayer, B.; Chae, G.T.; Kim, K.H.; Kim, K.; Koh, Y.K. Identification of groundwater recharge sources and processes in a heterogeneous alluvial aquifer: Results from multi-level monitoring of hydrochemistry and environmental isotopes in a riverside agricultural area in Korea. *Hydrol. Proc.* **2010**, *24*, 317–330. [CrossRef]

47. Daewoo Construction Co. *Survey Report for Bank Filtration Project at Gimhae*, Unpublished Report. 2006. (In Korean)

48. Paik, S.; Cheong, D.; Shin, S.; Kim, J.C.; Park, Y.-H.; Lim, H.S. A paleoenvironmental study of Holocene delta sediments in Nakdong River Estuary. *J. Geol. Soc. Korea* **2016**, *52*, 15–30. [CrossRef]

49. Kelly, W.R. Heterogeneities in ground-water geochemistry in a sand aquifer beneath an irrigated field. *J. Hydrol.* **1997**, *198*, 154–176. [CrossRef]

50. Owens, L.B.; van Keuren, R.W.; Edward, W.M. Budgets of non-nitrogen nutrients in a high fertility pasture system. *Agric. Ecosyst. Environ.* **1998**, *70*, 7–18. [CrossRef]

51. Böhlke, J.K. Groundwater recharge and agricultural contamination. *Hydrogeol. J.* **2002**, *10*, 153–179. [CrossRef]

52. Chae, G.T.; Kim, K.; Yun, S.T.; Kim, K.; Kim, S.; Choi, B.; Kim, H.; Rhee, C.W. Hydrogeochemistry of alluvial groundwaters in an agricultural area: An implication for groundwater contamination susceptibility. *Chemosphere* **2004**, *55*, 369–378. [CrossRef] [PubMed]

53. Kim, K.H.; Yun, S.T.; Choi, B.Y.; Chae, G.T.; Joo, Y.; Kim, K.; Kim, H.S. Hydrochemical and multivariate statistical interpretations of spatial controls of nitrate concentrations in a shallow alluvial aquifer around oxbow lakes (Osong area, central Korea). *J. Contam. Hydrol.* **2009**, *107*, 114–127. [CrossRef]

54. Wright, R.F.; Lotse, E.; Semb, A. Reversibility of acidification shown by whole-catchment experiments. *Nature* **1988**, *334*, 670–675. [CrossRef]

55. Vitousek, P.M.; Aber, J.D.; Howarth, R.W.; Likens, G.E.; Matson, P.A.; Schindler, D.W.; Schlesinger, W.H.; Tilman, D.G. Human alteration of the global nitrogen cycle: Sources and consequences. *Ecol. Appl.* **1997**, *7*, 737–750. [CrossRef]

56. Chae, G.T.; Yun, S.T.; Kwon, M.J.; Kim, Y.S.; Mayer, B. Batch dissolution of granite and biotite in water: Implication for fluorine geochemistry in groundwater. *Geochem. J.* **2006**, *40*, 95–102. [CrossRef]

57. Pierso n-Wickmann, A.C.; Aquilina, L.; Weyer, C.; Molénat, J.; Lischeid, G. Acidification processes and soil leaching influenced by agricultural practices revealed by strontium isotopic ratios. *Geochim. Cosmochim. Acta* **2009**, *73*, 4688–4704. [CrossRef]

58. Kim, K.; Jeong, G.Y. Factors influencing natural occurrence of fluoride-rich groundwaters: A case study in the southeastern part of the Korean Peninsula. *Chemosphere* **2005**, *58*, 1399–1408. [CrossRef] [PubMed]

59. Chae, G.T.; Yun, S.T.; Mayer, B.; Kim, K.H.; Kim, S.Y.; Kwon, J.S.; Kim, K.; Koh, Y.K. Fluorine geochemistry in bedrock groundwater of South Korea. *Sci. Total Environ.* **2007**, *385*, 272–283. [CrossRef] [PubMed]

60. Choi, B.Y.; Yun, S.T.; Kim, K.H.; Kim, J.W.; Kim, H.M.; Koh, Y.K. Hydrogeochemical interpretation of South Korean groundwater monitoring data using Self-Organizing Maps. *J. Geochem. Explor.* **2014**, *137*, 73–84. [CrossRef]

61. Lee, S.; Currell, M.; Cendón, D.I. Marine water from mid-Holocene sea level highstand trapped in a coastal aquifer: Evidence from groundwater isotopes, and environmental significance. *Sci. Total Environ.* **2016**, *544*, 995–1007. [CrossRef] [PubMed]

62. Thomas, N.E.; Kan, K.T.; Bray, D.I.; MacQuarrie, K.T.B. Temporal changes in manganese concentrations in water from the Fredericton Aquifer, New Brunswick. *Ground Water* **1994**, *32*, 650–656. [CrossRef]

63. Brown, C.J.; Schoonen, M.A.A.; Candela, J.L. Geochemical modeling of iron, sulfur, oxygen and carbon in a coastal plain aquifer. *J. Hydrol.* **2000**, *237*, 147–168. [CrossRef]

64. Liaghati, T.; Preda, M.; Cox, M. Heavy metal distribution and controlling factors within coastal plain sediments, Bells Creek catchment, southeast Queensland, Australia. *Environ. Int.* **2004**, *29*, 935–948. [CrossRef]

65. Bencala, K.E. Hyporheic zone hydrological processes. *Hydrol. Proc.* **2000**, *14*, 2797–2798. [CrossRef]

water

MDPI

Article

Riverbank Filtration for the Water Supply on the Nakdong River, South Korea

Sung Kyu Maeng [1,*] and Kyung-Hyuk Lee [2]

[1]   Department of Civil and Environmental Engineering, Sejong University, 209 Neungdong-ro, Gwangjin-Gu, Seoul 05006, Korea
[2]   K-water Research Institute, Korea Water Resources Corporation, Daejeon 34045, Korea; kh.lee@kwater.or.kr
*   Correspondence: smaeng@sejong.ac.kr; Tel.: +82-2-3408-3858

Received: 11 November 2018; Accepted: 7 January 2019; Published: 12 January 2019

Abstract: A field study was carried out to investigate the feasibility of a riverbank filtration site using two vertical wells on the Nakdong River, South Korea. The riverbank filtration site was designed to have eleven horizontal collector wells in order to supply 280,000 m³/day. This field study provided more insight into the fate of the dissolved organic matter's characteristics during soil passage. The vertical production wells (PWs) were located in different aquifer materials (PW-Sand and PW-Gravel) in order to determine the depth of the laterals for the horizontal collector wells. The turbidity of the riverbank filtrates from the PW-Sand (0.9 NTU) and PW-Gravel (0.7 NTU) was less than 1 NTU, which was the target turbidity of the riverbank filtrate in this study. The iron concentrations were 18.1 ± 0.8 and 25.9 ± 1.3 mg/L for PW-Sand and PW-Gravel respectively, and were higher than those of the land-side groundwater. The biodegradable organic matter-determined biochemical oxygen demand in the river water was reduced by more than 40% during soil passage, indicating that less microbial growth in the riverbank filtrate could be possible. Moreover, the influence of the pumping rates of the vertical wells on the removal of dissolved organic matter and the turbidity was not significant.

Keywords: dissolved organic matter; fluorescence excitation-emission matrix; LC-OCD; Nakdong River; riverbank filtration

1. Introduction

Climate change influences both water availability and water quality through floods and droughts [1]. In Korea, the characteristics of water sources and their availability have been affected by economic growth, insufficient water management, and uncertainties due to climate change [2]. Therefore, it will become more difficult to secure clean water during extreme meteorological events. Korea is also included among the world's water-stressed nations between 2000 and 2025 [3]. South Korea is heavily dependent on its surface water for sources of drinking water, and approximately more than 90% of its drinking water comes from a river or man-made reservoir. The rainfall from June to September provides nearly 70% of the regional drinking water supply [4]. The mean annual rainfall is 1274 mm, and heavy rainfall that occurs during the summer leads to water shortages during the dry season (spring). Environmental accidents, including the contamination of tap water sources in the 1990s, raised many concerns and caused people in Korea to be reluctant to use tap water as drinking water [2]. Therefore, there is a need to find other water resources that are safe and to improve the public view of tap water quality.

When surface water's characteristics change due to extreme weather events, conventional water treatments have difficulty securing high quality water resources. To secure high quality water resources, alternative water resources such as managed aquifer recharge (MAR) systems were investigated [5]. MAR systems use natural water treatments and are effective at removing biodegradable organic matter.

MAR systems such as riverbank filtration (RBF) are emerging in Korea as an alternative solution [2]. RBF is a water treatment process that uses the physical, chemical and biological degradation processes of aquifers [6,7], and it is a nature-friendly water treatment process that removes pollutants without using chemicals [8]. RBF is also effective at alleviating the production of disinfection by-products and reducing trace organic contaminants [9–11]. In addition, it is also suitable for water safety and management [8,12]. In the early 2000s, a RBF system was first introduced to improve the water quality of drinking water resources in South Korea, especially in the regions where there was poor water quality for a decade due to the wastewater effluents discharged from local industries. A number of chemical spills had occurred in the river, which caused people to lose confidence in tap water quality [5].

A number of cities located downstream of the Nakdong River (Busan, Korea) are vulnerable to various water pollution sources and the seasonal water quality changes. Therefore, it is necessary to improve their water treatment system by improving the water source's quality. Currently, there are three drinking water treatment plants that are currently providing water via RBF using the Nakdong River (Table 1). The city of Changwon, which is on the Nakdong River in Korea, has been providing 80,000 m^3/day of drinking water since 2006 using RBF systems with vertical and horizontal collector wells. This system was the first RBF site installed to supply drinking water in South Korea. The city of Gimhae, South Korea, is currently providing 127,000 m^3/day via RBF (designed capacity: 180,000 m^3/day). Moreover, the Korea Water Resources Corporation (K-Water, Daejeon) in South Korea is currently investigating potential RBF sites that can supply 680,000 m^3/day to cities including Busan that are located near the lower part of the Nakdong River. This field study of two vertical wells will be used for the design of the horizontal collector wells that contribute part of the water supply to the city of Busan.

Table 1. Riverbank filtration sites located along the Nakdong River, South Korea.

City	Capacity (m^3/day)	Since	Specification
Changwon	70,000	2006	43 vertical wells
Changwon	10,000	1998	7 vertical wells
Haman county	20,000	2005	18 vertical wells
Gimhae	180,000	2017	9 horizontal collector wells

Before the installation of eleven horizontal collector wells, which provide 280,000 m^3/day, the field study was carried out using two vertical wells to investigate the quality of RBF. To improve the post-treatment requirements after the RBF, there is a need to investigate the water quality of RBF filtrates, including the dissolved organic matter's characteristics. Previously, Lee et al. [5] investigated the performance of RBF filtrates by comparing river water quality. However, there were no studies conducted on the fate of dissolved organic matter characteristics during RBF from a field study in South Korea. Moreover, there has been no report on using vertical wells to determine the depths of the laterals for horizontal collector wells.

The objective of this study was to conduct a detailed investigation of the characteristics of dissolved organic matter during soil passage using two vertical wells along the Nakdong River, South Korea (January to June, 2011). This study also investigated the removal efficiency of dissolved organic matter and the turbidity of two vertical wells whose screens are located at different depths (i.e., sand and gravel layers). The performances of the vertical wells at different pumping rates were also investigated. The water quality characteristics from two different vertical wells helped to determine the depth of the laterals for horizontal collector wells, even when there were other factors that needed to be considered (such as quantity).

2. Materials and Methods

2.1. Field Site

Figure 1 shows the schematic diagram of a vertical well site, which consisted of two production wells (PWs) and land-side groundwater monitoring wells (GMWs). The pilot RBF site consisted of two PWs and two GMWs at the sand and gravel layers. The two PWs were located at 30 m from the river, and the well screens were 6 m long. The well depths for the PW-Sand and PW-Gravel were 18 and 27 m below the land surface, respectively. Horizontal collector wells were planned to be installed at the site; therefore, two vertical wells were located at two different depths/aquifer layers that consisted of different materials (e.g., sand or gravel) to investigate the water quality characteristics. Two GMWs at the same depths as the PWs screens were located 380 m from the PWs (PW-Sand and PW-Gravel) in order to compare the riverbank filtrates. A bank filtration (BF) simulator, developed as a part of the NASRI project (Germany), was used to estimate the shortest (i.e., minimum) travel time at different pumping rates (Table 2) [13]. It was reported that there was a small discrepancy, below 5%, when the numerically computed shortest travel time from the bank filtration simulator was compared to that of the MODFLOW model [14]. The shortest travel times estimated in the bank filtration simulator were used to determine the shortest travel time of the flow paths from the surface water to the well. Further information on the mathematical algorithms behind the BF simulator is given elsewhere [15].

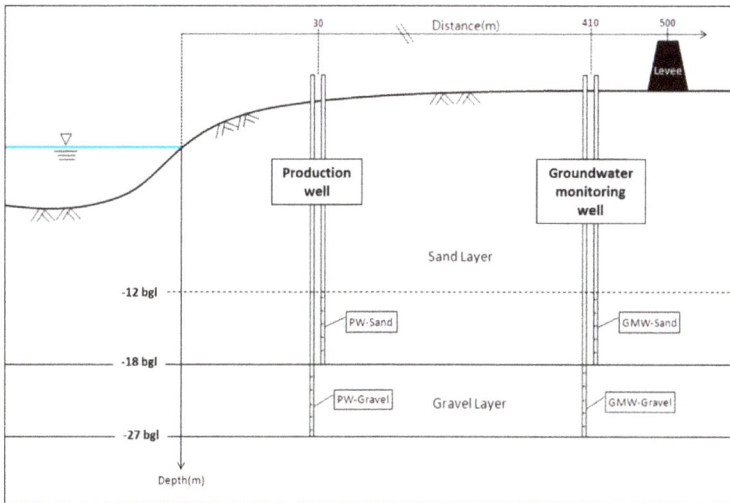

Figure 1. Schematic diagram of the vertical wells installed at different depths and the groundwater monitoring wells.

Table 2. Estimation of the shortest travel times (days) and the shares of riverbank filtrate and groundwater estimated by the bank filtration (BF) simulator.

Pumping Rate (m³/day)	Surface: Groundwater (%)	The Shortest Travel Time (days)
2000	63:33	1
1300	60:40	2
1000	54:46	3
700	46:54	4

2.2. Analytical Methods

The dissolved organic carbon (DOC) concentrations were analyzed using a total organic carbon (TOC) analyzer (TOC-V CPN, Shimadzu, Kyoto, Japan). For the DOC analysis, the samples were filtered using a 0.45-μm membrane filter (Whatman, Clifton, NJ, USA) and then analyzed using the TOC analyzer.

The ultraviolet absorbance at the 254 nm wavelength (UVA$_{254}$) is a useful indicator for predicting the trihalomethane (THM) formation potential [16] and characterizing the aromaticity of natural organic matter [17]. UVA$_{254}$ was measured using a spectrophotometer (DR5000, Hach, Loveland, CO, USA). The biochemical oxygen demand (BOD$_5$) was analyzed according to the standard methods [18]. The fluorescence excitation-emission matrix (EEM) spectra and liquid chromatography-organic carbon detection (LC-OCD) were used to determine the dissolved organic matter's characteristics. For the fluorescence EEM spectra, the sample was filled with a quartz cuvette (Hellma, Plainview, NY, USA) and the fluorescence intensity was measured using a fluorescence spectrophotometer (LS45, Perkin Elmer, Waltham, MA, USA). The spectrophotometer scanned emission wavelengths between 280 nm and 600 nm with excitation wavelengths between 200 nm and 400 nm at 10-nm intervals. The four selected peak regions that were found in this study were aromatic protein-like substances T1 (excitation 220–240 nm and emission 330–360 nm), tryptophan protein-like substances T2 (excitation 270–280 nm and emission 330–360 nm), fulvic-like substances A (excitation 230–260 nm and emission 400–450 nm), and humic-like substances C (excitation 300–340 nm and emission 400–450 nm) [19]. LC-OCD (DOC-LABOR, Karlsruhe, Germany), which consists of a size-exclusion chromatography column (HW-55S, GROM Analytik + HPLC GmbH, Herrenberg, Germany) with ultraviolet and organic carbon detectors, was used to classify the dissolved organic matter into five different organic matter fractions (biopolymers, humic substances, building blocks, low molecular weight neutrals, and low molecular weight acids) according to their molecular weights. Further details of the fractions that were determined via LC-OCD are reported elsewhere [20].

3. Results and Discussion

3.1. Turbidity

The turbidity of river water significantly varied during the time of the study, but the turbidity of the riverbank filtrates from PW-Sand and PW-Gravel was fairly stable with levels less than 1.0 NTU (Figure 2, one-way variance (ANOVA) $p < 0.05$). The high turbidity removal has been proven to be effective when using RBF [21]. Even when the turbidity of the river was higher than 60 NTU (data are not shown) in this study, the turbidity of the riverbank filtrate was below 1 NTU, which was the target turbidity of the riverbank filtrate in this study, indicating fairly stable riverbank filtrates.

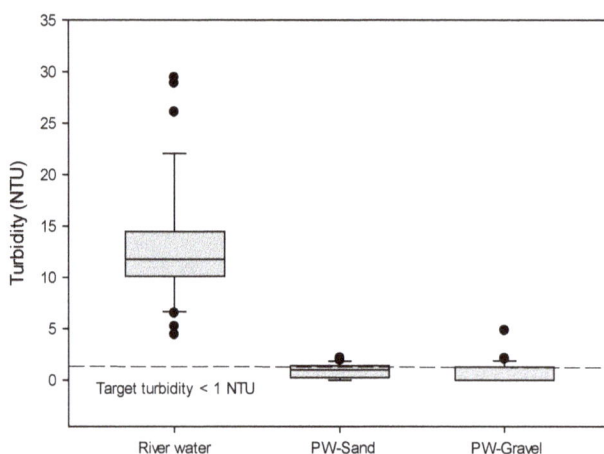

Figure 2. The turbidity of the vertical wells installed at different depths (production well (PW)-Sand and PW-Gravel).

3.2. Iron and Manganese

Iron and manganese were investigated during RBF and compared to the river water and groundwater monitoring wells. The iron concentration in the river water was 1.2 ± 0.9 mg/L, which was lower than the samples collected from the PW-Sand (18.1 ± 0.8 mg/L) and the PW-Gravel (25.9 ± 1.3 mg/L) during the time of the study (Figure 3). The iron concentrations between the two production wells were different because the PW-Sand and the PW-Gravel were located at different layers (e.g., sand or gravel). This study showed that the gravel layer had higher iron concentrations compared to those of the sand layer. The iron concentrations from production wells (PW-Sand and PW-Gravel) were higher than those of the groundwater (GMW-Sand: 14.1 ± 1.1 mg/L and GMW-Gravel: 10.3 ± 0.9 mg/L). Previous studies reported the occurrence of high iron concentrations in MAR [22,23]. The amount of manganese in the river water was 0.2 mg/L, but it was higher in the RBF filtrate samples (PW-Sand (1.5 ± 0.1 mg/L) and PW-Gravel (2.6 ± 0.8 mg/L)). Oxide-forming metals such as iron and manganese are easily mobilized as result of the reduced zone occurring during soil passage. We observed that dissolved oxygen (DO) dropped from 12 mg/L (Nakdong River) to below 0.5 mg/L (PW-Sand and PW-Gravel) during RBF (Figure 4). The high concentrations of iron and manganese that were observed in the RBF filtrates were due to the reduced conditions that occurred during soil passage. It was necessary to treat the riverbank filtrates that have very high levels of iron and manganese, and the riverbank filtrate needs to be treated at below 0.3 mg/L for iron in order to meet South Korean Guidelines for drinking water. The high concentrations of iron and manganese in the riverbank filtrates must be reduced by existing drinking water treatment plants, even though they were not designed to treat iron. The most common method in the ex situ removal of iron is oxidation followed by filtration, and the removal can be simply carried out by aeration followed by rapid sand filtration.

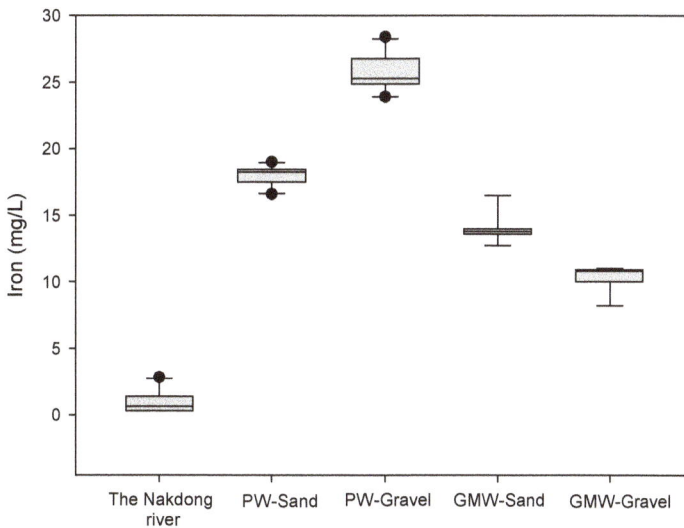

Figure 3. Iron concentrations of the vertical wells (PW-Sand and PW-Gravel).

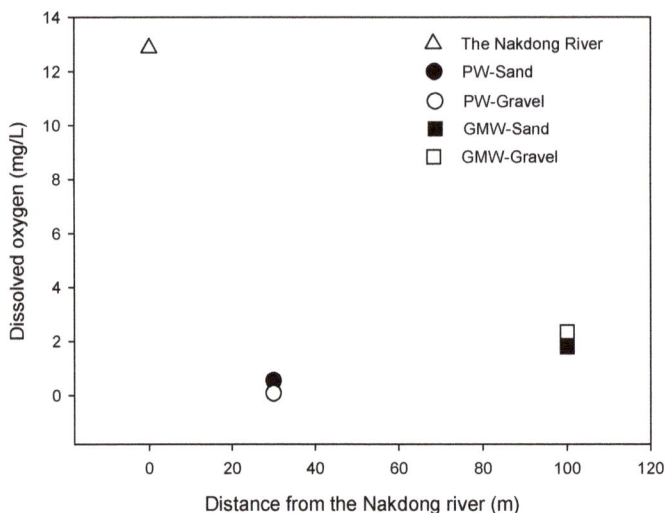

Figure 4. Changes in the dissolved oxygen during riverbank filtration.

3.3. Bulk Organic Matter

3.3.1. Biochemical Oxygen Demand (BOD_5)

The dissolved organic matter consisted of non-biodegradable and biodegradable organic matter fractions. In this study, BOD_5 was used to investigate the changes in the biodegradable organic matter during RBF. BOD_5 decreased during soil passage and was lower than that of the river. The removal efficiencies of BOD_5 were 64 and 40% for PW-Sand and PW-Gravel, respectively. The significant reduction of DO during soil passage reflected the biodegradation of bulk organic matter during soil passage. The DO reduction as a result of biodegradation was also reported during soil passage [24]. Biodegradable organic matter determined by BOD_5 was effectively removed during RBF which can reduce the occurrence of bio-regrowth and disinfection byproducts using less chlorine in distribution systems. There are limited treatments available (e.g., biological activated carbon, slow sand filter, etc.) to remove biodegradable organic matter during drinking water treatment processes. Therefore, RBF is an efficient treatment technology suitable for providing biostable water.

3.3.2. Dissolved Organic Carbon and UVA_{254}

The DOC was used to investigate the fate of dissolved organic matter during RBF. As shown in Table 3, DOC was effectively reduced. The reduction was not different from BOD and DO. The DOC is known to be the fraction that is most often removed by biodegradation during soil passage and it corresponds to the DO reduction [24]. The removal of DOC during RBF was 50 and 57% for PW-Sand and PW-Gravel, respectively. It is clear that the removal of DOC was effective during RBF. UVA_{254} is a useful tool to characterize the dissolved organic matter characteristics and can be used as an indicator of its aromaticity and hydrophobicity. The reduction of UVA_{254} was clearly observed during RBF, thus indicating the attenuation of the aromaticity in dissolved organic matter. The UVA_{254} levels were significantly high at 0.36 and 0.23 cm^{-1} for GMW-Sand and GMW-Gravel, respectively, indicating the high amount of aromatic compounds in the organic matter. Allochthonous dissolved organic matter, which consists of more humic-like compounds including polycyclic aromatics, is strongly associated with the formation of THMs [25]. Therefore, the mixing effects from groundwater that contained high concentrations of humic substances should be minimized at this site in order to reduce the THM formation potential during treatment of the riverbank filtrate. In this study, it is important to investigate the effect of the portion of groundwater in the pumped water, which is a mixture of riverbank filtrate and land-side groundwater.

Table 3. UV254 and dissolved organic carbon (n = 10).

	UVA_{254} (cm^{-1})	DOC (mg/L)
Nakdong River	0.038 ± 0.005	1.9 ± 0.1
PW-Sand	0.010 ± 0.008	0.7 ± 0.1
PW-Gravel	0.012 ± 0.004	0.6 ± 0.1
GMW-Sand	0.36 ± 0.09	0.5 ± 0.1
GMW-Gravel	0.23 ± 0.03	0.7 ± 0.2

3.3.3. Dissolved Organic Matter's Characteristics

The characteristics of the dissolved organic matter between surface water (Nakdong River), riverbank filtrates, and land-side groundwater were different. The fluorescence EEM spectra of the Nakdong River, PW-Sand and PW-Gravel were used to investigate dissolved organic matter's characteristics during soil passage. As previously mentioned, there are peaks at known wavelengths that represent protein-like substances and humic-like substances. The protein-like substances were also reported to indicate the presence of biodegradable organic matter [11]. The protein-like substances that were contained in the river water were attenuated in the RBF filtrate. Figure 5 demonstrated that the Fluorescence EEM intensities of the protein-like substances (T1 and T2) in the river water were reduced by more than 50% for the PW-Sand and PW-Gravel during soil passage. In addition, the reductions in the intensities of protein-like substances were similar to those of the DOC removals that are shown in Table 3 (PW-Sand: 63% and PW-Gravel: 68%). The fulvic-like and humic-like substances in groundwater were relatively high compared to those of PW-Sand and PW-Gravel. It was found that the humic-like substances in the Nakdong River were relatively low. In the case of land-side groundwater, it was confirmed that humic-like substances were dominant. The origins of aquatic humic-like substances are different from one another, and dissimilar properties could also result in high DOC [26]. The high fulvic-like and humic-like substances that were detected in PW-Gravel may be due to the influence of land-side groundwater, which contained high amounts of humic-like substances.

LC-OCD was used to investigate the changes in the dissolved organic matter's characteristics in the RBF filtrates according to their molecular weights. The biopolymers in the river water were significantly degraded compared to other DOM fractions (Figure 6). A previous study reported that biodegradation was found to be an important mechanism for removing biodegradable organic matter including biopolymers during soil passage [10]. As shown in Figure 6, biopolymers were relatively high in the Nakdong River. In the RBF filtrates taken from PW-Sand and PW-Gravel, the dissolved organic matter fractions with relatively high molecular weights, such as biopolymers and humics, were effectively removed during RBF. In particular, the biopolymers can be easily used as a carbon source for microorganisms; therefore, the biodegradation during soil passage is an important mechanism for supplying biologically stable water. It is important to understand the dissolved organic matter characteristics in riverbank filtrates using advanced organic matter characterization tools such as fluorescence EEM and LC-OCD. The humic-like substances were relatively high in the riverbank filtrate; therefore, post-treatments should focus on the removal of humics and biological filtration may not be necessary since most of the biodegradable organic matter was effectively removed. Based on the fluorescence EEM and LC-OCD results, RBF could effectively remove biodegradable organic matter, such as biopolymers, including protein-like substances.

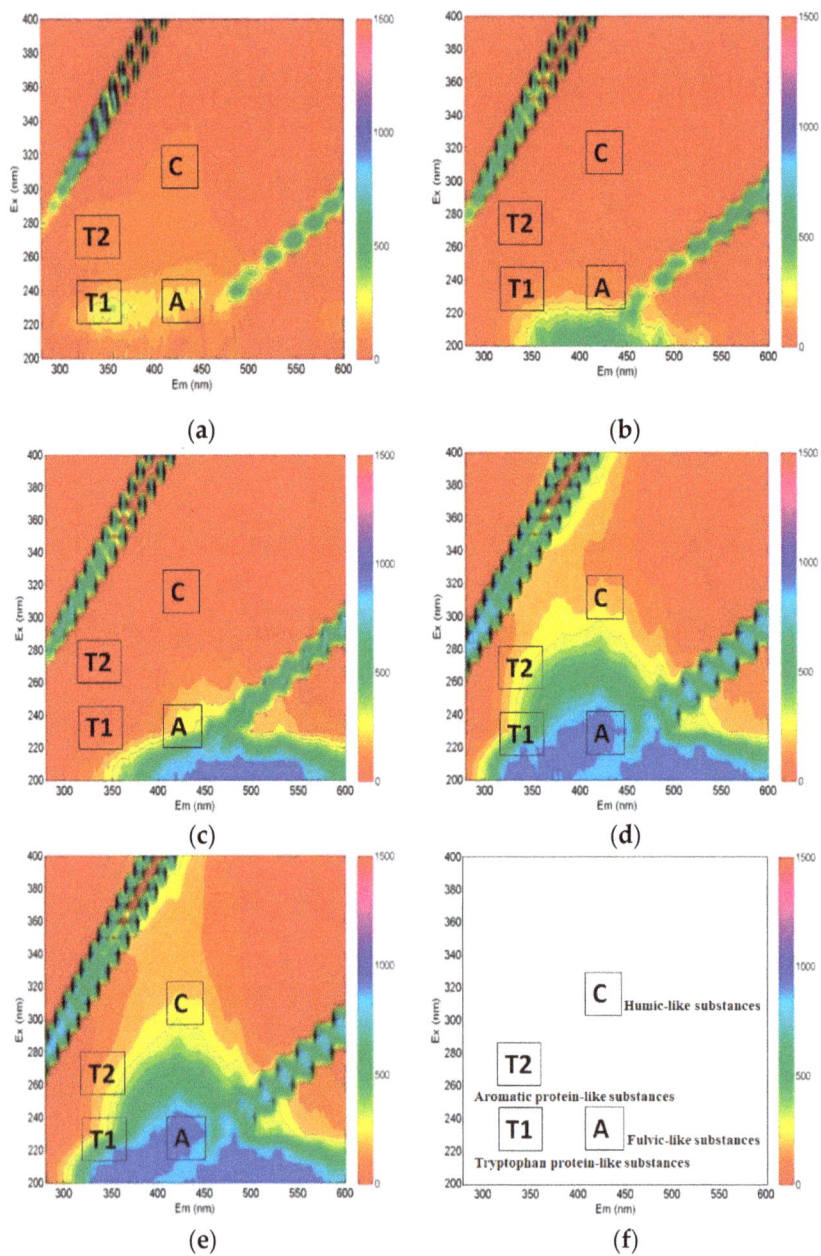

Figure 5. Fluorescence excitation-emission matrix (EEM) spectra: (**a**) river water, (**b**) PW-Sand, (**c**) PW-Gravel, (**d**) Groundwater (GW)-Sand, (**e**) GW-Gravel, and (**f**) the location of the EEM peaks for four groups of substances.

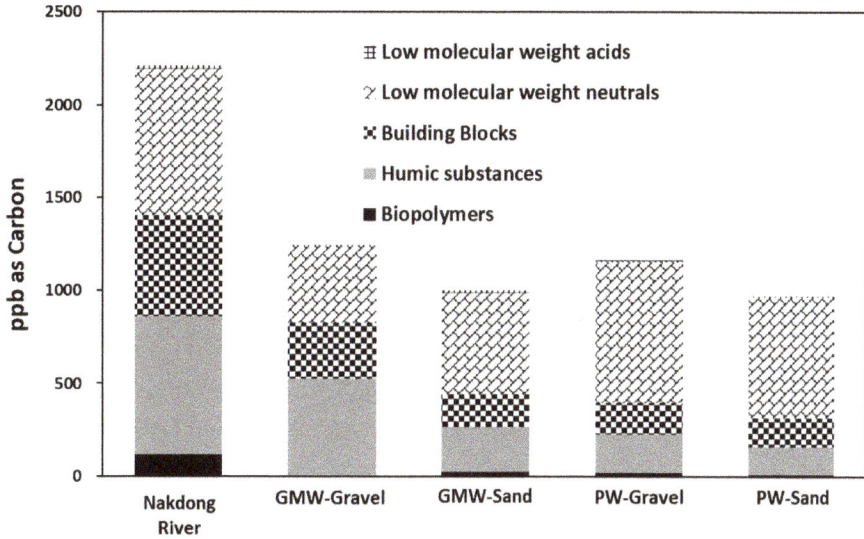

Figure 6. The changes of the organic matter fractions (biopolymers, humic substances, building blocks, neutrals, and acids) determined using liquid chromatography-organic carbon detection (LC-OCD) for river water, PW-Sand, PW-Gravel, GW-Sand, and GW-Gravel.

3.4. Effect of Different Pumping Rates

The pumping rates of the two vertical wells were varied for two weeks in order to investigate the impact of the shortest travel times on the removal of dissolved organic matter. The BF simulator, part of the NASRI Project (Germany), was used to estimate the shortest travel times at different pumping rates (Table 2) [13–15]. The different pumping rates (700, 1000, 1300 and 2000 m^3/day) that were tested in this study showed that there were no significant changes in the removal of turbidity, ammonia (data not shown), and DOC (Figure 7).

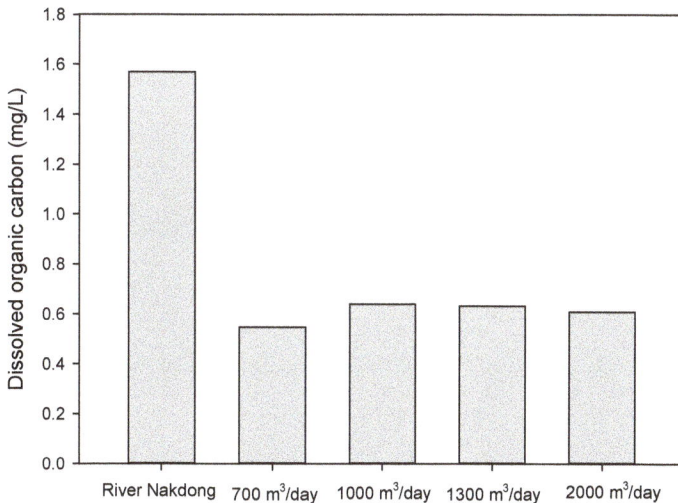

Figure 7. Changes in the dissolved organic carbon (PW-Sand) at different pumping rates (700, 1000, 1300 and 2000 m^3/day, n = 2).

The removal of ammonia and DOC are associated with biodegradation, which occurs predominantly within a few meters of infiltration [27]. Therefore, longer travel times may enhance the removal of slowly biodegradable organic matter that requires a longer time to degrade. However, there were no significant changes in the removal of dissolved organic matter when the estimated travel time increased from 1 to 3 d.

The iron concentrations with different pumping rates were also investigated. The iron concentrations were 18.1 ± 0.8, 17.2± 0.5, 14.7 ± 0.9, and 14.5 ± 0.6 mg/L at 2000, 1300, 1000 and 700 m^3/day, respectively. As the pumping rates decrease, there is more land-side groundwater in the pumped water, containing relatively lower iron concentrations (Table 2), indicating the higher fraction of land-side groundwater in the riverbank filtrate. As the pumping rates were changed, the water quality in the riverbank filtrates could also be influenced by the groundwater portion.

4. Conclusions

RBF is a robust water treatment method, which provides a degree of bulk organic matter removal and protection against water contamination. This study was carried out to investigate the feasibility of horizontal collector wells, and two vertical PWs were used to assess the water quality of filtrates during soil passage and the influence of land-side groundwater. Based on the field studies described in this paper, the following conclusions can be drawn:

- The turbidity was effectively removed via RBF, although there were significant turbidity increases in the river water.
- The iron concentrations from the RBF wells (i.e., PW-Sand and PW-Gravel) were higher than those of the groundwater monitoring wells (GMW-Sand and GMW-Gravel) and were 18.1 ± 0.8 and 25.9 ± 1.3 mg/L for the PW-Sand and PW-Gravel, respectively, during the study. The occurrence of a high iron concentration was due to the biodegradation of dissolved organic matter, which led to the reduced redox potential during soil passage. The reduced zone occurred between the river and RBF wells, which enhanced the mobilization of iron under more reducing conditions compared to that of land-side groundwater.
- As a result of the dissolved organic matter characteristics via LC-OCD and fluorescence EEM, the biopolymers contained in the river were effectively removed while passing through the aquifer. It was also confirmed that most of the humic components, which are difficult to reduce biologically and were detected from the land-side groundwater, could influence the quality of the RBF filtrate.
- The RBF wells (PW-Sand and PW-Gravel) in this study did not show any changes with respect to turbidity and DOC at different pumping rates (700, 1000, 1300 and 2000 m^3/day).
- Vertical wells at different layers (sand and gravel layers in the aquifer) were tested in this study in order to determine the depth of the laterals for the horizontal wells. This study used vertical wells in order to investigate the RBF site before the construction of the horizontal collectors, which usually cost much more than vertical wells.

Author Contributions: S.K.M. and K.-H.L. prepared the article. S.K.M. was responsible for the water quality data collection.

Funding: This research was supported by the Basic Science Research Program through the National Research Foundation of Korea (NRF), which was funded by the Ministry of Science, ICT & Future Planning [Grant number: 2017R1A1A1A05001231].

Conflicts of Interest: The authors declare no conflict of interest. The funders had no role in the design of the study; in the collection, analyses, or interpretation of data; in the writing of the manuscript; or in the decision to publish the results.

References

1. Delpla, I.; Jung, A.-V.; Baures, E.; Clement, M.; Thomas, O. Impacts of climate change on surface water quality in relation to drinking water production. *Environ. Int.* **2009**, *35*, 1225–1233. [CrossRef] [PubMed]
2. Choi, I.-C.; Shin, H.-J.; Nguyen, T.T.; Tenhunen, J. Water policy reforms in South Korea: A historical review and ongoing challenges for sustainable water governance and management. *Water* **2017**, *9*, 717. [CrossRef]
3. Pryor, F.L. Water stress and water wars. *Econ. Peace Secur. J.* **2007**, *2*, 7–18. [CrossRef]
4. Kim, J.-S.; Jain, S. Precipitation trends over the Korean peninsula: Typhoon-induced changes and a typology for characterizing climate-related risk. *Environ. Res. Lett.* **2011**, *6*, 034033. [CrossRef]
5. Lee, J.-H.; Hamm, S.-Y.; Cheong, J.-Y.; Kim, H.-S.; Ko, E.-J.; Lee, K.-S.; Lee, S.-I. Characterizing riverbank-filtered water and river water qualities at a site in the lower Nakdong River basin, Republic of Korea. *J. Hydrol.* **2009**, *376*, 209–220. [CrossRef]
6. Greskowiak, J.; Prommer, H.; Massmann, G.; Johnston, C.D.; Nützmann, G.; Pekdeger, A. The impact of variably saturated conditions on hydrogeochemical changes during artificial recharge of groundwater. *Appl. Geochem.* **2005**, *20*, 1409–1426. [CrossRef]
7. Henzler, A.F.; Greskowiak, J.; Massmann, G. Modeling the fate of organic micropollutants during river bank filtration (Berlin, Germany). *J. Contam. Hydrol.* **2014**, *56*, 78–92. [CrossRef]
8. Kim, H.-C.; Lee, W.M.; Lee, S.; Choi, J.; Maeng, S.K. Characterization of organic precursors in DBP formation and AOC in urban surface water and their fate during managed aquifer recharge. *Water Res.* **2017**, *123* (Suppl. C), 75–85. [CrossRef]
9. Bertelkamp, C.; Reungoat, J.; Cornelissen, E.R.; Singhal, N.; Reynisson, J.; Cabo, A.J.; van der Hoek, J.P.; Verliefde, A.R.D. Sorption and biodegradation of organic micropollutants during river bank filtration: A laboratory column study. *Water Res.* **2014**, *52*, 231–241. [CrossRef]
10. Maeng, S.K.; Sharma, S.K.; Abel, C.D.T.; Magic-Knezev, A.; Amy, G.L. Role of biodegradation in the removal of pharmaceutically active compounds with different bulk organic matter characteristics through managed aquifer recharge: Batch and column studies. *Water Res.* **2011**, *45*, 4722–4736. [CrossRef]
11. Maeng, S.K.; Sharma, S.K.; Abel, C.D.T.; Magic-Knezev, A.; Song, K.-G.; Amy, G.L. Effects of effluent organic matter characteristics on the removal of bulk organic matter and selected pharmaceutically active compounds during managed aquifer recharge: Column study. *J. Contam. Hydrol.* **2012**, *140–141*, 139–149. [CrossRef] [PubMed]
12. Dillon, P. Future management of aquifer recharge. *Hydrogeol. J.* **2005**, *13*, 313–316. [CrossRef]
13. Sharma, S.K.; Chaweza, D.; Bosuben, N.; Holzbecher, E.; Amy, G. Framework for feasibility assessment and performance analysis of riverbank filtration systems for water treatment. *J. Water Supply Res. Technol.* **2012**, *61*, 73–81. [CrossRef]
14. Rustler, M.; Grützmacher, G. Bank Filtration Simulator-Manual, TECHNEAU Executive Summary. 2009. Available online: https://www.kompetenz-wasser.de/wp-content/uploads/2017/05/d5-2-5.pdf (accessed on 4 April 2018).
15. Holzbecher, E.; Grützmacher, G.; Amy, G.; Wiese, B.; Sharma, S.K. The Bank Filtration Simulator—A MATLAB GUI. In *Environmental Informatics and Industrial Ecology, Proceedings of the 22nd International Conference on Informatics for Environmental Protection (Enviroinfo 2008), Lüneburg, Germany, 10–12 September 2008*; Möller, A., Page, B., Schreiber, M., Eds.; Shaker: Aachen, Germany, 2008.
16. Chow, A.T.; Dahlgren, R.A.; Zhang, Q.; Wong, P.K. Relationships between specific ultraviolet absorbance and trihalomethane precursors of different carbon sources. *J. Water Supply Res. Technol.* **2008**, *57*, 471–480. [CrossRef]
17. Ma, N.; Zhang, Y.; Quan, X.; Fan, X.; Zhao, H. Performing a microfiltration integrated with photocatalysis using an Ag-TiO2/HAP/Al2O3 composite membrane for water treatment: Evaluating effectiveness for humic acid removal and anti-fouling properties. *Water Res.* **2010**, *44*, 6104–6114. [CrossRef]
18. Korean Ministry of Environment. *Water Quality Standard and Test*; Korean Ministry of Environment: Sejong City, Korea, 2011.
19. Park, J.W.; Kim, H.C.; Meyer, A.S.; Kim, S.; Maeng, S.K. Influences of NOM composition and bacteriological characteristics on biological stability in a full-scale drinking water treatment plant. *Chemosphere* **2016**, *160*, 189–198. [CrossRef]

20. Huber, S.A.; Balz, A.; Abert, M.; Pronk, W. Characterisation of aquatic humic and non-humic matter with size-exclusion chromatography–organic carbon detection–organic nitrogen detection (LC-OCD-OND). *Water Res.* **2011**, *45*, 879–885. [CrossRef]
21. Kumar, P.; Mehrotra, I.; Gupta, A.; Kumar, S. Riverbank Filtration: A Sustainable Process to Attenuate Contaminants during Drinking Water Production. *J. Sustain. Dev. Energy Water Environ. Syst.* **2018**, *6*, 150–161. [CrossRef]
22. Vanderzalm, J.L.; Page, D.W.; Barry, K.E.; Dillon, P.J. A comparison of the geochemical response to different managed aquifer recharge operations for injection of urban storm water in a carbonate aquifer. *Appl. Geochem.* **2010**, *25*, 1350–1360. [CrossRef]
23. Vanderzalm, J.L.; Page, D.W.; Barry, K.E.; Scheiderich, K.; Gonzalez, D.; Dillon, P.J. Probabilistic approach to evaluation of metal(loid) fate during stormwater aquifer storage and recovery. *Clean Soil Air Water* **2016**, *44*, 1672–1684. [CrossRef]
24. Maeng, S.K.; Sharma, S.K.; Amy, G.; Magic-Knezev, A. Fate of effluent organic matter (EfOM) and natural organic matter (NOM) through riverbank filtration. *Water Sci. Technol.* **2008**, *57*, 1999–2007. [CrossRef] [PubMed]
25. Williams, C.J.; Conrad, D.; Kothawala, D.N.; Baulch, H.M. Selective removal of dissolved organic matter affects the production and speciation of disinfection byproducts. *Sci. Total Environ.* **2019**, *652*, 75–84. [CrossRef] [PubMed]
26. Artinger, R.; Buckau, G.; Geyer, S.; Fritz, P.; Wolf, M.; Kim, J.I. Characterization of groundwater humic substances: Influence of sedimentary organic carbon. *Appl. Geochem.* **2000**, *15*, 97–116. [CrossRef]
27. Schmidt, C.; Lange, F.; Brauch, H.; Kuhn, W. Experiences with riverbank filtration and infiltration in Germany. Proceeding of the International Symposium of Artificial Recharge of Groundwater, Daejon, Korea, 7–8 July 2003.

water

MDPI

Article

Riverbank Filtration Impacts on Post Disinfection Water Quality in Small Systems—A Case Study from Auburn and Nebraska City, Nebraska

Matteo D'Alessio [1], Bruce Dvorak [2] and Chittaranjan Ray [1,*]

[1]   Nebraska Water Center, University of Nebraska Lincoln, Lincoln, NE 68588-6204, USA; mdalessio2@unl.edu
[2]   Department of Civil Engineering, University of Nebraska Lincoln, Lincoln, NE 68588-6105, USA; bdvorak1@unl.edu
*   Correspondence: cray@nebraska.edu; Tel.: +1-402-472-8427

Received: 26 September 2018; Accepted: 13 December 2018; Published: 15 December 2018

Abstract: Small water systems can experience a fluctuating quality of water in the distribution system after disinfection. As chlorine is the most common disinfectant for small systems, the occurrence of disinfection byproducts (DBPs) represents a common problem for these systems. Riverbank filtration (RBF) can be a valuable solution for small communities located on riverbanks. The objectives of this study were to evaluate (i) the improvements in water quality at two selected RBF systems, and (ii) the potential lower concentrations of DBPs, in particular, trihalomethanes (THMs), in small systems that use RBF. Two small communities in Nebraska, Auburn and Nebraska City, using RBF were selected. Results from this study highlight the ability of RBF systems to consistently improve the quality of the source water and reduce the occurrence of THMs in the distribution water. However, the relative removal of THMs was directly impacted by the dissolved organic carbon (DOC) removal. Different THM concentrations and different DOC removals were observed at the two RBF sites due to the different travel distances between the river and the extractions wells.

Keywords: riverbank filtration; small communities; disinfection by-products; trihalomethanes

1. Introduction

Small water systems (served population <10,000) represent more than 97% of the USA public water systems and often experience a fluctuating quality of water in the distribution system after disinfection [1]. Dissolved organic carbon (DOC) present in surface waters contributes to the formation of several disinfection byproducts (DBPs) when chlorine is used as the disinfectant [2–5]. Systems that use chloramine also experience the depletion of chlorine residuals due to nitrification in summer months [6–8]. DBPs have been detected in concentration up to few mg L^{-1} and many of them are suspected or known human carcinogens [9–11]. Among halogenated DBPs, trihalomethanes (THMs) have been widely detected [2,10,12,13]. In order to reduce general public exposure to DBPs and lower the potential of cancer and reproductive and development risks, in 2005, the U.S. Environmental Protection Agency (EPA) issued a Stage 2 Disinfectants and Disinfection Byproducts Rule (DBPR) requiring all drinking water treatment plants to maintain levels of total THMs (TTHM) below the annual average maximum contaminant level (MCL) (80 μg L^{-1}) on the location running annual average (EPA Stage 1 and 2) [14]. Small water systems, especially in rural communities, may struggle to comply with the USEPA Stage 2 DBPR due to source water variation, limited resources, aging infrastructures, and low-cost efficiency [2]. For example, Hua et al., investigating three small drinking water systems in rural communities in Missouri, observed consistently higher THMs in finished water than the MCL (80 μg L^{-1}) [2]. Nationally, it has been identified that small systems using surface waters experience three times the rate of Stage 1 DBP violations than systems that serve populations above

10,000 [15]. Therefore, there is a need to provide cost-effective solutions to improve the source of water by reducing its variations and lowering its level of organic carbon. Natural filtration, a technology that has been used for communities of various sizes to fully treat or pretreat surface water before supply, can represent a valuable solution [16,17]. Natural filtration includes two primary technologies: Riverbank filtration (RBF) and slow sand filtration (SSF) [18,19]. Both of these types of technologies have been shown to produce water of consistent quality and remove a significant amount of organic carbon and microorganisms present in surface water. Additionally, natural filtration is resistant to rapid contamination [18–23]. Particularly, RBF is potentially ideal for small communities that are located on riverbanks. The use of groundwater in many areas (e.g., Nebraska) can be an issue due to high levels of nitrate (>MCL, 10 mg L^{-1} [24]) [25–28]. State regulatory agencies frequently request case studies for the application of technologies to address specific regulations before approving plans and funding [29]. There is a scarcity of research in small communities about how water produced from RBF wells or SSF affects the formation and subsequent fate of DBPs when chlorine or chloramines are used as disinfectants. Preliminary investigations have shown the ability of RBF to reduce DBP precursors (i.e., DOC) [30–34]. Increased understanding of the formation and dispersal of DBPs in the disinfection system and other contaminants of concern will provide greater protection of public health.

This study investigates (i) the improvements in water quality (i.e., total organic carbon, TOC, DOC, total coliforms, and *Escherichia coli*) at two selected RBF systems and (ii) the potential lower concentrations of DBPs, in particular, THMs, in small systems that use natural filtration compared to systems that directly use surface water. Improvements in water quality were measured by comparing the quality of river waters and the filtrate and examining the response of the systems to hydrologic forcing, such as spring runoff or low flow events in rivers.

2. Materials and Methods

2.1. Description of the RBF Sites

Auburn and Nebraska City, two small towns in Nebraska, were selected. The two cities have a population of 3460 and 7289, respectively [35]. Temperature ranged between −27 °C and 38 °C in both towns [36]. At the time of the study, the town of Auburn drew its drinking water from eight wells (1, 2, 3, 4, 5, 7, 13, and 19) located on the bank of the Little Nemaha River (distance well–river >88 m). The wells are placed in different locations along the river on straight stretches of the river as well as on limited curves (Figure S1). The outside well diameter ranges between 0.46 and 0.91 m, the well screen length is about 4.64 m (ranging between 4.58 and 4.72 m), and the well depth between 13.41 and 18.75 m (average: 15.72 m) (Table S1).

The aquifer sediments at the RBF site in Auburn consist of a superficial layer of brown/gray clay (1.5 to ~7.0 m below ground level, bgl) followed by fine sand to coarse (~7.0 to ~16.0 m bgl) with traces of gravel and boulders overlaying blue shale (>~16.0 m bgl) (Table S2).

After the water is collected, the current treatment practice for municipal use consists of aeration, adsorption clarification, high-rate gravity sand filters, fluoridation, and chlorination. Well 4, due to the high level of nitrate (~7.5 mg NO$_3$-N L^{-1}), is not being used. The water treatment plant is capable of producing 2 million gallons per day of drinking water.

At the time of the study, the town of Nebraska City drew its drinking water from eleven wells (1–11) located near the Missouri River (distance well–river >15.2 m). The outside well diameter ranges between 0.46 and 0.64 m, the well screen length is about 11 m (ranging between 8.53 and 13.41 m), and the well depth between 25.3 and 29.26 m (average: 26.72 m) (Table S1). In contrast with Auburn, the wells are placed in a single location along the river on a limited curve of the river (Figure S1). The aquifer sediments at the RBF site in Nebraska City show a more heterogeneous distribution compared to those of Auburn (Tables S2 and S3). In fact, aquifer sediments near wells 1, 2, 5, 6, and 7 mostly consist of sand with blue clay (~3.0 to ~4.5 m bgl) and gravel (~18.3 to ~23.0 m bgl), while superficial occurrence of clay (0 to ~5 m bgl) was observed followed by fine and coarse sand and limestone at the bottom near Wells 4, 8, 10, and 11 (Table S3). The aquifer sediments near Wells 3 and 9 mostly consist of fine and coarse sand with gravel and traces of clays (~15.0 m bgl) (Table S3). After the water is collected, the current treatment practice for municipal use consists of aeration, filtration, addition of lime, and chlorination.

The monitoring program started in May 2016 and continued until June 2017. Monthly samples were collected throughout the study (low flow period), while biweekly samples were collected in May 2016 (moderately high river flow period). River water samples were collected approximately 50 cm deep and 2 m from the riverbed of the Little Nemaha River and Missouri River, at the same locations throughout the study. Well water samples were collected from available well house spigots at the identified wells after water quality parameters (i.e., temperature, pH, dissolved oxygen, etc.) had stabilized. Water samples were also collected at the two water treatment plants, before additional treatment (referred herein as "pre") and after chlorination (water entering the distribution system; referred herein as "post").

2.2. Water Quality Analysis

Certified glass vials for TOC with Teflon lined, and septa (VWR, Thorofare, NJ, USA) were used to collect samples for TOC, DOC, ultraviolet absorbance (UVA), and THM analysis. DOC samples were filtered through a 0.45 μm glass microfiber (VWR, Thorofare, NJ, USA). Sulfuric acid was added to preserve the TOC and DOC samples. The collected samples were stored at 4 °C before analysis. Major anions, THMs, total coliforms, and *E. coli* were analyzed within a few hours of collection. Major anions (bromide, chloride, fluoride, nitrate, nitrite, orthophosphate, and sulfate) were measured using a Dionex ICS-90 ion chromatograph with a Dionex IonPac AS14 column (diameter: 4 mm and length: 250 mm) (Dionex, Bannockburn, IL, USA). TOC and DOC were measured by hot persulfate oxidation using an OI 1010 carbon analyzer (OI Analytical, College Station, TX, USA). The four most common THMs (bromodichloromethane, bromoform, chlorodibromomethane, and chloroform) [37,38] were measured using purge and trap gas chromatography/mass spectrometry with an OI 4552 Analytical autosampler and Eclipse Purge-and-Trap Sample Concentrator OI 4660 (OI Analytical, College Station, TX, USA) coupled with a 6890N GC/MS (Agilent Technologies, Santa Clara, CA, USA). Specific ultraviolet absorbance (SUVA), defined as the ratio between DOC and UVA at 254 nm, was used to estimate the nature of the organic matter present in the natural water [38–40]. The absorbance at 254 nm was measured with a Lambda 25 UV/VIS spectrometer (PerkinElmer Instruments, Akron, OH, USA). Table S4 shows the method detection limits of the different analytes.

Total coliform and *E. coli* were quantified using a commercial most probable number (MPN) test, Colilert 18, with a Quanti-Tray 2000 from IDEXX Laboratories (Westbrook, ME, USA) [41,42]. Due to its simplicity, the IDEXX method has been used in RBF investigations [32,43,44]. Samples were collected aseptically from the rivers as well as after RBF wells. 100 mL or an appropriate dilution of the sample was mixed with the reagent, poured into sterile trays, heat sealed, and incubated at 35 °C for 18 h to detect total coliform and *E. coli*.

3. Results

3.1. Improvement in Water Quality: Auburn RBF Site

The Little Nemaha River flows through S–E Nebraska and drains into the Missouri River. During the investigation, the Little Nemaha River, in Auburn, had a discharge ranging between 4 and 63 m^3 s^{-1} (Figure S2) [45]. At the Auburn RBF site, pH ranged between 6.96 and 8.41 in the Little Nemaha River water samples, and between 6.44 and 8.45 at the investigated wells (Figure S3a, Table S5). Electrical conductivity (EC) in the river ranged between 304 and 551 μS cm^{-1} and between 441 and 687 μS cm^{-1} at the investigated wells (Figure S4a, Table S5). Among the different anions, fluoride (MCL: 4 mg L^{-1} [24]) and phosphate were consistently below 0.4 mg L^{-1} in the river as well as at the wells (data not shown). Nitrate (measured as NO$_3$-N) (MCL: 10 mg L^{-1} [24]) ranged between 2.3 and

7.0 mg L^{-1} in the river and between 0.2 mg L^{-1} and 1.5 mg L^{-1} at Wells 6 and 13. High levels of nitrate (7.8 to 8.7 mg L^{-1}) were observed at Well 4 (Figure 1a). Due to the high levels of nitrate, Well 4 was not being used. Higher values of sulfate (14.8 to 180 mg L^{-1}) and chloride (5.9 to 31.9 mg L^{-1}) were observed at the wells than in the river (15.8 to 78.0 mg L^{-1} and 4.1 and 14.3 mg L^{-1}, respectively) (Figures S5a and S6a). Nitrite and bromide were below 0.0239 mg L^{-1} throughout the study (data not shown).

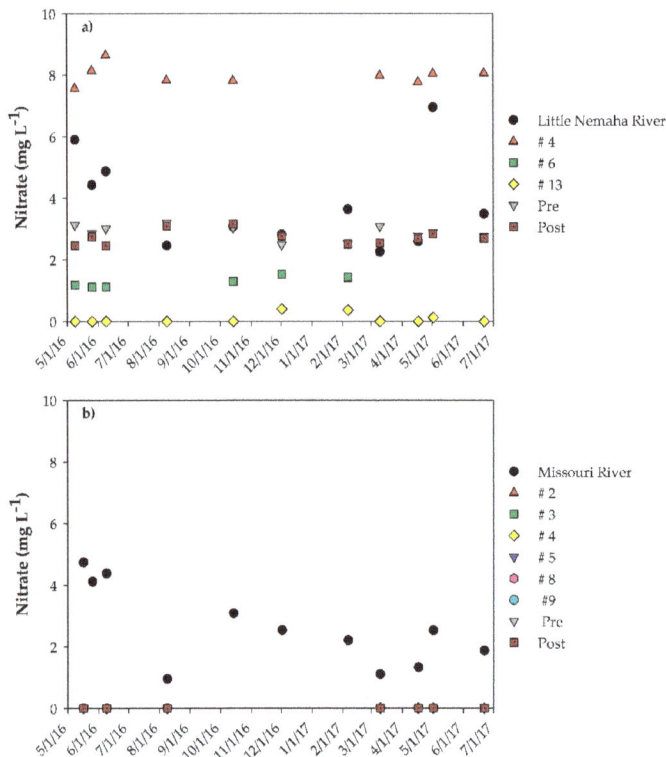

Figure 1. Nitrate (in NO$_3$–N) in (**a**) Auburn and (**b**) Nebraska City. LNR: Little Nemaha River; MR: Missouri River; Pre: Inflow water to the water treatment plants prior to any additional treatment; Post: Water collected at the two water treatment plants after chlorination.

TOC and DOC along the Little Nemaha River ranged between 1.8 and 17.6 mg L^{-1} and between 1.6 and 7.2 mg L^{-1}, respectively. In RBF well water, TOC and DOC ranged between 0.3 and 2.7 mg L^{-1} and 0.2 and 2.1 mg L^{-1} (Figure 2a,c, Table S5). TOC removal through RBF (water collected at investigated wells vs. stream water) ranged between 58 and 96% throughout the study. Similarly, DOC removals ranged between 53 and 90% (Table S6). Removal from RBF of TOC and DOC was not impacted by TOC and/or DOC values in the Little Nemaha River (Table S6). SUVA ranged between 1.9 and 16.0 L mg^{-1} m^{-1} at the Little Nemaha River and between 1.3 and 8.8 L mg^{-1} m^{-1} in extracted water from RBF wells. Lower values were observed at the water facility. After chlorination, SUVA ranged between 0.6 and 5.0 L mg^{-1} m^{-1} (Figure S7a). TTHM ranged between 11.2 and 30.5 µg L^{-1}. Among the THMs investigated, chloroform (1.16 to 4.03 µg L^{-1}) and dichlorobromomethane (3.24 to 14.97 µg L^{-1}) showed the lowest and highest concentrations, respectively (Figure 3a, Table S7). The RBF facility was also able to consistently remove bacteria. Total coliforms, 2.06 × 10^3 to 8.30 × 10^6 MPN/100 mL, and *E. coli*, 1.34×10^2 to 6.31×10^2 MPN/100 mL, present in the Little Nemaha River, decreased to <1 MPN/100 mL (IDEXX detection limit) in RBF well water throughout the study.

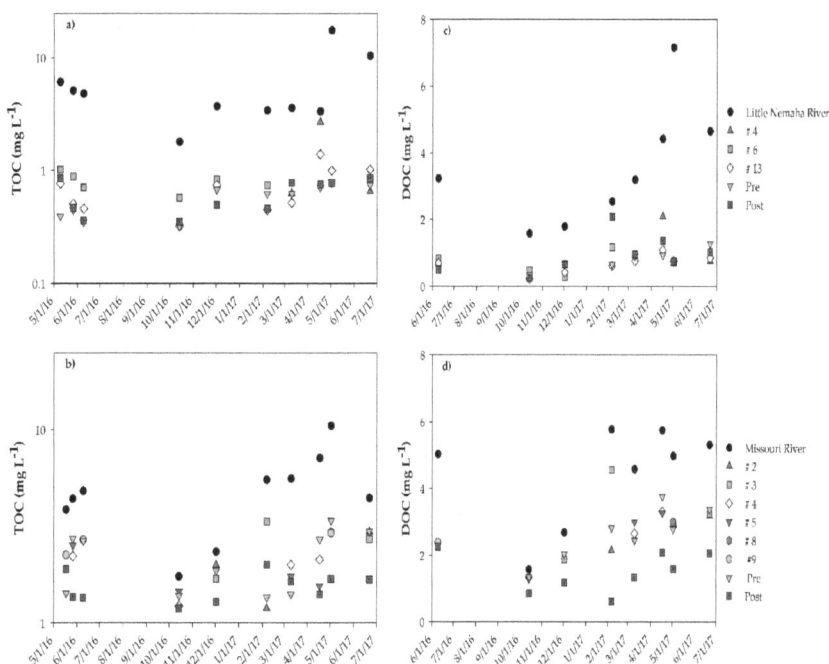

Figure 2. Total organic carbon (TOC) in (**a**) Auburn and (**b**) Nebraska City and dissolved organic carbon (DOC) in (**c**) Auburn and (**d**) Nebraska City. LNR: Little Nemaha River; MR: Missouri River; Pre: Inflow water to the water treatment plant prior to any additional treatment; Post: Water collected at the two water treatment plants after chlorination.

3.2. Improvement in Water Quality: Nebraska City RBF Site

The Missouri River flows through S–E Nebraska and receives the Little Nemaha River south of Auburn. During the investigation, in Nebraska City, the discharge of the Missouri River ranged between 960 and 1910 $m^3 s^{-1}$ (Figure S2) and the water temperature between 0 and 30 °C [45]. pH ranged between 7.10 and 8.30 in the Missouri River water samples, and between 6.68 and 8.00 at the investigated wells (Figure S3b, Table S5). EC in the river ranged between 579 and 763 $\mu S cm^{-1}$ and between 548 and 856 $\mu S cm^{-1}$ at the investigated wells (Figure S4b, Table S5). Among the different anions, phosphate was consistently below 0.0517 mg L^{-1}, while fluoride was consistently below 0.5 mg L^{-1} in the river as well as at the wells (data not shown). Nitrate ranged between 0.9 and 4.7 mg L^{-1} in the river (consistently <MCL), while no detection occurred at the investigated wells throughout the study (Figure 1b). Slightly higher values of sulfate (86.7 to 399 mg L^{-1}) and chloride (10.9 to 25.5 mg L^{-1}) were observed in the river than at the wells (80.4 to 353 mg L^{-1} and 5.3 and 22.3 mg L^{-1}, respectively) (Figures S5b and S6b). Nitrite and bromide were below 0.0239 mg L^{-1} throughout the study (data not shown).

TOC and DOC in the Missouri River water samples ranged between 1.7 and 10.5 mg L^{-1} and between 1.6 and 5.8 mg L^{-1} respectively. In the RBF well water, TOC and DOC ranged between 1.2 and 3.4 mg L^{-1} and between 1.3 and 4.6 mg L^{-1} (Figure 2b,d, Table S5). TOC removal ranged between 14.2 and 78.2%. DOC removal ranged between 15.2 and 62.8% (Table S7). Low removal of TOC and DOC was achieved in the presence of low TOC and DOC values (<2.7 mg L^{-1}) in the Missouri River. SUVA ranged between 1.51 and 6.69 L mg^{-1} m^{-1} at the Missouri River and between 1.53 and 6.16 L mg^{-1} m^{-1} in RBF well water. Slightly lower values were observed at the water facility. After chlorination, SUVA ranged between 0.82 and 5.46 L mg^{-1} m^{-1} (Figure S7b). The total concentration of THMs ranged between 28.9 and 98.6 µg L^{-1}. Among the THMs investigated, bromoform (0.4 to 11.4 µg L^{-1}) and chloroform (15.6 to 53.8 µg L^{-1}) showed the lowest and highest concentrations, respectively (Figure 3a, Table S7). By the results obtained in RBF well water in Auburn, the RBF facility was also able to consistently remove bacteria. Total coliforms, 2.60×10^3 to 4.35×10^4 MPN/100 mL, and *E. coli*, 24.6 to 2.00×10^3 MPN/100 mL, present in the Missouri River, decreased to <1 MPN/100 mL (IDEXX detection limit) in RBF well water throughout the study.

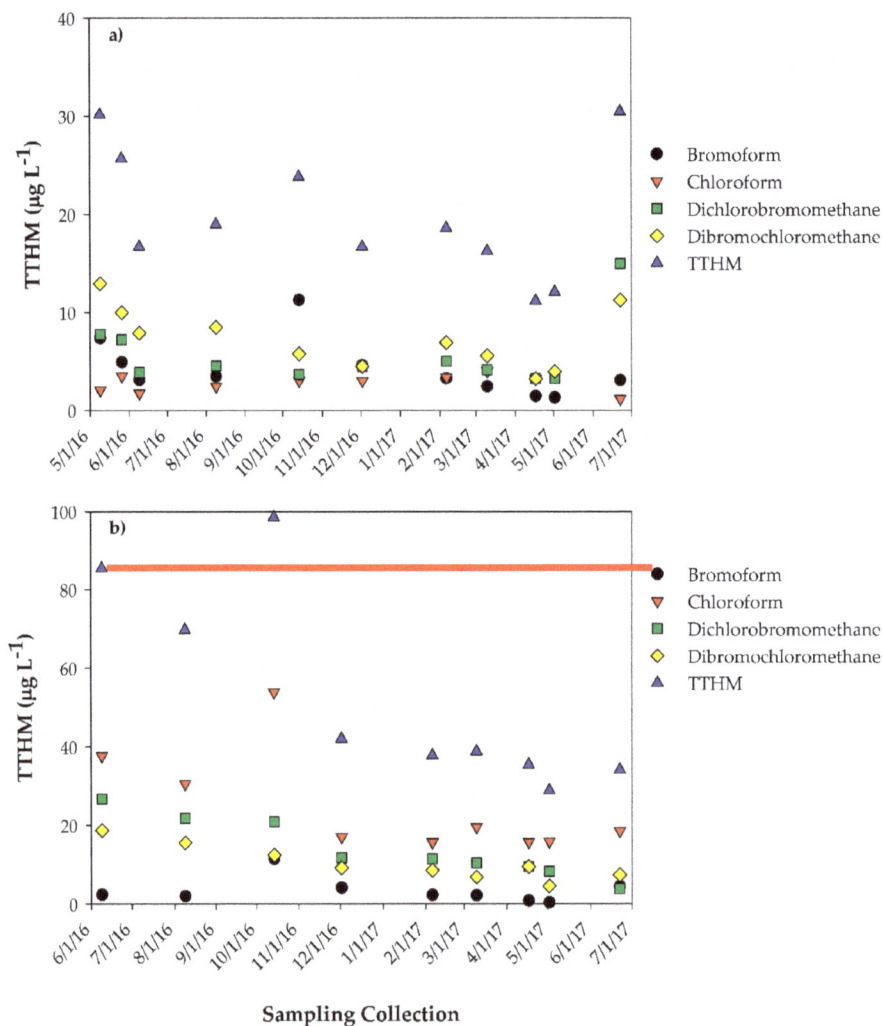

Figure 3. Total Trihalomethanes (TTHM) in (**a**) Auburn and (**b**) Nebraska City. Water samples were collected at the two water treatment plants after chlorination. Total TTHM maximum contaminant level (MCL): 80 µg L^{-1} (Stage 1 and Stage 2 Disinfectants and Disinfection Byproducts Rules. Available online: http://water.epa.gov/lawsregs/rulesregs/sdwa/stage2/regs_factsheet.cfm).

4. Discussion

Results from our study highlight the ability of the two RBF systems to consistently improve the quality of the source water. Total coliforms and *E. coli* were consistently removed (below detection limit) throughout the study regardless of the environmental conditions (summer vs. winter and dry vs. rainy period) and the starting counts. High bacterial removals (up to 4 logs) were also observed elsewhere [32–34,46]. RBF also represents a valuable option to provide treated water with low levels of nitrate even in agriculturally intensive areas. However, local conditions may impact the occurrence of nitrate in the treated water. While similar nitrate concentrations, up to 6 mg L^{-1}, were detected at the two rivers, different concentrations occurred in RBF well water at both sites. Nitrate was consistently absent in RBF well water in Nebraska City, while it ranged between 0 and 8.7 mg L^{-1} in RBF well water in Auburn. In Auburn, the occurrence of nitrate varied in different wells. Nitrate was almost absent in Well 13 (the newest well), while it was consistently detected at approximately 8 mg L^{-1} in Well 4. The different nitrate behavior may be related to the location of the two RBF sites. The RBF site in Nebraska City is adjacent to a conservation area, while the RBF site in Auburn is located in an agricultural area. The identification of a conservation area (ideal habitat for birds and animals and no agriculture) for the wellfield by the city will provide an ideal buffer around the wells. Therefore, local land use control can help small communities using RBF in dealing with nitrate. If a significant amount of flow for an RBF well is derived from the river and the well is located in a zone that receives limited or no nitrate, the expected concentrations will be low in the well as the background concentration will be low and denitrification is expected to remove a substantial part of the nitrate present in the river water. TOC and DOC removal efficiencies are similar to those observed around the world [25,32–34,47]; this is probably achieved within the first few meters of infiltration and can be due to a combined effect of biodegradation [23,48] and mixing with landside groundwater. The Missouri River is a far greater river than the Little Nemaha River. Its discharge is between 30 and 230 times the discharge of the Little Nemaha River. No to limited correlation was observed between discharge and TOC/DOC (*p*-value > 0.05) throughout the study at the two locations (Figure S8). In fact, except the Little Nemaha River on June 2017, TOC was fairly stable in the presence of low as well as high discharge (Figure S8). Similarly, to the limited impact on TOC/DOC, limited to no correlation (*p*-value > 0.05) was also observed between discharge and nitrate in the two rivers throughout the study (Figure S8). These findings suggest limited to no impact on the levels of nitrate in the two rivers due to increasing discharge associated with snowmelt observed in May 2016 and May 2017 (Figure S2). In cold climates, much of the springtime runoff and streamflow in rivers is associated with snowmelt.

RBF represented a potentially effective option to reduce the production of THMs in small systems. The occurrence of THMs in small systems represents a challenge for local water utilities [2,15,49]. For example, Hua et al. [2], investigating three small drinking water systems in rural Missouri using groundwater, surface water, and reservoir water as source water reported high levels of THMs (>80 µg L^{-1}) in finished water. The three systems used chlorine as final treatment. The low removal efficiency of DOC was consistently observed at the three water systems. However, after enhancing DOC removal in one of the three systems by adding powdered activated carbon, THM concentrations in finished water were lowered to approximately 40 µg L^{-1} [2]. In our study, THM values were mostly lower than the MCL. However, the occurrence of THMs was different at the two RBF sites. Historical data collected between 2012 and 2017, showed a higher concentration of THMs in Nebraska City (54.56 ± 27.76 µg L^{-1}) compared to Auburn (22.29 ± 6.92 µg L^{-1}) [50]. A similar trend was also observed in our study. A higher concentration of THMs was observed in Nebraska City (57.48 ± 25.62 µg L^{-1}) compared to Auburn (20.05 ± 6.63 µg L^{-1}). The small difference between historical data and our results at the two facilities can be related to the sampling location. THMs increased with increasing residence time [51]. Results from our study showed that the levels of THMs in the post-treatment water were strongly linked to the DOC removal (p-values < 0.005), and poorly linked to the DOC level (*p*-value > 0.05) as well as to SUVA (*p*-value > 0.005). The statistical analysis revealed a negative correlation between the levels of THMs in the post-treatment water and the DOC removal (correlation coefficient, r: -0.684). The highest THMs' concentration (98.54 μg L^{-1}) was observed in Nebraska City on October 14, corresponding to the lowest DOC removal (15.2%). While the two rivers had similar DOC (Little Nemaha River, Auburn: 3.57 ± 1.82 mg L^{-1}; Missouri River, Nebraska City: 4.46 ± 1.52 mg L^{-1}), the two RBF sites showed different DOC removal (Auburn: 76.17 ± 10.77%; Nebraska City: 37.82 ± 13.19%) (Table S5). The significantly higher DOC reductions observed in Auburn could be directed related to the longer extraction wells' distance from the river due to longer travel time (>88 m at the Auburn RBF site vs. >15 m at the Nebraska City RBF site). The results of our study were in agreement with previous findings suggesting that DOC reductions by RBF were site-specific, with higher reductions/removal correlating with greater travel distances between the river and the extractions wells [32,52] even in the presence of large capacity collector wells [53]. Both RBF sites showed a similar trend for SUVA throughout the study. SUVA was >4 L mg^{-1} m^{-1} in both rivers and in RBF well water at both sites between June 2016 and February 2017; after that SUVA was <3 L mg^{-1} m^{-1}. Samples collected after chlorination had SUVA <3 L mg^{-1} m^{-1}. High SUVA values suggested the occurrence of hydrophobic and especially aromatic matter, while SUVA <3 L mg^{-1} m^{-1} suggested the dominance of hydrophilic matter [54]. The results from our study confirmed the weak correlation between SUVA and THMs [55]. The presence of brominated THMs can be linked to the possible occurrence of a very low amount of naturally occurring bromide. During our study, bromide was constantly below the analytical detection limit. However, even at concentrations in the range of μg L^{-1} bromide can generate brominated THMs [56]. Naturally occurring bromide can be rapidly oxidized by chlorine to hypobromous acid and hypoiodous acid and consequently react with natural organic matter to form brominated DBPs [57].

Supplementary Materials: The following are available online at http://www.mdpi.com/2073-4441/10/12/1865/s1, Figure S1: RBF wells' sites: Auburn (left) and Nebraska City (Right). Figure S2: Discharge, Little Nemaha River (Auburn) and Missouri River (Nebraska City). Figure S3: pH in (a) Auburn and (b) Nebraska City. LNR: Little Nemaha River; MR: Missouri River; Pre: Inflow water to the water treatment plant prior to any additional treatment; Post: Water collected at the two water treatment plants after chlorination. Figure S4: Electrical conductivity (EC) in (a) Auburn and (b) Nebraska City. LNR: Little Nemaha River; MR: Missouri River; Pre: Inflow water to the water treatment plant prior to any additional treatment; Post: Water collected at the two water treatment plants after chlorination. Figure S5: Chloride in (a) Auburn and (b) Nebraska City. LNR: Little Nemaha River; MR: Missouri River; Pre: Inflow water to the water treatment plant prior to any additional treatment; Post: Water collected at the two water treatment plants after chlorination. Figure S6: Sulfate in (a) Auburn and (b) Nebraska City. LNR: Little Nemaha River; MR: Missouri River; Pre: Inflow water to the water treatment plant prior to any additional treatment; Post: Water collected at the two water treatment plants after chlorination. Figure S7: Specific ultraviolet absorbance (SUVA) in (a) Auburn and (b) Nebraska City. LNR: Little Nemaha River; MR: Missouri River; Pre: Inflow water to the water treatment plant prior to any additional treatment; Post: Water collected at the two water treatment plants after chlorination. Figure S8: Discharge, total organic carbon (TOC), and nitrate at the two sampling locations along the Little Nemaha River (Auburn) and Missouri River (Nebraska City). Table S1: Characteristics of wells at the two RBF sites. Table S2: Auburn—Geological formations at the riverbank filtration site. Table S3: Nebraska City—Geological formations at the riverbank filtration site. Table S4: Method detection limits. Table S5: Basic water quality properties (pH, electrical conductivity, EC, total organic carbon, TOC, dissolved organic carbon, DOC) at the two RBF sites and water utilities. Table S6: Removal (%) of dissolved organic carbon (DOC) at the two water treatment facilities. Table S7: Trihalomethanes at the two water treatment plants. Water samples.

Author Contributions: Methodology, M.D. and C.R.; Investigation, M.D.; Data Curation, M.D.; Writing—Original Draft Preparation, M.D.; Writing—Review and Editing, B.D. and C.R.; Supervision, C.R.; Project Administration, B.D.; Funding Acquisition, B.D. and C.R.

Funding: This research was funded by the U.S. EPA National Center for Innovation in Small Drinking Water Systems, grant number 15008462A02.

Acknowledgments: The authors are especially grateful to D. Hunter, K. Swanson, and their staff at the Auburn Water Work and C. Mayer, M. Lant, and their staff at the Nebraska Water Works for allowing this study. The authors would also like to thank C.A. Olson, P. Juntakut, and N. Schumacher for their support during water samples' collection. The authors are also grateful for the assistance of David Cassada, Separations Chemist at the Nebraska Water Center.

Conflicts of Interest: The authors declare no conflict of interest. The founding sponsors had no role in the design of the study; in the collection, analyses, or interpretation of data; in the writing of the manuscript, and in the decision to publish the results.

References

1. U.S. Environmental Protection Agency. Building the Capacity of Drinking Water Systems. Available online: https://www.epa.gov/dwcapacity/learn-about-small-drinking-water-systems (accessed on 14 October 2018).
2. Hua, B.; Mu, R.; Shi, H.; Inniss, E.; Yang, J. Water quality in selected small drinking water systems of Missouri rural communities. *Beverages* **2016**, *2*, 10. [CrossRef]
3. Allard, S.; Tan, J.; Joll, C.A.; Von Gunten, U. Mechanistic study on the formation of Cl-/Br-/I-trihalomethanes during chlorination/chloramination combined with a theoretical cytotoxicity evaluation. *Environ. Sci. Technol.* **2015**, *49*, 11105–11114. [CrossRef] [PubMed]
4. Roccaro, P.; Vagliasindi, F.; Korshin, G. Relationships between trihalomethanes, haloacetic acids, and haloacetonitriles formed by the chlorination of raw, treated, and fractionated surface waters. *J. Water Supply Res. Tech. AQUA* **2014**, *63*, 21–30. [CrossRef]
5. Hua, G.; Reckhow, D.A. Characterization of disinfection byproduct precursors based on hydrophobicity and molecular size. *Environ. Sci. Technol.* **2007**, *41*, 3309–3315. [CrossRef] [PubMed]
6. Zhang, Y.; Love, N.; Edwards, M. Nitrification in drinking water systems. *Crit. Rev. Environ. Control* **2009**, *39*, 153–208. [CrossRef]
7. Odell, L.H.; Kirmeyer, G.J.; Wilczak, A.; Jacangelo, J.G.; Marcinko, J.P.; Wolfe, R.L. Controlling nitrification in chloraminated systems. *J. AWWA* **1996**, *88*, 86–98. [CrossRef]
8. Wilczak, A.; Jacangelo, J.G.; Marcinko, J.P.; Odell, L.H.; Kirmeyer, G.J. Occurrence of nitrification in chloraminated distribution systems. *J. AWWA* **1996**, *88*, 74–85. [CrossRef]
9. Grellier, J.; Rushton, L.; Briggs, D.J.; Nieuwenhuijsen, M.J. Assessing the human health impacts of exposure to disinfection by-products—A critical review of concepts and methods. *Environ. Int.* **2015**, *78*, 61–81. [CrossRef]
10. Richardson, S.D.; Plewa, M.J.; Wagner, E.D.; Schoeny, R.; DeMarini, D.M. Occurrence, genotoxicity, and carcinogenicity of regulated and emerging disinfection by-products in drinking water: A review and roadmap for research. *Mutat. Res. Rev. Mutat. Res.* **2007**, *636*, 178–242. [CrossRef]
11. Singer, R.C. Human substances as precursors for potentially harmful disinfection by-products. *Water Sci. Technol.* **1999**, *40*, 25–30. [CrossRef]
12. Krasner, S.W.; Kostopoulou, M.; Toledano, M.B.; Wright, J.; Patelarou, E.; Kogevinas, M.; Villanueva, C.M.; Carrasco-Turigas, G.; Marina, L.S.; Fernández-Somoano, A.; et al. Occurrence of DBPs in drinking water of European regions for epidemiology studies. *J. AWWA* **2016**, *108*, E501–E512. [CrossRef]
13. Guilherme, S.; Rodriguez, M.J. Occurrence of regulated and non-regulated disinfection by-products in small drinking water systems. *Chemosphere* **2014**, *117*, 425–432. [CrossRef] [PubMed]
14. Stage 1 and Stage 2 Disinfectants and Disinfection Byproducts Rules. Available online: http://water.epa.gov/lawsregs/rulesregs/sdwa/stage2/regs_factsheet.cfm (accessed on 3 September 2018).
15. Ringenberg, D. Regulatory Barriers to Approval of New Technologies for Small Drinking Water Systems. Master's Thesis, University of Nebraska, Lincoln, NE, USA, 2017. Available online: http://digitalcommons.unl.edu/envengdiss/13/ (accessed on 3 September 2018).
16. Ray, C. Worldwide potential of riverbank filtration. *Clean Technol. Environ. Policy* **2008**, *10*, 223–225. [CrossRef]
17. Ray, C.; Melin, G.; Linsky, R.B. *River Bank Filtration—Improving Source-Water Quality*; Kluwer: Dordrecht, The Netherlands, 2003; 364p.
18. D'Alessio, M.; Yoneyama, B.; Ray, C. Fate of selected pharmaceutically active compounds during simulated riverbank filtration. *Sci. Total Environ.* **2015**, *505*, 615–622. [CrossRef] [PubMed]
19. D'Alessio, M.; Yoneyama, B.; Kirs, M.; Kisand, V.; Ray, C. Pharmaceutically active compounds: Their removal during slow sand filtration and their impact on slow sand filtration bacterial removal. *Sci. Total Environ.* **2015**, *524–525*, 124–135. [CrossRef]
20. Sudhakaran, S.; Lattemann, S.; Amy, G.L. Appropriate drinking water treatment processes for organic micropollutants removal based on experimental and model studies—A multi-criteria analysis study. *Sci. Total Environ.* **2013**, *442*, 478–488. [CrossRef] [PubMed]
21. Storck, F.R.; Schmidt, C.K.; Lange, F.T.; Henson, J.W.; Hahn, K. Factors controlling micropollutant removal during riverbank filtration. *J. AWWA* **2012**, *4*, E643–E652. [CrossRef]
22. Hoppe-Jones, C.; Oldham, G.; Drewes, J.E. Attenuation of total organic carbon and unregulated trace organic chemicals in U.S. riverbank filtration systems. *Water Res.* **2010**, *44*, 4643–4659. [CrossRef]

23. Hiscock, K.M.; Grischeck, T. Attenuation of groundwater pollution by bank filtration. *J. Hydrol.* **2002**, *266*, 139–144. [CrossRef]
24. U.S. Environmental Protection Agency. National Primary Drinking Water Regulations. Available online: https://www.epa.gov/ground-water-and-drinking-water/national-primary-drinking-water-regulations#one (accessed on 15 October 2018).
25. Wells, M.J.; Gilmore, T.E.; Mittelstet, A.R.; Snow, D.D.; Sibray, S.S. Assessing decadal trends of a nitrate-contaminated shallow aquifer in Western Nebraska using groundwater isotopes, age-dating, and monitoring. *Water* **2018**, *10*, 1047. [CrossRef]
26. Pennino, M.J.; Compton, J.E.; Leibowitz, S.G. Trends in drinking water nitrate violations across the United States. *Environ. Sci. Technol.* **2017**, *51*, 13450–13460. [CrossRef] [PubMed]
27. Exner, M.E.; Hirsh, A.J.; Spalding, R.F. Nebraska's groundwater legacy: Nitrate contamination beneath irrigated cropland. *Water Resour. Res.* **2014**, *50*, 4474–4489. [CrossRef] [PubMed]
28. Spalding, R.F.; Exner, M.E. Occurrence of nitrate in groundwater—A review. *J. Environ. Qual.* **1993**, *22*, 392–402. [CrossRef]
29. Ringenberg, D.; Wilson, S.; Dvorak, B. State barriers to approval of drinking water technologies for small systems. *J. AWWA* **2017**, *109*, E343–E352. [CrossRef]
30. Kim, H.-C.; Lee, W.M.; Lee, S.; Choi, J.; Maeng, S.K. Characterization of organic precursors in DBP formation and AOC in urban surface water and their fate during managed aquifer recharge. *Water Res.* **2017**, *123*, 75–85. [CrossRef] [PubMed]
31. Liu, P.; Farré, M.J.; Keller, J.; Gernjak, W. Reducing natural organic matter and disinfection by-product precursors by alternating oxic and anoxic conditions during engineered short residence time riverbank filtration: A laboratory-scale column study. *Sci. Total Environ.* **2016**, *565*, 616–625. [CrossRef] [PubMed]
32. Partinoudi, V.; Collins, M.R. Assessing RBF reduction/removal mechanisms for microbial and organic DBP precursors. *J. AWWA* **2007**, *99*, 61–71. [CrossRef]
33. Weiss, W.J.; Bouwer, E.J.; Ball, W.P.; O'Melia, C.R.; Lechevallier, M.W.; Arora, H.; Speth, T.F. Riverbank filtration—Fate of DBP precursors and selected microorganisms. *J. AWWA* **2003**, *95*, 68–81. [CrossRef]
34. Weiss, W.J.; Bouwer, E.J.; Ball, W.P.; O'Melia, C.R.; LeChevallier, M.W.; Arora, H.; Aboytes, R.; Speth, T.F. Study of water quality improvements during riverbank filtration at three Midwestern United States drinking water utilities. *Geophys. Res. Abstr.* **2003**, *5*, 04297.
35. United States Census Bureau, Census 2010. Available online: https://factfinder.census.gov/faces/tableservices/jsf/pages/productview.xhtml?src=bkmk (accessed on 9 September 2018).
36. High Plains Regional Climate Center CLIMOD. Available online: http://climod.unl.edu/ (accessed on 18 September 2018).
37. U.S. Environmental Protection Agency (U.S. EPA). Drinking water guidance on disinfection by-products. In *Disinfection By-Products in Drinking Water*; Advice Note. No. 4 Version 2; U.S. Environmental Protection Agency: Washington, DC, USA, 2012; p. 27.
38. WHO. *Guidelines for Drinking-Water Quality*; WHO: Geneva, Switzerland, 2011; p. 518.
39. Baghoth, S.A.; Sharma, S.K.; Amy, G.L. Tracking natural organic matter (NOM) in a drinking water treatment plant using fluorescence excitation–emission matrices and PARAFAC. *Water Res.* **2011**, *45*, 797–809. [CrossRef]
40. Chen, W.; Westerhoff, P.; Leenheer, J.A.; Booksh, K. Fluorescence excitation–emission matrix regional integration to quantify spectra for dissolved organic matter. *Environ. Sci. Technol.* **2003**, *37*, 5701–5710. [CrossRef] [PubMed]
41. U.S. Environmental Protection Agency (U.S. EPA). *Method 1623: Cryptosporidium and Giardia in Water by Filtration/IMS/FA*; EPA 815-R-05-002; Office of Water: Washington, DC, USA, 2005; p. 68.
42. ISO. *Water Quality—Enumeration of Escherichia coli and Coliform Bacteria—Part 2: Most Probable Number Method*; ISO 9308-2: 2012; International Organization for Standardization: Geneva, Switzerland, 2012.
43. Cady, P.; Boving, T.B.; Choudri, B.S.; Cording, A.; Patil, K.; Reddy, V. Attenuation of bacteria at a riverbank filtration site in rural India. *Water Environ. Res.* **2013**, *85*, 2164–2174. [CrossRef]
44. Balzer, M.; Witt, N.; Flemming, H.C.; Wingender, J. Faecal indicator bacteria in river biofilms. *Water Sci. Technol.* **2010**, *61*, 1105–1111. [CrossRef] [PubMed]
45. USGS Water Data for the Nation. Available online: https://waterdata.usgs.gov/nwis (accessed on 18 September 2018).

46. Dash, R.R.; Prakash, E.B.; Kumar, P.; Mehrotra, I.; Sandhu, C.; Grischek, T. River bank filtration in Haridwar, India: Removal of turbidity, organics and bacteria. *Hydrogeol. J.* **2010**, *18*, 973–983. [CrossRef]

47. Maeng, S.K.; Sharma, S.K.; Lekkerkerker-Teunissen, K.; Amy, G.L. Occurrence and fate of bulk organic matter and pharmaceutically active compounds in managed aquifer recharge: A review. *Water Res.* **2011**, *45*, 3015–3033. [CrossRef] [PubMed]

48. Quanrud, D.M.; Hafer, J.; Karpiscak, M.M.; Zhang, J.; Lansey, K.E.; Arnold, R.G. Fate of organics during soil-aquifer treatment: Sustainability of removals in the field. *Water Res.* **2003**, *37*, 3401–3411. [CrossRef]

49. U.S. Environmental Protection Agency (U.S. EPA). *Occurrence Assessment for the Final Stage 2 Disinfectants and Disinfection Byproducts Rule*; EPA 815-R-05-011; U.S. EPA: Washington, DC, USA, 2005.

50. Nebraska Drinking Water Watch. Available online: https://sdwis-dhhs.ne.gov:8443/DWW/ (accessed on 9 September 2018).

51. Chen, W.J.; Weisel, C.P. Halogenated DBP concentrations in a distribution system. *J. AWWA* **1998**, *90*, 151–163. [CrossRef]

52. Skark, C.; Remmler, F.; Zullei-Seibert, N. Classification of Riverbank Filtration Sites and Removal Capacity. In *Recent Progress in Slow Sand and Alternative Biofiltration Processes*; Gimbel, R., Graham, N.J.D., Collins, M.R., Eds.; IWA Publishing: London, UK, 2006.

53. Wang, J. Riverbank Filtration Case Study at Louisville, Kentucky. In *Riverbank Filtration*; Ray, C., Melin, G., Linsky, R.B., Eds.; Kluwer Academic Publishers: Dordrecht, The Netherlands, 2002; pp. 117–145.

54. Edzwald, J.K.; Tobiason, J.E. Enhanced coagulations: US requirements and a broader view. *Water Sci. Technol.* **1999**, *40*, 63–70. [CrossRef]

55. Hua, G.; Reckhow, D.A.; Abusallout, I. Correlation between SUVA and DBP formation during chlorination and chloramination of NOM fractions from different sources. *Chemosphere* **2015**, *130*, 82–89. [CrossRef]

56. Kitis, M.; Yigita, N.O.; Harmana, B.I.; Muhammetoglu, H.; Muhammetoglu, A.; Karadirek, I.E.; Demirel, I.; Ozdenc, T.; Palancic, I. Occurrence of trihalomethanes in chlorinated groundwaters with very low natural organic matter and bromide concentrations. *Environ. Forensics* **2010**, *11*, 264–274. [CrossRef]

57. Hua, G.; Reckhow, D.A.; Kim, J. Effect of bromide and iodide ions on the formation and speciation of disinfection byproducts during chlorination. *Environ. Sci. Technol. Lett.* **2006**, *40*, 3050–3056. [CrossRef]

Article

Flow Analysis through Collector Well Laterals: A Case Study from Sonoma County Water Agency, California

Matteo D'Alessio [1], John Lucio [2], Ernest Williams [3], James Warner [2], Donald Seymour [4], Jay Jasperse [4] and Chittaranjan Ray [1,*]

[1] Nebraska Water Center, University of Nebraska Lincoln, Lincoln, NE 68588-6204, USA; mdalessio2@unl.edu

[2] ERM, Inc., 1277 Treat Boulevards, Walnut Creek, CA 94597, USA; john.lucio@erm.com (J.L.); jim.warner@erm.com (J.W.)

[3] Bennet & Williams Environmental Consultants, Inc., 98 County Line Road West, Westerville, OH 43082, USA; ebwilliams@bennettandwilliams.com

[4] Sonoma County Water Agency, 404 Aviation Blvd, Santa Rosa, CA 95403, USA; don.seymour@scwa.ca.gov (D.S.); jay.jasperse@scwa.ca.gov (J.J.)

* Correspondence: cray@nebraska.edu; Tel.: +1-402-472-8427

Received: 3 November 2018; Accepted: 1 December 2018; Published: 13 December 2018

Abstract: The Sonoma County Water Agency (SWCA) uses six radial collector wells along the Russian River west of Santa Rosa, to provide water for several municipalities and water districts in north-western California. Three collector wells (1, 2, and 6) are located in the Wohler area, and three collector wells (3, 4, and 5) are located in the Mirabel area. The objective of this paper is to highlight the performance of the three collector wells located in the Mirabel area since their construction. The 2015 investigation showed a lower performance of Collectors 3 and 4 compared to their original performances after construction in 1975, while the performance of Collector 5 was relatively stable since 1982. The potential change in capacity could be due to the increase in encrustation observed during the visual inspection of laterals in all three collector wells. Overall, the three collectors are still within the optimal design parameters (screen entrance velocity < 0.305 m min^{-1} and axial flow velocity of lateral screens < 1.524 m s^{-1}).

Keywords: riverbank filtration; collector wells; performance; entrance velocity

---

1. Introduction

Several municipalities (i.e., the cities of Santa Rosa, Sonoma, Cotati, Rohnert Park, and Petaluma) and water districts (i.e., the Forestville Water District, Valley of the Moon Water District, North Marin Water District, and Marin Municipal Water District) in Sonoma and Marin Counties receive water from the Sonoma County Water Agency (SWCA). The SCWA water system has an estimated peak production capacity of 4.907 m^3 s^{-1}. The SWCA uses six radial collector wells, along the Russian River west of Santa Rosa, to provide water for approximately 570,000 people [1]. Three collector wells (1, 2, and 6) are located in the Wohler area, and three collector wells (3, 4, and 5) are located in the Mirabel area [1].

In 1998, a preliminary investigation highlighted declined capacities (−24 to −77%) of the collector wells compared to their original capacities. The declines were more pronounced in the oldest collector wells (e.g., collector wells 1 and 2) [2]. Clogging of lateral well screens, clogging of the aquifer adjacent to the lateral well screens, compaction of the alluvial aquifer material due to long-term pumping, problems with pumping equipment in the collectors, decreased recharge from the ponds and/or river due to long-term silt/organic material build-up or changes in the operation of the inflatable dam, and regional declines in groundwater levels due to changes in precipitation, river discharge, and/or groundwater extraction were among the possible reasons [1]. However, data evaluation was highly impacted by the operations of nearby collectors during the testing.

To have a better understanding of the status (i.e., magnitude, rate, and causes of the loss of capacity) of each collector well, SCWA developed a program to evaluate flow to the collector wells in fixed time intervals (about five years). The collector wells located in the Mirabel area were investigated in 2010 and 2015 [1,3], while the collector wells in the Wohler area were investigated in 2010 and will be investigated in 2018–2019.

The objective of this paper is to show the performance of three collector wells (3, 4, and 5) located in the Mirabel area since their construction. Additionally, we also show the flow variations along the laterals (along the length and among themselves). We also compare the design parameters such as theoretical screen entrance velocity and axial flow velocity for the lateral screens as well as comparing fluxes through individual laterals. The study is unique in the sense that it attempts to examine flow variation through lateral screens.

2. Materials and Methods

2.1. Description of the Riverbank Filtration (RBF) Sites

The Mirabel area is located approximately one mile south of the west bank of the Russian River (Figure 1). The wells extract water from the unconsolidated alluvial aquifer adjacent to and beneath the Russian River using large-volume Ranney-type (lateral) collector wells. The pumping wells induce large vertical fluxes from the river and nearby the infiltration ponds [4]. An inflatable dam and four infiltration ponds are present in this area (Figure 1).

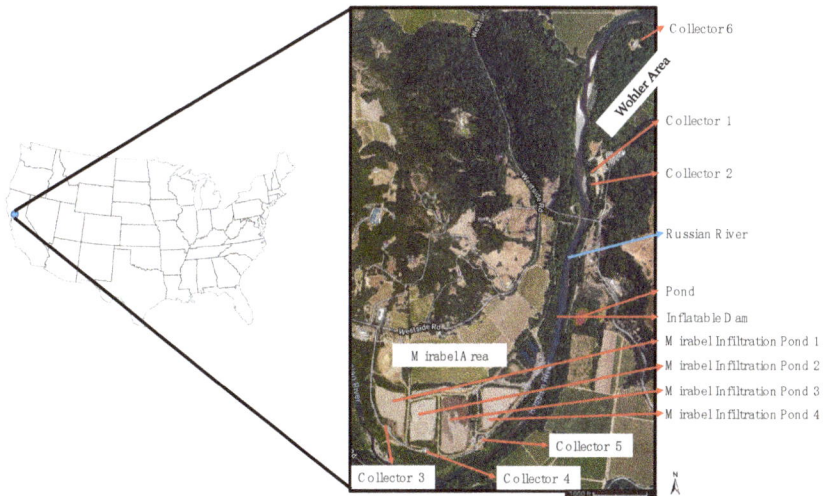

Figure 1. Project area map. Wohler (Collectors 1, 2, and 6) and Mirabel Areas (Collectors 3, 4, 5), Sonoma County Water Agency, Sonoma County, California. Modified from [3].

To account for the low flow periods, May to November, SCWA raises the inflatable dam which creates a low-velocity pool of water that extends approximately 2.5 km upstream and raises the stage of the river. A higher river stage produces a pressure gradient that forces water into the streambed and recharges the alluvial aquifer. The dam also diverts water to infiltration ponds that flank the river and water quickly enters the underlying aquifer [5]. Collector wells 3 and 4 were constructed in 1975, while collector well 5 was constructed in 1982 (Table S1). The three collector wells consist of

3.96 m inside diameter steel-reinforced concrete caissons. Collector wells 3, 4, and 5 have 6, 8, and 10 laterals (25.4-cm diameter mild steel), respectively (Table S1). Laterals range between 21.34 and 53.34 m (Table S2). Additional details are included in Tables S1 and S2 (Figure 2) [1,3].

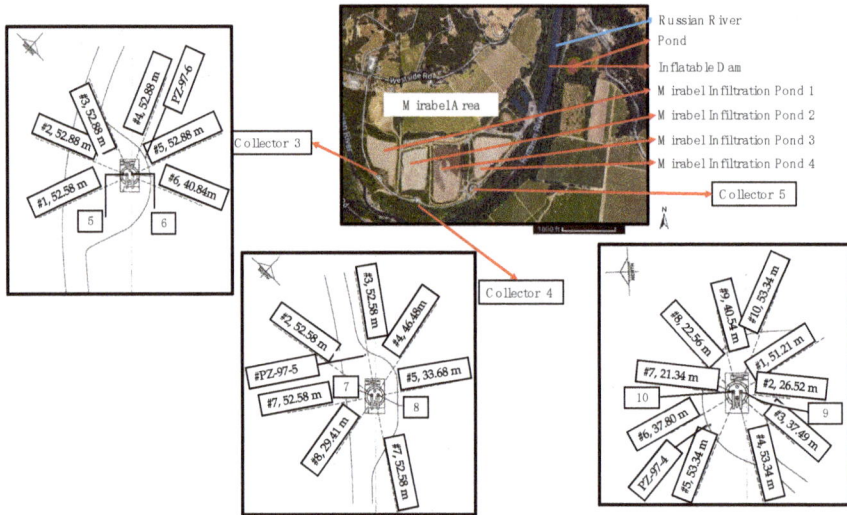

Figure 2. Mirabel Area, Sonoma County Water Agency, Sonoma County, California. Modified from [3].

The alluvium along the Russian River is the primary source of water production for SCWA. The Russian River is approximately 180 km long, originating from the Laughlin Range of California and draining to the Pacific Ocean near Jenner, California. The river drains a basin of 3866 km². The west coast of California receives most of its precipitation in the winter months. The US Geological Survey gage at Guerneville, California (CA) indicates a long-term mean flow of about 64 m³ s⁻¹ with a maximum exceeding 2888 m³ s⁻¹ during peak flow events. The minimum flow recorded at the gage is 0.02 m³ s⁻¹ [6]. As the Russian River is home to certain species of salmonid fish that migrate upstream for spawning, SCWA has installed fish ladders around the inflatable rubber dam for ease of fish migration. The ponded water behind the dam as well as in the recharge ponds encourages weed and algae growth during summer and also allows fine particles to settle to the bottom. This is also speculated to be one of the reasons for decreasing recharge capacity of the riverbed as well as the recharge ponds. The river is underlain primarily by alluvium and river channel deposits consisting of unconsolidated sands and gravels, with thin layers of silt and sand [7]. In the investigated area, the alluvial aquifer is bounded by metamorphic bedrock and is considered impermeable relative to the alluvial materials [7]. The shallow aquifer sediments in the investigated site have a measured hydraulic conductivity between 5.5×10^{-5} to 2.0×10^{-4} m s⁻¹ and from 1.4×10^{-5} to 2.6×10^{-4} m s⁻¹ within the same area using seepage meter methods [5,7].

2.2. Evaluation Procedures

One week prior to the initiation of the constant rate test pumping, the collector wells were shut off to allow for recovery of the water table. Pressure transducers equipped with data loggers (In-Situ Inc., Fort Collins, CO, USA) were installed before shutting down. The bottom floor and interior walls of the caisson, pump intakes, gate valves, and stem riser assemblies, if present, were visually inspected by a diver in October/November 2010 and October 2015 for collector wells 3, 4, and 5 [1,3]. Before diving in each of the collector well caissons for inspection and testing, chlorination of the collector water was temporarily ceased as a health and safety consideration.

The wall thickness of each of the lateral screens was estimated using an underwater ultrasonic digital thickness gauge (Cygnus Instrument Inc. Annapolis, MD, USA). The gauge (accuracy: ±0.05 mm) was inserted into the section of the lateral nearest the caisson and thickness measurements were obtained at 0°, 90°, 180°, and 270° from the vertical position for the three collector wells. To evaluate capacity, the collector wells were separately pumped continuously for approximately five days. During the constant rate capacity test, the collector well undergoing testing was placed back on-line at a controlled pumping rate roughly comparable to typical operating conditions [1,3].

Periodically, water levels within the collector well and five site monitoring wells were measured using an electric tape (accuracy: ±0.3 cm). Measurement of pH (accuracy: ±0.2), oxygen reduction potential, redox potential (ORP), (accuracy: ±20 mV), dissolved oxygen, DO, (accuracy: ±0.1 mg L^{-1}), specific conductance (accuracy: ±0.5% of reading plus 0.001 mS cm^{-1}), salinity (accuracy: ±0.1‰), total dissolved solids (TDS), turbidity (accuracy: ±0.1 NTU), and temperature (accuracy: ±0.15 °C) of the pumped water were done using a multi-parameters probe (YSI, Yellow Springs, OH, USA) [1,3]. The probe was inserted by the diver into each lateral, and equilibrated for approximately three minutes before data collection (Figure S1).

Lateral flow was measured using a mechanical flow meter (Gurley Precision Instrument, Troy, NY, USA) attached to an approximately 3.05-m long rod. The diver inserted the flow meter at the mouth of each lateral for a minimum of 1 min, after that the data was transmitted to the surface and read using a digital indicator (Gurley Precision Instrument, Troy, NY, USA). Once the data was recorded, the diver moved to the next lateral and repeated the same process. Upon completion of the lateral flow testing, an underwater video camera was inserted into the first 3 m of every lateral within each of the collector wells to provide preliminary information on the condition of the laterals adjacent to the collector caisson. Based on the results of the initial video inspection and the flow testing, laterals were prioritized for full-accessible length video inspection and lateral flow profiling along the entire accessible length. Video inspection and lateral flow testing were completed in a total of 22, 20, and 18 laterals within the three collector wells during the 2008, 2010, and 2015 monitoring campaigns. The video camera vehicle was controlled remotely and was used to position the flow meter within the lateral, and flow velocity was measured and recorded at 3 m increments along the accessible length of the lateral. The flow meter remained at each position within the lateral for a minimum of 1 min. In laterals with high velocity, an aluminum bull float rod approximately 2 m in length was used in conjunction with a cable and slip-fit ring to hold the flow meter and camera vehicle in place [1,3].

Entrance velocity was calculated by dividing the incremental flow measured approximately every 3 m (10 feet) by the screen open area. The screen open area was estimated using the diameter and the length of the screen as well as the estimated open area (45%) [6]. The axial flow was calculated by dividing the measured flow along the lateral by the cross-sectional area of the lateral [6,7].

3. Results and Discussion

3.1. Caisson and Lateral Condition

During both inspections, the caissons of the three collectors as well as all underwater structures appeared to be in good structural condition. No evidence of fracturing or spalling in the caissons was observed [1,3]. However, surface corrosion was observed on several valve stem risers and brackets and on the ladder [1,3]. The pumps and pump columns were in good conditions even if they contain surface corrosion. The laterals from the three caissons appeared to be in good condition with varying amounts of surface corrosion (Figure S2). Compared to the 2010 inspection, the 2015 inspection highlighted the presence of more corrosion along the internal steel pipe surfaces and screen slots [3]. The video inspection of the laterals showed signs of progressive encrustation along the lateral pipe interior and within the slot openings (Figure S2). Gravel piles were detected at the end of many laterals and were probably related to the high screen entrance velocity. In addition, the average thickness of the screen metal in the laterals slightly increased during the two inspections (Figure 3). In the laterals from Caisson 3, it ranged between 0.46 and 0.84 cm during the 2010 inspection and between 0.71 and 0.91 cm during the 2015 inspection (Figure 3). In the laterals from Caisson 4, it ranged between 0.56 and 0.73 cm during the 2010 inspection and between 0.66 and 0.98 cm during the 2015 inspection. Similarly, in the laterals from Caisson 5, it ranged between 0.44 and 1.03 cm during the 2010 inspection and between 0.66 and 0.98 cm during the 2015 inspection (Figure 3).

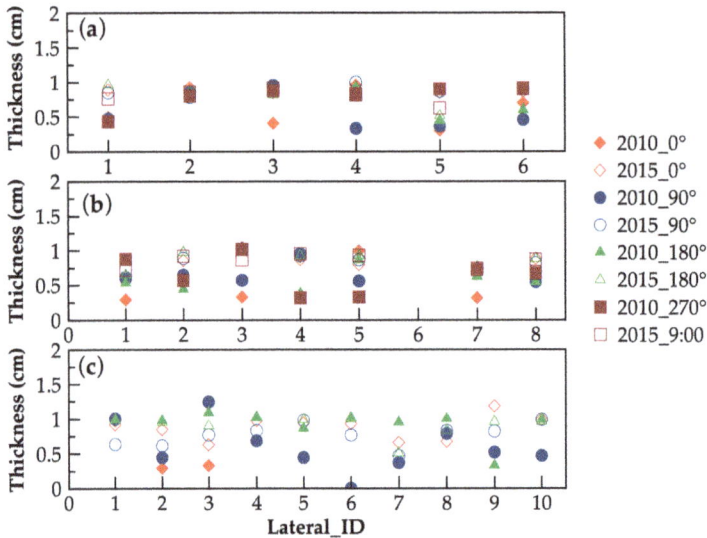

Figure 3. Lateral structural integrity in (**a**) Collector 3, (**b**) Collector 4, and (**c**) Collector 5 during the two inspections. The gauge was inserted into the section of the lateral nearest the caisson and obtained thickness measurements at 0°, 90°, 180°, and 270° from the vertical position for the three collector wells.

3.2. Lateral Flow Testing

The pumping levels in Collectors 3 and 4 had not stabilized after pumping (average rate ~0.6 m^3 s^{-1}) for 1 week and for 4 days, respectively, and steady state conditions were not achieved. At the conclusion of this test pumping, Collector 3 was producing approximately 0.1 m^3 s^{-1} per m of observed drawdown. In contrast, at the end of testing in 2010, the collector well was producing about 0.14 m^3 s^{-1} per m of drawdown, suggesting a decline in performance of one-third in the intervening 5-year period. By adjusting the 2015 results and those from previous testings to be equivalent in terms of static water level and pumping water level, and without interference from the nearby Mirabel wellfield, the 2015 performance of Collector 3 is 3% lower than when it was originally constructed and tested in 1975, and approximately 10% lower than for the last previous inspection in 2010. By using similar adjustments, the 2015 performance of Collector 4 is about 23% better than it was in 2010, but 11% lower than its original performance after construction in 1975. On the other hand, the 2015 performance of Collector 5 has not changed substantially (±1%) [1,3].

Throughout the three investigations (1998, 2010, and 2015), minimal changes in relative percentage flow were observed (Figure 4). In Collector 3, the laterals (1, 2, and 3) closest to the river had the largest percentages of flow (Figure 4). The collective gain in flow in these three riverward laterals balances the collective loss of flow in the three landward laterals. During the 2015 evaluation, the relative distribution of flow among Collector 3 laterals ranged between 12.5 (Lateral 4) and 20.7% (Lateral 1). Lateral 6 had the largest decline in flow (−2.7%) compared to the 2010 results [1,3]. The limited changes observed may be related due to varying influence of recharge from the Russian River, as well as Infiltration Ponds 2 and 3 northeast and east of Collector 3 (Figure 2), respectively.

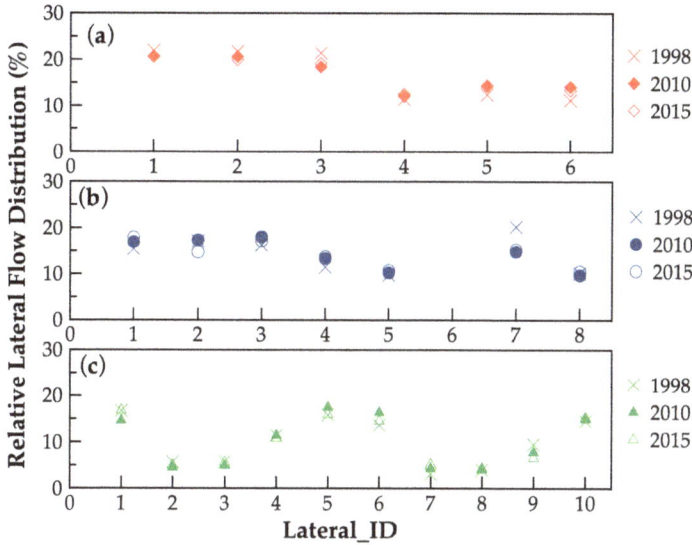

Figure 4. Relative lateral flow distribution (%) in (**a**) Collector 3, (**b**) Collector 4, and (**c**) Collector 5.

In Collector 4, during the three investigations, the productions of most of the laterals slightly improved over time. Laterals 1 and 2 (oriented toward the river) and Laterals 7 and 3 (parallel to the river) showed the highest percentages of flow. On the other hand, the lateral with the lowest percentage of flow (Lateral 8) is oriented on about a 45-degree angle towards the river, similar to Lateral 2, but it is also the shortest lateral (29.41 m) (Figure 2). The two landward laterals projected towards the infiltration ponds collectively provided 24.4% of the well's total production. During the 2015 evaluation the relative flow percentages ranged between 10.5% (Lateral 8) to 17.9% (Lateral 1). Lateral 2 showed a decline (−2.5%) in production between 2010 and 2015 (Figure 2) [1,3]. Based on the video obtained, this decline was probably due to the presence of sand and gravel within the lateral.

In contrast with the trend observed in Collectors 3 and 4, in Collector 5, the laterals showed constrasting results. The production slightly increased in Laterals 1 and 7, slightly decreased in Laterals 4, 5, and 9, and remained constant in the remaining laterals. In addition, the impact of the 10 laterals is different. Four laterals individually procuded 14.5% or more of the total capacity, while five laterals individually produced less than 6.6% (Figure 4). The orientation of the laterals had no impact on the percentage of flow distributions. On the other hand, the length of the laterals impacted the flow distribution. The four longest laterals (>51.20 m) were also four of the five highest producing laterals, while the three shortest laterals (<26.52 m) were also the three lowest producing laterals. The three remaining laterals were of intermediate lengths (37.49 to 40.54 m), as well as in producing capability [1,3].

3.3. Lateral Flow Profiling

Non-uniform flow occurred along the laterals from the three collectors (Figure 5). In fact, for uniform flow along the lateral, the trend for each lateral would be a straight line beginning with zero m min^{-1} at the outer end of the lateral and concluding with the total flow for the lateral where it enters the caisson. However, none of the laterals follow this straight-line trend of uniform distribution. Each lateral displays steeper gaining trends in production in the outermost segments of their length where they are obtaining most of their flow. After these steep gains in flow in the outer segments, the remaining trends of flow while moving progressively closer to the caisson were more gradual because of the generally slower production in those segments. Similar overall trends were observed during the 2010 and 2015 monitoring events. However, for Collector 3, regardless of the lateral, higher flow was observed during the 2015 campaign. On the other hand, for Collectors 4 and 5, a slightly higher flow was observed during the 2010 campaign [1,3].

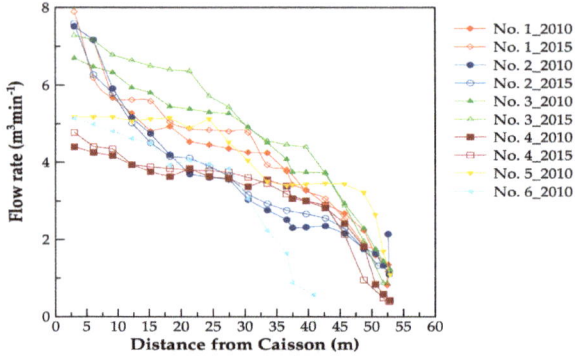

(a)

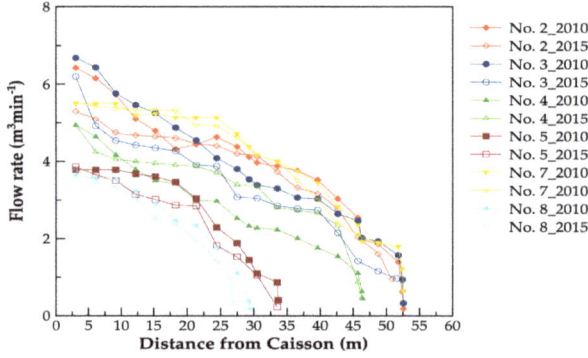

(b)

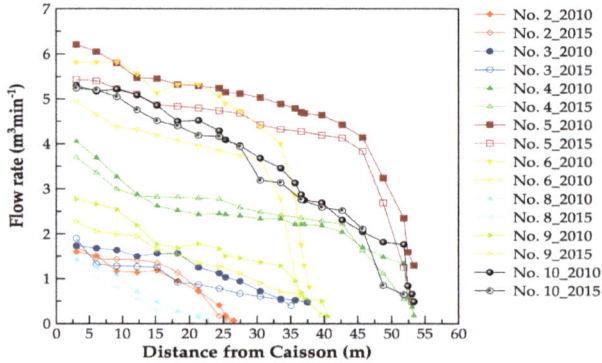

(c)

Figure 5. Flow versus distance at (**a**) Collector 3, (**b**) Collector 4, and (**c**) Collector 5.

Average unit flow capacity of each lateral, which is also greater at the end of lateral flow, effects large gains in production within only a few meters. This is particularly noticeable in Laterals 3 and 4 from Collector 3. In these outermost few feet, unit flow capacities can approach or even exceed $2 \, m^3 \, min^{-1}$ per m. On the other hand, while moving along the lateral toward the caisson, the gains in flow progressively decrease and the average unit flow capacity begins to assume a more consistent trend (Figure 6).

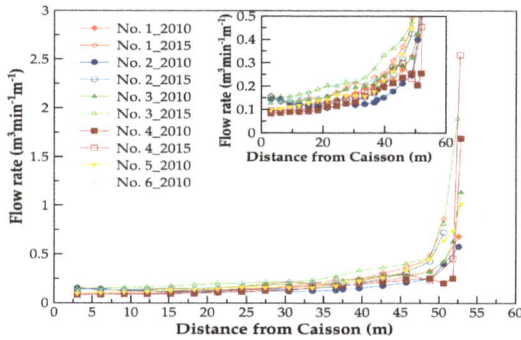

(a)

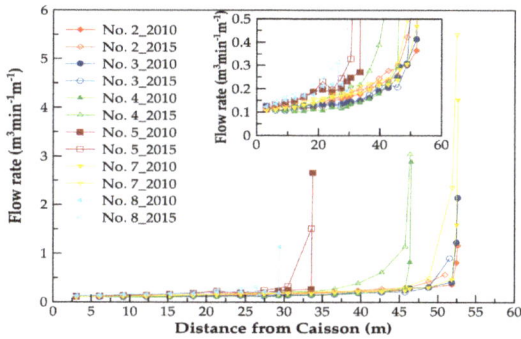

(b)

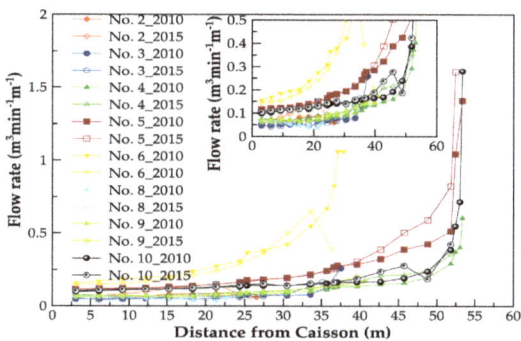

(c)

Figure 6. Flow along different laterals at increasing distances between the screen length and the caisson for (**a**) Collector 3, (**b**) Collector 4, and (**c**) Collector 5.

Overall, the three collectors are still within the optimal design parameters (entrance velocity < 0.305 m min^{-1} and axial velocity < 1.524 m s^{-1} [8,9]) (Figure 7 and Figure S3). Collector 6, constructed in 2002, consisted of a larger steel reinforced concrete caisson (5.49 m vs. 3.96 m inside diameter) with larger laterals (30.48 cm vs. 24. cm) than the Collectors discussed in this investigation, showed a lower entrance velocity (consistently < 0.610 m min^{-1}) and axial velocity (<1.524 m s^{-1}) during the 2010 monitoring campaign. This can also be related to lower presence of deposited materials and rust in Collector 6 (Figure S6) compared to the older collectors (Figure S2).

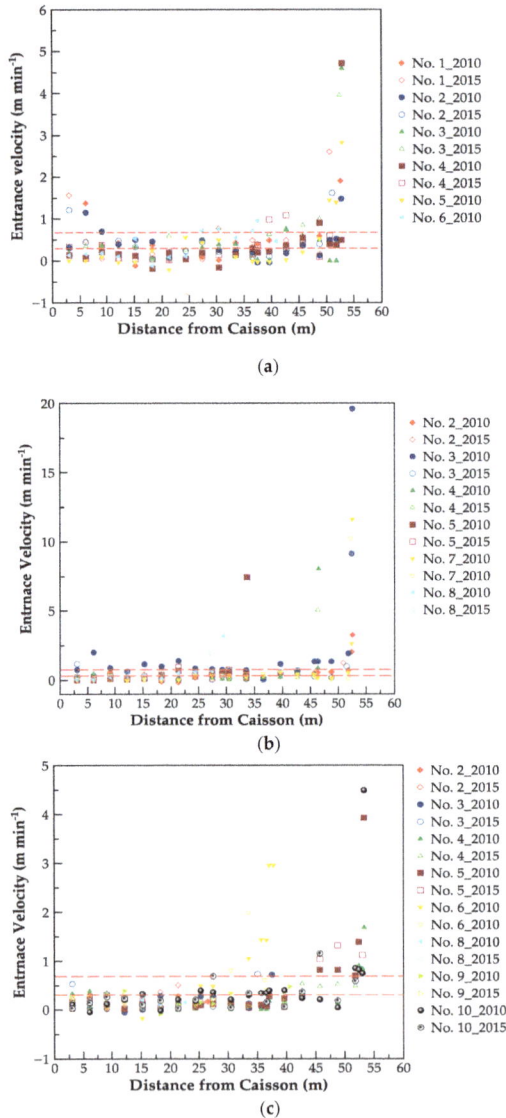

Figure 7. Entrance velocity through different laterals with increasing distances for their ends for (**a**) Collector 3, (**b**) Collector 4, and (**c**) Collector 5. Ideal and optimal design < 0.305 and < 0.610 m min^{-1} (bottom and top red line).

3.4. Water Quality

During the 2010 field campaign, the Russian River showed pH, EC, and DO values ranging between 7.5 and 8.5, between 190 and 240 μS cm^{-1} and between 9 and 10 mg L^{-1}, respectively. During the same field campaign, similar pH and EC values were observed along the different laterals (Figure S4), while DO was consistently lower (Figure S5). Limited changes in terms of basic water quality parameters (i.e., pH, electrical conductivity, EC) were observed at three collectors during the two sampling campaigns, with more significant changes in dissolved oxygen (DO) and redox potential (ORP) (Figures S4 and S5). The pH ranged between 6.67 (Collector 3) and 7.31 (Collector 5) during the 2010 campaign, and between 6.78 (Collector 5) and 7.25 (Collector 3) during the 2015 campaign (Figure S5). Low EC (~220 μs cm^{-1}) were observed at the three collectors throughout the study (Figure S5). The DO was consistently lower during the 2015 campaign compared to the 2010 campaign. For example, at Collector 3, DO ranged from 3.5 to 7 mg L^{-1} during the 2010 campaign and decreased during the 2015 campaign to a range between 1.5 and 6 3.5 to 7 mg L^{-1} (Figure S5a). A similar but more pronounced trend was observed at Collectors 4 and 5 (Figure S5b,c). The ORP was also significantly different between the two sampling campaigns. In fact, ORP decreased over time at Collectors 4 and 5, and increased at Collector 3 (Figure S4). This different behavior in terms of DO and ORP may be related to the changes in temperature observed during the two investigations at the three collectors. Even if the two sampling campaigns were conducted during the same time of year (late October-early November), there was a difference in the river temperature. In particular, higher temperatures (~20.6 °C vs. 19.2 °C) were observed at Collector 4 during the 2015 campaing compared to the 2010 campaign. On the other hand, slightly warmer temperatures were observed during the 2015 campaign at Collectors 4 and 5 compared to the 2010 campaign [1,3].

4. Conclusions

This study highlights the performance of the three collector wells located over four decades. The field methods used during the 2010 and 2015 campaigns represented a valuable tool to evaluate the performance of collector wells regarding overall conditions, specific capacities, and entrance and axial velocities. Also, the impact of precipitation and consequently the variability of the river stage represent a key component for temporal comparisons. While water quality monitoring at the different laterals during the different campaigns are valuable information, basic river water quality parameters should also be monitored during these campaigns.

While Collectors 3 and 4 achieved lower performances compared with their original performances (1975), Collector 5 was relatively stable since 1982. The potential change in capacity could be due to the increase in encrustation observed during the visual inspection of laterals in all three collector wells. Overall, the three collectors are still within the optimal design parameters (screen entrance velocity < 0.305 m min^{-1} and axial flow velocity of lateral screens < 1.524 m s^{-1}). The underwater structures in three collector wells are in generally good condition, early stages of rusting and encrustation are present. A more frequent cleaning and/or replacement of some of the rusted units may be required to further improve the efficiency of the collector wells.

Supplementary Materials: The following are available online at http://www.mdpi.com/2073-4441/10/12/1848/s1, Figure S1: Diver preparing to enter one of the collector caisson (left) and diver climbing down one of the collector caisson ladder (right) (Source: [3]). Figure S2. Collector Well 3, Lateral 4: 10 ft (3.048 m) progression from the lateral video inspection, 2015 (Source: [3]). Figure S3. Axial velocity in different laterals at increasing distances between the screen length and the caisson for Collector 3 (top), Collector 4 (middle), and Collector 5 (bottom). Optimal design < 1.524 m s^{-1} (red line). Figure S4: pH at (a) Collector 3, (b) Collector 4, and (c) Collector 5; electrical conductivity (EC) in (d) Collector 3, (e) Collector 4, and (f) Collector 5 (Source: [1,3]). Daily average pH and EC values associated to the Russian River were collected between 4 October (day 0) and 11 October (day 8) 2010, between 8 November (day 0) and 15 November (day 8), and between 18 October (day 0) and 25 October (day 8) 2010 during the capacity testing for Collector 3 (S4a and S4d), Collector 4 (S4b and S4e), and Collector 5 (S4c and S4f), respectively. Figure S5: Dissolved oxygen (DO) at (a) Collector 3, (b) Collector 4, and (c) Collector 5; electrical conductivity (EC) in (d) Collector 3, (e) Collector 4, and (f) Collector 5 (Source: [1,3]). Daily average DO values associated to the Russian River were collected between 4 October (day 0) and 11 October (day 8)

2010, between 8 November (day 0) and 15 November (day 8), and between 18 October (day 0) and 25 October (day 8) 2010 during the capacity testing for Collector 3 (S5a), Collector 4 (S5b), and Collector 5 (S5c), respectively. Figure S6: Collector Well 6, Lateral 4: 10 ft (3.048 m) progression from the lateral video inspection, 2008 (Source: Sonoma County Water Agency). Table S1: Summary of collector wells and construction parameters. Table S2: Summary of laterals' construction parameters.

Author Contributions: Methodology, E.W., J.J., J.W., and C.R.; Investigation, E.W. and J.L.; Data Curation, J.L.; Writing-Original Draft Preparation, M.D. and J.W.; Writing-Review & Editing, C.R.; Supervision, J.J. and J.W.; Project Administration, J.W. and D.S.; Funding Acquisition, J.J. and J.W.

Acknowledgments: The authors are especially grateful to Sonoma County Water Agency personnel for working with ERM, Inc. for making this project successful. Additionally, we thank S. Stowe and H. Hunt of Granite Construction for providing design criteria for screens.

Conflicts of Interest: The authors declare no conflict of interest.

References and Note

1. Sonoma County Water Agency. *Capacity Analysis of Sonoma County Water Agency Mirabel Radial Collector Wells 3, 4, and 5*; Sonoma County Water Agency: Santa Rosa, CA, USA, 2011; pp. 1–310.

2. Collector Wells International. *Inspection of Collector Wells 1 and 2 at Wohler and 3, 4, and 5 at Mirabel*; Collector Wells International: Columbus, OH, USA, 1998.

3. Sonoma County Water Agency. *Capacity Analysis Report—Sonoma County Water Agency, Mirabel Radial Collector Wells 3, 4, and 5*; Sonoma County Water Agency: Santa Rosa, CA, USA, 2018; pp. 1–271.

4. Metge, D.W.; Harvey, R.W.; Aiken, G.R.; Lincoln, G.; Jasperse, J. Influence of organic carbon loading, sediment associated metal oxide content and sediment grain size distributions upon Cryptosporidium parvum removal during riverbank filtration operations, Sonoma County, Ca. *Water Res.* **2010**, *44*, 1126–1137. [CrossRef] [PubMed]

5. Gorman, P.D.; Constantz, J.; Laforce, M.J. *Spatial and Temporal Variability of Hydraulic Properties in the Russian River Streambed, Central Sonoma County, California*; American Geophysical Union (AGU): Washington, DC, USA, 2007.

6. USGS Water Data for the Nation. Available online: https://waterdata.usgs.gov/ca/nwis/uv/?site_no=11467000&PARAmeter_cd=00065,00060 (accessed on 26 October 2018).

7. Su, G.; Jasperse, J.; Seymour, D.; Constantz, J. Estimation of hydraulic conductivity in an alluvial system using temperatures. *Gr. Water* **2004**, *42*, 890–901.

8. Private email with Granite Foundation.

9. Kim, S.-H.; Ahn, K.-H.; Ray, C. Distribution of discharge intensity along small-diameter collector well laterals in a model riverbed filtration. *ASCE J. Irrigat. Drain. Eng.* **2008**, *134*, 493–500. [CrossRef]

water

MDPI

Article

The Fate of Dissolved Organic Matter (DOM) During Bank Filtration under Different Environmental Conditions: Batch and Column Studies

Ahmed Abdelrady [1,2,*], Saroj Sharma [2], Ahmed Sefelnasr [3] and Maria Kennedy [1,2]

[1] Department of Water Management, Faculty of Civil Engineering and Geoscience, Delft University of Technology, Stevinweg 1, 2628 CN, Delft, the Netherlands; m.kennedy@un-ihe.org

[2] Department of Environmental Engineering and Water Technology, IHE Delft Institute for Water Education, Westvest 7, 2611 AX, Delft, the Netherlands; s.sharma@un-ihe.org

[3] Geology Department, Faculty of Science, Assiut University, 71516 Assiut, Egypt; ahmed.sefelnasr@yahoo.com

* Corresponding author: A.R.A.Mahmoud@tudelft.nl; Tel.: +0031-633821855

Received: 29 October 2018; Accepted: 22 November 2018; Published: 26 November 2018

Abstract: Dissolved organic matter (DOM) in source water highly influences the removal of different contaminants and the dissolution of aquifer materials during bank filtration (BF). The fate of DOM during BF processes under arid climate conditions was analysed by conducting laboratory—scale batch and column studies under different environmental conditions with varying temperature (20–30 °C), redox, and feed water organic matter composition. The behaviour of the DOM fractions was monitored using various analytical techniques: fluorescence excitation-emission matrix spectroscopy coupled with parallel factor analysis (PARAFAC-EEM), and size exclusion liquid chromatography with organic carbon detection (LC-OCD). The results revealed that DOM attenuation is highly dependent ($p < 0.05$) on redox conditions and temperature, with higher removal at lower temperatures and oxic conditions. Biopolymers were the fraction most amenable to removal by biodegradation (>80%) in oxic environments irrespective of temperature and feed water organic composition. This removal was 20–24% lower under sub-oxic conditions. In contrast, the removal of humic compounds exhibited a higher dependency on temperature. PARAFAC-EEM revealed that terrestrial humic components are the most temperature critical fractions during the BF processes as their sorption characteristics are negatively correlated with temperature. In general, it can be concluded that BF is capable of removing labile compounds under oxic conditions at all water temperatures; however, its efficiency is lower for humic compounds at higher temperatures.

Keywords: dissolved organic matter; high temperature; sub-oxic conditions; organic matter composition; PARAFAC-EEM; LC-OCD

---

1. Introduction

Pollution of surface water systems and the high cost of treatment have obliged water authorities to extend the use of cost-effective treatment techniques. Therefore, bank filtration (BF) has gained widespread interest in recent years as an economic surrogate for traditional drinking water treatment [1]. This technique has been employed in many European countries as a common method to supply drinking water. Many cities around the Rhine, Elbe, and Danube Rivers were primarily supplied with bank filtrate water for hundreds of years [2,3]. In recent years, BF has been utilized to contribute to the overall drinking water production in many developing countries: e.g., Egypt [4] and India [5], with variable hydrological and environmental conditions. Thus, there is a need to evaluate the effectiveness of the BF process under these hot-semi arid climates conditions. BF is a natural water treatment system in which surface water is induced to flow through a porous media towards a vertical or horizontal pumped well in response to a hydraulic gradient [2]. The riverbed and the underlying aquifer have been proven to act as a natural filter to remove chemical and biological pollutants from the surface water system and thereby improve the pumped water quality. Moreover, the biochemical and physical processes (i.e., adsorption) that occur during subsurface flow have a substantial role in pollutant attenuation [6]. The biochemical process taking place during infiltration is mainly controlled by the abundance and composition of dissolved organic matter (DOM) during the filtration process.

Natural water bodies contain a multitude of DOM types which determine the efficacy of the treatment processes in engineered and natural treatment systems [7]. The organic matter present in surface water systems can be divided into two main categories: (I) non-biodegradable matter (e.g., humics HS), which is mainly formed from the decay of animals and plants in the environment; and (II) biodegradable matter (e.g., protein-like compounds), which principally discharges into the water system from wastewater treatment plants [8]. Although DOM does not have an adverse effect on human health, it negatively impacts the physical properties of the water (e.g., odour, taste, and colour). In addition, it is considered the precursor for disinfection by-products (DBP) carcinogenic compounds formation [9]. Furthermore, DOM components play major roles in the removal of pollutants during the treatment processes [10]. Ma et al. [11] reported that HS has an influential role in the biodegradation of organic micropollutants (e.g., estrogen) in the treatment systems. Due to its high shuttle-electrons capacity, HS might enhance the bacterial growth and thereby the biotransformation of these micropollutants in treatment systems. Moreover, it can act as a redox mediator, thereby stimulating the iron and manganese microbial reduction process and enhancing the release of toxic metals (e.g., As and Cd) from sediment into the filtrate water in natural treatment systems [12]. Recently, Chianese et al. [13] stated that HS absorbs a wide range of wavelengths of UV radiation and thus reducing the available energy for photo-degradation of organic micropollutants. Biodegradable matter, on the other hand, takes part in the following processes: (I) it enhances biofouling in reverse osmosis, nanofiltration, and ultrafiltration membranes [14]; (II) it is used as a substrate for microorganism regrowth in distribution systems; and (III) it serves as a precursor to nitrogenous DBP (N-DBPs) formation in conventional treatment plants [15].

BF is reportedly effective at reducing the labile organic compounds during infiltration, thus increasing the biological stability of drinking water in distribution systems by >60%, as well as reducing the potential for disinfection by-product formation by 40–80% [16]. The natural attenuation of (DOM) during BF processes is primarily due to initial adsorption followed by biodegradation [17]. These processes are highly influenced by subsurface flow area environmental conditions (i.e., temperature, redox conditions, travel time, raw water quality) [18]. Maeng et al. [19] found that more than 50% of the DOM is principally removed during the first 50 cm of infiltration and thus it is highly controlled by raw water temperature. Temperature may affect the DOM behaviour directly by altering the associated soil microbial activity and changing the pollutant adsorption character. Indirectly, DOM may reduce the dissolved oxygen in the infiltrate water and thus increase the potential for developing anoxic and even anaerobic environments in the adjacent aquifer. Adversely, redox alteration may impact the DOM biodegradation rate [20]. Hoehn et al. [21] reported the redox environment turning to Mn(III/IV)—and Fe(III)—reducing conditions during the hot summer of 2003 along the Thur River. Derx et al. [22] observed that a rising water temperature will lead to a lower water viscosity, thereby increasing the infiltration capacity and shortening the travel time, which inversely affects the chemical pollutant removal efficiency. Ray et al. [6] reported that the impact of temperature on water viscosity doubled the infiltration capacity during summer along the Ohio and Danube Rivers. However, this research focussed on the direct influence of temperature and redox conditions on DOM removal during BF processes.

Several field- and lab-scale studies have tracked the behaviour of DOM during BF processes [20,22,23]. However, most research was conducted under cold and moderate-temperature (5–25 °C) conditions. The bank filtrate temperature was recently recorded as 26.4 °C along the Nile River in Egypt [24] and 30 °C along the Yamuna River in India [25]. Moreover, recent climate models predict an increase in average global temperature of 1.4–5.8 °C by 2099 [26]. Therefore, it is highly important to assess the effectiveness of BF to remove DOM under these extreme hot climate conditions. The main objectives of this research are: (1) to study the impact of high temperature (20–30 °C) on bulk organic matter removal during BF processes; (2) to track the behaviour of the DOM fractions during BF processes using innovative analytical tools (i.e., fluorescence spectroscopy coupled with parallel factor analysis (PARAFAC) and liquid chromatography with an on-line organic carbon detection (LC-OCD); (3) to determine which DOM fraction is more impacted by the temperature change and redox conditions; and (4) to quantify the role of biodegradation in DOM removal. To achieve these objectives, laboratory-scale batch studies were conducted to assess the impact of temperature (20, 25, and 30 °C) on DOM behaviour using different influent water sources. Additionally, the impact of redox conditions on the reduction of DOM during BF was tracked in laboratory-scale soil columns at a controlled room temperature (30 ± 2 °C).

2. Materials and Methods

2.1. Batch Experiments

Batch experiments were conducted to study the impact of temperature on effluent and DOM behaviour in a saturated subsurface flow system. The batch reactors were operated (in duplicate) using 0.5 L glass bottles filled with 100 g of sand (grain size 0.8–1.25 mm) and fed with 400 mL of Delft canal water. The reactors were placed on a horizontal reciprocal shaker (shaking speed 100 rpm). Three sets of batch reactors were used at three different temperatures (20, 25, and 30) ± 2 °C. Initially, the reactors were acclimated (with respect to DOC removal) at their respective temperature for 90 days. After the acclimation period, the reactors were fed with four different water types that had a different organic matter composition: (1) Delft canal water, the Netherlands (DC); (2) Delft canal water spiked with secondary treated wastewater effluent from Hoek van Holland, The Netherlands (DCWW); (3) secondary treated wastewater effluent (WW); and (4) water extractable organic matter (WEOM). WEOM was used to simulate the DOM water with a high concentration of humic aromatic compounds. It was prepared using 100 g of clay (obtained from Delftse Hout, Delft, Netherlands) in a 0.5 L glass bottle filled with 400 mL of DC water and placed on a shaker at 150 rpm for 24 h. Then, the extracted solution was centrifuged at 4800 rpm for 30 min, and filtered with 0.45-μm pore-size cellulose acetate filters [27]. Samples were taken from the influent and effluent water and analysed to determine their chemical and physical characteristics. Control samples were taken by filling the glass bottles with the same amount of each influent (without silica sand). Another series of batch reactor studies were performed to estimate the role of biodegradation in the removal of organic matter and to what extent it may be affected by temperature. Maeng et al. [28] suggested sodium azide as a biocide to suppress biological activity. However, this research found that sodium azide enhances fluorescence intensity and UV-absorbance measurements, thus reducing their reliability, as also reported by Park et al. [29]. Alternatively, the batch reactors were spiked with mercuric chloride (20 mm) to develop an abiotic environment inside the reactors [30].

2.2. Column Experiments

A laboratory-scale column study was conducted to assess the impact of redox conditions on the removal of DOM at a high temperature (30 °C) during the BF process. Six columns were established and run under three different redox conditions (oxic, anoxic, and anaerobic). Each column was made of a PVC pipe with a 0.05 m internal diameter and 0.5 m height. The column bottom was packed with a support layer of graded gravel (7 cm), and then with cleaned silica sand (size 0.8–1.25 mm), allowing the media to settle in deionized water and thus ensuring packing homogeneity. The columns were operated in up-flow mode (saturated flow), where a variable speed peristaltic pump was connected to the bottom of each column to introduce the influent water from the tank into the column at a constant hydraulic loading rate of 0.5 m·day^{-1}. Two valves were attached at the inlet and outlet of each column, which allowed the air to dissipate from the system, as well as to collect samples of the influent and effluent water. The oxic environment was maintained through continuous aeration of the influent tanks to keep the dissolved oxygen level at 7 mg·L^{-1}. Anaerobic conditions were developed in the second two columns by degassing the influent tanks with nitrogen to dissipate the air. Anoxic conditions were created through the degassing processes, followed by spiking 5 mg·L^{-1} of nitrate into the influent tank. The columns were acclimated for 70 days until the removal of DOC for three successive measurements was ±1%. Then, three columns were fed with DC and run under the identified redox conditions. The other three columns were fed with WEOM and run under the same redox conditions. All influents were filtered through a microsieve (38 μm) to avoid physically clogging the column inlets. The experiment lasted 30 days, and influent and effluent samples were taken daily.

2.3. Analytical Methods

The collected samples were filtered using 0.45 μm filtration (Whatman, Dassel, Germany) and analysed within three days to avoid organic matter degradation. DOC (in mg·L^{-1}) was measured through the combustion technique using a total organic carbon analyser (TOC-VCPN (TN), Shimadzu, Japan). UV-Absorbance at 254 nm UV254 (cm^{-1}) was measured using a UV/Vis spectrophotometer (UV-2501 PC, Shimadzu, Japan). Specific ultraviolet absorbance SUVA$_{254}$ (L·mg^{-1}·m^{-1}) was used as an indicator for the aromaticity degree and unsaturated structures of the bulk organic matter. It was determined by dividing the UV254 by its corresponding DOC measurement. Adenosine triphosphate (ATP) was measured as an indicator for microbial activity associated with the sand. The sampling and preparation protocols of ATP measurements were explained in [19]. Details of the ATP extraction procedures and the detection method employed are described in Abushaban et al. [31].

The constituents of bulk organic matter were elucidated using different analytical methods, including: Liquid chromatography–organic carbon and nitrogen detection (LC-OCD-OND) (manufacturer DOC-LABOR Dr. Huber, Karlsruhe, Germany) and fluorescence excitation-emission spectrophotometry. LC-OCD is used to separate the pool of DOC into five major fractions: biopolymers BP, humic substances (humic and building blocks) HS, low molecular weight (LMW) acids (LMWa), neutrals (LMWn), and hydrophobic organic carbon (HOC), based on their molecular weight distribution. The measurement procedures were described in detail by Huber et al. [32].

The Fluorescence Emission Excitation Matrices (EEMs) technique was widely used to characterize the bulk organic matter into three main components (humic-, fulvic-, and protein-like fractions) [10]. EEM measurements were conducted at excitation wavelengths from 240 to 452 nm with 4 nm intervals and emission wavelengths ranging between 290 and 500 nm with 2 nm intervals using a Fluoromax-3 spectrofluorometer (HORIBA Jobin Yvon, Edison, NJ, USA). The EEMs were corrected and recorded in Raman units (RU) using MATLAB (version 8.3, R2014a, The MathWorks, Natick, MA, USA).

2.4. PARAFAC Modelling

Fluorescence excitation-emission matrix spectroscopy coupled with parallel factor analysis (PARAFAC-EEMs) is used to decompose the EEMs to independent fluorescent components representing different DOM compositions. PARAFAC-EEMs have been extensively developed to characterize DOM behaviour in natural and treatment systems [10]. PARAFAC is based on decomposing the fluorescence signals into tri-linear components and a residual array using an alternating least squares algorithm [33]:

$$X_{ijk} = \sum_{f=1}^{f} a_{if} b_{jf} c_{kf} + \varepsilon_{ijk}, \; i = 1,\ldots\ldots,I; \; j = 1,\ldots\ldots,J; \; k = 1,\ldots\ldots,k; \; f = 1,\ldots\ldots,F \quad (1)$$

where X_{ijk} represents the fluorescence intensity of the *i*th sample at the *k*th excitation and *j*th emission wavelength; f is the number of model components; a_{if} is the score for the *f*th component and is proportional to the fluorophore f concentration in sample *i*; b_{jf} is the scaled estimates of the emission spectrum for the *f*th component; c_{kf} is linearly related to the specific absorption coefficient at excitation wavelength *k*th; and ε_{ijk} is the residual term representing the unaccounted variation of the model [34].

To further assess the behaviour of different DOM components during the filtration process, a PARAFAC model was developed and validated using the complete measured EEMs dataset (184 samples) from the influent and effluent water of the batch and column experiments. The PARAFAC model with three to seven components was implemented using the N-Way and drEEM MATLAB toolboxes developed by Murphy et al. [35]. The right number of PARAFAC components was selected and validated using diagnostic tools such as split-half validation [36], Tucker's congruence coefficients [37], and the residual error technique [38,39].

2.5. Statistical Analysis

A two-way analysis of variance (ANOVA) test was applied to assess if an environmental parameter's influence on the DOM constituent behaviour during the BF process was statistically significant, in which a significant difference was ($p < 0.05$).

3. Results

3.1. PARAFAC Components

PARAFAC analysis successfully decomposed the fluorescence measurements into five components. The validated model explained more than 99.6% of the data variance. The excitation and emission loadings, as well as the contour plots of these fluorescent components in RU, are shown in Figure 1 and Figure S1. The spectral slopes of the identified components were successfully cross-referenced with the OpenFluor database [40] (Table S1).

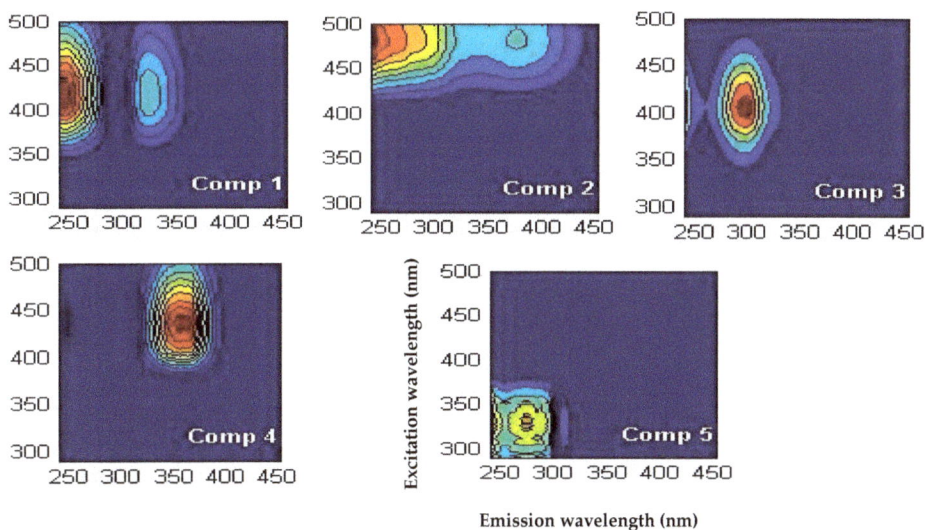

Figure 1. Contour plots of the five components identified from the complete measured F-EEMs dataset for the influent and effluent water of the batch and column experiments.

Four of the PARAFAC components were identified previously as humics: (1) Component 1 (C1) found at (maximum excitation wavelength (λ_{ex})~240 and 320, maximum emission wavelength (λ_{em})~410 nm) and Component 2 (C2) (λ_{ex}~244 and 376, λ_{em}~480 nm(are both associated with humic-like fluorophore substances originating from terrestrial resources, as reported previously in Shutova et al. [41]. It can be seen that component 2 (C2) appeared at longer excitation and emission wavelengths, suggesting it possesses a more condensed and conjugated structure. According to Baghoth et al. [10], these components are characterized by a high molecular weight (>1000 Da).

Moreover, they have low biodegradable matter and are thus principally removed by adsorption and coagulation in water treatment systems. (2) Component 3 (C3) (λ_{ex}~300, λ_{em}~400 nm) mimics microbial humic components in surface water systems [42]. This component is highly related to recent biologically produced fluorescent compounds. It is characterized by an intermediate molecular weight (650 < C3 < 1000 Da). (3) Component 4 (C4) (λ_{ex}~360, λ_{em}~440 nm) is related to a humic-like component derived from agricultural activity and it is common in freshwater environments, as reported in Osburn et al. [43]. These compounds mainly contain carboxylic and phenolic moieties in their structures [44]. (4) Component 5 (C5) (λ_{ex}~240 and 270, λ_{em}~320 nm) is spectrally similar to a protein-like fluorophore (tyrosine and tryptophan compounds) identified in Kulkarni et al. [45]. These components are highly correlated with microbial activity in water systems and principally their removal in engineered water treatment systems is attributed to biodegradation [10]. Therefore, it can be used as a surrogate for tracking the manner of bioavailable matter during filtration.

To further investigate the behaviour of the DOM fractions during the filtration process, the maximum fluorescence intensity (F_{max}) was used to characterize the influent and effluent water and to track the behaviour of PARAFAC components during the infiltration process under different environmental conditions. F_{max} fluorescence intensities give an estimation of the proportional contribution of each component to the full fluorescence spectra. This contribution highly relies on the DOM source and the behaviour of the fluorescent components during the filtration process [10].

3.2. Batch Experiments

During this research, laboratory-scale batch studies were employed to assess the impact of temperature (20, 25, 30 ± 2 °C) on the removal of organic matter during the filtration process.

3.2.1. Characteristics of Influent Water DOM

The feed water quality has a clear impact on microbial activity and thus on DOM behaviour during the filtration process [19]. Four different water types were prepared and applied to the batch reactors. The average values of the chemical and physical water quality parameters are presented in Table 1. The results show that WEOM influent water had the highest DOC concentration (14.6 ± 1.6 mg·L^{-1}), followed by DC (11.6 ± 0.7 mg·L^{-1}), DCWW (10.5 ± 0.4 mg·L^{-1}), and WW (9.7 ± 0.6 mg·L^{-1}). Furthermore, the WEOM had a relatively higher SUVA$_{254}$ value (3.56 ± 0.71 L·mg^{-1}·m^{-1}) compared to DC (2.84 ± 0.33 L·mg^{-1}·m^{-1}). This implies that the DOC of the WEOM influent was composed of higher aromatic compounds (i.e., humic substances) than the DC influent water DOC. The average SUVA$_{254}$ values of DCWW and WW were 2.37 ± 0.28 and 2.56 ± 0.42 L·mg^{-1}·m^{-1}, respectively, indicating the relatively low aromatic character of their organic matter composition.

Table 1. Chemical and physical characteristics of the influent water.

	Unit	DC	DCWW	WW	WEOM
pH	-	7.87	7.79	7.66	7.65
DOC	mg·L^{-1}	11.6 ± 0.7	10.5 ± 0.4	9.7 ± 0.6	14.6 ± 1.6
SUVA$_{254}$	L·mg^{-1}·m^{-1}	2.84 ± 0.33	2.37 ± 0.28	2.56 ± 0.42	3.56 ± 0.71
NO$_3$-N	mg-N·L^{-1}	2.06 ± 0.27	2.10 ± 0.19	1.87 ± 0.15	4.03 ± 0.43
NH$_4$-N	mg-N·L^{-1}	0.24 ± 0.04	0.21 ± 0.07	0.17 ± 0.03	0.31 ± 0.06
Mn	μg·L^{-1}	46.8	14	14.03	86.74
Fe	μg·L^{-1}	175	87.4	37.6	109.6
Zn	μg·L^{-1}	20.9	30.1	36.6	36.6

Co, Cd, and Pb values were below the limit of detection.

LC-OCD results showed that humic substances (HS) are the dominant fraction of DOC in all influent water. The contributions of HS to total DOC were 74%, 73%, 68%, and 75%, respectively, for DC, DCWW, WW, and WEOM influent. The hydrophobic fraction (HOC) was only 5.7% of the DOC in WEOM influent water, a typical value for surface water systems. However, the HOC of DC,

DCWW, and WW influent was 10.3, 9.8, and 13.7% of the total DOC, respectively. This indicates the impact of effluent organic matter (EfOM) on their organic compositions [32]. Though WEOM had the highest concentration of BP, only 42% can be considered protein (assuming the C:N is 3, and all organic nitrogen in BP originates from protein) [46]. However, the protein represents 51, 65, and 82% of the BP for DC, DCWW, and WW influent, respectively, which also reflects the impact of EfOM. Furthermore, WW and DCWW influent contain more LMW (acids and neutrals), which are more subject to biological treatment. However, DC and WEOM contain relatively lower concentrations of LMW.

PARAFAC components (C1–C5) were recorded for all the influent water. F_{max} was lower for protein component C5 than the humic/fulvic components (C1–C4). The maximum and minimum F_{max} of component C5 were observed for WEOM (1.09 ± 0.05 RU) and DC (0.38 ± 0.03 RU), respectively. A humic-like component (C4) exhibited a comparable contribution with a protein-like component to the DOM fluorescence of the influent. The F_{max} of C4 ranged between 0.33 ± 0.04 RU and 0.89 ± 0.1 RU. However, the terrestrial humic-like component (C1) contributed much more highly than other humic/fulvic components. An exception was the WEOM influent, which possessed the highest concentration of conjugated humic component (C2). Microbial humic (C3) contributed moderately to the fluorescence spectrums of the influent water, with a higher contribution (1.39 ± 0.28 RU) observed for the WW influent and a lower contribution (0.99 ± 0.15 RU) for WEOM.

3.2.2. Bulk Organic Matter Parameters

The results demonstrated that the DOC removal during the filtration process is highly dependent ($p < 0.001$) on its concentration in the feed water. Table 2 showed that the DOC removal values for the DC, DCWW, and WW influent were 9.5, 11.4, and 14.7%, respectively, at 30 °C. However, WEOM influent water exhibited the highest DOC removal (44%) at the same temperature and that may be attributed to the higher feed water DOC concentration promoting biomass formation associated with sand. The ATP values of reactor media were measured to be 4.69, 5.21, 5.39, and 7.95 ng·g^{-1} sand at 30 °C for DC, DCWW, WW, and WEOM, respectively. These values increased by 7–9% at 25 °C and 8–16% at 20 °C (Figure 2). However, the statistical analysis revealed that there is no significant ($p > 0.05$) effect of temperature on biological activity (ATP concentration) and thereby DOM biodegradation is not significantly affected by temperature in the range of 20–30 °C. Nevertheless, the results showed a higher DOC removal efficiency at a lower temperature (20 °C) ($p < 0.05$). For instance, the DOC removal for DC increased from 9.5 ± 2.3% at 30 °C to 20.3 ± 3.7% at 20 °C (Table 2). On the other hand, the results of abiotic batch reactors revealed that adsorption mechanisms contributed to the overall removal of DOC for DC influent by 18 ± 2.1% at 30 °C, 38.5 ± 5.4% at 25 °C, and 51 ± 4.7% at 20 °C, and for WEOM influent by 27 ± 3.7% at 30 °C, 42 ± 5.1% at 25 °C, and 58 ± 6.8% at 20 °C (Table 2). In the same regard, SUVA$_{254}$ values exhibited a positive relationship with temperature, increasing from 2.84 ± 0.3 L·mg^{-1}·m^{-1} for the DC influent to 3.71 ± 0.3, 3.59 ± 0.5, and 3.57 ± 0.2 L·mg^{-1}·m^{-1} for the effluent water at 30, 25, and 20 °C, respectively (Table S2). This implies that aromatic compounds are favourably removed at lower temperatures, considering that there is no significant change in the removal of aliphatic compounds at respective temperatures.

Table 2. DOC (mg·L^{-1}) values of the batch effluents at different temperatures (20, 25, 30 °C) and biotic/abiotic conditions under oxic conditions.

	30 °C		25 °C		20 °C	
	Biotic	Abiotic	Biotic	Abiotic	Biotic	Abiotic
DC	10.5 ± 0.21	11.4 ± 0.35	9.76 ± 0.0.28	10.89 ± 0.61	9.24 ± 0.32	10.4 ± 0.43
DCWW	9.3 ± 0.18	10.29 ± 0.33	9.24 ± 0.37	10.1 ± 0.23	8.18 ± 0.17	9.4 ± 0.29
WW	8.27 ± 0.24	9.42 ± 0.51	7.95 ± 0.27	9.37 ± 0.23	7.37 ± 0.19	8.96 ± 0.18
WEOM	8.18 ± 0.26	12.87 ± 0.37	7.74 ± 0.41	11.7 ± 0.53	6.8 ± 0.27	10.1 ± 0.41

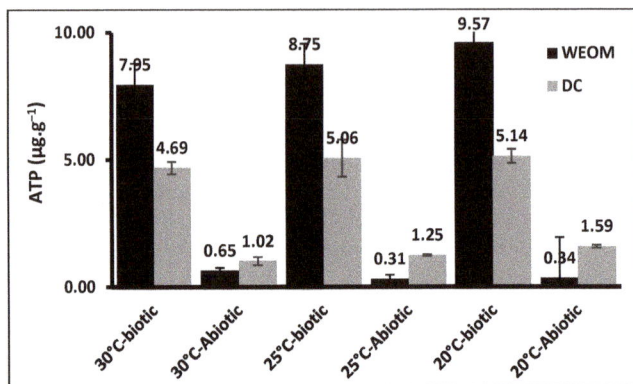

Figure 2. Changes of ATP concentrations ($\mu g \cdot g^{-1}$) in a function of temperature and biotic/abiotic conditions during the batch experiment.

3.2.3. LC-OCD Analysis

Total DOC measured by LC-OCD is largely well-matched with the measured values of a conventional TOC analyser to within 0.5 mg·L^{-1}. The changes of DOM fractions at different temperatures are presented in Figure 3 and Table S3. The removal of BP values for DC influent were 87, 94, and 95%; 96, 91, and 88% for DCWW; 94, 86, and 83% for WW; and 98, 97, and 97% for WEOM, at 30, 25, and 20 °C, respectively. However, the statistical analysis revealed that this process is independent of temperature ($p > 0.05$). Other biogenic organic matter fractions (LMWn and LMWa) exhibited lower removal for all the influent water compared to BP. The removal rates of LMWn for DC, DCWW, WW, and WEOM were 20, 16, 6, and 47%, respectively, at 30 °C. This removal increased by 5–16% at 20 °C. Likewise, the LMWa removal ranged between 10–47% for all the influents water at 20 °C. This removal decreased by (6–25%) when the temperature went up by 10 °C. This indicates that the decomposition of higher molecular weight compounds (BP and HS) into LMW hydrophilic compounds is lower than the removal of LWM compounds during filtration. However, this removal is also independent ($p > 0.05$) of temperature. An exception was the LMW acid of WEOM, which increased by 24% at 30 °C.

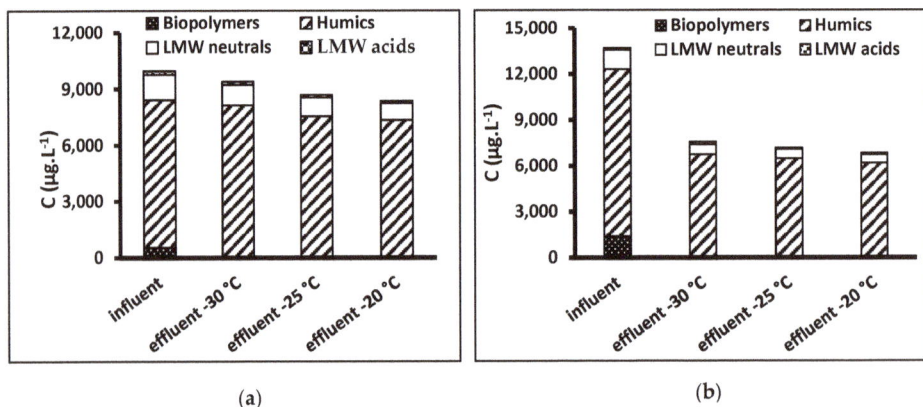

(a)

(b)

Figure 3. Changes in the LC-OCD fractions concentration in batch reactors at different temperatures under oxic conditions: (**a**) DC and (**b**) WEOM.

Humic compound removal exhibited a significant dependency on temperature ($p < 0.05$). Figure 3 displays a highly reduced HS concentration at lower temperatures (20, 25 °C). The HS removal varied between 7–44% at 20 °C and 4–41% at 25 °C for all the influents water. However, an increase in the HS concentration was observed for DC, DCWW, and WW at 30 °C.

3.2.4. PARAFAC-EEM Analysis

The protein-like (C5) component exhibited the highest reduction rate at all three temperatures (Figure 4). The protein-like component removal increased consistently with increasing influent concentration ($p < 0.001$). The highest F_{max} reduction (93.6 ± 2.6%) was recorded for WEOM at 20 °C, followed by WW (60.2 ± 3.2%), DCWW (43 ± 3.8%), and DC (36.1 ± 1.7%), respectively, at the same temperature. Similar to BP removal, these labile compounds exhibited independent behaviour upon temperature variation ($p > 0.05$). The removal percentage was reduced by 23.9 ± 4.8%, 15.9 ± 1.6%, 15.08 ± 5.3%, and 4.5 ± 1.87%, respectively, for DC, DCWW, WW, and WEOM at 30 °C. On the contrary, humic components (C1–C4) removals were impacted significantly by variations in temperature and feed water characteristics ($p < 0.05$). An exception was microbial humic, which showed independent behaviour with temperature variations during the filtration process ($p = 0.09$). Figure 4 illustrates the attenuation of humic components decreasing with rising temperature. The average removal of C1 was 48.6 ± 12.3%, 47 ± 16% for C2, 49.8 ± 13.5% for C3, and 56.2 ± 8.4% for C4, at 20 °C. These removals decreased by 28.4, 26, 19, and 30.4% for C1–C4, respectively, at 30 °C.

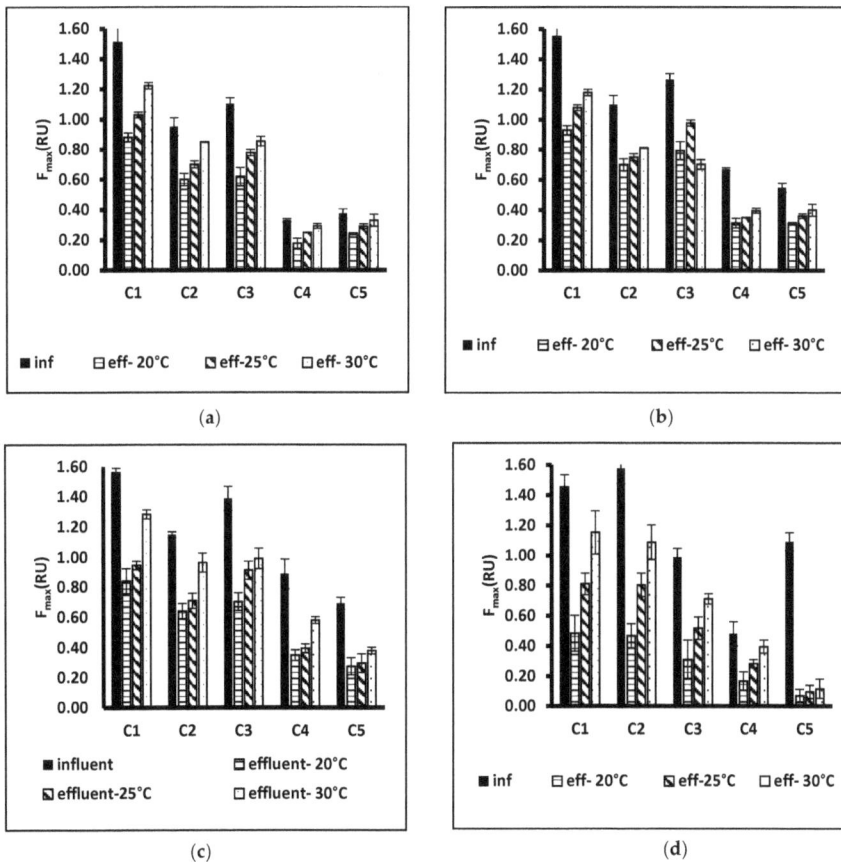

Figure 4. Changes of PARAFAC components (F_{max}) in the batch reactors at different temperatures (20, 25 and 30 °C) under oxic conditions: (**a**) DC, (**b**) DCWW, (**c**) WW, and (**d**) WEOM (influent = inf and effluent = eff).

3.3. Column Experiments

A column experiment was conducted to assess the impact of redox conditions (oxic, anoxic, and anaerobic) on the behaviour of DOM constituents during the filtration process. The experiment was conducted in a controlled temperature room ($30 \pm 2\,°C$) using two different feed water types (DC and WEOM).

3.3.1. Characteristics of Influent Water DOM

The bulk organic characteristics of the feed water are presented in Table 3. It can be shown that WEOM had a higher DOC concentration (14.16 ± 0.73 mg·L^{-1}) than DC influent (10.80 ± 0.51 mg·L^{-1}). Moreover, WEOM possessed higher aromatic characteristics (SUVA$_{254}$ = 3.67 ± 0.21 L·mg^{-1}·m^{-1}) compared to DC influent (SUVA$_{254}$ = 3.05 ± 0.31 L·mg^{-1}·m^{-1}). These results were confirmed with PARAFAC-EEM results, which demonstrated that WEOM had a higher concentration of terrestrial-derived; the average F$_{max}$ values of C1, C2, and C4 for WEOM were 1.52 ± 0.06, 1.61 ± 0.1, and 0.51 ± 0.03 RU, and 1.31 ± 0.04, 0.99 ± 0.07, and 0.41 ± 0.02 for DC influent, respectively. Furthermore, LC-OCD results revealed that the humic fraction was the dominant fraction in the feed water, representing 72 and 73% of the DOM pool for DC and WEOM influents, respectively. However, biogenic fractions (BP, LMWn, and LMWa) represent only 5.9, 11, and 2.3% of DOM for DC influent and 10.7, 8.7, and 1.1% of DOM for WEOM influent, respectively. In addition, PARAFAC-EEM results revealed that DC influent possessed a higher concentration of microbial humic-like component (C3) and lower concentration of protein-like component (C5); the F$_{max}$ values of C3 and C5 were 0.98 ± 0.07 and 0.58 ± 0.03 RU for DC and 0.78 ± 0.06 and 0.92 ± 0.04 RU for WEOM influent, respectively.

Table 3. Characteristics of the influents and effluents water of the columns under different redox conditions.

	DC			WEOM		
	pH	DOC	SUVA$_{254}$	pH	DOC	SUVA
	-	(mg·L^{-1})	(L·mg^{-1}·m^{-1})	-	(mg·L^{-1})	(L·mg^{-1}·m^{-1})
Influent	7.82	10.80 ± 0.51	3.05 ± 0.31	7.73	14.16 ± 0.73	3.67 ± 0.21
effluent-oxic	7.91	9.49 ± 0.36	4.01 ± 0.24	7.88	7.91 ± 0.17	3.92 ± 0.14
effluent-anoxic	8.08	10.12 ± 0.25	3.06 ± 0.19	8.16	9.37 ± 0.28	3.73 ± 0.33
effluent-anaerobic	8.13	10.26 ± 0.55	3.21 ± 0.27	7.95	10.08 ± 0.39	3.67 ± 0.37

3.3.2. Bulk Organic Matter Parameters

The redox environment significantly ($p < 0.05$) impacts the removal of DOC during filtration. Table 3 shows that the removal of DOC decreased by 5–10% under anoxic, and 7–15% under anaerobic, conditions, compared to oxic conditions. This is highly linked to the biological activity associated with the sand. ATP from active microbial biomass associated with sand was higher for oxic conditions. The average concentrations of ATP in the oxic, anoxic, and anaerobic columns were 5.44 ± 0.64, 3.68 ± 0.37, and 3.21 ± 0.47 for DC and 7.32 ± 0.51, 4.1 ± 0.15, and 4.69 ± 0.21 ng·g^{-1} sand for WEOM, respectively (Figure 5). In the same regard, SUVA$_{254}$ increased from 3.05 ± 0.31 to 4.01 ± 0.24, 3.06 ± 0.19, and 3.22 ± 0.27 L·mg^{-1}·m^{-1} for DC, and from 3.67 ± 0.21 to 3.92 ± 0.14, 3.73 ± 0.33, and 3.67 ± 0.37 L·mg^{-1}·m^{-1} for WEOM, respectively, under oxic, anoxic, and anaerobic conditions.

3.3.3. LC-OCD Analysis

LC-OCD results showed that BP is preferentially removed during soil passage; its removal under anoxic and anaerobic conditions was less than oxic conditions by 20–24% (Figure 6). Similarly, LMWn compounds exhibited higher removal during oxic filtration. The removal was decreased under anoxic and anaerobic conditions by 21 and 50% for WEOM and by 15 and 17% for DC, respectively. The same behaviour was observed for LMWa of DC influent, where its removal decreased by 25–32% under sub-oxic conditions. However, LMWa of WEOM influent exhibited inconsistent behaviour, where its concentration was increased by 15–21% under sub-oxic conditions compared to its concentration in the feed water. In the same way, HS demonstrated a higher removal efficiency under oxic conditions. For WEOM, the average removal of HS under oxic, anoxic, and anaerobic conditions was 36, 29, and 24%, respectively. However, HS removal for DC was only decreased by 2–4% when the environment turned into sub-oxic conditions.

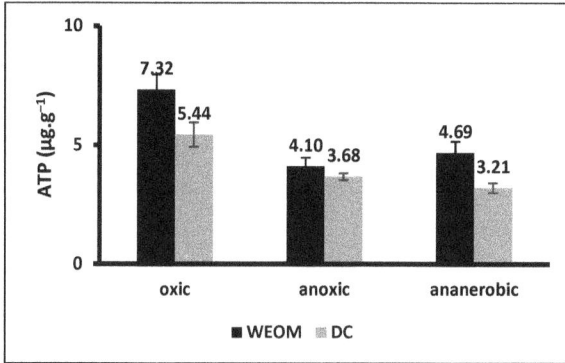

Figure 5. Changes of ATP concentrations ($\mu g \cdot g^{-1}$) in a function of redox conditions during the column experiment for DC and WEOM influent.

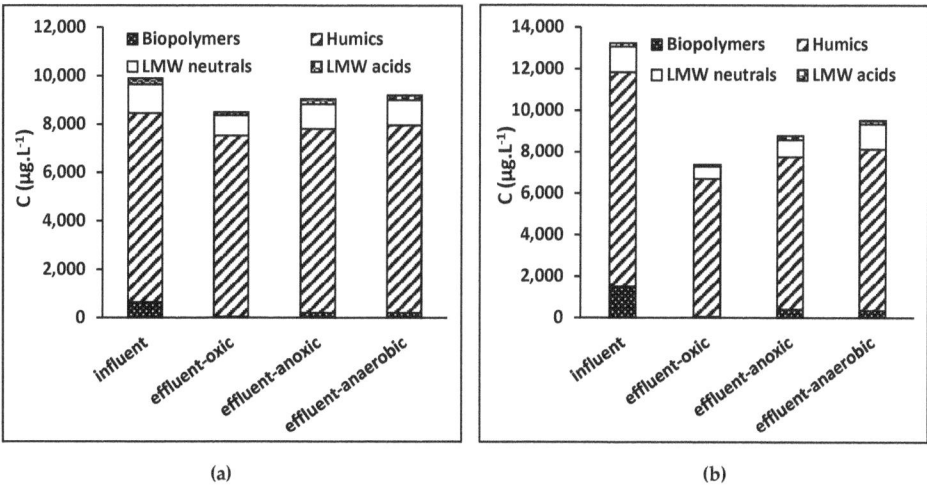

(a)　　　　　　　　　　　(b)

Figure 6. Changes of LC-OCDND fractions in batch reactors under different redox conditions: (a) DC and (b) WEOM (temperature = 30 °C, column study).

3.3.4. PARAFAC-EEM Analysis

The fate of PARAFAC components under different redox conditions was examined using their maximum fluorescence intensity (F_{max}). The results reveal that the redox environment plays a substantial role in the removal efficiency of fluorescence components during soil passage (Figure 7). The removal of protein-like components was decreased by 15–22% under sub-oxic conditions. In the same manner, the microbial humic component (C3) displayed redox-dependent behaviour, with higher reduction under oxic conditions. The removal of microbial humics was reduced by 20–22% for WEOM and 2–5% for DC influent under sub-oxic conditions. Similar to protein-like components, terrestrial-derived humic components showed a higher removal under oxic conditions; the removals of C1, C2, and C4 were reduced by 14–19%, 9–14%, and 10–18%, respectively, during sub-oxic filtration.

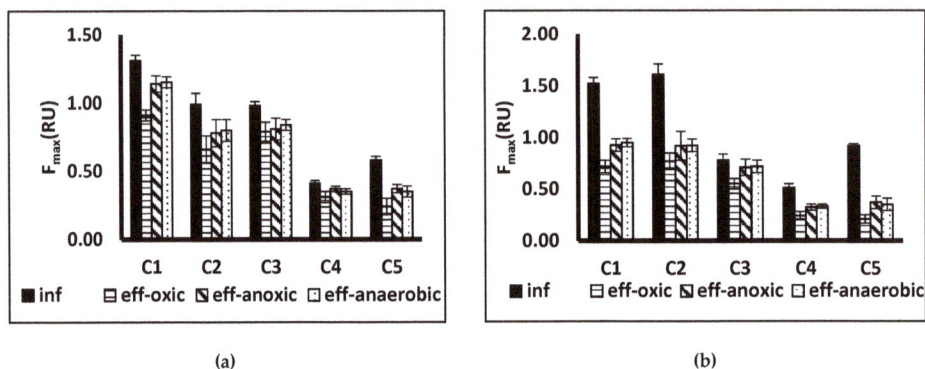

(a)   (b)

Figure 7. Changes of PARAFAC components (F_{max}) under different redox conditions: (**a**) DC and (**b**) WEOM (temperature = 30 °C) (column study) (influent = inf and effluent = eff).

4. Discussion

4.1. Impact of Temperature and Influent Organic Composition on DOM Behaviour

DOM removal during BF is principally due to a combination of DOM sorption to the sand media and biodegradation through bacteria in biofilms associated with the media. In this study, DOC decreased during the filtration process, showing a high dependence on the feed water DOC composition, which is highly correlated with the biomass activity associated with the sand. This is in agreement with Li et al. [47], who reported a positive correlation between biofilm density and influent DOC concentration. However, the results infer that there is no effect of the temperature on the biomass activity associated with the sand and thus the impact of temperature on DOC biodegradation is low. On the other hand, the abiotic results indicate preferential DOM adsorption at a lower temperature. Thus, it can be concluded that the relatively higher removal of DOM at lower temperatures in the range of 20–30 °C during the BF process is mainly ascribed to adsorption. These results are inconsistent with Massmann et al. [48], who reported DOM attenuation independent of 0–24 °C temperature, based on a field study conducted at an operational artificial recharge site over Tegel Lake in Germany. In contrast, Abel et al. [49] found a positive relationship between DOC removal and temperature (5–25 °C) during the filtration process. Alidina et al. [50], on the other hand, reported a minor temperature (10–30 °C) effect on DOM removal during a column filtration process, with higher removal at lower temperatures. These contradictions in DOM behaviour during infiltration at different temperatures could be attributed to different feed water DOM characteristics. This goes in line with the conclusion of Chen et al. [51] who reported that the organic composition of the raw water determines its behaviour during filtration. Thus, it is important to assess the behaviour of DOM fractions individually during filtration.

BPs (Molecular weight MW > 20,000 Da) are the most readily biodegradable DOM fraction, thus they are preferentially removed during the filtration process [52]. According to So et al. [53], the BP was reported to be highly degraded at higher temperatures; however, in this research, there is no significant effect of temperature (20–30 °C) on BP removal observed during filtration. The removal of BP reached >80% at all temperatures during filtration, and this is mainly attributed to the prolonged filtration period. Moreover, the high ratio between organic nitrogen BP and DOC in the feed water indicated that they are principally composed of proteinaceous matter that is highly degraded during filtration. This is in consonance with the PARAFAC-EEM results, where labile compounds (i.e., protein-like compounds) exhibited independent behaviour upon temperature variation. However, the results demonstrated that protein-like component removal is more sensitive to temperature variation when the protein content of the feed water is low. This specifies the vital role of co-metabolism in the removal of this biodegradable matter. These results are consistent with Maeng et al. [23], who reported the total removal of BP during bank filtration over Tegel Lake (Germany). Likewise, other biogenic organic matter fractions (LMWn and LMWa) exhibited temperature-independent behaviour during the filtration process.

In contrast, the removal of humic compounds was highly dependent on the temperature of the feed water, with a favourable reduction at a lower temperature (20–25 °C). This reduction is mainly ascribed to the high ability of these refractory compounds to adsorb onto sand grains at a lower temperature. On the contrary, an increase in the concentration of HS was observed at a higher temperature (30 °C) for DC, DCWW, and WW influent. The HS concentration enrichment may be attributed to microorganisms and enzymes that are able to: (I) transform microbial matter to more refractory and conjugated matter (i.e., microbial humification process), that was reported in several laboratory-scale and field studies [54–56]; and (II) leach soil humic compounds into filtrate water [57]. In contrast, the humic of WEOM influent exhibited a unique behaviour, and its concentration decreased at all three temperatures. This may be due to: (I) the absence of microorganisms to transform labile matter or leach organics from soil, as mentioned above; and (II) the inability of these organic compounds to bio-transform into refractory compounds, which may be attributed to their higher aromaticity (SUVA$_{254}$ = 2.9 L·mg^{-1}·m^{-1}) compared to other influents. PARAFAC-EEM results revealed that the condensed structure humic compounds are the most impacted by changing temperature during the filtration process. These results are compatible with those of Abel et al. [49], who illustrated that the optimum temperature for the removal of these refractory compounds is 15 °C. The ratios between the PARAFAC components were used in many studies [10,45] to assess treatment efficacy. In this research, only the ratios between the F$_{max}$ of terrestrial humic components (C1, C2, and C4) and the protein-like component (C5) exhibited a clear increasing trend with rising temperature. These results also confirm the preferential removal of terrestrial humic components at lower temperatures.

4.2. Impact of Redox Conditions on DOM Behaviour

This research specifies the preferential removal of DOM under oxic conditions during the BF process, which is mainly attributed to oxygen as an electron acceptor for microorganism respiration to degrade the organic matter. Slower biodegradation of organic matter under sub-oxic conditions was also reported in previous studies [49,58]. Moreover, SUVA$_{254}$ values exhibited a lower increase during sub-oxic conditions, which refers to the preferential removal of aliphatic compounds during oxic filtration.

LC-OCD data revealed that BP is the most impacted DOM fraction by the alteration in the redox environment, with favourable removal under oxic conditions. In the same regard, the ratio between nitrogen and carbon BP exhibited higher values under oxic conditions, which infers lower biodegradation of protein compounds under sub-oxic conditions. This finding was confirmed with PARAFAC-EEM results, where F$_{max}$ of the protein-like component (C5) exhibited higher reduction (relative to influent F$_{max}$) under oxic conditions than other redox conditions. This reduction is mainly ascribed to the degradation of high molecular weight biodegradable organic matter into non-fluorescing material. This is in agreement with field data collected at the Tegel Lake (Berlin, Germany) BF site, where the partial removal of biopolymers was detected under sub-oxic conditions [58]. Furthermore, previous studies [19,49] also emphasized the superior removal of protein-like components under oxic environmental conditions by conducting laboratory-scale experiments. Likewise, LMW (acids and neutrals) exhibited a lower removal efficiency under sub-oxic conditions. An exception was the LMW acid of WEOM, which increased under sub-oxic conditions, likely due to the breakdown of larger molecular weight humic matter into lower molecular weight compounds under these conditions [59].

HS removal followed the same trend as BP (with less removal efficiency), in that higher removal was obtained under oxic than other redox conditions. Nonetheless, the HS removal efficiency is much lower for DC influent than WEOM influent, presumably attributed to the nature and molecular weight of the humic present. PARAFAC-EEM humic components (C1–C4) also exhibited higher reduction under oxic conditions. Terrestrial humic-like components (C1 and C2) exhibited the highest reduction of F_{max} among other humic components, followed by lower aromatic-humic, such as component (C4). According to Gerlach et al. [60], humic compounds with a higher molecular weight are preferentially removed during aerobic soil passage.

5. Conclusions

Based on the results of laboratory-scale batch and column studies, the following conclusions can be drawn:

- A positive correlation was found between DOM biodegradation and raw water concentration, which was likely due to the higher microbial activity associated with sand, as determined by ATP measurements of the biomass attached to the sand grains.
- The removal of DOM during filtration is significantly impacted by temperature variation, with higher removal at lower temperatures.
- LC-OCD results revealed that the labile compounds (i.e., biopolymers) are highly removed (>80%) under oxic filtration, regardless of the temperature and organic matter composition of the feed water. Likewise, the PARAFAC protein-like component exhibited the highest reduction at all temperatures studied.
- Humic compound removal exhibited a significant dependence on temperature, with higher removal at a lower temperature. PARAFAC analysis indicated that terrestrial humic components are the least persistent humic type adsorbed at a lower temperature. The contradictory behaviour of protein and humic compounds explains the positive relationship between SUVA and temperature.
- DOM was preferentially removed under oxic conditions; its removal decreased by 5–10% under anoxic, and by 7–15% under anaerobic conditions. LC-OCD results reveal that biopolymers are the most impacted fraction by altering the redox conditions. Humic compounds also exhibited a lower removal efficiency (with less extent) under sub-oxic conditions. Therefore, post-treatment steps should be considered in case of sub-oxic filtration.
- In general, this study revealed that the BF removal efficiency for DOM components under arid conditions (high temperature) is determined by the feed water organic composition and redox conditions in the infiltration area.
- Finally, this study shows that PARAFAC-EEM and LC-OCD can be promising tools to provide further insight into BF processes and for determining the treatment efficiency for DOM components.

Supplementary Materials: The following are available online at http://www.mdpi.com/2073-4441/10/12/1730/s1, Figure S1: The contour plots and the corresponding excitation (solid curves) and emission (dotted curves) loadings of the fluorescent components C1–C5; Table S1: The spectral slopes of the identified components and their corresponding components in previous studies from the OpenFluor database; Table S2: $SUVA_{254}$ ($L \cdot mg^{-1} \cdot m^{-1}$) values of the batch effluents at different temperatures (20, 25, 30 °C) and biotic/abiotic conditions under oxic conditions; Table S3: Changes in the LC-OCD fractions concentration in batch reactors at different temperatures under oxic conditions using different water types (DC, DCWW, WW, and WEOM).

Author Contributions: Conceptualization, A.A and S.S.; methodology, A.A. and S.S. and M.K.; software, A.A. and A.S.; validation, A.A.; formal analysis, A.A.; investigation, A.A.; resources, S.S. and M.K.; data curation, A.A.; writing—original draft preparation, A.A.; review and editing, S.S., A.S., and M.K.; visualization, A.A.; supervision, S.S., A.S., and M.K.; project administration, M.K.; funding acquisition, A.A.

Funding: This work was financially supported by the Netherlands Fellowship Programme NFP.

Acknowledgments: We would like to acknowledge the support of M. Abushaban (IHE Delft Institute for Water Education, Delft, The Netherlands) in laboratory ATP measurements.

Conflicts of Interest: The authors declare that they have no conflict of interest.

References

1. Stahlschmidt, M.; Regnery, J.; Campbell, A.; Drewes, J. Application of 3D-fluorescence/PARAFAC to monitor the performance of managed aquifer recharge facilities. *J. Water Reuse Desalin.* **2015**, *6*, 249–263. [CrossRef]
2. Hiscock, K.M.; Grischek, T. Attenuation of groundwater pollution by bank filtration. *J. Hydrol.* **2002**, *266*, 139–144. [CrossRef]
3. Tufenkji, N.; Ryan, J.N.; Elimelech, M. The Promise of Bank Filtration. *Environ. Sci. Technol.* **2002**, *36*, 422A–428A. [CrossRef] [PubMed]
4. Bartak, R.; Grischek, T.; Ghodeif, K.; Wahaab, R. Shortcomings of the RBF Pilot Site in Dishna, Egypt. *J. Hydrol. Eng.* **2014**. [CrossRef]
5. Boving, T.B.; Choudri, B.S.; Cady, P.; Cording, A.; Patil, K.; Reddy, V. Hydraulic and Hydrogeochemical Characteristics of a Riverbank Filtration Site in Rural India. *Water Environ. Res.* **2014**, *86*, 636–648. [CrossRef] [PubMed]
6. Ray, C.; Melin, G.; Linsky, R.B. *Riverbank Filtratio: Improving Source-Water Quality*; Kluwer Academic Publishers: Dordrecht, The Netherlands; Boston, MA, USA; NWRI, National Water Research Institute: Fountain Valley, CA, USA, 2002.
7. McKnight, D.M.; Boyer, E.W.; Westerhoff, P.K.; Doran, P.T.; Kulbe, T.; Andersen, D.T. Spectrofluorometric characterization of dissolved organic matter for indication of precursor organic material and aromaticity. *Limnol. Oceanogr.* **2001**, *46*, 38–48. [CrossRef]
8. Nam, S.-N.; Amy, G. Differentiation of wastewater effluent organic matter (EfOM) from natural organic matter (NOM) using multiple analytical techniques. *Water Science Technol.* **2008**, *57*, 1009–1015. [CrossRef] [PubMed]
9. Zhang, B.; Xian, Q.; Lu, J.; Gong, T.; Li, A.; Feng, J. Evaluation of DBPs formation from SMPs exposed to chlorine, chloramine and ozone. *J. Water Health* **2016**, *15*, 185–195. [CrossRef] [PubMed]
10. Baghoth, S.A.; Sharma, S.K.; Amy, G.L. Tracking natural organic matter (NOM) in a drinking water treatment plant using fluorescence excitation-emission matrices and PARAFAC. *Water Res.* **2011**, *45*, 797–809. [CrossRef] [PubMed]
11. Ma, L.; Yates, S.R. Dissolved organic matter and estrogen interactions regulate estrogen removal in the aqueous environment: A review. *Sci. Total Environ.* **2018**, *640*, 529–542. [CrossRef] [PubMed]
12. Vega, M.A.; Kulkarni, H.V.; Mladenov, N.; Johannesson, K.; Hettiarachchi, G.M.; Bhattacharya, P.; Kumar, N.; Weeks, J.; Galkaduwa, M.; Datta, S. Biogeochemical Controls on the Release and Accumulation of Mn and As in Shallow Aquifers, West Bengal, India. *Front. Environ. Sci.* **2017**, *5*, 29. [CrossRef]
13. Chianese, S.; Iovino, P.; Leone, V.; Musmarra, D.; Prisciandaro, M. Photodegradation of Diclofenac Sodium Salt in Water Solution: Effect of HA, NO_3- and TiO_2 on Photolysis Performance. *Water Air Soil Pollut.* **2017**, *228*, 270. [CrossRef]
14. Shi, Y.; Huang, J.; Zeng, G.; Gu, Y.; Hu, Y.; Tang, B.; Zhou, J.; Yang, Y.; Shi, L. Evaluation of soluble microbial products (SMP) on membrane fouling in membrane bioreactors (MBRs) at the fractional and overall level: A review. *Rev. Environ. Sci. Bio/Technol.* **2018**, *17*, 71–85. [CrossRef]
15. Rostad, C.E.; Leenheer, J.A.; Katz, B.; Martin, B.S.; Noyes, T.I. Characterization and Disinfection By-Product Formation Potential of Natural Organic Matter in Surface and Ground Waters from Northern Florida. *ACS Symp. Ser.* **2000**, *761*, 154–172. [CrossRef]
16. Drewes, J.E.; Summers, R.S. Natural Organic Matter Removal During Riverbank Filtration: Current Knowledge and Research Needs. In *Riverbank Filtration: Improving Source-Water Quality*; Ray, C., Melin, G., Linsky, R.B., Eds.; Springer: Dordrecht, The Nerthands, 2003; pp. 303–309.
17. Gross-Wittke, A.; Gunkel, G.; Hoffmann, A. Temperature effects on bank filtration: Redox conditions and physical-chemical parameters of pore water at Lake Tegel, Berlin, Germany. *J. Water Clim. Chang.* **2010**, *1*, 55–66. [CrossRef]
18. Pan, W.; Huang, Q.; Huang, G. Nitrogen and Organics Removal during Riverbank Filtration along a Reclaimed Water Restored River in Beijing, China. *Water* **2018**, *10*, 491. [CrossRef]

19. Maeng, S.K.; Sharma, S.K.; Magic-Knezev, A.; Amy, G. Fate of effluent organic matter (EfOM) and natural organic matter (NOM) through riverbank filtration. *Water Sci. Technol.* **2008**, *57*, 1999–2007. [CrossRef] [PubMed]

20. Diem, S.; von Rohr, M.R.; Hering, J.G.; Kohler, H.-P.E.; Schirmer, M.; von Gunten, U. NOM degradation during river infiltration: Effects of the climate variables temperature and discharge. *Water Res.* **2013**, *47*, 6585–6595. [CrossRef] [PubMed]

21. Hoehn, E.; Scholtis, A. Exchange between a river and groundwater, assessed with hydrochemical data. *Hydrol. Earth Syst. Sci.* **2011**, *15*, 983–988. [CrossRef]

22. Derx, J.; Andreas, H.F.; Matthias, Z.; Liping, P.; Jack, S.; Blaschke, A.P. Evaluating the effect of temperature induced water viscosity and density fl uctuations on virus and DOC removal during river bank fi ltration—A scenario analysis. *River Syst.* **2012**, *20*, 169–183. [CrossRef]

23. Maeng, S.K.; Ameda, E.; Sharma, S.K.; Grützmacher, G.; Amy, G.L. Organic micropollutant removal from wastewater effluent-impacted drinking water sources during bank filtration and artificial recharge. *Water Res.* **2010**, *44*, 4003–4014. [CrossRef] [PubMed]

24. Ghodeif, K.; Grischek, T.; Bartak, R.; Wahaab, R.; Herlitzius, J. Potential of river bank filtration (RBF) in Egypt. *Environ. Earth Sci.* **2016**, *75*, 671. [CrossRef]

25. Sprenger, C.; Lorenzen, G.; Asaf, P. Environmental Tracer Application and Purification Capacity at a Riverbank Filtration Well in Delhi (India). *J. Indian Water Works Assoc.* **2012**, *1*, 25–32.

26. Misra, A.K. Climate change and challenges of water and food security. *Int. J. Sustain. Built Environ.* **2014**, *3*, 153–165. [CrossRef]

27. Guigue, J.; Mathieu, O.; Lévêque, J.; Mounier, S.; Laffont, R.; Maron, P.A.; Navarro, N.; Chateau, C.; Amiotte-Suchet, P.; Lucas, Y. A comparison of extraction procedures for water-extractable organic matter in soils. *Eur. J. Soil Sci.* **2014**, *65*, 520–530. [CrossRef]

28. Maeng, S.K.; Sharma, S.K.; Abel, C.; Magic-Knezev, A.; Amy, G.L. Role of biodegradation in the removal of pharmaceutically active compounds with different bulk organic matter characteristics through managed aquifer recharge: Batch and column studies. *Water Res.* **2011**, *45*, 4722–4736. [CrossRef] [PubMed]

29. Park, M.; Snyder, S.A. Sample handling and data processing for fluorescent excitation-emission matrix (EEM) of dissolved organic matter (DOM). *Chemosphere* **2018**, *193*, 530–537. [CrossRef] [PubMed]

30. Wang, L.K.; Chen, J.P.; Hung, Y.T.; Shammas, N.K. *Membrane and Desalination Technologies*; Humana Press: New York, NY, USA, 2008.

31. Abushaban, A.; Mangal, M.; Salinas-Rodriguez, S.G.; Nnebuo, C.; Mondal, S.; Goueli, S.; Schippers, J.; Kennedy, M. Direct Measurement of ATP in Seawater and Application of ATP to Monitor Bacterial Growth Potential in SWRO Pre-Treatment Systems. *Desalin. Water Treat.* **2017**, *99*, 91–101. [CrossRef]

32. Huber, S.A.; Balz, A.; Abert, M.; Pronk, W. Characterisation of aquatic humic and non-humic matter with size-exclusion chromatography-Organic carbon detection-Organic nitrogen detection (LC-OCD-OND). *Water Res.* **2011**, *45*, 879–885. [CrossRef] [PubMed]

33. Andersen, C.M.; Bro, R. Practical aspects of PARAFAC modeling of fluorescence excitation-emission data. *J. Chemometr.* **2003**, *17*, 200–215. [CrossRef]

34. Stedmon, C.A.; Markager, S.; Bro, R. Tracing dissolved organic matter in aquatic environments using a new approach to fluorescence spectroscopy. *Mar. Chem.* **2003**, *82*, 239–254. [CrossRef]

35. Murphy, K.R.; Stedmon, C.A.; Graeber, D.; Bro, R. Fluorescence spectroscopy and multi-way techniques. PARAFAC. *Anal. Methods* **2013**, *5*, 6557–6566. [CrossRef]

36. Colin, A.S.; Rasmus, B. Characterizing dissolved organic matter fluorescence with parallel factor analysis: A tutorial. *Limnol. Oceanogr. Methods* **2008**, *6*, 572–579. [CrossRef]

37. Fellman, J.B.; Hood, E.; Spencer, R.G. Fluorescence spectroscopy opens new windows into dissolved organic matter dynamics in freshwater ecosystems: A review. *Limnol. Oceanogr.* **2010**, *55*, 2452–2462. [CrossRef]

38. Cuss, C.W.; Guéguen, C.; Andersson, P.; Porcelli, D.; Maximov, T.; Kutscher, L. Advanced Residuals Analysis for Determining the Number of PARAFAC Components in Dissolved Organic Matter. *Appl. Spectrosc.* **2016**, *70*, 334–346. [CrossRef] [PubMed]

39. Li, W.-T.; Chen, S.-Y.; Xu, Z.-X.; Li, Y.; Shuang, C.-D.; Li, A.-M. Characterization of Dissolved Organic Matter in Municipal Wastewater Using Fluorescence PARAFAC Analysis and Chromatography Multi-Excitation/Emission Scan: A Comparative Study. *Environ. Sci. Technol.* **2014**, *48*, 2603–2609. [CrossRef] [PubMed]

40. Murphy, K.R.; Stedmon, C.A.; Wenig, P.; Bro, R. OpenFluor—An online spectral library of auto-fluorescence by organic compounds in the environment. *Anal. Methods* **2014**, *6*, 658–661. [CrossRef]

41. Shutova, Y.; Baker, A.; Bridgeman, J.; Henderson, R.K. Spectroscopic characterisation of dissolved organic matter changes in drinking water treatment: From PARAFAC analysis to online monitoring wavelengths. *Water Res.* **2014**, *54*, 159–169. [CrossRef] [PubMed]

42. Walker, S.A.; Amon, R.M.W.; Stedmon, C.; Duan, S.; Louchouarn, P. The use of PARAFAC modeling to trace terrestrial dissolved organic matter and fingerprint water masses in coastal Canadian Arctic surface waters. *J. Geophys. Res. Biogeosci.* **2009**, *114*. [CrossRef]

43. Osburn, C.L.; Handsel, L.T.; Peierls, B.L.; Paerl, H.W. Predicting Sources of Dissolved Organic Nitrogen to an Estuary from an Agro-Urban Coastal Watershed. *Environ. Sci. Technol.* **2016**, *50*, 8473–8484. [CrossRef] [PubMed]

44. Singh, S.; Inamdar, S.; Scott, D. Comparison of Two PARAFAC Models of Dissolved Organic Matter Fluorescence for a Mid-Atlantic Forested Watershed in the USA. *J. Ecosyst.* **2013**, *2013*, 16. [CrossRef]

45. Kulkarni, H.V.; Mladenov, N.; Johannesson, K.H.; Datta, S. Contrasting dissolved organic matter quality in groundwater in Holocene and Pleistocene aquifers and implications for influencing arsenic mobility. *Appl. Geochem.* **2017**, *77*, 194–205. [CrossRef]

46. Rehman, Z.; Jeong, S.; Tabatabai, S.; Emwas, A.; Leiknes, T. Advanced characterization of dissolved organic matter released by bloom-forming marine algae. *Desalin. Water Treat.* **2017**, *69*, 1–11. [CrossRef]

47. Li, D.; Sharp, J.O.; Saikaly, P.E.; Ali, S.; Alidina, M.; Alarawi, M.S.; Keller, S.; Hoppe-Jones, C.; Drewes, J.E. Dissolved Organic Carbon Influences Microbial Community Composition and Diversity in Managed Aquifer Recharge Systems. *Appl. Environ. Microbiol.* **2012**, *78*, 6819. [CrossRef] [PubMed]

48. Massmann, G.; Greskowiak, J.; Dünnbier, U.; Zuehlke, S.; Knappe, A.; Pekdeger, A. The impact of variable temperatures on the redox conditions and the behaviour of pharmaceutical residues during artificial recharge. *J. Hydrol.* **2006**, *328*, 141–156. [CrossRef]

49. Abel, C.; Sharma, S.K.; Malolo, Y.N.; Maeng, S.K.; Kennedy, M.D.; Amy, G.L. Attenuation of Bulk Organic Matter, Nutrients (N and P), and Pathogen Indicators During Soil Passage: Effect of Temperature and Redox Conditions in Simulated Soil Aquifer Treatment (SAT). *Water Air Soil Pollut.* **2012**, *223*, 5205–5220. [CrossRef]

50. Alidina, M.; Shewchuk, J.; Drewes, J.E. Effect of temperature on removal of trace organic chemicals in managed aquifer recharge systems. *Chemosphere* **2015**, *122*, 23–31. [CrossRef] [PubMed]

51. Chen, F.; Peldszus, S.; Elhadidy, A.M.; Legge, R.L.; Van Dyke, M.I.; Huck, P.M. Kinetics of natural organic matter (NOM) removal during drinking water biofiltration using different NOM characterization approaches. *Water Res.* **2016**, *104*, 361–370. [CrossRef] [PubMed]

52. Vasyukova, E.; Proft, R.; Uhl, W. Evaluation of dissolved organic matter fractions removal due to biodegradation. In *Progress in Slow Sand and Alternative Biofiltration Processes: Further Developments and Applications*; IWA Publishing: London, UK, 2014; pp. 59–66.

53. So, S.H.; Choi, I.H.; Kim, H.C.; Maeng, S.K. Seasonally related effects on natural organic matter characteristics from source to tap in Korea. *Sci. Total Environ.* **2017**, *592*, 584–592. [CrossRef] [PubMed]

54. Jørgensen, L.; Stedmon, C.A.; Kragh, T.; Markager, S.; Middelboe, M.; Søndergaard, M. Global trends in the fluorescence characteristics and distribution of marine dissolved organic matter. *Mar. Chem.* **2011**, *126*, 139–148. [CrossRef]

55. Yang, L.; Guo, W.; Hong, H.; Wang, G. Non-conservative behaviors of chromophoric dissolved organic matter in a turbid estuary: Roles of multiple biogeochemical processes. *Estuar. Coast. Shelf Sci.* **2013**, *133*, 285–292. [CrossRef]

56. Yang, L.; Shin, H.-S.; Hur, J. Estimating the Concentration and Biodegradability of Organic Matter in 22 Wastewater Treatment Plants Using Fluorescence Excitation Emission Matrices and Parallel Factor Analysis. *Sensors* **2014**, *14*, 1771–1786. [CrossRef] [PubMed]

57. Sun, H.Y.; Koal, P.; Gerl, G.; Schroll, R.; Joergensen, R.G.; Munch, J.C. Water-extractable organic matter and its fluorescence fractions in response to minimum tillage and organic farming in a Cambisol. *Chem. Biol. Technol. Agric.* **2017**, *4*, 15. [CrossRef]

58. Gimbel, R.; Graham, N.J.D.; Collins, M.R. *Recent Progress in Slow sand and Alternative Biofiltration Processes*; IWA Publishing: London, UK, 2006.

59. Wang, Y. *Assessment of Ozonation and Biofiltration as a Membrane Pre-Treatment at a Full-Scale Drinking Water Treatment Plant*; UWSpace: Waterloo, ON, Canada, 2014.

60. Gerlach, M.; Gimbel, R. Influence of humic substance alteration during soil passage on their treatment behaviour. *Water Sci. Technol.* **1999**, *40*, 231–239. [CrossRef]

![water logo] *water*

MDPI

Article

Trace Organic Removal during River Bank Filtration for Two Types of Sediment

Victoria Burke [1,*], Laura Schneider [1], Janek Greskowiak [1], Patricia Zerball-van Baar [2], Alexander Sperlich [2], Uwe Dünnbier [2] and Gudrun Massmann [1]

[1]   Working Group Hydrogeology and Landscape Hydrology, Department of Biology and Environmental Sciences, Carl von Ossietzky University of Oldenburg, D-26111 Oldenburg, Germany; laura.schn@web.de (L.S.); janek.greskowiak@uni-oldenburg.de (J.G.); gudrun.massmann@uni-oldenburg.de (G.M.)

[2]   Berliner Wasserbetriebe, 10864 Berlin, Germany; Patricia.Zerball-Vanbaar@bwb.de (P.Z.-v.B.); Alexander.Sperlich@bwb.de (A.S.); Uwe.Duennbier@bwb.de (U.D.)

*   Correspondence: victoria.burke@uni-oldenburg.de; Tel.: +49-441-7984683

Received: 26 October 2018; Accepted: 21 November 2018; Published: 26 November 2018

Abstract: The process of bank filtration acts as a barrier against many anthropogenic micropollutants, such as pharmaceuticals and industrial products, leading to a substantial improvement of groundwater quality. The performance of this barrier is, however, affected by seasonal influences and subject to significant temporal changes, which have already been described in the literature. Much less is known about spatial differences when considering one field site. In order to investigate this issue, two undisturbed cores from a well-investigated bank filtration field site were sampled and operated in the course of a column study. The ultimate aim was the identification and quantification of heterogeneities with regard to the biodegradation of 14 wastewater derived micropollutants, amongst others acesulfame, gabapentin, metoprolol, oxypurinol, candesartan, and olmesartan. While six of the compounds entirely persisted, eight compounds were prone to degradation. For those compounds that were subject to degradation, degradation rate constants ranged between 0.2 day^{-1} (gabapentin) and 31 day^{-1} (valsartan acid). Further, the rate constants consistently diverged between the distinct cores. In case of the gabapentin metabolite gabapentin-lactam, observed removal rate constants differed by a factor of six between the cores. Experimental data were compared to values calculated according to two structure based prediction models.

Keywords: redox sensitivity; micropollutants; oxypurinol; gabapentin

1. Introduction

The process of induced bank filtration—defined as the extraction of groundwater near or under a river or lake to induce infiltration from the surface water body [1]—results in recharge of the implicated aquifer on the one hand, but is predominantly intended for water quality improvement. The elimination of suspended solids, particles, bacteria, and viruses due to filtration and the removal of biodegradable compounds contribute to an effective natural attenuation of the bank filtrate [2]. The significance of bank filtration as a measure of managed aquifer recharge is clarified by the fraction of bank filtrate in drinking water supplies of individual countries in Europe summarized in Dillon et al. [3] ranging from 7% in the Netherlands, 9% in Germany, 25% in Switzerland, to 50% in Slovakia and Hungary.

Especially in densely populated areas, bank filtration is used within the concept of a partly closed water cycle. Given these conditions, the infiltrated surface water is to some extent wastewater influenced as it acts as receiving water for wastewater treatment plant effluents. Hence, wastewater treatment plant effluents represent point-sources for organic micropollutants, as the removal of these compounds is often incomplete during treatment [4]. Accordingly, a large number of studies previously reported on elevated concentrations of organic micropollutants, such as pharmaceuticals and personal care products, industrial agents, or artificial sweeteners, in treated wastewater (e.g., [5–8]). Besides potential negative impacts on the aquatic ecosystem [9], these compounds may also influence the groundwater quality and eventually the drinking water quality by entering the groundwater body via bank filtration.

A number of studies have shown that many wastewater-derived organic micropollutants are (at least to some extent) attenuated during the process of bank filtration (e.g., [10–13]). As proven in the course of field and lab studies, the transport- and degradation behavior of organic micropollutants during subsurface flow highly depends on the prevailing (hydrochemical) conditions along the flow path. Parameters identified to be of special importance in that context are temperature [14,15], pH value [16], redox conditions [13,17,18], sediment characteristics (i.e., organic carbon content [19]), and availability of primary substrate in the infiltrating water [20,21].

Some of the above-mentioned parameters are highly transient during bank filtration due to seasonal variations in the source water and also closely interrelated [11,22]. Accordingly, the removal efficiency during bank filtration undergoes temporal fluctuations, which was demonstrated for example by Greskowiak et al. [17].

Though the temporal variability was repeatedly proven, investigations on the spatial variation of the attenuation efficiency during bank filtration are sparse. It is evident, that removal rates notably differ between individual field sites. Take the analgesic diclofenac as an example: a compilation of biodegradation rate constants published by Greskowiak et al. [23] revealed rate constants observed at bank filtration sites ranging between 0.0025 day^{-1} and 17 day^{-1}. Apart from the knowledge of differences between individual locations, the question of heterogeneities at one field site is rather unclear. In order to fill this gap, the presented study covers column experiments simulating the process of bank filtration using two undisturbed cores, which were sampled from the same bank filtration site (Berlin Tegel, Berlin, Germany), but differed with regard to vegetation cover.

The study included a set of 14 human pharmaceuticals from different pharmacological classes discharged by a wastewater treatment plant nearby the sampling site and hence present in the respective source water also used for the experiments. Among them the anticonvulsants gabapentin, carbamazepine, primidone, and the respective transformation products gabapentin-lactam, 10,11-dihydro-10,11-dihydroxy-carbamazepine (DiOH-CBZ) and phenylethylmalonamide (PEMA). Another anticonvulsant present in source water and thus considered during this study was pregabalin.

Antihypertensive medications belong to the most prescribed therapeutic groups in human medicines, as hypertension is a serious public health problem [24] and one of the main risks leading to death in the world [25]. This study considered the antihypertensives metoprolol, candesartan and olmesartan. Whereas metoprolol is one of the classic beta-blockers, candesartan and olmesartan belong to a relatively new generation of antihypertension medications called angiotensin II receptor blockers (ARBs). The market launch of the first ARB (losartan) was 1995 and followed by highly increasing prescription rates [26,27]. So far, only a few studies dealt with the entry into the environment and the environmental behavior of these two sartans. Gurke et al. [28] reported on candesartan concentrations >1 µg L^{-1} in treated wastewater effluents. Bayer et al. [29] detected candesartan and olmesartan in wastewater influenced surface waters with maximum concentrations of 1.1 µg L^{-1} and 2.2 µg L^{-1}, respectively.

As transformation product evolving from different sartans (e.g., valsartan, olmesartan, candesartan) during wastewater treatment, valsartan acid was also examined during this study. It was previously shown to occur in elevated concentrations in wastewater treatment plant effluents, surface water, groundwater, and even in tap water [30–32]. Recently, its biodegradation in pilot-scale granular activated carbon filters for drinking water treatment was shown [33]. However, information published on the environmental behavior of valsartan acid is sparse.

The list of compounds is completed by the phenazone-type metabolite formylaminoantipyrine (FAA), the artificial sweetener acesulfame and oxypurinol, which is the active metabolite of the widely prescribed anti-gout agent allopurinol. Whereas the first-mentioned compounds were already subject of various studies [14,34–36], oxypurinol, in comparison, has been insufficiently studied. Funke et al. [37] detected oxypurinol in surface water samples in concentrations up to 23 µg L^{-1} and even in drinking water in concentrations up to 0.4 µg L^{-1} caused by the discharge of treated wastewater. However, oxypurinol can efficiently be removed by adsorption onto activated carbon [33].

The overall objectives of the study were to quantitatively characterize the degradation behavior of the investigated compounds during bank filtration and to identify and quantify spatial variations in degradation. A comparison of the observed data with two different quantitative structure-activity relationships (QSAR) approaches targets an assessment according to existing models.

2. Materials and Methods

2.1. Core Sampling

Two undisturbed sediment cores were collected at a bank filtration site in Berlin, Germany, where groundwater is abstracted by the local water supplier for drinking water production (Latitude: 52.575384, Longitude: 13.262819). The study site located at 'Lake Tegel' is well investigated and described due to former research activities focusing on the fate and transport of organic micropollutants during bank filtration [11,38,39]. In order to identify local heterogeneities in micropollutant removal related to different sediment properties, one sediment core was taken from a sandy section of the lakeshore (hereinafter referred to as core A) and another one from a reed-covered shore section (hereinafter referred to as core B). Accordingly, these cores showed considerable differences regarding the fraction of sedimentary organic carbon. Information on the sediment properties of the cores are to be found in Section 3.1.

During core drilling, tubes with an inner diameter of 0.08 m were pushed into the sediment by application of the vibrocorer technique. Subsequently, tube and sediment core were drawn using a combination of tripod and pulley. For a detailed description of the sampling procedure refer to Burke et al. [14]. Thus, sediment cores of a length of 1.02 m (core A) and 0.81 m (core B) were retrieved and incorporated into the experimental setup described in Section 2.2. In order to determine the sedimentary organic carbon fraction via loss of ignition (combustion of sample aliquots at 430 °C for 3 h), a second core was sampled in parallel by using a common inliner-system. This parallel core was also used for sieving in order to characterize the grain size distribution.

2.2. Experimental Setup

The experimental setup used within this study was adapted from Burke et al. [14]. Briefly, each column was equipped with oxygen probes and rhizome samplers in different depths, enabling periodic O_2 measurements and sampling in order to define the hydrochemical conditions along the flow path (Figure 1). The sample ports of core A were placed in infiltration depths of 0, 3, 5, 13, 23, 38, 51, 76, 96, and 102 cm. The sample ports of core B were installed at infiltration depth of 0, 3, 5, 14, 24, 38, 51, 74, and 81 cm.

Turned upside down, the columns were operated in an upward mode, thus preventing the entrapment of air and ensuring fully saturated conditions. Each column was connected to a peristaltic pump injecting surface water sampled from Lake Tegel as column influent solution in order to design the experimental conditions as site specific as possible. The surface water quality of Lake Tegel is influenced by treated wastewater, as it is part of a surface water system receiving wastewater treatment plant effluents [40], establishing a semi-closed urban water cycle. Thus, all compounds targeted during this study were present in the lake water and not spiked during the course of the experiment. Information on the composition of the column inlet, averaged according to the duration of the experiment, are listed in the Supplementary Material (Table S1).

Conservative tracer tests conducted with bromide revealed longitudinal dispersion coefficients, D_l, of 3.2×10^{-7} m^2 s^{-1} and 6.12×10^{-7} m^2 s^{-1} and mean pore water velocities, v, of 1.3×10^{-5} m s^{-1} and 1.2×10^{-5} m s^{-1} for cores A and core B, respectively. Consequently, hydraulic retention times of 22 h (core A) and 19 h (core B) were calculated. Temperature measurements throughout the study yielded an average of 21 °C.

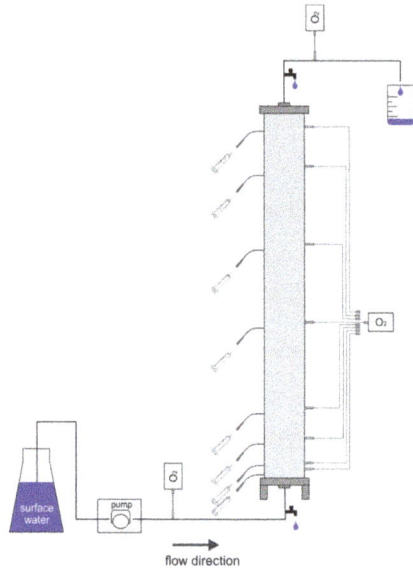

Figure 1. Experimental setup—adapted from Burke et al. [14].

2.3. Sampling and Sample Analysis

In the course of a conditioning period, the cores were operated for approximately six months in order to reach steady state conditions regarding the redox system. During this time, oxygen measurements and sample collection considering the entire depth profiles were performed once a week. Samples were analyzed for nitrate (NO_3^-), manganese (Mn^{2+}), iron (Fe^{2+}) and sulfate (SO_4^{2-}). Biweekly, additional samples for the determination of dissolved organic carbon (DOC) and pH value were abstracted at the in- and outlets of the columns.

The conditioning period was followed by the experimental period consisting of three individual sampling events with an interval of one week between the respective events. Apart from the aforementioned parameters, samples intended for trace organic analysis were abstracted from all sampling ports.

Samples meant for analysis of NO_3^- and SO_4^{2-} were filtered using 0.45 µm cellulose acetate filters (Sartorius Minisart®, Göttingen, Germany) and immediately analyzed by ion chromatography with a Basic IC plus (Metrohm, Filderstadt, Germany) according to DIN EN ISO 14911. Fe^{2+} and Mn^{2+} were determined photometrically using a compact photometer (PF-12, Macherey-Nagel, Düren, Germany). DOC samples were filtered using 0.7 µm glass fibre filters (Whatman, Maidstone, UK). Previously, sampling vessels and filters were combusted for five hours at 400 °C in order to remove possible residues of organic substances. Samples were acidified with hydrochlorid acid to a pH of 2 and stored dark and cool (~4 °C) until further processing. Analysis was carried out by application of high temperature combustion at a total organic carbon (TOC) analyzer (Shimadzu, Kyoto, Japan). For details on this method refer to Wurl [41].

Samples intended for trace organic analysis were filled into glass vessels and stored at −18 °C until further processing. Analysis was carried out in the laboratories of the Berliner Wasserbetriebe. Chromatographic separation was carried out on an ACQUITY UPLC HSS T3 column (100 mm × 2.1 mm; 1.8 μm) (Waters GmbH, Milford, MA, USA) using an ACQUITY Ultra Performance HPLC-system (Waters GmbH). The column oven was maintained at 40 °C. The mobile phase consisting of water (containing 0.05% acetic acid) and methanol with a linear gradient from 5% to 95% methanol in 8 min at a flow rate of 0.4 mL/min. The samples were kept at 5 °C in the autosampler. Samples were analyzed directly with an injection volume of 50 μL. The mass spectrometry was performed on a Xevo TQ-S mass spectrometer (Waters GmbH). Masslynx software was used for data acquisition and analysis (Waters GmbH).

Ionization was attained using an electrospray ionization (ESI) source in positive and negative ion mode. The ESI source was operated with a desolvation temperature of 600 °C. The desolvation gas and cone gas were 1000 L h^{-1} and 150 L h^{-1}, respectively. Nebulizer gas pressure was set by 6 bar, and the capillary voltage used was 3.5 V in positive mode and 2 V in negative mode. Target compounds were identified in the selected reaction monitoring (SRM) mode recording two transitions between precursor ions and the two most abundant product ions. Details on certain transitions, including cone voltages and collision energies applied, as well as retention times and limits of quantification (LOQs) are presented in the Supplementary Material (Table S2).

2.4. Data Evaluation

The concentration depth profiles presented within Section 3.2 display arithmetic mean values and standard deviations calculated from the individual experimental runs (n = 3). Values below the limit of quantification (LOQ) were set to LOQ/2. Degradation rate constants were fitted assuming 1st order degradation. After 6 months of adaption phase to the respective input concentrations, sorption was assumed to be negligible and disregarded.

In order to describe the prevailing redox conditions along the flow path, the classification scheme proposed by Regnery et al. [42] was slightly modified. Oxygen, nitrate, dissolved manganese and iron as well as sulfide served as redox indicators. Redox zones were classified as oxic, suboxic and anoxic, whereby anoxic conditions were further specified as nitrate reducing, iron/manganese reducing or sulfidic. The criteria used within this classification scheme are listed in Table 1.

Table 1. Classification scheme used for redox indication.

Redox Environment		Criteria
oxic		$DO > 1$ mg L^{-1}
suboxic		$0 < DO < 1$ mg L^{-1}
anoxic	nitrate reducing	$NO_3^- > 0$ mg L^{-1} and $Mn^{2+} < 0.05$ mg L^{-1} and $Fe^{2+} < 0.05$ mg L^{-1}
	mangnese/iron reducing	$Mn^{2+} > 0.05$ mg L^{-1} or/and $Fe^{2+} > 0.05$ mg L^{-1}
	sulfidic	$S^{2-} > 0$ mg L^{-1}

3. Results and Discussion

3.1. Sediment Properties and Hydrochemical Conditions

Both cores used in the course of this study consisted of sandy sediments, where fine sand is the dominating fraction (Figure 2). The fraction of medium to coarse sand tends to increase with depth along core A, with a maximum of 40% observed at the bottom of the core. With regard to core B, the fraction of medium to coarse sand varies between 6 and 39%. Except for a single part in the middle of core A (60–70 cm), where the silt fraction accounts 4%, the amount of silt is in general negligibly small (<1%).

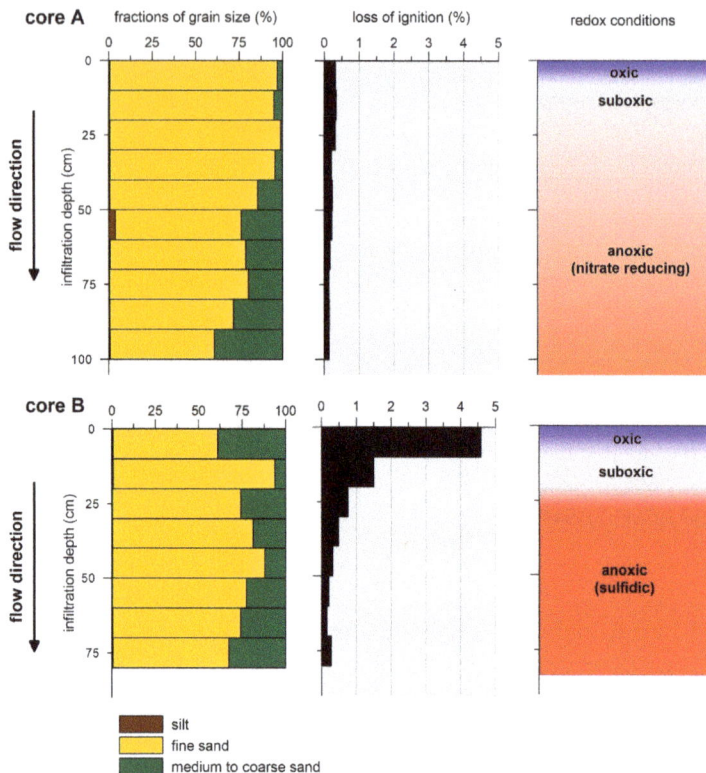

Figure 2. Information on sediment properties (fractions of grain sizes at the left; amount of particulate organic carbon (POC) in the middle) and hydrochemical conditions (at the right) observed along core A (upper part) and core B (lower part). With increasing infiltration depth (from top to bottom) the redox milieu changes from oxic (indicated by blue colours) to suboxic (indicated by grey colors) to anoxic (orange = manganese reducing; red = sulfidic).

The content of particulate organic carbon (POC) revealed differences between the two cores. While along core A nearly no variations in POC content over depth appeared, an elevated POC content was detected for the upper part of core B, where values decreased from 4.5% within the first 10 cm—which is 10 fold of the core A in the same depth—to 0.8% in a depth of 30 cm. The lower part of core B resembles the counterpart of core A.

For the description of the hydrochemical conditions along the flow path, we focused on pH, DOC concentration as well as on the redox relevant parameters described in Section 2.4, since these are strongly influencing the degradation behavior of organic micropollutants. The column influent solutions for both, core A and core B, were similar regarding pH value (8.2–8.3) and DOC concentration (7.2–7.3 mg L^{-1}). Along the flow path, pH values slightly decreased to 7.7 along core A and to 7.2 along core B. As it acts as primary source for the microbial community, also DOC concentrations dropped to 5.4 mg L^{-1} (core A) and 6.0 mg L^{-1} (core B).

Fundamental differences were observed with regard to the development of distinct redox zones. In both columns a thin oxic zone (~8–10 cm) at the inlet of the column was followed by a thin suboxic zone (~10–15 cm). At further depth, the main part of core A was characterized by nitrate reducing conditions, as NO_3^- concentrations decreased from 9.1 mg L^{-1} to 5.7 mg L^{-1} and neither Mn^{2+} nor Fe^{2+} were observed in notable concentrations. In contrast, core B originating from the reed-covered section became highly reducing at an infiltration depth of about 23 cm. This was indicated by increased S^{2-} concentrations of up to 0.4 mg L^{-1}.

3.2. Fate of Organic Micropollutants

3.2.1. Persistent Compounds

Six compounds including carbamazepine, candesartan, olmesartan, primidone, as well as the transformation products DiOH-CBZ and PEMA did not show attenuation when passing through the column, since concentrations measured in the column outlet were similar to those measured in the column inflow solution (see Figure S1 in the Supplementary Material). The fact that persistence prevailed in both columns is an indication for the general persistence of these compounds in the aquatic environment, apparently independent of the prevailing hydrochemical conditions and sediment composition.

In case of carbamazepine the environmental persistence has been frequently pointed out (e.g., [43–45]). The poor biodegradability is accompanied by a low tendency to sorb onto soils and sediment (e.g., [46,47]), pointing towards its high environmental relevance and ability to enter various aquatic compartments including groundwater. DiOH-CBZ, known as the predominant human metabolite evolving from carbamazepine and frequently detected in wastewater treatment plant effluents [48,49], likewise persisted along the flow path at concentrations around 0.5 µg L^{-1}.

In lab-scaled sewage treatment plants olmesartan and candesartan were shown to be rather poorly degradable with total elimination percentages <20% [29]. Similar to our results, Hellauer et al. [50] found candesartan and olmesartan to be biologically persistent in a column system simulating bank filtration conditions. In the same study, the process of ozonation led to an efficient removal of both compounds. Further, Khan and Nicell [51] expected candesartan to be highly mobile and persistent and suggested its prioritization for further studies.

Primidone and its metabolite PEMA were detected in rather small concentrations of 7 ng L^{-1} in the surface water used as column influent solution, and—independent of the prevailing redox conditions—no removal was observed along the flow path. The environmental persistence of primidone was already pointed out by other authors in the course of field studies (e.g., [11,52]) and lab studies (e.g., [21,53]). Although the number of studies on the environmental behavior of the primidone metabolite PEMA is small, its poor biodegradability has been revealed by Hass et al. [52] and Nham et al. [54].

3.2.2. Reactive Compounds

Due to high consumption rates, as well as largely unaffected passage through the human body and persistence during wastewater treatment [44,55], the artificial sweetener acesulfame is generally detected at elevated concentrations in treated wastewater [56,57] and therefore used as an indicator for wastewater influenced surface waters [58,59] and groundwater [60,61]. Presuming a low sorption affinity and recalcitrance to microbial degradation, acesulfame has formerly been proposed as an ideal anthropogenic marker [45,55,62]. However, other previous studies revealed that—under certain conditions—acesulfame is actually prone to microbial degradation [14,50,63]. Thereby, acesulfame degradation is largely affected by (i) temperature, (ii) redox conditions, and (iii) biodegradable carbon content [64]. This is supported by our findings, as acesulfame was attenuated under oxic conditions with degradation rate constants of 14.6 day^{-1} (core A) and 10.4 day^{-1} (core B), and under suboxic conditions ($\lambda = 2$ day^{-1}), but persisted under anoxic conditions (Figure 3a). According to Kahl et al. [64], first evidences questioning the recalcitrance of acesulfame came up in 2014. Based on their findings the same authors hypothesized, that acesulfame degrading species evolved during the last few years—for example, due to horizontal gene transfer.

The phenazone type metabolite FAA has formerly been stated to behave redox dependent [34,65]. This was also proven during this study, as fast degradation was recognized for the oxic zone while

FAA persisted under anoxic conditions (Figure 3b). Degradation rate constants observed under oxic conditions were 24.7 day^{-1} (core A) and 7.8 day^{-1} (core B), which are higher than those published elsewhere [23]. However, rate constants of similar magnitude (5.7 day^{-1} and 1.4 day^{-1}) have already been noticed under oxic conditions within sandy columns [14,66].

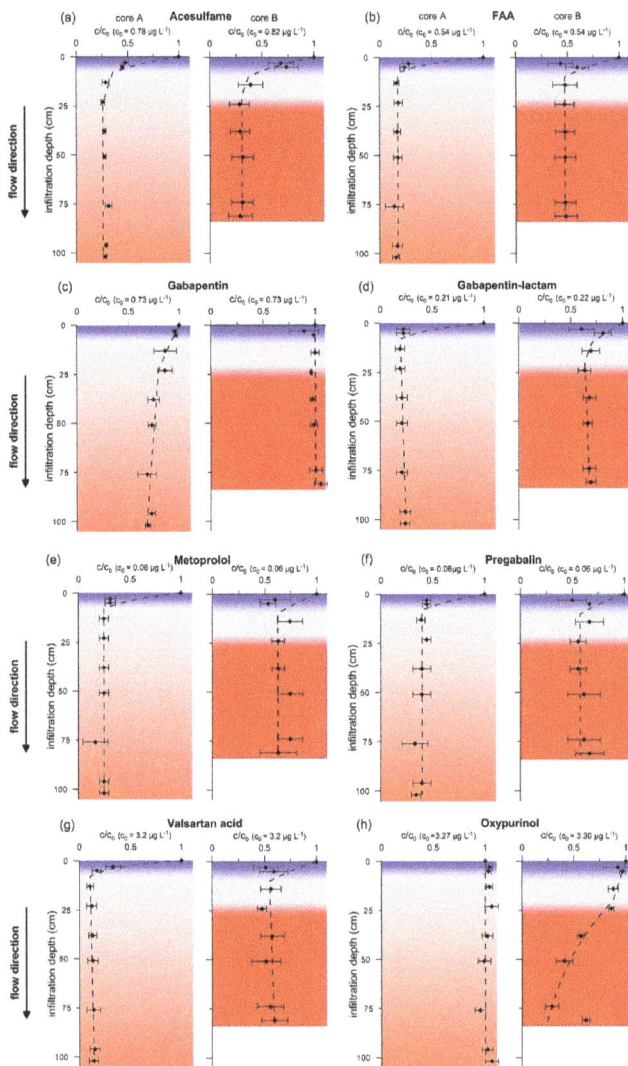

Figure 3. Concentration depth profiles (core A to the left, core B to the right) observed for (**a**) acesulfame, (**b**) FAA, (**c**) gabapentin, (**d**) gabapentin-lactam, (**e**) metoprolol, (**f**) pregabalin, (**g**) valsartan acid, and (**h**) oxypurinol. Black circles indicate arithmetic mean values of measured concentrations (n = 3) and standard deviations are given by error bars. Removal curves (dashed lines) were created assuming 1st order degradation. The colored background indicates the redox conditions prevalent in the respective part of the column as presented in Figure 2 (blue = oxic, grey = suboxic, orange/red = anoxic). The concentrations are presented here as c/c$_0$, i.e., normalized to the inlet concentration c$_0$ (shown in brackets).

A similar picture—characterized by an efficient removal within the upper oxic zone paired with persistence within the remaining suboxic to anoxic part of the column—emerges for metoprolol, pregabalin and valsartan acid (Figure 3e–g). Based on field data, Nödler et al. [31] suggested valsartan acid to behave persistent during bank filtration. However, evaluating the data for valsartan acid of this study yielded the highest degradation rate constant of all compounds (31.5 day^{-1}), corresponding to a half-life time of 0.5 h (core A). In accordance, Hellauer et al. [50] found valsartan acid to be efficiently attenuated after aeration of a test system, while persistence was noticed in an anoxic reference system.

With regard to metoprolol, for which degradation rate constants of 19.7 day^{-1} (core A) and 4.9 day^{-1} (core B) have been observed, our results concur with results reported by other authors, who found a strong redox dependency of metoprolol degradation by means of laboratory experiments [18,67], and also confirm our previous findings [66].

The anticonvulsant gabapentin behaved different along the distinct cores—while the concentration decreased along core A with degradation rate constants of 1.1 day^{-1} in the upper (oxic to suboxic) part and 0.2 day^{-1} in the lower (anoxic) part, persistence was observed along core B (Figure 3c). Hence, the availability of oxygen does not seem to be the controlling factor for gabapentin degradation. These findings differ from those reported by Henning et al. [68] and Hellauer et al. [50], who observed by means of batch and column experiments, respectively, a redox dependent degradation of gabapentin with enhanced removal under aerobic conditions. Since one major difference between both cores is the organic carbon content, which is higher in core B, low carbon contents may favor gabapentin removal. However, by investigating the influence of particular organic carbon on oxygen consumption and attenuation of organic trace compounds, Filter et al. [19] detected neither a distinct impact of the carbon content nor any correlation with the prevailing redox regime. Indications for biodegradation of gabapentin in GAC fixed-bed and tertiary filtration systems have also been found in pilot-scale studies on advanced water and wastewater treatment steps [33,69,70]. Further studies for clarification are needed here.

Gabapentin-lactam, the quantitatively most relevant transformation product evolving from Gabapentin [68], showed enhanced attenuation under oxic conditions whereas the concentration remained constant under reducing conditions (Figure 3d). The degradation rate constants obtained were 23.0 day^{-1} for core A and 3.6 day^{-1} for core B. These results share similarities with those of Henning et al. [68], who described also a redox dependent degradation of gabapentin-lactam, even though the reported degradation rate constants of 0.06 day^{-1} were clearly lower.

The attenuation pattern of oxypurinol appeared to be quite different. While persistence was noticed within core A, decreasing concentrations along core B were detected. The degradation rate constants ranged from 0.7 day^{-1} in the upper part to 2.2 day^{-1} in the lower part. Hence, oxypurinol was more efficiently attenuated under strongly reducing (sulfidic) conditions. By investigating its degradation during managed aquifer recharge, Hellauer et al. [71] found oxypurinol to be persistent during two meters of infiltration under oxic conditions.

By comparing the degradation rate constants observed during this study as shown in Figure 4, it becomes evident that highest removal rates mostly appeared under oxic conditions (blue bars). Four compounds, namely FAA, metoprolol, pregabalin and valsartan acid, where solely degraded under oxic conditions, while acesulfame and gabapentin-lactam also were prone to degradation under suboxic conditions. Deviating from that, the removal of oxypurinol and gabapentin seems not primarily to be controlled by the redox environment, as oxypurinol concentration decreased only along core B while gabapentin concentrations only decreased along core A.

Further, it is apparent from Figure 4 that degradation rate constants within the upper, oxic zone observed considering core A (blue, solid bars) are systematically higher than those observed in core B (blue, striped bars). Highest discrepancies appeared for gabapentin-lactam, for which the removal within the upper (oxic) part of core A was 6 times larger than in the upper (oxic) part of core B. By looking at the sediment characteristics, the main evident difference is the content of organic carbon determined by the loss of ignition, as it is by factor 10 higher in the upper zone of core B than of core A

(Figure 2). Thus, it seems reasonable that the organic carbon content of the sediment influenced the degradation processes. Consistently, Kahl et al. [64] suggested the removal of acesulfame to be most efficient when the availability of biodegradable organic carbon is low. However, during investigations the influence of a compost layer on the attenuation of organic micropollutants, Schaffer et al. [72] observed an enhanced degradation when levels of biodegradable dissolved organic carbon were higher. Since this relation remains unclear, further research needs to be done.

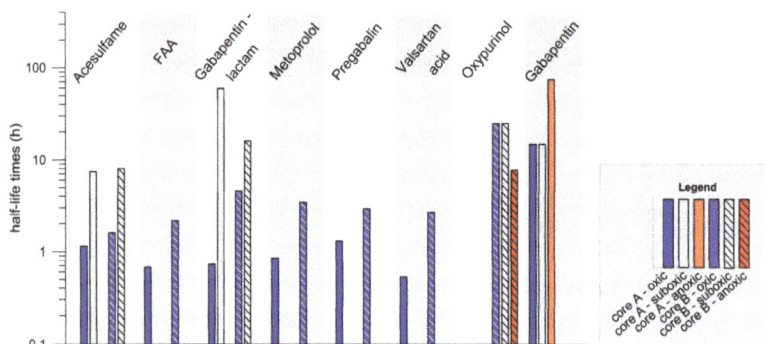

Figure 4. Compilation of observed half-life times (hours) along core A (solid filled bars) and as well as along core B (striped bars) under distinct redox conditions. FAA: formylaminoantipyrine.

3.3. Comparison with Recent QSAR Approaches

In order to classify the degradation rate constants observed during this study in relation to existing quantitative structure activity relation models, two QSAR models were picked and rates calculated accordingly and plotted against experimental findings (Figure 5). Firstly, the degradation expert survey BIOWIN 4 from the EPI Suite™ package (United States Environmental Protection Agency, Washington, DC, USA) [73] was used, which calculates time required for primary degradation, i.e., the change in molecular structure resulting in the formation of a new compound. Secondly, the predictive multi-linear regression model published by Bertelkamp et al. [74] was applied since it focusses especially on bank filtration data. Due to the fact that both models account for degradation under oxic conditions, only data derived from the upper, oxic layer of the columns were compared.

The BIOWIN 4 output indicates the degradation probability of a compound by means of half-life categories (hours, hours–days, days, days–weeks, weeks, weeks–months, months, and recalcitrant). Most of the target compounds investigated during this study, namely acesulfame, FAA, gabapentin-lactam, metoprolol and oxypurinol end up in the category 'days–weeks' in BIOWIN 4, which, after EPI Suite™, is converted to a degradation rate constant of 0.08 day^{-1}. Gabapentin is grouped in the category 'days', which is equivalent to a degradation rate constant of 0.29 day^{-1}. The comparison shows that the BIOWIN 4 half-lives are generally larger than those obtained experimentally, hence degradation is faster in the cores. This could be expected, since the data base used for this primary degradation survey primarily includes studies on well defined batch experiments in the absence of sediment. However, reactions at the water-sediment interface are of special importance considering physical and biological processes [75]. Additionally, it was repeatedly shown that the first decimeters of infiltration during bank filtration are highly reactive [14,76,77], leading to comparably high removal rates in this study.

Based on data from a column study with material from a riverbank filtration site and considering a set of 31 organic micropollutants, Bertelkamp et al. [74] derived a multi-linear regression model for predicting their degradation during bank filtration. Applying their model to compounds that behaved reactive during this study revealed degradation rate constants between 1.8 and 4.1 day^{-1}. Even though the calculated values are also lower than the experimental data, the deviations are obviously lower and data range in a similar order of magnitude. Hence, the model Bertelkamp et al. [74] calibrated for a specific field site was not fully able to predict the compounds behavior of the Berlin site considered in this study, but could have been used for first approximations. However, taking into account the variability in removal rates detected during this study between two cores from the same field site—one can roughly estimate the difficulties that still exist regarding the development of a holistic model aiming at predicting the degradation behavior of organic micropollutants in the aquatic environment.

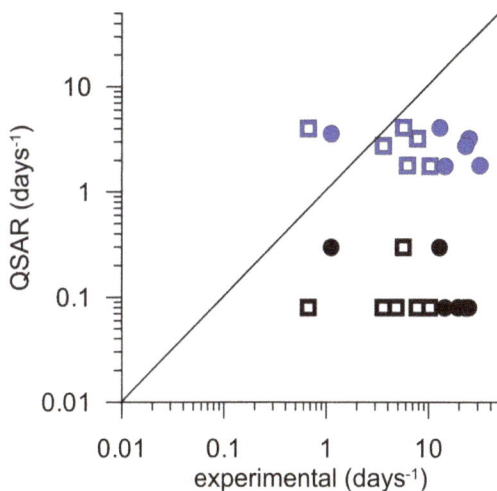

Figure 5. Comparison of data determined experimentally in core A (circles) and core B (rectangles) with data calculated according to Bertelkamp et al. ([74], blue) and the BIOWIN 4 model ([73], black).

4. Conclusions

The compounds acesulfame, FAA, gabapentin, gabapentin-lactam, metoprolol, pregabalin, valsartan acid, and oxypurinol are subject to primary degradation under certain conditions during bank filtration. Observed degradation rate constants ranged between 0.2 day^{-1} and 31.4 day^{-1}, resulting in half-life times between 3 days and 0.5 h, respectively. Compared to other investigations on the degradation behavior of organic micropollutants these rates are relatively high [23] and may tend to represent removal under most favorable conditions (i.e., in terms of oxygen availability, elevated temperatures, and the grade of nutrient availability). Moreover, our data documents the high reactivity of the initial part of the infiltration flow path, i.e., the lake base.

Hydrochemical conditions and removal patterns observed in the columns indicated notable differences between two cores from the same field site. Although the distance between the sampled cores was only a few meters, (completely) different redox conditions and concentration depth profiles of the organic trace compounds developed. The largest difference in compound removal was found for gabapentin-lactam, which was less degraded by a factor of six in one core compared to another. Since one major difference between the cores was the content of POC, we hypothesize that compound removal was more efficient where the carbon content was low. In order to confirm this assumption a systematic study on this issue is needed. It became obvious, that a robust characterization of a field site, which is intended to be used for process-oriented investigations, requires a closed network of sampling points.

The comparison of experimental data with two structure based prediction models elucidated that—due to the complex interaction of influencing parameters—it is still highly challenging to assess the biodegradation of organic micropollutants under environmental conditions based on compound specific properties. Moreover, reactions at the sediment-water interface are of central importance with regard to predictive models.

Supplementary Materials: The following are available online at http://www.mdpi.com/2073-4441/10/12/1736/s1, Table S1: Information on the water composition of the lake water used as column inlet (averaged over the duration of the experiment, n = 8), Table S2: Compilation of details describing the analytical method used for trace pollutant analysis (UHPLC-MSMS), Figure S1: Concentration depth profiles observed for compounds that behaved persistent during this study.

Author Contributions: Conceptualization, G.M., U.D., V.B., and J.G.; formal analysis, V.B. and P.Z.-v.B.; investigation, L.S. and V.B.; writing—original draft preparation, V.B. and P.Z.-v.B.; writing—review and editing, J.G., A.S., U.D. and G.M.; supervision, G.M.; project administration, V.B., G.M. and A.S.

Funding: This research was funded by the Berlin Water Company.

Acknowledgments: We would like to thank Britta Drude, Nina Gnuschke, Jessica Schnitger and Ulrike Kücks for practical assistance. Moreover the MPI research group 'Marine Geochemistry' is acknowledged for performing DOC analyses. We kindly acknowledge Helmo Nicolai from the Institute for Chemistry and Biology of the Marine Environment (ICBM) for helping us with the instrumentation used for core drilling.

Conflicts of Interest: The authors declare no conflict of interest.

References

1. Dillon, P. Future management of aquifer recharge. *Hydrogeol. J.* **2005**, *13*, 313–316. [CrossRef]
2. Hiscock, K.M.; Grischek, T. Attenuation of groundwater pollution by bank filtration. *J. Hydrol.* **2002**, *266*, 139–144. [CrossRef]
3. Dillon, P.; Stuyfzand, P.; Grischek, T.; Lluria, M.; Pyne, R.D.G.; Jain, R.C.; Bear, J.; Schwarz, J.; Wang, W.; Fernandez, E.; et al. Sixty years of global progress in managed aquifer recharge. *Hydrogeol. J.* **2018**. [CrossRef]
4. Luo, Y.; Guo, W.; Ngo, H.H.; Nghiem, L.D.; Hai, F.I.; Zhang, J.; Liang, S.; Wang, X.C. A review on the occurrence of micropollutants in the aquatic environment and their fate and removal during wastewater treatment. *Sci. Total Environ.* **2014**, *473–474*, 619–641. [CrossRef] [PubMed]
5. Vieno, N.; Tuhkanen, T.; Kronberg, L. Elimination of pharmaceuticals in sewage treatment plants in Finland. *Water Res.* **2007**, *41*, 1001–1012. [CrossRef] [PubMed]
6. Bartelt-Hunt, S.L.; Snow, D.D.; Damon, T.; Shockley, J.; Hoagland, K. The occurrence of illicit and therapeutic pharmaceuticals in wastewater effluent and surface waters in Nebraska. *Environ. Pollut.* **2009**, *157*, 786–791. [CrossRef] [PubMed]
7. Kasprzyk-Hordern, B.; Dinsdale, R.M.; Guwy, A.J. The removal of pharmaceuticals, personal care products, endocrine disruptors and illicit drugs during wastewater treatment and its impact on the quality of receiving waters. *Water Res.* **2009**, *43*, 363–380. [CrossRef] [PubMed]
8. Gros, M.; Petrovic, M.; Ginebreda, A.; Barcelo, D. Removal of pharmaceuticals during wastewater treatment and environmental risk assessment using hazard indexes. *Environ. Int.* **2010**, *36*, 15–26. [CrossRef] [PubMed]
9. Schwarzenbach, R.P.; Escher, B.I.; Fenner, K.; Hofstetter, T.B.; Johnson, C.A.; von Gunten, U.; Wehrli, B. The challenge of micropollutants in aquatic systems. *Science* **2006**, *313*, 1072–1077. [CrossRef] [PubMed]
10. Kovacevic, S.; Radisic, M.; Lausevic, M.; Dimkic, M. Occurrence and behavior of selected pharmaceuticals during riverbank filtration in The Republic of Serbia. *Environ. Sci. Pollut. Res.* **2017**, *24*, 2075–2088. [CrossRef]
11. Henzler, A.F.; Greskowiak, J.; Massmann, G. Modeling the fate of organic micropollutants during river bank filtration (Berlin, Germany). *J. Contam. Hydrol.* **2014**, *156*, 78–92. [CrossRef] [PubMed]
12. Hamann, E.; Stuyfzand, P.J.; Greskowiak, J.; Timmer, H.; Massmann, G. The fate of organic micropollutants during long-term/long-distance river bank filtration. *Sci. Total Environ.* **2016**, *545*, 629–640. [CrossRef] [PubMed]
13. Maeng, S.K.; Ameda, E.; Sharma, S.K.; Grutzmacher, G.; Amy, G.L. Organic micropollutant removal from wastewater effluent-impacted drinking water sources during bank filtration and artificial recharge. *Water Res.* **2010**, *44*, 4003–4014. [CrossRef] [PubMed]
14. Burke, V.; Greskowiak, J.; Asmuss, T.; Bremermann, R.; Taute, T.; Massmann, G. Temperature dependent redox zonation and attenuation of wastewater-derived organic micropollutants in the hyporheic zone. *Sci. Total Environ.* **2014**, *482*, 53–61. [CrossRef] [PubMed]

15. Trinh, T.; van den Akker, B.; Coleman, H.M.; Stuetz, R.M.; Drewes, J.E.; Le-Clech, P.; Khan, S.J. Seasonal variations in fate and removal of trace organic chemical contaminants while operating a full-scale membrane bioreactor. *Sci. Total Environ.* **2016**, *550*, 176–183. [CrossRef] [PubMed]
16. Gulde, R.; Helbling, D.E.; Scheidegger, A.; Fenner, K. pH-Dependent Biotransformation of Ionizable Organic Micropollutants in Activated Sludge. *Environ. Sci. Technol.* **2014**, *48*, 13760–13768. [CrossRef] [PubMed]
17. Greskowiak, J.; Prommer, H.; Massmann, G.; Nutzmann, G. Modeling seasonal redox dynamics and the corresponding fate of the pharmaceutical residue phenazone during artificial recharge of groundwater. *Environ. Sci. Technol.* **2006**, *40*, 6615–6621. [CrossRef]
18. Bertelkamp, C.; Verliefde, A.R.D.; Schoutteten, K.; Vanhaecke, L.; Bussche, J.; Singhal, N.; van der Hoek, J.P. The effect of redox conditions and adaptation time on organic micropollutant removal during river bank filtration: A laboratory-scale column study. *Sci. Total Environ.* **2016**, *544*, 309–318. [CrossRef]
19. Filter, J.; Jekel, M.; Ruhl, A.S. Impacts of Accumulated Particulate Organic Matter on Oxygen Consumption and Organic Micro-Pollutant Elimination in Bank Filtration and Soil Aquifer Treatment. *Water* **2017**, *9*, 12. [CrossRef]
20. Li, D.; Alidina, M.; Drewes, J.E. Role of primary substrate composition on microbial community structure and function and trace organic chemical attenuation in managed aquifer recharge systems. *Appl. Microbiol. Biotechnol.* **2014**, *98*, 5747–5756. [CrossRef]
21. Alidina, M.; Li, D.; Ouf, M.; Drewes, J.E. Role of primary substrate composition and concentration on attenuation of trace organic chemicals in managed aquifer recharge systems. *J. Environ. Manag.* **2014**, *144*, 58–66. [CrossRef]
22. Massmann, G.; Greskowiak, J.; Duennbier, U.; Zuehlke, S.; Knappe, A.; Pekdeger, A. The impact of variable temperatures on the redox conditions and the behaviour of pharmaceutical residues during artificial recharge. *J. Hydrol.* **2006**, *328*, 141–156. [CrossRef]
23. Greskowiak, J.; Hamann, E.; Burke, V.; Massmann, G. The uncertainty of biodegradation rate constants of emerging organic compounds in soil and groundwater—A compilation of literature values for 82 substances. *Water Res.* **2017**, *126*, 122–133. [CrossRef]
24. Whelton, P.K.; He, J.; Appel, L.J.; Cutler, J.A.; Havas, S.; Kotchen, T.A.; Roccella, E.J.; Stout, R.; Vallbona, C.; Winston, M.C.; et al. Primary prevention of hypertension: Clinical and public health advisory from The National High Blood Pressure Education Program. *JAMA* **2002**, *288*, 1882–1888. [CrossRef]
25. Murray, C.J.L.; Lopez, A.D. Evidence-based health policy–Lessons from the global burden of disease study. *Science* **1996**, *274*, 740–743. [CrossRef]
26. Stafford, R.S.; Monti, V.; Furberg, C.D.; Ma, J. Long-term and short-term changes in antihypertensive prescribing by office-based physicians in the United States. *Hypertension* **2006**, *48*, 213–218. [CrossRef]
27. Campbell, N.R.C.; McAlister, F.A.; Brant, R.; Levine, M.; Drouin, D.; Feldman, R.; Herman, R.; Zarnke, K.; Canadian Hypertension Eval, P. Temporal trends in anti hypertensive drug prescriptions in Canada before and after introduction of the Canadian Hypertension Education Program. *J. Hypertens.* **2003**, *21*, 1591–1597. [CrossRef]
28. Gurke, R.; Rossmann, J.; Schubert, S.; Sandmann, T.; Rossler, M.; Oertel, R.; Fauler, J. Development of a SPE-HPLC-MS/MS method for the determination of most prescribed pharmaceuticals and related metabolites in urban sewage samples. *J. Chromatogr. B* **2015**, *990*, 23–30. [CrossRef]
29. Bayer, A.; Asner, R.; Schussler, W.; Kopf, W.; Weiss, K.; Sengl, M.; Letzel, M. Behavior of sartans (antihypertensive drugs) in wastewater treatment plants, their occurrence and risk for the aquatic environment. *Environ. Sci Pollut. Res.* **2014**, *21*, 10830–10839. [CrossRef]
30. Letzel, T.; Bayer, A.; Schulz, W.; Heermann, A.; Lucke, T.; Greco, G.; Grosse, S.; Schüssler, W.; Sengl, M.; Letzel, M. LC-MS screening techniques for wastewater analysis and analytical data handling strategies: Sartans and their transformation products as an example. *Chemosphere* **2015**, *137*, 198–206. [CrossRef]
31. Nödler, K.; Hillebrand, O.; Idzik, K.; Strathmann, M.; Schiperski, F.; Zirlewagen, J.; Licha, T. Occurrence and fate of the angiotensin II receptor antagonist transformation product valsartan acid in the water cycle—A comparative study with selected beta-blockers and the persistent anthropogenic wastewater indicators carbamazepine and acesulfame. *Water Res.* **2013**, *47*, 6650–6659. [CrossRef]
32. Hermes, N.; Jewell, K.S.; Wick, A.; Ternes, T.A. Quantification of more than 150 micropollutants including transformation products in aqueous samples by liquid chromatography-tandem mass spectrometry using scheduled multiple reaction monitoring. *J. Chromatogr. A* **2018**, *1531*, 64–73. [CrossRef]

33. Sperlich, A.; Harder, M.; Zietzschmann, F.; Gnirss, R. Fate of trace organic compounds in Granular Activated Carbon (GAC) adsorbers for drinking water treatment. *Water* **2017**, *9*. [CrossRef]
34. Massmann, G.; Duennbier, U.; Heberer, T.; Taute, T. Behaviour and redox sensitivity of pharmaceutical residues during bank filtration—Investigation of residues of phenazone-type analgesics. *Chemosphere* **2008**, *71*, 1476–1485. [CrossRef]
35. Bichler, A.; Muellegger, C.; Brünjes, R.; Hofmann, T. Quantification of river water infiltration in shallow aquifers using acesulfame and anthropogenic gadolinium. *Hydrol. Process.* **2015**. [CrossRef]
36. Storck, F.R.; Skark, C.; Remmler, F.; Brauch, H.J. Environmental fate and behavior of acesulfame in laboratory experiments. *Water Sci. Technol.* **2016**, *74*, 2832–2842. [CrossRef]
37. Funke, J.; Prasse, C.; Eversloh, C.L.; Ternes, T.A. Oxypurinol—A novel marker for wastewater contamination of the aquatic environment. *Water Res.* **2015**, *74*, 257–265. [CrossRef]
38. Heberer, T.; Mechlinski, A.; Fanck, B.; Knappe, A.; Massmann, G.; Pekdeger, A.; Fritz, B. Field studies on the fate and transport of pharmaceutical residues in bank filtration. *Ground Water Monit. Remediat.* **2004**, *24*, 70–77. [CrossRef]
39. Massmann, G.; Sueltenfuss, J.; Duennbier, U.; Knappe, A.; Taute, T.; Pekdeger, A. Investigation of groundwater a residence times during bank filtration in Berlin: Multi-tracer approach. *Hydrol. Process.* **2008**, *22*, 788–801. [CrossRef]
40. Schimmelpfennig, S.; Kirillin, G.; Engelhardt, C.; Nutzmann, G.; Dunnbier, U. Seeking a compromise between pharmaceutical pollution and phosphorus load: Management strategies for Lake Tegel, Berlin. *Water Res.* **2012**, *46*, 4153–4163. [CrossRef]
41. Wurl, O. *Practical Guidelines for the Analysis of Seawater*; CRC Press: Boca Raton, FL, USA; Taylor & Francis Group: Abingdon-on-Thames, UK, 2009.
42. Regnery, J.; Wing, A.D.; Alidina, M.; Drewes, J.E. Biotransformation of trace organic chemicals during groundwater recharge: How useful are first-order rate constants? *J. Contam. Hydrol.* **2015**, *179*, 65–75. [CrossRef]
43. Fenz, R.; Blaschke, A.P.; Clara, M.; Kroiss, H.; Mascher, D.; Zessner, M. Monitoring of carbamazepine concentrations in wastewater and groundwater to quantify sewer leakage. *Water Sci. Technol.* **2005**, *52*, 205–213. [CrossRef]
44. Scheurer, M.; Storck, F.R.; Graf, C.; Brauch, H.J.; Ruck, W.; Lev, O.; Lange, F.T. Correlation of six anthropogenic markers in wastewater, surface water, bank filtrate, and soil aquifer treatment. *J. Environ. Monit.* **2011**, *13*, 966–973. [CrossRef]
45. Foolad, M.; Ong, S.L.; Hu, J. Transport of sewage molecular markers through saturated soil column and effect of easily biodegradable primary substrate on their removal. *Chemosphere* **2015**, *138*, 553–559. [CrossRef]
46. Calisto, V.; Esteves, V.I. Adsorption of the antiepileptic carbamazepine onto agricultural soils. *J. Environ. Monit.* **2012**, *14*, 1597–1603. [CrossRef]
47. Yu, L.; Fink, G.; Wintgens, T.; Melina, T.; Ternes, T.A. Sorption behavior of potential organic wastewater indicators with soils. *Water Res.* **2009**, *43*, 951–960. [CrossRef]
48. Kaiser, E.; Prasse, C.; Wagner, M.; Broder, K.; Ternes, T.A. Transformation of oxcarbazepine and human metabolites of carbamazepine and oxcarbazepine in wastewater treatment and sand filters. *Environ. Sci. Technol.* **2014**, *48*, 10208–10216. [CrossRef]
49. Brezina, E.; Prasse, C.; Meyer, J.; Mückter, H.; Ternes, T.A. Investigation and risk evaluation of the occurrence of carbamazepine, oxcarbazepine, their human metabolites and transformation products in the urban water cycle. *Environ. Pollut.* **2017**, *225*, 261–269. [CrossRef]
50. Hellauer, K.; Mergel, D.; Ruhl, A.S.; Filter, J.; Hubner, U.; Jekel, M.; Drewes, J.E. Advancing sequential managed aquifer recharge technology (SMART) using different intermediate oxidation processes. *Water* **2017**, *9*, 14. [CrossRef]
51. Khan, U.; Nicell, J. Human health relevance of pharmaceutically active compounds in drinking water. *AAPS J.* **2015**, *17*, 558–585. [CrossRef]
52. Hass, U.; Duennbier, U.; Massmann, G. Occurrence and distribution of psychoactive compounds and their metabolites in the urban water cycle of Berlin (Germany). *Water Res.* **2012**, *46*, 6013–6022. [CrossRef]
53. Lin, K.D.; Bondarenko, S.; Gan, J. Sorption and persistence of wastewater-borne psychoactive and antilipidemic drugs in soils. *J. Soils Sediments* **2011**, *11*, 1363–1372. [CrossRef]
54. Nham, H.T.T.; Greskowiak, J.; Hamann, E.; Meffe, R.; Hass, U.; Massmann, G. Long-term transport behavior of psychoactive compounds in sewage-affected groundwater. *Grundwasser* **2016**, *21*, 321–332. [CrossRef]

55. Buerge, I.J.; Buser, H.R.; Kahle, M.; Muller, M.D.; Poiger, T. Ubiquitous occurrence of the artificial sweetener acesulfame in the aquatic environment: An ideal chemical marker of domestic wastewater in groundwater. *Environ. Sci. Technol.* **2009**, *43*, 4381–4385. [CrossRef]

56. Scheurer, M.; Brauch, H.-J.; Lange, F.T. Analysis and occurrence of seven artificial sweeteners in German waste water and surface water and in soil aquifer treatment (SAT). *Anal. Bioanal. Chem.* **2009**, *394*, 1585–1594. [CrossRef]

57. Li, S.L.; Ren, Y.H.; Fu, Y.Y.; Gao, X.S.; Jiang, C.; Wu, G.; Ren, H.Q.; Geng, J.J. Fate of artificial sweeteners through wastewater treatment plants and water treatment processes. *PLoS ONE* **2018**, *13*, e16. [CrossRef]

58. Li, Z.; Sobek, A.; Radke, M. fate of pharmaceuticals and their transformation products in four small European rivers receiving treated wastewater. *Environ. Sci. Technol.* **2016**, *50*, 5614–5621. [CrossRef]

59. Müller, C.E.; Gerecke, A.C.; Alder, A.C.; Scheringer, M.; Hungerbühler, K. Identification of perfluoroalkyl acid sources in Swiss surface waters with the help of the artificial sweetener acesulfame. *Environ. Pollut.* **2011**, *159*, 1419–1426. [CrossRef]

60. Yang, Y.Y.; Zhao, J.L.; Liu, Y.S.; Liu, W.R.; Zhang, Q.Q.; Yao, L.; Hu, L.X.; Zhang, J.N.; Jiang, Y.X.; Ying, G.G. Pharmaceuticals and personal care products (PPCPs) and artificial sweeteners (ASs) in surface and ground waters and their application as indication of wastewater contamination. *Sci. Total Environ.* **2018**, *616*, 816–823. [CrossRef]

61. Spoelstra, J.; Senger, N.D.; Schiff, S.L. Artificial sweeteners reveal septic system effluent in rural groundwater. *J. Environ. Qual.* **2017**, *46*, 1434–1443. [CrossRef]

62. Robertson, W.D.; Van Stempvoort, D.R.; Solomon, D.K.; Homewood, J.; Brown, S.J.; Spoelstra, J.; Schiff, S.L. Persistence of artificial sweeteners in a 15-year-old septic system plume. *J. Hydrol.* **2013**, *477*, 43–54. [CrossRef]

63. Castronovo, S.; Wick, A.; Scheurer, M.; Nodler, K.; Schulz, M.; Ternes, T.A. Biodegradation of the artificial sweetener acesulfame in biological wastewater treatment and sandfilters. *Water Res.* **2017**, *110*, 342–353. [CrossRef]

64. Kahl, S.; Kleinsteuber, S.; Nivala, J.; van Afferden, M.; Reemtsma, T. Emerging biodegradation of the previously persistent Artificial Sweetener Acesulfame in biological wastewater treatment. *Environ. Sci. Technol.* **2018**, *52*, 2717–2725. [CrossRef]

65. Burke, V.; Richter, D.; Hass, U.; Duennbier, U.; Greskowiak, J.; Massmann, G. Redox-dependent removal of 27 organic trace pollutants: Compilation of results from tank aeration experiments. *Environ. Earth Sci.* **2014**, *71*, 3685–3695. [CrossRef]

66. Burke, V.; Greskowiak, J.; Grunenbaum, N.; Massmann, G. Redox and temperature dependent attenuation of twenty organic micropollutants—A systematic column study. *Water Environ. Res.* **2017**, *89*, 155–167. [CrossRef]

67. de Wilt, A.; He, Y.; Sutton, N.; Langenhoff, A.; Rijnaarts, H. Sorption and biodegradation of six pharmaceutically active compounds under four different redox conditions. *Chemosphere* **2018**, *193*, 811–819. [CrossRef]

68. Henning, N.; Kunkel, U.; Wick, A.; Ternes, T.A. Biotransformation of gabapentin in surface water matrices under different redox conditions and the occurrence of one major TP in the aquatic environment. *Water Res.* **2018**, *137*, 290–300. [CrossRef]

69. Altmann, J.; Rehfeld, D.; Trader, K.; Sperlich, A.; Jekel, M. Combination of granular activated carbon adsorption and deep-bed filtration as a single advanced wastewater treatment step for organic micropollutant and phosphorus removal. *Water Res.* **2016**, *92*, 131–139. [CrossRef]

70. Altmann, J.; Sperlich, A.; Jekel, M. Integrating organic micropollutant removal into tertiary filtration: Combining PAC adsorption with advanced phosphorus removal. *Water Res.* **2015**, *84*, 58–65. [CrossRef]

71. Hellauer, K.; Karakurt, S.; Sperlich, A.; Burke, V.; Massmann, G.; Hubner, U.; Drewes, J.E. Establishing sequential managed aquifer recharge technology (SMART) for enhanced removal of trace organic chemicals: Experiences from field studies in Berlin, Germany. *J. Hydrol.* **2018**, *563*, 1161–1168. [CrossRef]

72. Schaffer, M.; Kroger, K.F.; Nodler, K.; Ayora, C.; Carrera, J.; Hernandez, M.; Licha, T. Influence of a compost layer on the attenuation of 28 selected organic micropollutants under realistic soil aquifer treatment conditions: Insights from a large scale column experiment. *Water Res.* **2015**, *74*, 110–121. [CrossRef]

73. United States Environmental Protection Agency. *Estimation Programs Interface Suite™ for Microsoft® Windows, v 4.1*; United States Environmental Protection Agency: Washington, DC, USA, 2012.

74. Bertelkamp, C.; Verliefde, A.; Reynisson, J.; Singhal, N.; Cabo, A.; De Jonge, M.; van der Hoek, J.P. A predictive multi-linear regression model for organic micropollutants, based on a laboratory-scale column study simulating the river bank filtration process. *J. Hazard. Mater.* **2016**, *304*, 502–511. [CrossRef]

75. Santschi, P.; Hohener, P.; Benoit, G.; Buchholtztenbrink, M. Chemical processes at the sediment water interface. *Mar. Chem* **1990**, *30*, 269–315. [CrossRef]

76. Heberer, T.; Massmann, G.; Fanck, B.; Taute, T.; Dunnbier, U. Behaviour and redox sensitivity of antimicrobial residues during bank filtration. *Chemosphere* **2008**, *73*, 451–460. [CrossRef]

77. Jüttner, F. Efficacy of bank filtration for the removal of fragrance compounds and aromatic hydrocarbons. *Water Sci. Technol.* **1999**, *40*, 123–128. [CrossRef]

![water logo] *water*

MDPI

Article

Design and Optimization of a Fully-Penetrating Riverbank Filtration Well Scheme at a Fully-Penetrating River Based on Analytical Methods

Ya Jiang [1,2], Junjun Zhang [3], Yaguang Zhu [1], Qingqing Du [1], Yanguo Teng [1] and Yuanzheng Zhai [1,*]

[1] Engineering Research Center for Groundwater Pollution Control and Remediation of Ministry of Education of China, College of Water Sciences, Beijing Normal University, Beijing 100875, China; 15166587857@163.com (Y.J.); waterzyg@163.com (Y.Z.); 201621470033@mail.bnu.du.cn (Q.D.); teng1974@163.com (Y.T.)

[2] College of Water Conservancy and Civil Engineering, Shandong Agricultural University, Tai'an 271018, China

[3] Guangdong Geological Bureau, Guangzhou 510080, China; water_zjj@163.com

* Correspondence: diszyz@163.com; Tel.: +86-151-2009-8909

Received: 27 January 2019; Accepted: 21 February 2019; Published: 26 February 2019

Abstract: In order to maintain the sustainable development of pumping wells in riverbank filtration (RBF) and simultaneously minimize the possible negative effects induced, it is vital to design and subsequently optimize the engineering parameters scientifically. An optimizing method named Five-Step Optimizing Method was established by using analytic methods (Mirror-Image Method, Dupuit Equation and the Interference Well Group Method, etc.) systematically in this study considering both the maximum allowable drawdown of the groundwater level and the water demand as the constraint conditions, followed by a case study along the Songhua River of northeast China. It contained three parameters (number of wells, distance between wells, and distance between well and river) for optimizing in the method, in which the well type, depth and radius were beforehand designed and fixed, without the need of optimizing. The interference between wells was found to be a decisive factor that significantly impacts the optimizing effort of all the three parameters. The distance between the well and the river was another decisive factor impacting the recharge from the river and subsequently, the well water yield. There would be more than one optional scheme sometimes in the optimized result, while it's not yet difficult in practice to single out the optimal one considering both the field setting and the water demand. The established method proved to be applicable in the case study.

Keywords: riverbank filtration; riverside water source; analytical method; mirror-image method; optimization

---

1. Introduction

Riverside water source (RWS) refers to the water source where the wells are arranged close to the riverbank and mainly recharged by the adjacent river water through riverbank filtration (RBF) [1]. As a very important method in the development and utilization of water resources, the RWS has been widely valued and applied worldwide for its advantage of water pre-treatment and the regulatory capacity of water quantity [2]. In order to give full play to the advantages such as more sufficient and stable water supply, better water quality and more beneficial to centralized exploitation, etc., it is the most critical step for RWS to design the well group layout and the exploitation plan scientifically [3]. Different from the general surface water source and the general groundwater source, a set of parameters of

RWS should be considered systematically and skillfully, such as the hydrological and hydrogeological conditions, surface water and groundwater quality, the structure of the RBF and its physical and chemical properties, distance between wells, distance between well and river, well depth, location and length of the filter pipe, allowable drawdown of groundwater level and water yield [4]. Moreover, some parameters impact each other. Thus, the design of RWS is extraordinarily complex and difficult. In past decades, most of the studies on RWS and RBF focused on the surface water, groundwater interaction, pollutant migration and transformation, RBF clogging, numerical simulation, and an evaluation of the water resource, etc., while studies on the well layout optimizing are relatively insufficient considering its importance. How to design and optimize the layout of the pumping wells, and the exploitation schemes in RBF has become an urgent problem to be studied.

The well group layout and the exploitation schemes of the RWS play a decisive role in water yield, water quality and the impact on the geologic environment [5,6]. In addition to the surface water and groundwater level, the influence factors of the hydrodynamic process and water yield of the RWS include the integrity of the river (whether the river is disjointed is also included) [7], the topography [8] and silting [9] of the riverbed, the permeability of the riverbed and aquifer [10], and the river crossing seepage (partial penetrating river) [11], etc. Therefore, the above factors should be fully considered in the design of RWS. Moreover, the number of wells, well depth, distance between wells, and distance between well and river should be optimized in combination with the water demand and the allowable drawdown of the groundwater level [12]. At present, the study methods of RWS mainly include analytical methods [13] and numerical simulation methods [14]. As a means of obtaining hydrogeological parameters and verifying results, pumping tests are useful [15], and tracer tests have also been widely used [16].

The study on the RWS could be carried out in a number of ways, and the recharge rate of infiltration captured from the river water was usually determined through productive experiments in the early stage [17], which could provide a basis for the determination of the well location. Later, the iterative moving subdomain method [18] and fuzzy comprehensive evaluation model [19] based on the basic theory of fuzzy mathematics; were introduced to optimize the layout of pumping wells. The distance between well and river and the distance between wells could be determined by using the phreatic well equation of linear-arranged interferential well group [20], and the distance between well and river value could also be furtherly minimized by coupling riverbank filtration and reverse osmosis [21]. The optimal water yield can be determined by using the nonlinear optimizing method, evolutionary algorithm [22] and numerical simulation method using the Visual MODFLOW software. In addition, the sustainable water yield can be calculated by using the analytical method [23] while the optimization study of the water yield can be carried out by using analytical method [24], multi-objective optimizing model [25], and modelling method [26]. A large number of study cases showed that the distance between the well and river influenced the water yield profoundly. It could also pose an impact on the recharge from the river [27]. However, with a certain exploitation amount, the capture zone is less affected by the distance between well and river [28]. The construction and operation cost of an RBF scheme is also a factor impacting the selection of the types of pumping wells [29]. In addition, in terms of the study of the effectiveness of the RBF as a pre-treatment means of water, the methods of investigation and study are constantly being innovated and upgraded, and the joint application of multiple means is increasingly emphasized [30]. Although a lot of studies have been carried out on the design and optimization of well groups of RWS, critical issues that need to be coped with still exist. For example, it is worth further exploration to establish a popularizing method of optimizing the well layout from the perspective of river-groundwater dynamics.

In this study, both the maximum allowable drawdown of the groundwater level and the water demand are taken as the constraint conditions. On this basis, various well group layouts and exploitation schemes are firstly formed by combining some parameters with the well type, the number of wells, the distance between wells and the distance between well and river. Secondly, the interactions among engineering parameters are explored and the values of those parameters are compared and screened step-by-step, so as to form a modularized optimizing method for the well group layout and the exploitation plan of an RBF scheme. During this process, the sustainable water yield would be calculated by the analytical method. Further, a case study is carried out by using the established method and subsequently discussed.

2. Scenarios and Methods

2.1. Scenarios

In this study, the optimizing method was discussed by taking an RBF scheme with a condition of river fully penetrating the phreatic aquifer in the vertical dimension [31] as an example. The necessary parameters characterizing the geological and hydrogeological settings, such as hydraulic conductivity (K), aquifer thickness (H_0), etc., could be determined through hydrogeologic drilling, a pumping test, analogy and collecting previous data. After the hydrogeological conditions are identified, the characteristics of the aquifer, boundary conditions, initial conditions, hydraulic characteristics, and source sink term could be determined and subsequently generalized.

2.2. Argument Method of Water Supply Capacity of Rws

In order to reveal the differences in water supply capacity, the yields of single pumping wells under different conditions were calculated separately: an off-riverside well (known as non-riverside water source) and a riverside well (known as RWS). Besides, a pumping well group along a riverside was also considered here, as the third scenario considering the practice needs.

2.2.1. Scenario I: A Single Pumping Well Off-Riverside

Under the condition of a single pumping well off-riverside that was independent of the river, the well could be generalized as an incomplete well in an infinite phreatic aquifer in the horizontal dimension (excluding the upper sealing section). In order to facilitate the calculation, it was assumed that the well meets the Dupuit Hypothesis, that is, the flow line to the well is approximately horizontal, and the contour map of groundwater level was a coaxial cylinder, which was consistent with the passing water section. According to the Dupuit Equation, the water yield of the single well could be calculated as the following:

$$q = \pi K \frac{(2H_0 - s_w)s_w}{\ln \frac{R}{r_w}} \tag{1}$$

where q is the water yield (m³/d); K is the hydraulic conductivity (m/d); H_0 is the phreatic aquifer thickness (m); s_w is the drawdown of the groundwater level (m); R is the influence radius (m); and r_w is the well radius (m).

2.2.2. Scenario II: A Single Pumping Well along a Linear Riverside

Under the condition of a single pumping well along a linear riverside, the river could be determined as a recharge boundary with a specific water level, which could be coped with according to the Mirror-Image Method (Bear, 1979) [32]. To save space, the detailed description of the method was omitted here (the schematic diagram can be seen in Figure 1), which could be referenced to Bear (1979) [32], and the water yield of the single well could be calculated, as the following:

$$q = \pi K \frac{(2H_0 - s_w)s_w}{\lg \frac{2D_{wr}}{r_w}} \tag{2}$$

where D_{wr} is the distance between the well and the river (m); and the other symbols are the same as those in Equation (1).

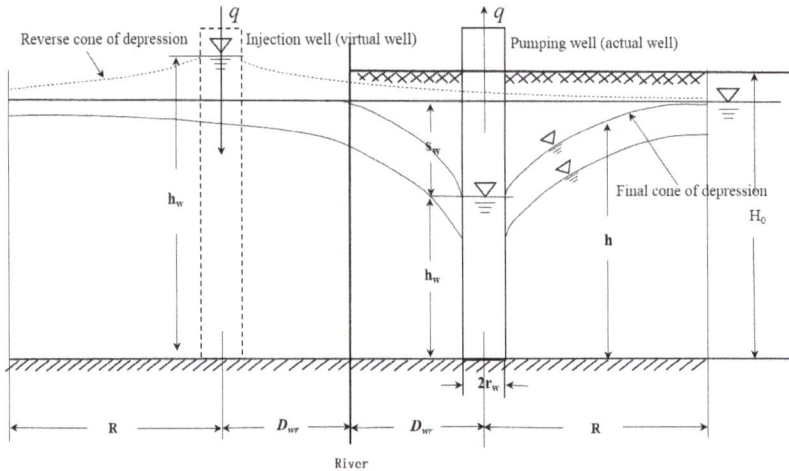

Figure 1. Diagram of the Mirror-Image method for a specified-head recharge boundary (river in special).

2.2.3. Scenario III: A Well Group along a Linear Riverside

In many practice cases, a set of pumping wells were arranged along a riverside replacing a single well due to the limited water supply capacity of the latter. Thus, the scenario where a well group along a linear riverside was considered in this study. The water supply capacity of this scenario was jointly impacted by the layout of the wells, and the exploitation plan. In this section, we deduced the calculating equations of the exploitation amount of a single well in the well group in addition to that of the well group. Those equations were convenient for the optimization of the layout of the wells and the exploitation plan in the following sections. The Mirror-Image Method was also applicable for this scenario, and the corresponding schematic diagram adapted from that of the scenario with a single well could be seen in Figure 2.

Figure 2. Schematic of Mirror-Image Method of the scenario of a well group paralleling to the river line.

The equation describing the interference between pumping wells could be deduced from Dupuit Equation:

$$H_0^2 - h_w^2 = \frac{q}{\pi K} \ln \frac{R}{r_w} \tag{3}$$

where h_w is the groundwater level of the pumping well compared with the bottom of the aquifer, which is valued as the difference value between the aquifer thickness (H_0) and the drawdown of the groundwater level (s_w) (m); and the other symbols are the same as those of the above equations. The water yield of a pumping well in the well group could be determined by this equation, based on which the total water yield of the well group could be calculated easily if the number of wells was specified.

Superposition calculation could be carried out based on Equation (3) when the total water yield of all the pumping wells were the same [33]:

$$H_0^2 - h_i^2 = \sum_{j=1}^{n} \frac{q}{\pi K} \ln \frac{R}{r_j^-} - \sum_{j=1}^{n} \frac{q}{\pi K} \ln \frac{R}{r_j^+} = \frac{q}{\pi K} \ln \frac{\prod_1^n r_j^+}{\prod_1^n r_j^-} \tag{4}$$

where r_j^+ is the distance between the j injection well and the calculated pumping well (the i well) (m); r_j^- is the distance between two adjacent pumping wells (the i and j well) (m); $r_j^- = r_w$ when the pumping well is the calculated pumping well; and the other symbols are the same as those of the above equations.

2.3. Method of Design and Optimization of Well Group of Rws

2.3.1. Constraint Conditions

Considering the sustainable exploitation and utilization of RWS, it was essential to avoid causing the persistent decline of the groundwater level and bring an unacceptable negative impact on the ecosystem at the site and around. In this regard, it is common to set one third of the aquifer thickness (H_0), as the maximum allowable drawdown of the groundwater level (s_{max}) [33] which should be the upper limit of the drawdown of the groundwater level (s_w). The planned water resource exploitation is usually determined jointly by the water demand (Q'), as well as the water supply capacity of the water source (Q) [34]. Sometimes, the construction and operational costs were also taken into account. Taking these factors into account, both s_w and Q were selected as the constraint conditions (objective function) as seen in Figure 3. In detail, the results of the design and optimization effort needed to satisfy the two conditions simultaneously: $s_w \leq s_{max}$, and $Q \geq Q'$.

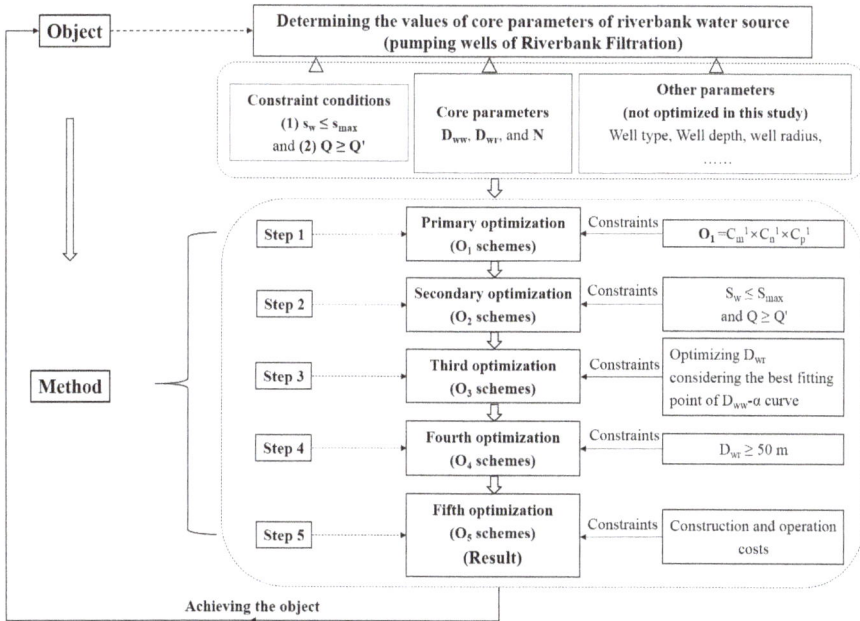

Note: s_w is the drawdown of the groundwater level (m); s_{max} is the maximum allowable drawdown of the groundwater level (m); Q is the total water yield of the water source (m³/d); Q' is the water demand (m³/d); D_{ww} is the distance between wells (m); D_{wr} is the distance between well and river (m); N is the number of wells (dimensionless); m, n and p are the numbers of the theoretical options for N, D_{ww} and D_{wr}, respectively (dimensionless); and α is the interference coefficient between wells (dimensionless).

Figure 3. Roadmap of the optimizing method (Five-Step Optimizing Method) established in the study.

2.3.2. Parameter Design

The layout of the well group directly determined the success of the establishment of the water source. The parameters involved included not only the well type, well depth and well radius (r_w), but also the number of wells (N), the distance between wells (D_{ww}) and the distance between well and river (D_{wr}). It is common to arrange tube wells in a straight line parallel to the river [33], considering it has better stimulation from the recharge from the river. In order to ensure the water intake efficiency, the pumping well was often designed as a completely penetrating well, which penetrated the whole phreatic aquifer in the vertical dimension. As to r_w, it is usually determined by the manufacturing technique, and 0.25 m was the most designed value in many regions.

Different from the fixed type and parameters discussed above, it is very complex and difficult to determine the values of N, D_{ww}, and D_{wr} due to their interactions with each other. Thus, N, D_{ww}, and D_{wr} were selected as the objective parameters for optimizing through a certain method in this study, and the numbers of the theoretical options for these three parameters were assumed to be m, n, and p, respectively.

2.3.3. Parameter Optimization

In order to facilitate the discussion, the method established in this study was named the "Five-Step Optimizing Method" (Figure 3), and the result schemes obtained by the latter step were the subset of those obtained by the former step ($O_5 \subset O_4 \subset O_3 \subset O_2 \subset O_1$).

(1) The First Step: The Establishment of all the Possible Schemes (O_1)

The first step was to establish the scheme set including all the possible schemes. Considering the numbers of the theoretical options for N, D_{ww}, and D_{wr} being m, n, and p, respectively, the number of all the possible schemes (O_1) equaled the result of $C_m^1 \times C_n^1 \times C_p^1$.

(2) The Second Step: Screening from N

The second step was to compare all the schemes established in O_1 and then screen the possible schemes satisfying $s_w \leq s_{max}$ and $Q \geq Q'$ simultaneously, forming a scheme set named O_2. N depended on Q' and the water yield of a single well, and the latter was determined by s_{max}. Thus, Q and s_w of each scheme should be calculated, followed by the screening of the favorable schemes.

(3) The Third Step: Screening from D_{ww}

The third step was to screen those schemes in O_2 with the favorable D_{ww}, which was carried out by establishing the relationship between D_{ww} and α (the interference coefficient between wells) and subsequently, finding the inflection point at the curve. The corresponding D_{ww} value of the inflection point or around was considered to be the favorable D_{ww} [35]. A strong relationship between D_{ww} and α existed, and the greater the D_{ww} was, the smaller α was, which could facilitate obtaining more water yield. However, the greater D_{ww} would increase, the cost including waterline, power transmission system and the corresponding management cost [33]. Thus, it is the designer's responsibility to minimize D_{ww} as much as possible under the premise of ensuring meeting the water demand [36]. After screening the favorable D_{ww} from O_2, the possible schemes further decreased temporarily to form O_3.

α referred to the percentage change of the water yield of a single well with disturbance relative to that of a single well without disturbance [35], which could be described as:

$$\alpha = (q' - q)/q' \tag{5}$$

where q is the water yield of a single well with disturbance calculated by Equation (4) (m^3/d); and q' is the water yield of a single well without disturbance calculated by Equation (2) (m^3/d).

(4) The Fourth Step: Screening from D_{wr}

The fourth step was to screen those schemes in O_3 with the favorable D_{wr}. Generally speaking, the smaller the D_{wr} was, the greater the water yield was, which could increase the efficiency of the water intake engineering. Inversely, the water yield would decrease with the D_{wr} increasing because the recharge from the river would decrease. Under certain conditions, the impact of river water quality on the well water quality should also be considered, especially in a circumstance of surface water pollution, as pollutants in the RBF usually decreased with the D_{wr} increasing [37]. After screening the favorable D_{wr} from O_3, the possible schemes further decreased temporarily to form O_4, which usually only included two or three schemes or less.

(5) The Fifth (Last) Step: Obtaining the Optimal Option (O_5) through Appropriate Consideration of Construction and Operation Costs

More than one scheme was often obtained after the accomplishment of the fourth optimizing step (O_4). If that happened, the optimal scheme usually could be screened by appropriately considering the construction and operation cost of the water intake engineering, which was usually not hard to accomplish. Besides, it is also common to encounter other factors that require consideration, such as the available land issue. Thus, it's occasionally necessary to go further to the fifth (last) step, while more details was omitted here considering the step being relatively strong subjective and maneuverable.

3. Case Study

3.1. Study Area and Generalization

The case study referred to an experimental riverside water source established by the Songhua River, which was located at Harbin City of Heilongjiang Province, northeastern China. The hydrological and hydrogeological investigation, including pumping tests had been carried out during 2017–2018, through which the study area setting had been identified in detail. Limited to the length of this paper, more details were omitted, which could be referenced to Zhu et al. (2019) [2]. Generally speaking, the conditions in the study area met the assumptions of the method with the main parameter values such as K, H_0, R, and r_w being 50 m/d, 42 m, 200 m and 0.5 m, respectively. As the designed requirement in the method, s_{max} equaled 14 m (one third of H_0).

3.2. Water Supply Capacity of the RWS

If a single pumping well was established at the study area without considering the impact of the river, the q (23,056 m^3/d ≈ 23,000 m^3/d) could be easily determined by the Equation (1). If a single pumping well was established at the study area with simultaneously considering the actual impact of the river, three q values (69,800 m^3/d, 61,400 m^3/d and 57,400 m^3/d) could be easily determined by Equation (2) separately with three designed D_{wr} values (20 m, 40 m and 60 m). These results showed that the water yields of the single pumping well in the riverside water source were 2.5–3 times as much as that of the single pumping well without the recharge from the river. The variation in multiples was caused by the differences of D_{wr}, and the smaller the D_{wr}, the greater both q and the corresponding multiple were. The results showed that D_{wr} had a great impact on q. As to the scenario of a well group along the riverside, the total water yield was impacted by many parameters complicatedly, which could be seen in the following section.

3.3. The Design and Optimization of the Well Group

3.3.1. Constraint Conditions

As discussed at the beginning of Section 3.1, s_{max} equaled 14 m. Q′ was designed as 2×10^5 m^3/d considering the water demand of Harbin City. Thus, the constraint conditions in this case study were:

$s_w \leq s_{max} = 14$ m, and $Q \geq Q' = 2 \times 10^5$ m^3/d. The subsequent design and optimization effort was carried out around this goal.

3.3.2. Parameter Design

In this case study, all the pumping wells of the well group of the RWS were designed with the same specifications, considering the convenience of construction and management. The well type (tube well), depth (50 m from the ground to the bottom) and radius (r_w, 0.25 m) were directly designed without optimizing. As for those parameters needing optimization (N, D_{ww} and D_{wr}), the alternative options of the parameter values for optimizing needed to be fixed, considering both the study area setting and the water demand.

3.3.3. Parameter Optimization

(1) The First Step: The Establishment of all the Possible Schemes (O_1)

The alternative values of N were designed as 7, 9, and 11 (m = 3) considering the ratio of Q'/q (8.7) and the possible interaction between wells. The alternative values of D_{ww} were designed as 20 m, 50 m, 100 m, 150 m, and 200 m (n = 5) considering the influence radius and the impact from the river. The alternative values of D_{wr} were designed as 20 m, 50 m, 100 m, 200 m and 250 m (p = 5) considering the influence radius and the impact from the river. That's to say, the optimizing effort was to screen the most favorable values set (scheme) of the three parameters from the 75 ($C_3^1 \times C_5^1 \times C_5^1$) established schemes ($O_1$) (Table 1) by using the established method. N, D_{ww}, D_{wr}, q, and Q in Table 1 refer to number of wells, distance between wells, distance between well and river, single well yield, and total yield, respectively, and q was calculated by Equation (4).

Table 1. All possible schemes and the corresponding results of each optimizing step.

No.	N	D_{ww} (m)	D_{wr} (m)	q (m^3/d)	Q ($\times 10^5$ m^3/d)	Result *
1	11	20	20	22,754	2.50	AB
2	11	50	20	28,022	3.08	AB
3	11	100	20	29,629	3.26	ABC
4	11	200	20	30,135	3.32	AB
5	11	250	20	30,199	3.32	AB
6	11	20	50	14,021	1.54	A
7	11	50	50	20,038	2.20	AB
8	11	100	50	23,330	2.57	ABCD
9	11	200	50	24,935	2.74	AB
10	11	250	50	25,185	2.77	AB
11	11	20	100	9555	1.05	A
12	11	50	100	14,499	1.60	A
13	11	100	100	18,379	2.02	ABCDE
14	11	200	100	21,111	2.32	AB
15	11	250	100	21,670	2.38	AB
16	11	20	150	7763	0.85	A
17	11	50	150	11,772	1.30	A
18	11	100	150	15,523	1.71	A
19	11	200	150	18,723	2.06	AB
20	11	250	150	19,485	2.14	AB
21	11	20	200	6785	0.75	A
22	11	50	200	10,135	1.12	A
23	11	100	200	13,610	14.97	A
24	11	200	200	16,974	1.87	A
25	11	250	200	17,858	1.96	A
26	9	20	20	22,902	2.06	AB
27	9	50	20	28,058	2.53	AB
28	9	100	20	29,639	2.67	ABC
29	9	200	20	30,137	2.71	AB
30	9	250	20	30,201	2.72	AB

Table 1. *Cont.*

No.	N	D_{ww} (m)	D_{wr} (m)	q (m³/d)	Q (×10⁵ m³/d)	Result *
31	9	20	50	14,343	1.29	A
32	9	50	50	20,153	1.81	A
33	9	100	50	23,370	2.10	ABCDE
34	9	200	50	24,946	2.25	AB
35	9	250	50	25,193	2.27	AB
36	9	20	100	10,020	0.90	A
37	9	50	100	14,727	1.33	A
38	9	100	100	18,476	1.66	A
39	9	200	100	21,144	1.90	A
40	9	250	100	21,692	1.95	A
41	9	20	150	8287	0.75	A
42	9	50	150	12,085	1.09	A
43	9	100	150	15,675	1.41	A
44	9	200	150	18,779	1.70	A
45	9	250	150	19,525	1.76	A
46	9	20	200	7333	0.66	A
47	9	50	200	10,507	0.95	A
48	9	100	200	13,811	1.24	A
49	9	200	200	17,056	1.54	A
50	9	250	200	17,917	1.61	A
51	7	20	20	23,141	1.62	A
52	7	50	20	28,117	1.97	A
53	7	100	20	29,656	2.08	AB
54	7	200	20	30,142	2.11	ABC
55	7	250	20	30,204	2.11	AB
56	7	20	50	14,857	1.04	A
57	7	50	50	20,338	1.42	A
58	7	100	50	23,433	1.64	A
59	7	200	50	24,964	1.75	A
60	7	250	50	25,204	1.76	A
61	7	20	100	10,738	0.75	A
62	7	50	100	15,092	1.06	A
63	7	100	100	18,631	1.30	A
64	7	200	100	21,196	1.48	A
65	7	250	100	21,727	1.52	A
66	7	20	150	9077	0.64	A
67	7	50	150	12,577	0.88	A
68	7	100	150	15,918	1.11	A
69	7	200	150	18,871	1.32	A
70	7	250	150	19,588	1.37	A
71	7	20	200	8148	0.57	A
72	7	50	200	11,086	0.78	A
73	7	100	200	14,132	0.99	A
74	7	200	200	17,188	1.20	A
75	7	250	200	18,011	1.26	A

* A: the result of the first optimizing step; AB: the result of the second optimizing step; ABC: the result of the third optimizing step; ABCD: the result of the fourth optimizing step; and ABCDE: the result of the fifth (last) optimizing step.

(2) The Second Step: Screening from N

The results (Table 1) showed that the favorable schemes ($Q \geq Q' = 2 \times 10^5$ m³/d) for N of 11, 9, and 7 were 14, 8, and 3, respectively, which constituted a new scheme set with 25 possible schemes (O_2) (Table 1).

(3) The Third Step: Screening from D_{ww}

All the interference coefficients (α) between wells of schemes in O_2 were calculated (Table 2), and correlation diagrams (Figure 4) were drawn to illustrate the relationships between D_{ww} and α

with different N and different D_{wr}. The results showed that α decreased with D_{ww} increasing at the beginning, while decreased almost no longer after the inflection points appeared at the top left corners of all the curves. It suggested that the favorable value of D_{ww} should be around the inflection point, because smaller value would bring more disturbance between wells while greater value would bring more construction, and operation costs considering the occupying space of the wells. Thus, the favorable value of D_{ww} should be 100 m for both eleven-wells and nine-wells, while the same value should be 200 m for seven-wells. The favorable schemes decreased to six (O_3) after this optimizing step (Table 2). N, D_{ww}, D_{wr}, q, q' and α in Table 2 refer to number of wells, distance between wells, distance between well and river, single well yield, single well yield without interference, and interference coefficient, respectively.

Table 2. Interference coefficient of wells.

No.	N	D_{ww} (m)	D_{wr} (m)	q (m³/d)	q' (m³/d)	α (%)
1	11	20	20	22,753.5	69,805.6	67.40
2	11	50	20	28,021.8	69,805.6	59.86
3	11	100	20	29,629.2	69,805.6	57.55
4	11	200	20	30,134.7	69,805.6	56.83
5	11	250	20	30,199.2	69,805.6	56.74
6	11	50	50	20,038.2	59,130.1	66.11
7	11	100	50	23,330.2	59,130.1	60.54
8	11	200	50	24,934.7	59,130.1	57.83
9	11	250	50	25,185.2	59,130.1	57.41
10	11	100	100	18,379.0	52,998.7	65.32
11	11	200	100	21,111.3	52,998.7	60.17
12	11	250	100	21,669.9	52,998.7	59.11
13	11	200	150	18,722.6	49,967.8	62.53
14	11	250	150	19,485.1	49,967.8	61.00
15	9	20	20	22,901.5	69,805.6	67.19
16	9	50	20	28,058.2	69,805.6	59.81
17	9	100	20	29,639.4	69,805.6	57.54
18	9	200	20	30,137.3	69,805.6	56.83
19	9	250	20	30,200.9	69,805.6	56.74
20	9	100	50	23,369.6	59,130.1	60.48
21	9	200	50	24,946.0	59,130.1	57.81
22	9	250	50	25,192.6	59,130.1	57.39
23	7	100	20	29,655.8	69,805.6	57.52
24	7	200	20	30,141.6	69,805.6	56.82
25	7	250	20	30,203.6	69,805.6	56.73

(4) The Fourth Step: Screening from D_{wr}

In O_3 with six schemes (Table 1), the three schemes with D_{wr} = 20 m were not a wise choice after considering the study area setting carefully, because the distance between the pumping wells and the river was too small to protect the wells from occasional flooding happening in the river. In addition, the safety of the river levee and the possible clogging of the RBF should be also considered. Thus, the schemes numbered 3, 28 and 54 in Table 1 had to be abandoned. So far, only three schemes left in Table 1 constituting O_4, which left very little scope to decision maker to select. That is to say, the optimal scheme in fact could be easily singled out from the three options (Figure 5). The drawdown of the groundwater level of the middle pumping well of each scheme is 14.0 m, while the drawdowns of the well at each edge of the three schemes are 12.7 m, 11.7 m, and 12.8 m respectively satisfy the constraint condition. Thus, sometimes the optimizing process could be terminated after this step.

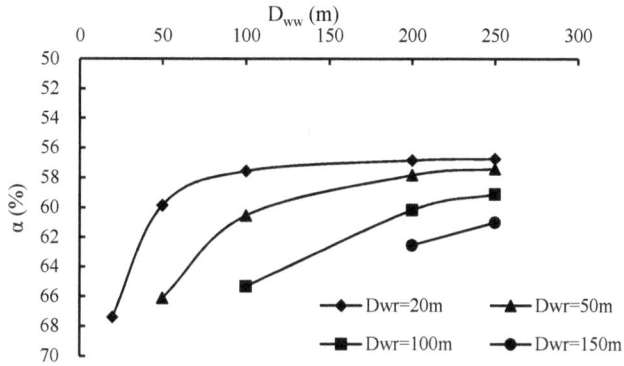

(a) Eleven-wells

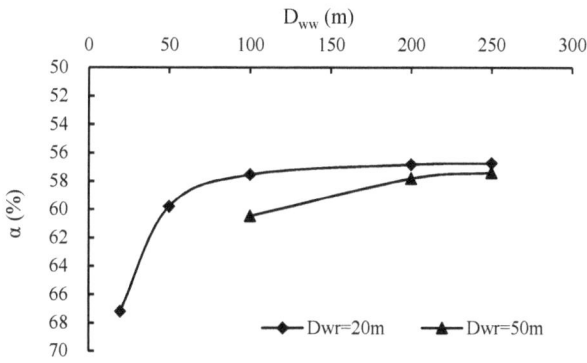

(b) Nine-wells

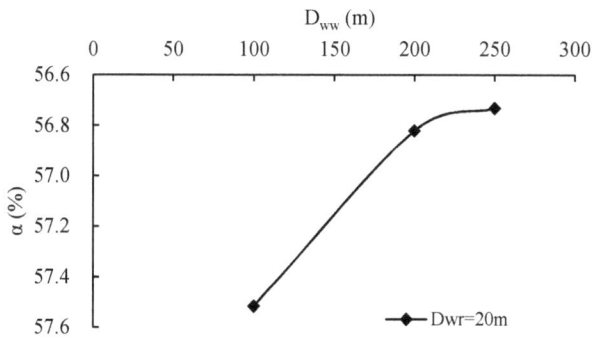

(c) Seven-wells

Figure 4. Relationships between D_{ww} and α of pumping wells.

(a) Optimized result A (eleven wells, numbered 8 in Table 1)

(b) Optimized result B (eleven wells, numbered 13 in Table 1)

(c) Optimized result C (nine wells, numbered 33 in Table 1)

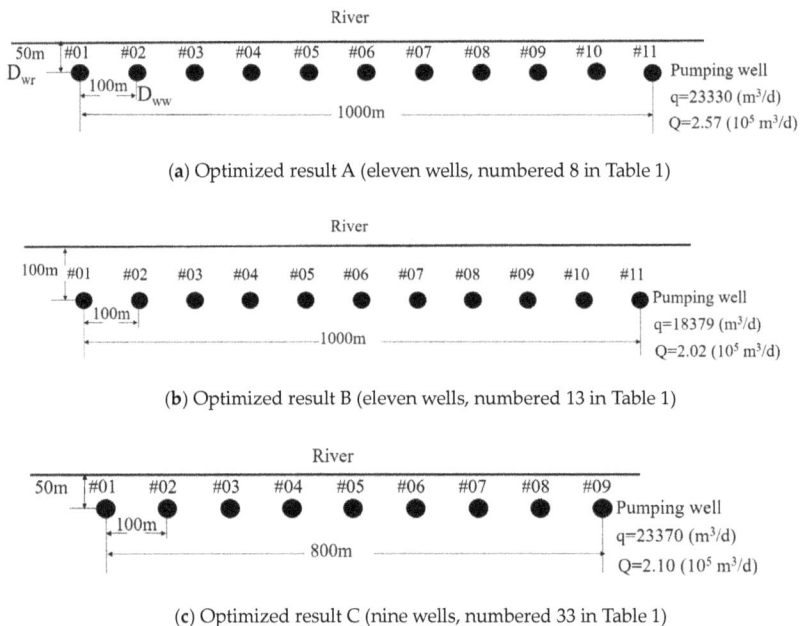

Figure 5. Schemes obtained from the fourth optimizing step.

(5) The Fifth (Last) Step: Obtaining the Optimal Option through the Appropriate Consideration of Construction and Operation Costs

In fact, the need to go on to the fifth step is obvious, on occasion, especially when the budget is not adequate, or the riverbed clogging occurs. Out of the three left schemes (Figure 5), the scheme (a) would have to be abandoned considering saving water resource and energy, as the total water yield ($Q = 2.57 \times 10^5$ m^3/d) of it is much more than needed ($Q' = 2 \times 10^5$ m^3/d). Nevertheless, scheme (a) could also become a choice if the water demand increases. And the scheme (b) would also have to be abandoned considering saving investment, as it has two more pumping wells compared with the scheme (c). Thus, the scheme (c) could be singled out as the final result of the optimizing effort, if other factors are not considered.

However, the scheme (b) is the best choice if 50 m as the distance between the well and the river is considered to be too small to protect against floods and water pollution incidents, or 23,370 m^3/d as the pumping yield of each well is considered to be too large to avoid the possible riverbed clogging in practice. That is to say, the final result may switch among the three scenarios considering the establishing investment, the energy consumption, the disaster risk, and the riverbed clogging, etc. Thus, all the three optimized schemes are valuable for the decision maker to choose.

4. Discussion

In this study, we tentatively established a method of design and optimization of pumping wells in riverbank filtration (RBF) by using analytical methods, in which the traditional methods such as the Mirror-Image Method, Dupuit Equation and Interference Well Group Method were jointly adopted. The method assumed that the phreatic aquifer was fully penetrated by the linear river in the vertical dimension, and there was a close hydraulic relation between the surface water and groundwater, both of which had nearly the same natural hydrological curve. The aquifer extended indefinitely in the other three directions and was bounded by an impervious rock at the bottom. The RBF was a homogeneous and isotropic medium, and there was no aquitard or aquiclude around the river.

Thus, the model assumed was generally the same as that established by Theis (1941) [38], which was usually considered the simplest mode of surface water-groundwater interactions and rare in practice [31]. The main purpose of this simplification was to facilitate calculation, so we used the Mirror-Image Method in this study, after we proved the applicability [2]. This method was established by Jacob (1979) [32] to calculate the extending of the groundwater level, which is induced by pumping groundwater near the river and the corresponding water supply capacity of the pumping wells. Considering their simplicity, the mode and the corresponding method were often adopted in order to obtain the approximate solutions.

Similar to many other optimizing efforts, the parameters are not independent [39], which makes the optimizing efforts more difficult, especially in the use of analytical methods [40]. The basic hydrogeology tells us that the cone of depression of the groundwater level will continually extend both in the vertical and horizontal dimensions when pumping, unless the aquifer could capture as much as water recharge relative to the water yield [41]. Thus, the cones of depression of the different pumping wells will interact with each other if the distance between wells is not far enough [42], which will inevitably affect the efficiency of the pumping wells through decreasing the water capacity of a single well [43]. However, the cost will increase considering the water supply network and the power supply line if the distance is too far, especially when the number of the pumping wells are great [44]. Thus, it is an art to design the distance between wells, especially when there are many wells [45]. More than that, the recharge of the pumping well of RBF is usually much more than that of off-riverside because the former could capture a portion of river water [2]. Thus, the interaction of cones of depression of different wells will be more complex, which further complicates the design art of the distance of wells. As to the river-groundwater interaction, the river water captured by the pumping well is directly determined by the distance of the well and the river if the other parameters are fixed [46]. It is better to design a smaller value to the distance between well and river, only if the water quantity is considered in the engineering. However, it is not favorable for the effectiveness of the RBF as a pre-treatment measure of water [47]. Furthermore, it is also not favorable for protecting the wells from possible river flooding and water pollution, which will pose significant impact on the safety of the social water supply. The water quality is obviously also an important issue considering both the pre-treatment function of the RBF and the subsequent design of the post-treatment of the plant. The water quality issue was not considered quantitatively in this study, mainly because we considered that, the optimal scheme considering both the water quantity and quality of the pumping wells should only be singled out from the optional schemes only considering the water quantity, which was obtained from the optimizing method established in this study. That is to say, the design and optimization effort made in this study is the basis of considering the water quality issue, based on which the optimal scheme can be determined for both water quantity and quality.

River-groundwater interactions are at the core of a wide range of major contemporary challenges [48], out of which the provision of high-quality drinking water in sufficient quantities is without doubt the most important because it involves water demand of human society. Riverbank filtration (RBF) has been used for many decades widely worldwide especially in Europe, the United States and some Asian countries (China, India, Japan, South Korea, etc.) to provide drinking water [1,49]. However, the existing RBF comprehension mainly depends on practical understandings, and no standards have been developed to guide the optimization of the RBF design [50]. Thus, we have good reason to believe that more design and optimization efforts on RBF considering water quantity and quality or both will be carried out in the following periods.

5. Conclusions

An optimizing method named Five-Step Optimizing Method was established systematically by us aiming to improve the design effort of engineering parameters of water pumping wells in riverbank filtration. The maximum allowable drawdown of groundwater level (s_{max}) and the water demand of society (Q') jointly constituted the constraint conditions. Three parameters including the number of wells (N), the distance between wells (D_{ww}) and the distance between well and river (D_{wr}) could be optimized through the established method step-by-step by screening the alternative values beforehand designed for the parameters. The interference between wells was found to be a decisive factor, which had a significant impact on the design and optimization effort of all the three parameters. D_{wr} was another decisive factor impacting the recharge from the river, and subsequently the well water yield. The optimized result would sometimes supply the decision maker with more than one optional scheme for selecting, while it is not yet difficult in practice to single out the optimal one considering both the field setting and the water demand.

A case study was carried out along the Songhua River in Harbin City of Heilongjiang Province of northeast China, whose setting generally met the assumptions of the method, aiming to illustrate the application and simultaneously verify the applicability of the method. Three schemes with different parameter values obtained from the optimizing method could meet the constraint conditions (s_{max} = 14 m; and Q' = 2 × 10^5 m^3/d), out of which the optimal option had the parameters: N = 9; D_{ww} = 100 m; and D_{wr} = 50 m. Meanwhile, the other two schemes were valuable in certain circumstances, which should also be referred to the decision maker. This case study proved that the established method was applicable.

Admittedly, the method established in this study has some limitations, especially in the limitations of the assumptions. At the very least, however, this effort provides a relatively simple and operational approach to relevant practices and decisions.

Author Contributions: Conceptualization, Y.Z. (Yuanzheng Zhai); Methodology, Y.Z. (Yuanzheng Zhai) and J.Z.; Formal analysis, Y.J. and J.Z.; Investigation, Y.Z. (Yaguang Zhu) and Q.D.; Data curation, J.Z.; Writing—original draft, Y.J. and J.Z.; Writing—review & editing, Y.Z. (Yuanzheng Zhai); Supervision, Y.T.; Funding acquisition, Y.Z. (Yuanzheng Zhai) and Y.T.

Funding: This work was supported by the National Natural Science Foundation of China (No. 41877174 and 41877355), and the Major Science and Technology Program for Water Pollution Control and Treatment (2018ZX07101005-04).

Conflicts of Interest: The authors declare no conflict of interest. The funders had no role in the design of the study; in the collection, analyses, or interpretation of data; in the writing of the manuscript; or in the decision to publish the results.

References

1. Hu, B.; Teng, Y.; Zhai, Y.; Zuo, R.; Li, J.; Chen, H. Riverbank filtration in China: A review and perspective. *J. Hydrol.* **2016**, *541*, 914–927. [CrossRef]

2. Zhu, Y.; Zhai, Y.; Du, Q.; Teng, Y.; Wang, J.; Yang, G. The impact of well drawdowns on the mixing process of river water and groundwater and water quality in a riverside well field, Northeast China. *Hydrol. Process.* **2019**. [CrossRef]

3. Yin, W.; Teng, Y.; Zhai, Y.; Hu, L.; Zhao, X.; Zhang, M. Suitability for developing riverside groundwater sources along Songhua River, Northeast China. *Hum. Ecol. Risk Assess.* **2018**, *8*, 2088–2100. [CrossRef]

4. Hester, E.T.; Cardenas, M.B.; Haggerty, R.; Apte, S.V. The importance and challenge of hyporheic mixing. *Water Resour. Res.* **2017**, *53*, 3565–3575. [CrossRef]

5. Steen, C.; Vitaly, A.Z.; Daniel, M.T. On the use of analytical solutions to design pumping tests in leaky aquifers connected to a stream. *J. Hydrol.* **2010**, *381*, 341–351.

6. Baalousha, H.M. Drawdown and stream depletion induced by a nearby pumping well. *J. Hydrol.* **2012**, *466–467*, 47–59. [CrossRef]

7. Jin, M.; Xian, Y.; Liu, Y. Disconnected stream and groundwater interaction: A review. *Adv. Water Sci.* **2017**, *28*, 149–160, (In Chinese with English abstract).

8. Lu, C.; Shu, C.; Chen, X. Numerical analysis of the impacts of bedform on hyporheic exchange. *Adv. Water Sci.* **2012**, *23*, 789–795, (In Chinese with English abstract).

9. Pholkern, K.; Srisuk, K.; Grischek, T.; Soares, M.; Schäfer, S.; Archwichai, L.; Saraphirom, P.; Pavelic, P.; Wirojanagud, W. Riverbed clogging experiments at potential river bank filtration sites along the Ping River, Chiang Mai, Thailand. *Environ. Earth Sci.* **2015**, *73*, 7699–7709. [CrossRef]

10. Grischek, T.; Bartak, R. Riverbed clogging and sustainability of riverbank filtration. *Water* **2016**, *12*, 604. [CrossRef]
11. Shaymaa, M.; Arifah, B.; Zainal, A.A.; Saim, S. Review of the role of analytical modelling methods in riverbank filtration system. *J. Teknol.* **2014**, *1*, 59–69.
12. Knapp, J.L.A.; Cirpka, O.A. Determination of hyporheic travel-time distributions and other parameters from concurrent conservative and reactive tracer tests by local-in-global optimization. *Water Resour. Res.* **2017**, *53*, 4984–5001. [CrossRef]
13. Xie, Y.; Peter, G.C.; Craig, T.S.; Zheng, C. On the limits of heat as a tracer to estimate reach-scale river-aquifer exchange flux. *Water Resour. Res.* **2015**, *51*, 7401–7416. [CrossRef]
14. Stefania, G.A.; Rotiroti, M.; Fumagalli, L.; Simonetto, F.; Capodaglio, P.; Zanotti, C.; Bonomi, T. Modeling groundwater/surface-water interactions in an Alpine valley (the Aosta Plain, NW Italy): The effect of groundwater abstraction on surface-water resources. *Hydrogeol. J.* **2018**, *26*, 147–162. [CrossRef]
15. Wen, Z.; Zhan, H.; Wang, Q.; Liang, X.; Ma, T.; Chen, C. Well hydraulics in pumping tests with exponentially decayed rates of abstraction in confined aquifers. *J. Hydrol.* **2017**, *548*, 40–45. [CrossRef]
16. Xie, X.; Johnson, T.M.; Wang, Y.; Lundstrom, C.C.; Ellis, A.; Wang, X.; Duan, M.; Li, J. Pathways of arsenic from sediments to groundwater in the hyporheic zone: Evidence from an iron isotope study. *J. Hydrol.* **2014**, *511*, 509–517. [CrossRef]
17. Cardenas, M.B. Hyporheic zone hydrologic science: A historical account of its emergence and a prospectus. *Water Resour. Res.* **2015**, *51*, 3601–3616. [CrossRef]
18. Gökçe, Ş.; Ayvaz, M.T. *Evaluation of Harmony Search and Differential Evolution Optimization Algorithms on Solving the Booster Station Optimization Problems in Water Distribution Networks*; Springer: Cham, Switzerland, 2015; pp. 245–261.
19. Miracapillo, C.; Morel-Seytoux, H.J. Analytical solutions for stream-aquifer flow exchange under varying head asymmetry and river penetration: Comparison to numerical solutions and use in regional groundwater models. *Water Resour. Res.* **2014**, *50*, 7430–7444. [CrossRef]
20. Hamann, E.; Stuyfzand, P.J.; Greskowiak, J.; Timmer, H.; Massmann, G. The fate of organic micropollutants during long-term/long-distance river bank filtration. *Sci. Total Environ.* **2015**, *545–546*, 629–640. [CrossRef] [PubMed]
21. Salamon, E.; Goda, Z. Coupling riverbank filtration with reverse osmosis may favor short distances between wells and riverbanks at RBF sites on the River Danube in Hungary. *Water* **2019**, *11*, 113. [CrossRef]
22. Mantoglou, A.; Papantoniou, M.; Giannoulopoulos, P. Management of coastal aquifers based on nonlinear optimization and evolutionary algorithms. *J. Hydrol.* **2004**, *297*, 209–228. [CrossRef]
23. Jin-Yong, L.; Kang-Kun, L.; Se-Yeong, H.; Yongcheol, K. Fifty years of groundwater science in Korea: A review and perspective. *Geosci. J.* **2017**, *6*, 951–969.
24. Tian, G.L.; Chang, J.B.; Wang, W. Application of analytical method to the calculation of groundwater permissible mining capacity. *J. Pearl River* **2016**, *10*, 1–7, (In Chinese with English abstract).
25. Hansen, A.K.; Franssen, H.J.H.; Bauer-Gottwein, P.; Madsen, H.; Rosbjerg, D.; Kaiser, H.P. Well field management using multi-objective optimization. *Water Resour. Manag.* **2013**, *27*, 629–648. [CrossRef]
26. Polomčić, D.; Hajdin, B.; Stevanović, Z.; Bajić, D.; Hajdin, K. Groundwater management by riverbank filtration and an infiltration channel: The case of Obrenovac, Serbia. *Hydrogeol. J.* **2013**, *21*, 1519–1530. [CrossRef]
27. Lee, E.; Hyun, Y.; Lee, K.K.; Shin, J. Hydraulic analysis of a radial collector well for riverbank filtration near Nakdong River, South Korea. *Hydrogeol. J.* **2012**, *20*, 575–589. [CrossRef]
28. AbdelFattah, A.; Langford, R.; SchulzeMakuch, D. Applications of particle-tracking techniques to bank infiltration: A case study from El Paso, Texas, USA. *Environ. Geol.* **2008**, *55*, 505–515. [CrossRef]
29. Fragoso, T.; Cunha, M.D.C.; LoboFerreira, J.P. Optimal pumping from Palmela water supply wells (Portugal) using simulated annealing. *Hydrogeol. J.* **2009**, *17*, 1935–1948. [CrossRef]
30. Haas, R.; Opitz, R.; Grischek, T.; Otter, P. The AquaNES Project: Coupling riverbank filtration and ultrafiltration in drinking water treatment. *Water* **2019**, *11*, 18. [CrossRef]
31. Hunt, B. Unsteady stream depletion from ground water pumping. *Ground Water* **1999**, *37*, 98–102. [CrossRef]
32. Bear, J. *Hydraulics of Groundwater*; McGraw-Hill: New York, NY, USA, 1979.

33. Wen, C.; Dong, W.; Cui, G.; Liu, Y.; Su, X. Application of analytical methods to determination of the exploitation scheme of a wellfield at riverside. *Hydrogeol. Eng. Geol.* **2017**, *3*, 19–26, (In Chinese with English abstract).

34. Gaur, S.; Mimoun, D.; Graillot, D. Advantages of the analytic element method for the solution of groundwater management problems. *Hydrol. Process.* **2011**, *25*, 3426–3436. [CrossRef]

35. Li, Y.; Li, Z. The analytical method for rationally distributing interference well group. *J. Xi'an Coll. Geol.* **1997**, *S1*, 47–50, (In Chinese with English abstract).

36. Katsifarakis, K.L. Groundwater pumping cost minimization-an analytical approach. *Water Resour. Manag.* **2008**, *22*, 1089–1099. [CrossRef]

37. Hoehn, E.; Scholtis, A. Exchange between a river and groundwater, assessed with hydrochemical data. *Hydrol. Earth Syst. Sci.* **2011**, *15*, 983–988. [CrossRef]

38. Theis, C.V. The effect of a well on the flow of a nearby stream. *Trans. Am. Geophys. Union* **1941**, *22*, 734–738. [CrossRef]

39. Wang, P.; Sergey, P.P.; Vsevolod, M.S. Optimum experimental design of a monitoring network for parameter identification at riverbank well fields. *J. Hydrol.* **2015**, *523*, 531–541. [CrossRef]

40. Hund-Der, Y.; Ya-Chi, C. Recent advances in modeling of well hydraulics. *Adv. Water Resour.* **2013**, *51*, 27–51.

41. Wang, Y.; Zheng, C.; Ma, R. Review: Safe and sustainable groundwater supply in China. *Hydrogeol. J.* **2018**, *26*, 1301–1324. [CrossRef]

42. Holzbecher, E. Analytical solution for well design with respect to discharge ratio. *Groundwater* **2013**, *1*, 128–134. [CrossRef] [PubMed]

43. Zhang, Y.; Susan, H.; Finsterle, S. Factors governing sustainable groundwater pumping near a river. *Groundwater* **2011**, *3*, 432–444. [CrossRef] [PubMed]

44. Mojtaba, S.; Javad, D.S.M. Optimum pumping well placement and capacity design for a groundwater lowering system in urban areas with the minimum cost objective. *Water Resour. Manag.* **2017**, *13*, 4207–4225.

45. Rao, S.V.N.; Sudhir, K.; Shashank, S.; Sinha, S.K.; Manju, S. Optimal pumping from skimming wells from the Yamuna River flood plain in north India. *Hydrogeol. J.* **2007**, *6*, 1157–1167. [CrossRef]

46. Hantush, M.S. Wells near streams with semipervious beds. *J. Geophys. Res.* **1965**, *70*, 2829–2838. [CrossRef]

47. Van Driezum, I.H.; Derx, J.; Oudega, T.J.; Zessner, M.; Naus, F.L.; Saracevic, E.; Kirschner, A.K.T.; Sommer, R.; Farnleitner, A.H.; Blaschke, A.P. Spatiotemporal resolved sampling for the interpretation of micropollutant removal during riverbank filtration. *Sci. Total Environ.* **2019**, *649*, 212–223. [CrossRef] [PubMed]

48. Brunner, P.R.; Therrien, P.R.; Simmons, C.T.; Franssen, H.J.H. Advances in understanding river-groundwater interactions. *Rev. Geophys.* **2017**, *3*, 818–854. [CrossRef]

49. Ray, C. Worldwide potential of riverbank filtration. *Clean Technol. Environ. Policy* **2008**, *10*, 223–225. [CrossRef]

50. Ahmed, K.A.A. Review on river bank filtration as an in situ water treatment process. *Clean Technol. Environ. Policy* **2017**, *2*, 349–359. [CrossRef]

water

MDPI

Article

Water Quality Changes during the Initial Operating Phase of Riverbank Filtration Sites in Upper Egypt

Rifaat Abdel Wahaab [1,2], Ahmed Salah [1] and Thomas Grischek [3,*

[1] Holding Company for Water and Waste Water, 1200 Corniche El Nile, Rod-El-Farag, 12622 Cairo, Egypt;
rawahaab@yahoo.com (R.A.W.); ahmedsalahosman49@gmail.com (A.S.)
[2] Environmental Science Division, National Research Centre, 12622 Cairo, Egypt
[3] Division of Water Sciences, University of Applied Sciences Dresden, 01069 Dresden, Germany
* Correspondence: grischek@htw-dresden.de; Tel.: +49-351-4623350

Received: 15 May 2019; Accepted: 13 June 2019; Published: 15 June 2019

Abstract: To meet the increasing water demand and to provide safe drinking water in Egypt, the Holding Company for Water and Wastewater (HCWW) and its affiliated companies have started a program to develop riverbank filtration (RBF) sites in all Egyptian governorates. The paper gives an overview of water quality changes as a result of RBF, during the initial phase of operation at three sites in Upper Egypt, between 2015 and 2018. Significant changes were observed for chloride, sulfate, iron, manganese, ammonium, and in the bacterial counts. After the initiation of pumping from the RBF wells, it took 2 to 8 months until stable water quality was observed for the hydrochemical parameters and 2 to 14 months for the microbiological parameters. The results showed that RBF wells should be operated continuously, to maintain the advantage of lower Fe and Mn concentrations achieved by the wash-out effect in the aquifer zone, between the river bank and the RBF wells.

Keywords: riverbank filtration; water quality; bank filtrate portion; iron; manganese; microorganisms; system costs

---

1. Introduction

Egypt's freshwater consumption is growing ten times faster than its freshwater production [1], and the current ballooning demographics show no sign of abating any time soon. The total population of Egypt increased from 22 million in 1950 to around 88 million in 2015 [2]. This rapid increase in population growth will continue for decades to come and it is likely to increase to between 120–150 million, by 2050. A high population growth rate would exaggerate the problems associated with water allocation.

Egypt is not only facing problems with an increase in water demand but also with pollution of the River Nile water through industrial, agricultural and municipal inflows. In Egypt, large amounts of untreated or poorly treated sewage are discharged into surface waters [3]. Along the River Nile, there are a total of 56 large drains that discharge water and transfer pollutants from industrial and settlement areas, and 72 drains that discharge water mainly from agricultural areas. The exports from agricultural areas and domestic wastes are considered to be the main sources of water pollutants in the River Nile [4]. The water quality of the River Nile mainly depends on the water quality in the Lake Nasser reservoir and the volume of water released from it. Despite the overall water quality of the River Nile being suitable for drinking water production using conventional treatment, accidental (oil) spills and flash floods occur frequently, which affect the operations of water treatment plants [5]. Additionally, from December to January, irrigation canals are put under maintenance (winter closure) and the water released from the Aswan dam is reduced, such that less dilution of sewage inputs occur, and some large water treatment plants suffer from higher siltation at the intake points. During this period, several small surface water abstraction units suffer from the lower river water level and use that time for maintenance, resulting in a decrease in drinking water supply [6].

Egypt relies on the River Nile for 95% of its freshwater needs for irrigation, drinking, and industrial purposes. Egypt's aquifers, which contain large amounts of fossil water that experiences little to no replenishment, cannot be abstracted easily. Additionally, while desalination of seawater is slowly picking up in the country, it still represents a very negligible amount of freshwater production overall, and comes with its own set of environmental issues. New strategies have to be developed by the governorates to overcome the water shortage. One strategy is to opt for riverbank filtration (RBF), which has been used for over 150 years in Germany and other European countries, to produce large quantities of drinking and industrial water with low cost and high quality, even during floods and droughts [7].

Riverbank filtration is the abstraction of water from aquifers that are hydraulically connected to the river, through pumping wells adjacent to the river [8]. The pumping lowers the groundwater table, such that the river water infiltrates into the aquifer. The bank filtrate percolates through the aquifer sediments towards the production wells, where it mixes with groundwater. Figure 1 shows an RBF cross-section with typical conditions from Upper Egypt. Favorable conditions include a good hydraulic connection between the river and the aquifer, erosive river flow conditions to prevent riverbed clogging, sufficient aquifer thickness (>10 m) and hydraulic conductivity ($K > 1 \times 10^{-4}$ m/s), and a low natural (pre-RBF) gradient of groundwater flow towards the river [8–10]. Such favorable hydrogeological conditions for RBF have been identified for Upper Egypt [11].

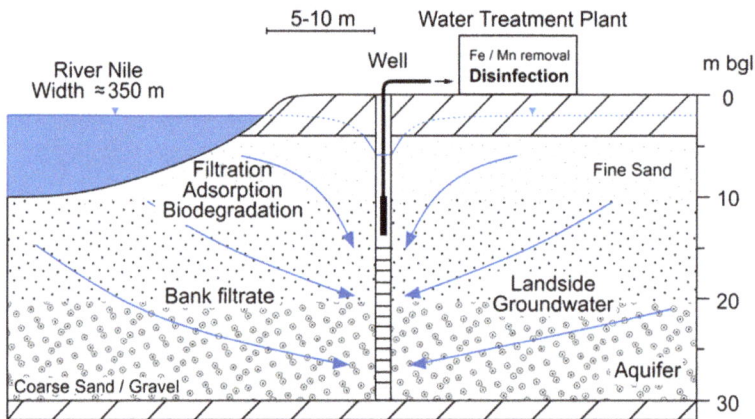

Figure 1. A generalized riverbank filtration (RBF) cross-section with typical conditions for Upper Egypt ©Grischek, HTW Dresden.

The technology in itself is quite simple, is cheaper than conventional water treatment systems, and requires little maintenance. Identifying the right location for an RBF site is a key issue. Therefore, water quality tests of river and groundwater need to be conducted at each specific site, and the composition of the riverbed and thickness and hydraulic conductivity of the adjacent aquifer need to be examined to assess the viability of a site. If an RBF scheme is properly designed, the subsurface passage of surface water through the riverbed and aquifer material provides several natural treatment processes, including filtration, biodegradation, adsorption, chemical precipitation, and improvement of water quality through redox processes [12].

According to [13], four stages of site investigation should be followed:

1. Initial site assessment, including visual reconnaissance by site visits, documentation of verbal and archived information, and in-situ sampling of river water and groundwater.

2. Basic site survey and installation of basic infrastructure: Identifying possible well locations, determining ground elevations and datum, river and groundwater monitoring locations, and construction of exploratory and monitoring wells.

3. Monitoring and determining aquifer parameters: Monitoring of river and groundwater levels and quality, river channel geometry and grain size analysis, and pumping tests.

4. Analytical or numerical groundwater flow modeling: Determining flow paths, travel times, and portions of bank filtrate and groundwater in the extracted water.

The aim of this study was to give an overview of the initiation of RBF processes at three recently developed RBF sites near Luxor and Sohag, Upper Egypt. While water quality changes during RBF, redox-zonation, and removal rates have been discussed by many authors (e.g., [14–16]), little information is available on the initial operation phase of new RBF schemes. For example, the wash-out effect of iron and manganese in aquifers between the riverbank and the production wells has been described for longer periods (in terms of years) [17,18], based on long-term monitoring, but no description has yet been provided for the initial phase. The new RBF sites in Upper Egypt make it possible to study the behavior of organic compounds and redox sensitive parameters during the initial operation phase and to draw conclusions on the required post-treatment.

2. Materials and Methods

2.1. Site Description

Results from investigations at three sites in Upper Egypt have been presented—Alsaayda near Luxor (Site L), and Eltawael (Site T) and Al-maragha (Site M) near Sohag (Figure 2).

Luxor Company for Water and Wastewater (LCWW) is an affiliated body of the Holding Company for Water and Wastewater (HCWW) that provides drinking water (according to the Egyptian standards) from 9 water treatment plants (WTPs), with a total design capacity of 3.5 Mm³/day. In the Governorate Luxor, 57% of the drinking water production is based on surface water from the River Nile and the irrigation canals, and the remaining 43% is abstracted from shallow groundwater. The Alsaayda site is located in Northern Luxor (Figure 2). One exploratory well was drilled in March 2018 and three production wells were drilled in May 2018. The production wells were located at a distance of 5 to 13 m from the right river bank (Table 1). The distance between well L1 and L2 was 25 m and the distance between well L2 and L3 was 19 m. The borehole diameter was 20 inches and the diameter of the well casing was 14 inches.

Table 1. Design parameters of RBF wells near Luxor and Sohag.

Site	Alsaayda, Luxor			Eltawael, Sohag			Al-Maragha, Sohag	
Well No.	L1	L2	L3	T1	T2	T3	M1	M2
Depth of well (mbgs)	31	31	31	36	36	36	36	36
Location of filter screen (mbgs)	10–25	10–25	6–21	18–35	18–35	18–35	18–35	18–35
Borehole diameter (inch)	20	20	20	20	20	20	20	20
Well diameter (inch)	14	14	14	14	14	14	14	14
Distance from river bank (m)	13	10	5	6	6	6	5	5
Distance from neighboring well (m)	25	25/19	19	10	10/10	10	12	12
Pumping rate (L/s)	35	35	35	35	35	35	35	35
Static groundwater level (mbgs)	2.0 *	1.8 *	1.5 *	2.2 **	2.4 **	2.4 **	2.5 ***	2.5 ***
Drawdown (m)	4.0	3.6	3.5	3.7	3.7	4.1	3.6	3.6

* May 2018, ** May 2015, *** February 2018.

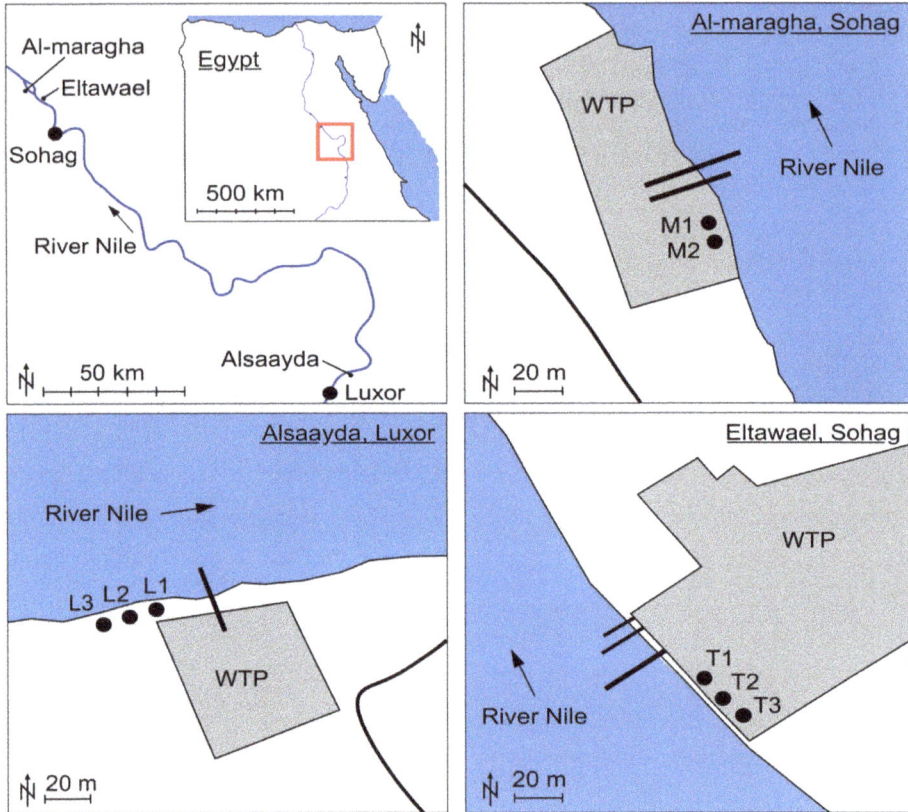

Figure 2. Location of the RBF sites near Luxor and Sohag, Upper Egypt.

The submersible pumps were operated at a rate of about 35 L/s each. The static groundwater depth was 1.5 to 2.0 mbgs (meter below ground surface), when pumping 5 to 6 mbgs. The aquifer had a thickness of about 30 m. During drilling, soil profiles were taken; the results are summarized in Table 2.

The Sohag Company for Water and Wastewater (SCWW) is another affiliated body of the HCWW that produces drinking water from 12 large WTPs, 60 small WTPs, and 450 groundwater wells, all over the governorate, with a total design capacity of 0.9 Mm³/day. Here, 60% of the drinking water production is based on surface water and 40% on groundwater. The aquifer has a thickness of about 36 m, lithological profiles from drilling are summarized in Table 2.

At the RBF site in Eltawael (Site T) (26°38′51 N, 31°38′51 E), one exploratory well was drilled in March 2015 and three production wells were drilled in May 2015. The wells were located at a distance of 6 m from the right river bank (Table 1, Figure 2). The distances between the wells T1, T2, and T3 were 10 m. The borehole diameter was 20 inches; the diameter of the well casing was 14 inches. The submersible pumps were operated at a rate of about 35 L/s each. The static groundwater depth was 2.2 to 2.4 mbgs, when pumping 5 to 6 mbgs.

At the RBF site in Al-maragha (Site M) (26°41′28 N, 31°36′48 E), one exploratory was drilled in December 2017 and two production wells were drilled in February 2018. The wells were located at a distance of 5 m from the right river bank (Table 1, Figure 2). The distance between wells M1 and M2 was 12 m. The static groundwater depth was 2.5 mbgs, when pumping 6 mbgs.

Table 2. Thickness of the sediment layers with depth (mbgs) at the RBF sites near Luxor and Sohag.

| Site | Alsaayda Near Luxor | | | Eltawael | | | Al-maragha | |
Well No.	L1	L2	L3	T1	T2	T3	M1	M2
Clay top	0–2	0–4	0–4	-	-	-	0–8	0–8
Fine sand	2–7	4–9	4–13	0–10	0–10	0–10	-	-
Medium sand	-	9–16	13–16	-	-	-	8–12	8–12
Coarse sand	7–18	-	-	10–22	10–22	10–22	12–17	12–17
Coarse sand with gravel	18–19	16–30	16–30	22–30	22–30	22–30	17–28	17–28
Gravel	19–25	-	-	30–35	30–35	30–35	28–35	28–35
Shale	>25	Not reached	Not reached	-	-	-	-	-

The depth of River Nile at Luxor is about 8 m, with a maximum of 14 m. Figure 3 shows a cross-section of River Nile, which was prepared in May 2018.

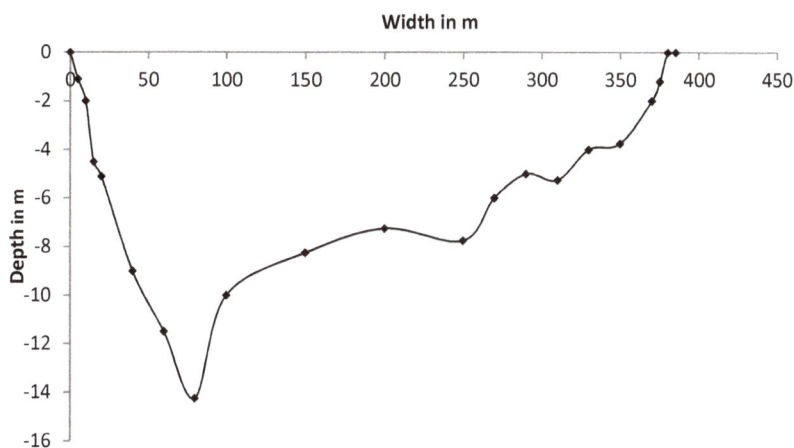

Figure 3. Cross-section of the River Nile at the RBF site Alsaayda near Luxor, May 2018.

2.2. Water Sampling and Analysis

Water sampling from the river and the wells was carried out following the Egyptian guidelines and standards [19,20]. The River Nile was sampled daily for its major parameters, such as pH, EC, turbidity, and microbiological parameters, and weekly for further parameters such as major ions. Every well in operation was sampled regularly (at least weekly) from its sampling tap. Table 3 provides all analytical methods for determining the discussed parameters.

The River Nile was sampled at the intake of several water treatment plants, along its flowpath from Luxor to Sohag. The river water quality decreased towards Sohag and further downstream, due to industrial and domestic wastewater input. Water analysis was conducted by different laboratories (local company labs and central labs). Groundwater was sampled from different wells in the catchments of the three sites, at different depths and distances from the RBF wells. Here, the differences in water quality were mainly driven by spatial differences.

From the collected dataset, the minimum, median, and maximum values were extracted. This was done, so that the outliers and short-term events such as spills or flash floods would not seriously affect the median values that were used to discuss the general effect of RBF processes and water quality.

Table 3. Parameters and analytical methods, according to the Standard Methods for Examination of Water and Wastewater [20].

Parameter	Abbreviation	Unit	Method, Equipment, Method No.
		Physical Parameters	
Electric Conductivity	EC	μS/cm	Conductivity (2510)/Laboratory method/WTW Cond. Meter, 2–55
Turbidity	Turb	NTU	Turbidity (2130)/Nephelometric method/Turbidimeter (Hach), 2–12
		Chemical Parameters	
pH	pH	-	pH (4500-H^+)/Electrometric method/ThermoScientific (Orion 3 STAR), 4–95
Alkalinity	Alk	mg/L	Alkalinity (2320)/Titrimetric method, 2–36
Total organic carbon	TOC	mg/L	TOC (5310)/C. Persulfate–Ultraviolet or Heated-Persulfate Oxidation Method, 5–29
Total Hardness	$CaCO_3$	mg/L	EDTA titrimetric method, 3–69
Ammonium	NH_4^+	mg/L	Ammonium (4500-NH_3)/Phenate method, 4–114
Chloride	Cl^-	mg/L	Chloride (4500-Cl^-)/Argentometric method, 4–75
Sulfate	SO_4^{2-}	mg/L	Sulfate (4500-SO_4^{2-})/Turbidimetric method, 4–197
Nitrate	NO_3^-	mg/L	Nitrate (4500-NO_3)/Ultraviolet spectrophotometric method/Cecil 2041 UV/VIS, 4–126
Iron	Fe	mg/L	Iron (3500-Fe)/Phenanthroline method/Cecil 2041 UV/VIS, 3–79
Manganese	Mn	mg/L	Manganese (3500-Mn)/Persulfate method/Cecil 2041 UV/VIS, 3–87
		Microbiological Parameters	
Heterotrophic Plate Count 35 °C	HPC 35 °C	count/mL	HPC (9215)/B-Pour Plate Method, 9–53
Total coliform	TC	count/100 mL	MFT (9222)/B-D, endo agar method, 9–81 for drinking water, MTFT 9221 B-C-E for intake water
Fecal coliforms	FCC	count/100 mL	MFT (9222)/Membrane filter procedure for coliform group D, thermotolerant (fecal) coliforms
		Biological Parameters	
Total Algae Count	Algae	cells/mL	Plankton (10200)/C, E and F, 10–11, 10–15, 10–17

Due to the limited budget, no extra monitoring wells could be installed. All samples were taken from the production wells. Thus, it has to be taken into account that the sampled well water was a mixture of the bank filtrates (of different age depending on the location of infiltration in the riverbed, the depth of the flow path in the aquifer, and the pumping rate of the well) and the land-side groundwater.

3. Results and Discussion

3.1. The Quality of River Nile and the Ambient Groundwater in Luxor and Sohag

Mean river water temperatures varied from 19.5–28.6 °C. The salinity of River Nile water showed an increase in the downstream direction; the electric conductivity (EC) ranged from 273–461 μS/cm (median 322 μS/cm) and the chloride concentrations were 11–27 mg/L (median 15 mg/L). Water in the main canals (Nag Hammadi West and Nag Hammadi East) had the same chemical composition as water in the river. Nevertheless, water quality deteriorated in the downstream direction, due to the disposal of municipal and industrial effluents, the inflow of agricultural drainage, as well as the decreasing water flow. Turbidity was relatively low but was above the drinking water limits, ranging from 0.3–8.5 NTU. During flash floods, turbidity could increase by a hundred-fold. This happened during a flash flood in March 2014 [21] (Figure 4) and in November 2016, when a shock load extended for about 15 days and severely affected all water treatment plants in Sohag. During those periods, the level of turbidity in River Nile reached a maximum of about 100 NTU in 2014 and 300 NTU in 2016. This caused the temporary closure of all conventional surface water treatment plants in Sohag, while the RBF wells at the Eltawael site could still be operated during flash floods periods. The surface water abstraction of the Elmonsha WTP in the Elmonsha city was shut down for 7 days during the flash flood.

The significant influence of heavy rains on the water turbidity of river water is shown in Figure 4, where the spike in March was caused by heavy rain.

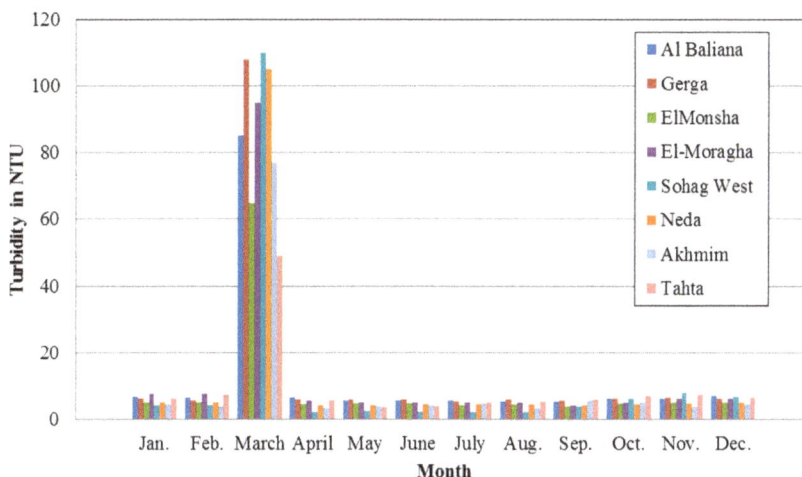

Figure 4. Impact of precipitation in March 2014 on the turbidity of the River Nile measured at the intakes of eight water treatment plants around Sohag, in 2014.

The unprecedented level of turbidity exceeded the tolerance capacity of the WTPs. As a result, all WTPs had to be shut down to a maximum of 72 h (conventional WTPs) and 100 h (compact water treatment plants).

All major ions in Nile water were generally within the limits for drinking water supply. The Nile water was alkaline and predominantly of the bicarbonate type. Total hardness ranged from 88 to 192 mg/L (as $CaCO_3$), with calcium as the main hardness constituent. The main problem in drinking water treatments is the high load of microbiological contaminants.

The ambient groundwater in the River Nile valley, near the current River Nile track, has EC values ranging from 341–1039 µS/cm with a median value of 634 µS/cm and, correspondingly, a TDS (Total Dissolved Solids) ranging from 218–665 mg/L with a median value of 406 mg/L (Table 4). The EC of the groundwater is more than twice as high as the EC of the river water. Additionally, chloride concentrations are also much higher, with a median of 30 mg/L.

In Egypt, risks might arise from the land-side part of the aquifer as it is subject to contamination, mainly from the unsecure conventional systems for sewage disposal in villages (latrines and septic tanks) but also from seepage of irrigation water containing nitrogen fertilizers [22]. However, median concentrations for ammonium and nitrate in the groundwater were found to be 0.68 mg/L and 1 mg/L, respectively, and did not indicate groundwater pollution. Maximum ammonium and nitrate concentrations found in the groundwater, at the studied sites, were 1.38 mg/L and 5.7 mg/L, respectively. Total coliform counts in some wells exceeded the drinking water standards. The median value for total coliforms in the River Nile valley aquifer was <1 CFU/100 mL. Besides pathogens and ammonium, primarily iron and manganese concentrations are relevant for water treatment design. Iron and manganese concentrations were found to be 0.012–0.76 mg/L and 0.07–1.0 mg/L, respectively.

Table 4. Standard water quality parameters, median (min–max) values, River Nile water, bank filtrate (BF), and groundwater (GW), GW: 2016–2018; Nile at Sohag and BF Sohag site T: 2015–2018; Sohag site M and Luxor site L: 2018 (*n* = number of samples).

Parameter	Unit	Standard Egypt	Ambient GW	Nile Water, Sohag	BF Luxor L * (2018)	BF Sohag T **	BF Sohag M ***
EC	μS/cm	-	634 341–1039 (*n* = 333)	322 273–461 (*n* = 290)	350 322–594 (*n* = 57)	507 388–1213 (*n* = 402)	547 452–734 (*n* = 82)
pH	-	6.5–8.5	7.5 6.9–7.9 (*n* = 329)	8.1 7.5–8.6 (*n* = 592)	7.5 7.4–7.7 (*n* = 57)	7.5 7.3–7.8 (*n* = 386)	7.7 7.5–8.02 (*n* = 82)
Turbidity	NTU	1	0.7 0.1–3.5 (*n* = 333)	3.3 0.3–300 (*n* = 517)	0.24 0.1–0.65 (*n* = 57)	0.38 0.1–0.99 (*n* = 385)	0.36 0.2–1.5 (*n* = 82)
Alkalinity	mg/L	500	297 139–477 (*n* = 327)	134 110–213 (*n* = 544)	n.d.	238 135–350 (*n* = 265)	n.d.
TDS	mg/L	-	406 218–665 (*n* = 333)	206 175–295 (*n* = 290)	224 206–380 (*n* = 57)	324 248–776 (*n* = 402)	350 289–470 (*n* = 82)
Total HardnessCaCO$_3$	mg/L	-	230 103–823 (*n* = 326)	115 88–192 (*n* = 552)	n.d.	209 150–443 (*n* = 383)	n.d.
Fe	mg/L	0.3	0.23 0.012–0.76 (*n* = 257)	<0.001 <0.001–0.19 (*n* = 205)	0.11 0.09–0.20 (*n* = 57)	0.25 0.1–0.51 (*n* = 393)	0.14 0.06–0.46 (*n* = 82)
Mn	mg/L	0.4	0.34 0.07–1.0 (*n* = 260)	<0.001 <0.001–0.16 (*n* = 199)	0.42 0.3–0.61 (*n* = 57)	0.55 0.38–1.2 (*n* = 388)	0.45 0.4–0.65 (*n* = 82)
NH$_4^+$	mg/L	0.5	0.68 0.002–1.38 (*n* = 239)	0.018 <0.002–0.5 (*n* = 188)	0.19 0.09–0.33 (*n* = 57)	0.12 0.01–0.52 (*n* = 367)	0.2 0.07–0.54 (*n* = 81)
Cl$^-$	mg/L	250	30 16–75 (*n* = 327)	15 11–27 (*n* = 289)	n.d.	25 16–64 (*n* = 381)	n.d.
SO$_4^{2-}$	mg/L	250	31 17–76 (*n* = 318)	22 14–37 (*n* = 289)	n.d.	27 17–81 (*n* = 375)	n.d.
NO$_3^-$	mg/L	45	1.0 <1–5.7 (*n* = 247)	1.5 <1–4.6 (*n* = 181)	n.d.	1.2 <1–4.6 (*n* = 263)	n.d.
TOC	mg/L	-	n.d.	2.7 2.1–3.2 (*n* = 11)	n.d.	1.2 1.1–1.4 (*n* = 17)	n.d.
HPC	CFU/mL	50	34 1–6500 (*n* = 300)	1900 220–9550 (*n* = 245)	8 0–720 (*n*=57)	2 0–1200 (*n*=378)	20 2–800 (*n*=82)
Total coliform	CFU/100 mL	<1	0 0–355 (*n* = 302)	2700 45–54,000 (*n* = 257)	10 0–240 (*n* = 57)	0 0–2410 (*n* = 369)	0 0–200 (*n* = 82)
Fecal coliform	CFU/100 mL	<1	0 0–16 (*n* = 302)	180 20–790 (*n* = 221)	0 0–120 (*n* = 57)	0 0–100 (*n* = 399)	0 0–6 (*n* = 82)

n.d.—not determined, * L—Alsaayda, Luxor, ** T—Eltawael, Sohag, *** M—Al-maragha, Sohag.

3.2. Quality of the Pumped Water from RBF Units Near Luxor and Sohag

At many RBF sites worldwide, seasonal temperature changes in river water and bank filtrates and flow-related changes in EC could be used to estimate the travel times of bank filtrate [23,24]. The temperature curve of River Nile has a comparatively low amplitude, which is not feasible for using temperature as a tracer to determine the travel times of bank filtrates. EC also does not show much fluctuation, except in the maximum values during the low-flow period of December/January. During this period, the water levels change, so the travel times are also affected by the changing gradients in the aquifer. However, the EC values of groundwater and river water differ significantly, offering the potential to use EC measurements to determine the portion of bank filtrates and land-side groundwater in the pumped water and residence times. Chloride concentration measurements might also be used, as described in [25,26]. In any case, installation of an observation well, land-side of the pumping well, is required. In this article, initial EC values and anion concentrations at the start of the well operation were assumed to be representative for groundwater quality and were used for the mixing calculations, as shown in Equation (1), where C_{PW} is the concentration in the pumped well water, C_{River} is the concentration in river water, and C_{GW} is the concentration in groundwater.

$$\text{Portion of bank filtrate in \%} = (C_{PW} - C_{River})/(C_{GW} - C_{River}) \times 100 \qquad (1)$$

The removal of organic compounds is relevant for the dissolution and release of iron and manganese along the flow path and for the required post-treatment and to inhibit the potential formation of disinfection by-products. The total organic carbon (TOC) concentration in River Nile ranges from 2.1 to 3.2 mg/L (Table 4), and has a median similar to that of the River Danube at Budapest [24], and about half of the concentration in the River Elbe [23]. Despite the low input concentration, the removal rate for TOC is relatively high, with about a 56% median and up to 62% at the RBF site Eltawael (site T), Sohag. As the samples were taken from the production well, the decrease in the TOC concentration could also be affected by a portion of land-side groundwater, presumably having a lower TOC than river water.

3.3. Water Quality Changes during the Initial Phase of RBF near Luxor and Sohag

Figures 5–14 show water quality changes during the initial phase of RBF operation for the three sites in Upper Egypt.

According to electric conductivity (EC) values (Table 4) the portion of bank filtrate is highest at the RBF site Alsaayda near Luxor (site L). Assuming that the EC value of 594 µS/cm in the first portion of the pumped water in May 2018, after the installation of wells, and an EC for River Nile water of 283 µS/cm (median from 2018, $n = 12$) were sufficiently representative to be used in the mixing calculations, the bank filtrate portion was more than 80% in the pumped RBF well water. At the other two sites at Sohag, the portion of bank filtrate was lower on average, with up to 80% at site T and about 50% at site M. Figures 5 and 6 show that the EC of the pumped water decreased with time, indicating an increasing portion of bank filtrate. This effect was most obvious for site T, where first, the groundwater was pumped and later a mixture of the groundwater and bank filtrate was pumped. A short peak in June 2016 indicated the effect of a short pause in pumping, which caused land-side groundwater flow towards the river, resulting in a high portion of groundwater with a high EC, after a restart of pumping (Figure 5).

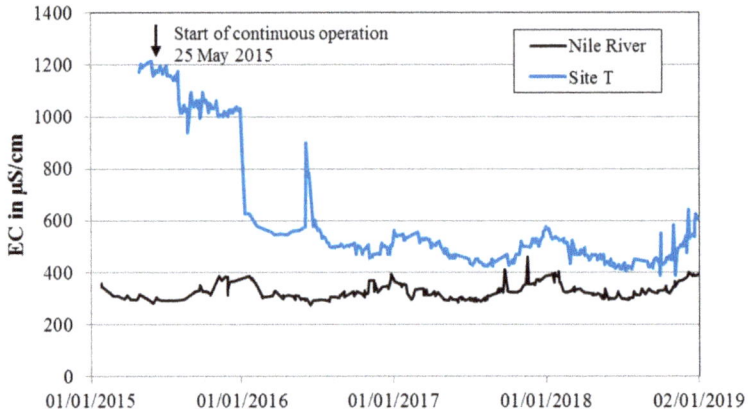

Figure 5. EC readings during the initial phase of RBF at Sohag, site T, 2015–2018.

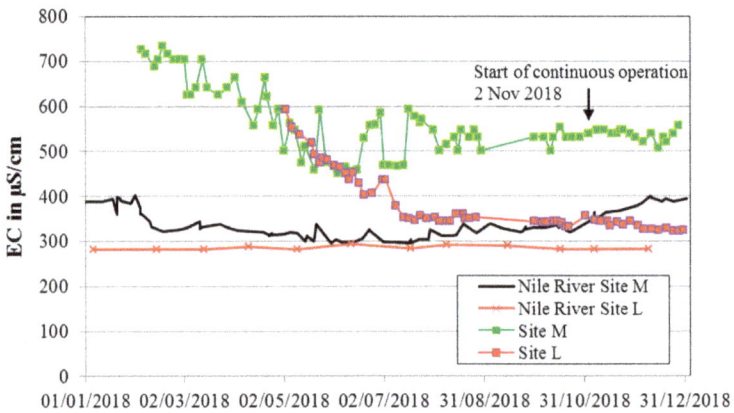

Figure 6. EC readings during the initial phase of RBF near Luxor, site L, and Sohag, site M, 2018.

Chloride, bromide, and iodide are conservative water constituents and can be used for the mixing calculations and to identify the water origin, with less ambiguity than other dissolved species [25]. The median chloride concentration of the river water was 15 mg/L. The initial chloride concentration of about 62 mg/L showed that the median chloride concentration given in Table 4 for ambient groundwater did not apply for site T. The chloride concentration in the pumped water at site T started at about 62 mg/L, and decreased after 30 weeks to about 30 mg/L, and later to a minimum of 16 mg/L (Figure 7, Table 4). This was in agreement with the observed changes in EC (Figure 5) and sulfate concentration (Figure 7). When the bank filtrate arrived at the well and comprised a major portion of the pumped water, the EC, chloride, and sulfate concentrations dropped. The time period of 30 weeks underlined that it might take months until the wells can truly abstract a certain portion of bank filtrate. The calculated portion of the pumped bank filtrate at site T (based on the calculations using the EC values) was supported by a calculation based on chloride concentrations. Taking 62 mg/L chloride for ambient groundwater, 15 mg/L as a reliable median for the river water (Table 4), and about 20 mg/L for the pumped water (Figure 7), the portion of bank filtrate would be about 90% at site T. This, in turn, underlined >50% stable removal of TOC during riverbank filtration at site T

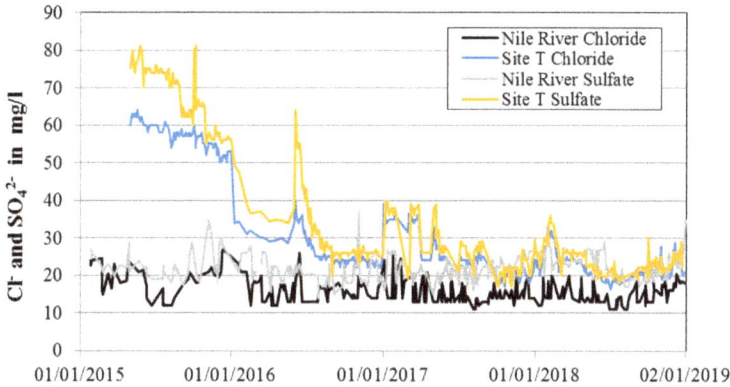

Figure 7. Chloride and sulfate concentration during the initial phase of RBF at Sohag, site T, 2015–2018.

The observed decrease in EC was in agreement with a delayed decrease in Fe and Mn concentration in the pumped water (Figures 8–10). The delay could be caused by cation exchange processes along the flow path, but was not too long because the distance between the wells and the river was quite short. There were many little peaks where Fe was about 0.1 mg/L higher than the average base line. This might be caused by the discontinuous operations of some wells at the sites. If the pumping was stopped, the land-side groundwater flowed towards the river, filling part of the aquifer between the wells and the river bank. If the pumps were switched on again, the groundwater was pumped back, causing a short-term increase in Fe and Mn concentration in the pumped water, before the average mixing determined the metal concentration. Thus, it is of utmost importance to continuously operate the RBF units (24/7).

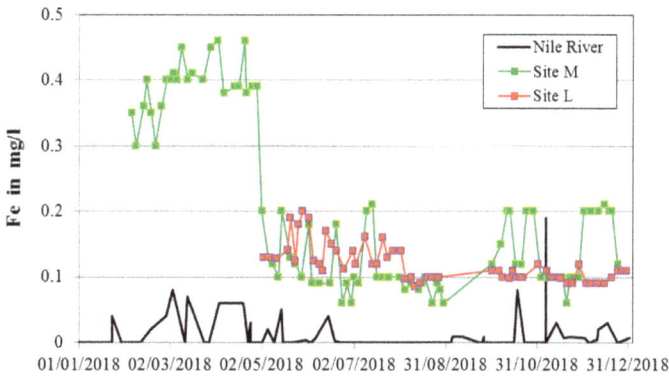

Figure 8. Fe concentration in the pumped RBF well water, during the initial phase of RBF near Luxor, site L, and Sohag, site M, 2015–2018.

The final Fe and Mn concentration depended on the mixing ratio (portion of pumped groundwater commonly having high Fe and Mn concentration) and the redox-dependent Fe and Mn concentration of the bank filtrate. The latter is determined by the river water quality, the composition of the riverbed and the redox conditions within the aquifer [17,18,27,28]. After starting the operations of new RBF wells, the Fe and Mn concentration in the pumped water changed. An optimal case would be a low portion of land-side groundwater and oxic conditions in the aquifer between the river and the wells. The TOC concentration in River Nile water was not very high and would not result in strongly anaerobic conditions in the aquifer, if e.g., 50% of the TOC was oxidized, resulting in an equivalent oxygen demand. At many RBF sites, fine particles settling on the riverbed contributed to the degradable organic carbon and oxygen demand [29]. As there was very little nitrate in the river water (median 1.5 mg/L, Table 4), denitrification was of minor importance for the redox conditions at the RBF sites in Upper Egypt. Consequently, Mn concentrations in the bank filtrate might have increased. Mn concentrations at Sohag, site T (Figure 9) were slightly above the permissible limit of 0.4 mg/L, according to the Egyptian drinking water standard, but Fe concentrations were below the permissible limit of 0.3 mg/L. The WHO (2006) has no specific standards for iron concentration but has a guideline value of 0.4 mg/L for Mn.

The lowest final Mn concentration was observed at site L near Luxor (Figure 10), where the portion of bank filtrate was the highest and nearly continuous operation was ensured. This indicated that continuous operations and a short residence time of bank filtrates in the riverbed and the aquifer prevented the release of Mn due to redox processes along the flow path.

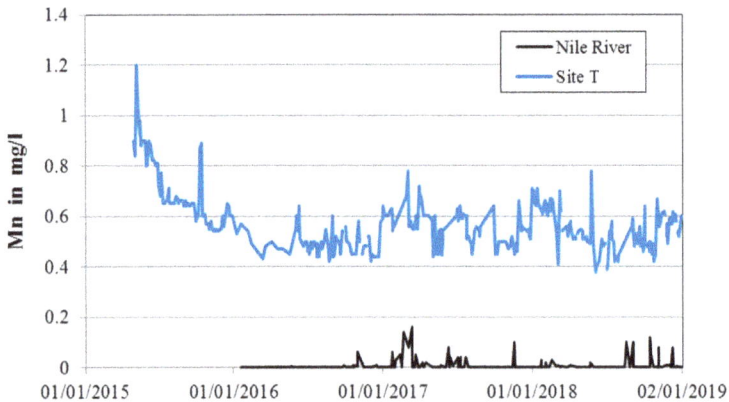

Figure 9. Mn concentration in the pumped RBF well water during the initial phase of RBF at Sohag, site T, 2015–2018.

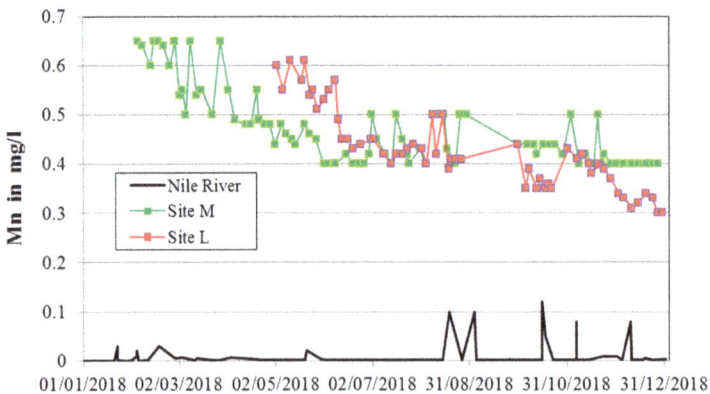

Figure 10. Mn concentration in the pumped RBF well water during the initial phase of RBF near Luxor, site L, and Sohag, site M, 2018.

Figure 11 shows a decrease in ammonium concentration in the pumped well water at Sohag, site M. Here, RBF resulted in lower ammonium concentrations compared to the abstraction of groundwater. Ammonium concentration in groundwater at site L near Luxor was already lower. At Sohag, site T, the initial ammonium concentration was 0.5 mg/L, the same as at site M, decreasing to about 0.1 mg/L in 2017, and afterwards fluctuating between 0.05 and 0.4 mg/L, with a median of 0.12 mg/L in 2018. After the initial phase, all data were below the permissible limit of 0.5 mg/L for ammonium, according to the Egyptian Drinking Water Standard. The final ammonium concentrations of 0.1–0.2 mg/L did not cause problems during disinfection.

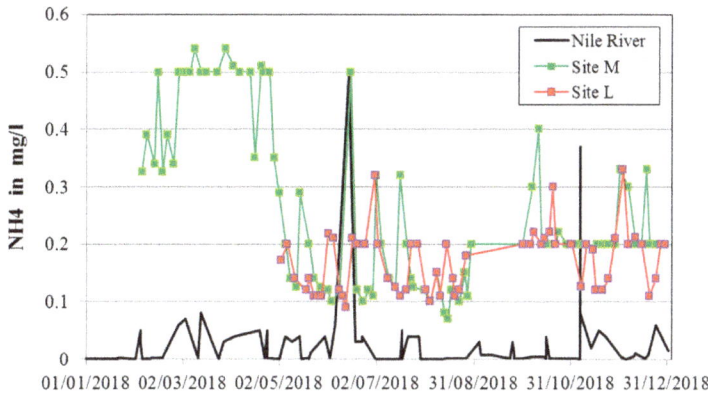

Figure 11. Ammonium concentration in River Nile water and the pumped RBF well water during the initial phase of RBF near Luxor and Sohag, 2018.

Figure 12 shows remarkable changes in the bacteria (heterotrophic plate counts (HPC)) found in the pumped RBF well water. The high levels at the start of the operation, at sites M and L, could be a result of disturbance of conditions in the subsurface, and contamination during drilling and well construction. If the upper soil enters the borehole or is even used as a filling material, (non-)fecal bacteria are released and can cause a higher HPC, until a microbiological equilibrium is achieved and biofilms are stabilized in the aquifer and the vicinity of the well. The results from site T were different in the beginning but did not indicate any contamination by sewage, as the coliform and fecal coliform counts were very low after one year of no fully continuous operation. After the initial phase, the HPC decreased to median values of 2–20 CFU/mL, compared to a median of 1900 CFU/mL in river water. The permissible limit of the HPC bacteria, according to the Egyptian Drinking Water Standards, is 50 CFU/mL.

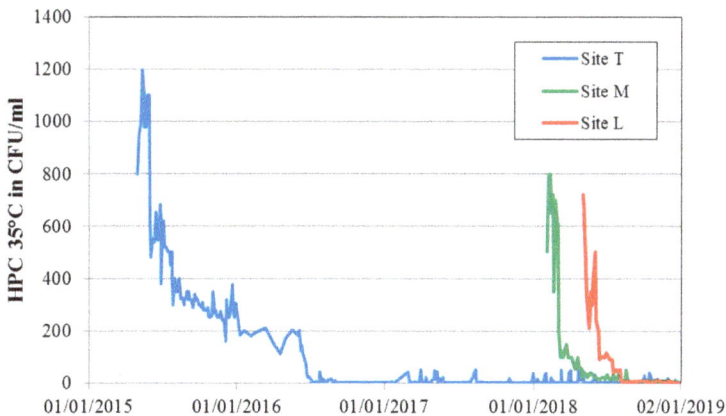

Figure 12. Heterotrophic plate counts in the pumped RBF well water during the initial phase of RBF near Luxor and Sohag, 2015–2018.

The contamination during the drilling and well installation was also obvious from the observed total and fecal coliforms, during the initial phase shown in Figures 13 and 14. The long-lasting effect at site T was unacceptable whereas at site M (low contamination) and site L the pumping resulted in a clean-up of the disturbed aquifer within 2 to 3 months. After the initial phase, an effective removal of coliforms and fecal coliforms was proved, with a very low number of positive detections of coliforms (Table 4). Despite this, disinfection is a required post-treatment step. During the initial phase, the dosage of disinfectants (e.g., chlorine) should be higher to ensure safe drinking water supply. This might also be required during floods, when the riverbed is eroded and the travel time of bank filtrate is shorter, due to a higher gradient.

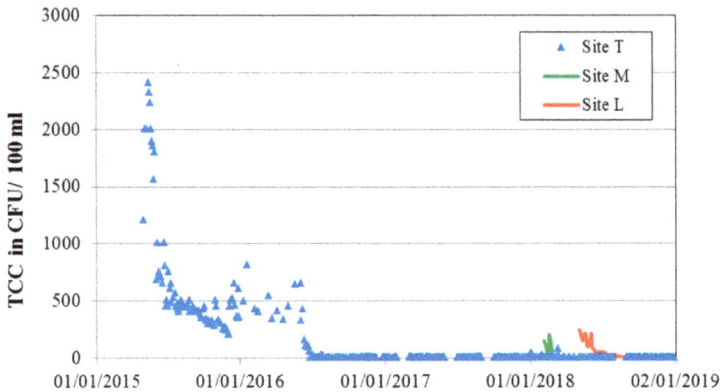

Figure 13. Total coliform counts in pumped RBF well water during the initial phase of RBF near Luxor and Sohag, 2015–2018.

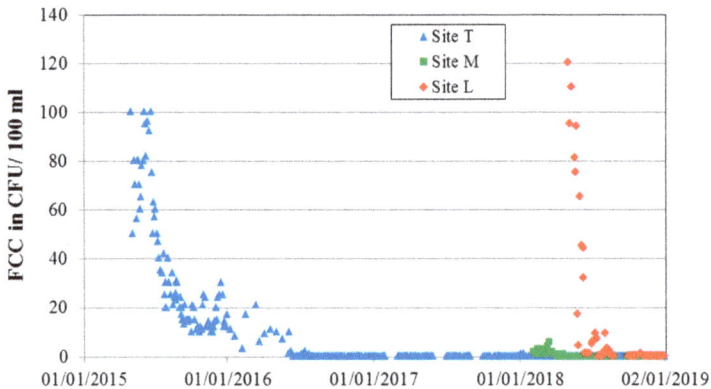

Figure 14. Fecal coliforms in the pumped RBF well water during the initial phase of RBF near Luxor and Sohag, 2015–2018.

Table 5 shows the observed time until nearly constant values were observed, for relevant parameters at the three RBF sites. For most parameters, it took between 2 and 4 months until a stable water quality was achieved at sites L and M. At site T, it took much longer. Looking at the available data from drilling and well design, no obvious reason was found to explain the longer time for site T. One reason could be a stronger gradient of groundwater flow towards the river. This would also explain the stronger fluctuation in Fe concentration and chloride concentration (Figure 8)—if the pumps are switched off for only a short time, groundwater fills the cone of depression and dominates the quality of the pumped water, again. This again underlines the need for a continuous operation of the pumps, to keep Fe and Mn concentrations low.

Table 5. Time (in months) of RBF unit operation until stable water quality was achieved for the pumped water.

Parameter	Site L	Site T	Site M
EC	2	7	3
Cl⁻	n.d.	7	n.d.
Fe	3	n.a.	3
Mn	>7*	8	4
NH_4^+	n.a.	6	3–4
HPC	3.5	14	4
TCC	3.5	13	2
FC	2.5	13	2

n.d.—not determined, n.a.—not applicable, * still decreasing.

4. Sustainability and Cost of RBF

RBF is a sustainable low-cost technology applicable in Egypt (Table 6) and could be integrated into the conventional WTPs allocated on the Nile River banks as an additional source to secure water supply during accidental oil spills and extreme climate events. The increased water demand, together with the limited available surface water resources, make RBF a potential alternative for drinking water supply, especially in Upper Egypt [11,21].

Many types of water treatment technology are used in Upper Egypt (Table 6), depending on criteria such as population number, water use (drinking or industries), and cost. To cover a large population, e.g. >1 million consumers, water companies tend to establish large water treatment plants with capacities >17,000 m³/day to 1 Mm³/day or more.

RBF units commonly consist of abstraction wells and a disinfection unit to provide safe drinking water. If Mn and/or Fe concentrations are above the threshold set by the drinking water standard, aeration and sand filtration have to be added as treatment steps.

Small conventional WTPs (rapid sand filters) are plants similar to the large ones, with same treatment steps and chemicals but with a capacity less than 17,000 m³/day. The small conventional WTPs are used for remote areas with limited population or for industrial water supply. This type has some disadvantages, such as high initial cost, high running cost, and a higher energy consumption.

Direct infiltration WTPs (slow sand filters) use large sand and gravel filters to treat water without any coagulant dosage. Chlorine gas is used for disinfection in most cases. These WTPs only produce small quantities of drinking water at a high operation cost.

Treatment of water from groundwater wells commonly requires Fe and Mn removal and disinfection, as in many parts of Egypt, people have to dispose-off sewage in the ground, resulting in groundwater quality deterioration. Fe and Mn removal is based on aeration or dosage of potassium permanganate and subsequent filtration, which affects the operation costs.

Compact units (built-in units) produce up to 2160 m³/day/unit. These mobile units can be moved to any place and are typically installed on small canal banks to provide drinking water for small communities. The basic process is sand filtration in small closed metal cylinders, each unit having three of them, and treatment processes are the same as in the conventional WTPs (coagulation/flocculation).

Table 6 indicates low capital costs for RBF units (without Fe/Mn removal) and small conventional WTPs. High capital costs for (deep) groundwater wells and subsequent treatment are affected by the deep drilling of the wells and commonly require manganese removal. Additionally, the operational costs are lowest for the RBF units, as compared to the groundwater wells, due to a lower drawdown and associated lower energy demand. If Fe/Mn removal is required, the operational costs would be more similar to those of groundwater wells.

Table 6. Cost estimates for different abstraction and treatment schemes in Sohag, Egypt [30].

Capacity and Cost	RBF Unit	Small Conventional WTP	Direct Infiltration WTP	Groundwater Well	Compact Unit
Capacity (m³/day)	3024	8640	3465	2160	2160
Capital cost (Million EGP/Unit)	0.6 *	30–40	60	0.75 *–10	15
Operational cost (Million EGP/Unit)	≈0.05 *	≈0.5	≈1.5	≈0.1 *	≈0.7

* without Fe/Mn removal, only disinfection with chlorine.

5. Conclusions

RBF can serve as a pre-treatment for waterworks; at some sites only disinfection is required as a further treatment step for drinking water production. Results from water quality monitoring during the initial phase of the RBF operation in Upper Egypt have demonstrated a good surface and groundwater interaction, and favorable hydrogeological conditions. Data from all investigated sites showed an efficient removal of turbidity and bacteria during RBF. Special care should be taken to prevent well contamination during the drilling and installation process. TOC removal was calculated to about 60%, based on measurements at a site at Sohag. Fe and Mn concentration decreased at all sites, compared to the initial concentrations in groundwater. At some sites, removal of Mn is required to meet the drinking water quality standard.

After initiation of pumping from the RBF wells, it might take 2 to 12 months until stable water quality is gained for the pumped water. Decisions on adequate further treatment—especially if Fe/Mn removal is required—should be made only after a monitoring period of a few months, to allow an optimal design for post-treatment. The RBF units should be operated continuously to prevent fluctuations in water quality and to limit the pumped portion of land-side groundwater, which commonly has higher Fe and Mn concentrations.

RBF can provide large volumes of drinking water with a high quality at low cost. The capital and operating costs of the RBF units are lower, compared to conventional water treatment plants. The highlighted conditions and advantages of RBF in Upper Egypt underline that RBF should be considered as an option for water supply, without requiring any further treatment besides disinfection or as a pre-treatment step at some sites, especially if the capacity of the existing water treatment plants needs to be increased.

Author Contributions: R.A.W., A.S., and T.G. prepared the article draft. A.S. collected site data and was responsible for the monitoring program, including water analysis. All authors reviewed and edited the article.

Funding: All primary data was collected within the UN Human Settlements Programme of the UN-Habitat, HCWW, SCWW, and LCWW. Guidance on well design and interpretation of water quality data was supported within the framework of the AquaNES project, which has received funding from the European Union's Horizon 2020 Research and Innovation Program under Grant No. 689450.

Acknowledgments: The authors would like to thank staff members of the Holding Company for Water and Wastewater, SCWW, and LCWW for their support in water sampling, analysis, and site management.

Conflicts of Interest: The authors declare no conflict of interest. The funding sponsors had no role in the design of the study; in the collection, analyses, or interpretation of data; in the writing of the manuscript, and in the decision to publish the results.

References

1. Deutsches Wissenschaftszentrum (DWZ) Cairo. *River Bank Filtration Expert Workshop under the Framework of the Egyptian German Water Cluster Report*; Deutsches Wissenschaftszentrum (DWZ) Cairo: Cairo, Egypt, 26 September 2016.

2. Central Agency for Public Mobilization and Statistics. Census Population. September 2017. Available online: https://censusinfo.capmas.gov.eg (accessed on 16 February 2019).
3. Wahaab, R.A. *One System Inventory: Environment Theme—Egypt Nile Basin Initiative*; Eastern Technical Regional Office: Addis Ababa, Ethiopia, 2006.
4. Yousry, M.; El-Sherbini, A.; Heikal, M.; Salem, T. Suitability of water quality status of Rosetta branch for west Delta water conservation and irrigation rehabilitation project. *Water Sci.* **2009**, *46*, 47–60.
5. Yehia, A.G.; Fahmy, K.M.; Mehany, M.A.S.; Mohamed, G.G. Impact of extreme climate events on water supply sustainability in Egypt—Case studies in Alexandria region and Upper Egypt. *J. Water Clim. Chang.* **2017**, *8*, 484–494. [CrossRef]
6. HCWW. RBF in Egypt—Strategy and Policy. In *Holding Company for Water and Wastewater Report*; HCWW: Cairo, Egypt, 2018.
7. Grischek, T.; Bartak, R. Riverbed clogging and sustainability of riverbank filtration. *J. Water* **2016**, *8*, 604. [CrossRef]
8. Ray, C.; Melin, G.; Linsky, R.B. *Riverbank Filtration: Improving Source Water Quality*; Springer: Dordrecht, The Netherlands, 2002.
9. Grischek, T.; Schubert, J.; Jasperse, J.L.; Stowe, S.M.; Collins, M.R. What is the appropriate site for RBF. In *Management of Aquifer Recharge for Sustainability, Proceedings of the ISMAR 6, Phoenix, AZ, USA, 28 October–2 November 2007*; Fox, P., Ed.; Acacia: Phoenix, AZ, USA, 2007; pp. 466–474.
10. Grischek, T.; Ray, C. Bank filtration as managed surface—groundwater interaction. *J. Water* **2009**, *5*, 125–139. [CrossRef]
11. Ghodeif, K.; Grischek, T.; Bartak, R.; Wahaab, R.; Herlitzius, J. Potential of river bank filtration (RBF) in Egypt. *Environ. Earth Sci.* **2016**, *75*, 671. [CrossRef]
12. Hiscock, K.; Grischek, T. Attenuation of groundwater pollution by bank filtration. *J. Hydrol.* **2002**, *266*, 139–144. [CrossRef]
13. Sandhu, C. A Concept for the Investigation of Riverbank Filtration Sites for Potable Water Supply in India. Ph.D. Thesis, Faculty of Environmental Sciences, TU Dresden, Division of Water Sciences, HTW Dresden, Dresden, Germany, 2015.
14. Henzler, A.F.; Greskowiak, J.; Massmann, G. Seasonality of temperatures and redox zonation during bank filtration—A modeling approach. *J. Hydrol.* **2016**, *535*, 282–292. [CrossRef]
15. Stuyfzand, P.J. Fate of pollutants during artificial recharge and bank filtration in the Netherlands. In *Artificial Recharge of Groundwater*; Peters, J.H., Ed.; Balkema: Rotterdam, The Netherlands, 1998; pp. 119–125.
16. Sandhu, C.; Grischek, T.; Börnick, H.; Feller, J.; Sharma, S.K. A water quality appraisal of some existing and potential riverbank filtration sites in India. *J. Water* **2019**, *11*, 215. [CrossRef]
17. Grischek, T.; Paufler, S. Prediction of iron release during riverbank filtration. *J. Water* **2017**, *9*, 317. [CrossRef]
18. Paufler, S.; Grischek, T.; Benso, M.; Seidel, N.; Fischer, T. The impact of river discharge and water temperature on manganese release from the riverbed during riverbank filtration—A case study from Dresden, Germany. *J. Water* **2018**, *10*, 1476. [CrossRef]
19. *Egyptian Standard Specifications for Potable Drinking Water*; Decision of the Minister of Health and Population No. 458; Minister of Health and Population: Cairo, Egypt, 2007.
20. American Water Works Association (AWWA); Water Environment Federation (WEF); American Public Health Association (APHA). *Standard Methods for the Examination of Water and Wastewater (SMWW)*, 23rd ed. 2018. Available online: https://doi.org/10.2105/SMWW.2882.219 (accessed on 1 May 2019).
21. Ghodeif, K.; Paufler, S.; Grischek, T.; Wahaab, R.; Souaya, E.; Bakr, M.; Abogabal, A. Riverbank filtration in Cairo, Egypt—Part I: Installation of a new riverbank filtration site and first monitoring results. *Environ. Earth Sci.* **2018**, *77*, 270. [CrossRef]
22. Hoehn, E.; Cirpka, O.A. Assessing residence times of hyporheic ground water in two alluvial flood plains of the Southern Alps using water temperature and tracers. *Hydrol. Earth Syst. Sci.* **2006**, *10*, 553–563. [CrossRef]
23. Grischek, T. Management of RBF along the Elbe River. Ph.D. Thesis, Department of Water Sciences, Dresden University of Technology, Dresden, Germany, 2003. (In German).
24. Nagy-Kovács, Z.; Davidesz, J.; Czihat-Mártonné, K.; Grischek, T. Water quality changes during riverbank filtration in Budapest, Hungary. *J. Water* **2019**, *11*, 302. [CrossRef]

25. Trettin, R.; Grischek, T.; Strauch, G.; Mallen, G.; Nestler, W. The suitability and usage of ^{18}O and chloride as natural tracers for bank filtrate at the Middle River Elbe. *Isot. Environ. Health Stud.* **1999**, *35*, 331–350. [CrossRef]

26. Davis, S.N.; Whittemore, D.O.; Fabryka-Martin, J. Uses of chloride/bromide ratios in studies of potable water. *Ground Water* **1998**, *36*, 338–350. [CrossRef]

27. Paufler, S.; Grischek, T.; Bartak, R.; Ghodeif, K.; Wahaab, R.; Boernick, H. Riverbank filtration in Cairo, Egypt—Part II: Detailed investigation of a new riverbank filtration site with a focus on manganese. *Environ. Earth Sci.* **2018**, *77*, 318. [CrossRef]

28. Schoenheinz, D.; Grischek, T. Behavior of dissolved organic carbon during bank filtration under extreme climate conditions. In *Riverbank Filtration for Water Security in Desert Countries*; Ray, C., Shamrukh, M., Eds.; Springer Science + Business Media B.V.: Berlin, Germany, 2011; pp. 51–67.

29. Grischek, T.; Hiscock, K.M.; Metschies, T.; Dennis, P.; Nestler, W. Factors affecting denitrification during infiltration of river water into a sand and gravel aquifer in Saxony, Germany. *Water Res.* **1998**, *32*, 450–460. [CrossRef]

30. SCWW. Internal Report. In *Sohag Company for Water and Wastewater*; SCWW: Sohag, Egypt, 2016.

MDPI

Article

Microturbines at Drinking Water Tanks Fed by Gravity Pipelines: A Method and Excel Tool for Maximizing Annual Energy Generation Based on Historical Tank Outflow Data

Thomas John Voltz * and Thomas Grischek

Division of Water Sciences, University of Applied Sciences Dresden (HTWD), PF 120701,
D-01008 Dresden, Germany
* Correspondence: thomas.voltz@htw-dresden.de

Received: 6 May 2019; Accepted: 29 June 2019; Published: 9 July 2019

Abstract: Wherever the flow of water in a gravity pipeline is regulated by a pressure control valve, hydraulic energy in the form of water pressure can instead be converted into useful mechanical and electrical energy via a turbine. Two classes of potential turbine sites exist—those with (class 1, "buffered") and those without (class 2, "non-buffered") a storage tank that decouples inflow from outflow, allowing the inflow regime to be modified to better suit turbine operation. A new method and Excel tool (freely downloadable, at no cost) were developed for determining the optimal hydraulic parameters of a turbine at class 1 sites that maximize annual energy generation. The method assumes a single microturbine with a narrow operating range and determines the optimal design flow rate based on the characteristic site curve and a historical time series of outflow data from the tank, simulating tank operation with a numerical model as it creates a new inflow regime. While no direct alternative methods could be found in the scientific literature or on the internet, three hypothetically applicable methods were gleaned from the German guidelines (published by the German Technical and Scientific Association for Gas and Water (DVGW)) and used as a basis of comparison. The tool and alternative methods were tested for nine sites in Germany.

Keywords: water supply; storage tank; drinking water hydropower; turbine; energy generation; renewable energy

---

1. Introduction

1.1. Advantages and Characteristics of Potential Drinking Water Hydropower Facilities

Hydropower is among the renewable energy sources with the best life-cycle energy balance—the ratio between the energy output (e.g., electricity) and the energy input required to manufacture, install, operate and dispose of the infrastructure and equipment needed for energy generation [1,2]. Drinking water hydropower has further advantages compared to much larger hydropower facilities at dams, weirs and run-of-river schemes, including high water quality (enabling long lifetime and low-maintenance operation), minimal to no environmental impact (due to a previously existing, closed system) and relatively good economic viability [2–4]. Thanks to technological progress, capitalizing on drinking water hydropower potential is increasingly becoming economically viable even at the bottom of the capacity spectrum, below 10 kW of available hydraulic power [4–9]. Practical methods and easy-to-use design tools play an important role in facilitating the implementation of such projects.

A gravity pipeline is the central element of most drinking water hydropower schemes, connecting a higher-elevation water source to lower-elevation water storage tanks and water users. With respect to hydropower development, there are two relevant types of gravity pipelines: those with and those without pressure control at points downstream, such as transfer stations or outlets into storage tanks. Gravity pipelines with pressure control valves can be thought of as possessing "surplus" potential energy, which can be "harvested" by using a turbine rather than "wasted" by using a pressure control valve, which "throttles" the flow of water (see Appendix A.1 for more background on gravity pipelines and the hydraulic explanation for surplus energy).

There are two main classes of eligible turbine sites for drinking water hydropower (as illustrated by Figure 1), in which the flow rate is either

1. decoupled from uncontrolled downstream water use through a storage tank ("buffered"), or
2. determined by uncontrolled water use in the downstream supply zone(s) ("non-buffered").

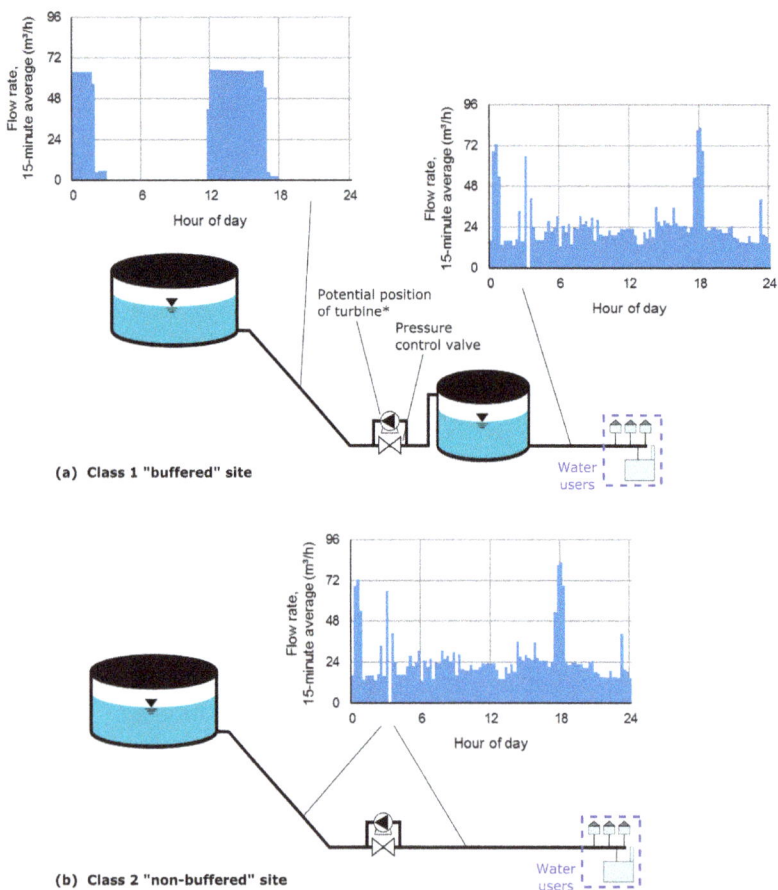

(a) Class 1 "buffered" site

(b) Class 2 "non-buffered" site

Figure 1. Key difference between class 1 ("buffered") and class 2 ("non-buffered") sites, based on data from 27 April 2017 from the class 1 site "Voigtsgrün" in Saxony, Germany (the isolated, very high peaks occurring between 00:00 and 04:00 and at 18:00 are due to a beer brewery). *This illustration only intends to show the turbine in parallel to the control valve; the turbine does not need to be located above the control valve and is typically located at nearly the same elevation.

Table 1 summarizes the most relevant consequences for drinking water hydropower development that follow from the presence or absence of pressure control and a storage tank. As sites without pre-existing pressure control are not promising for hydropower development, they will not be further considered in this paper.

Table 1. Matrix of consequences for drinking water hydropower development, depending on the type of gravity pipeline and site.

Water Supply Site / Gravity Pipeline	with Storage Tank (Class 1, "Buffered")	without Storage Tank (Class 2, "Non-Buffered")
with pressure control	Hydropower is very practical, as the storage tank provides flexibility in re-defining the inflow regime	Hydropower is possible, but may require a complex design to accommodate high variability in flow rate and pressure due to uncontrolled downstream water use
without pressure control	Hydropower is theoretically possible, but would reduce inflow rate if installed at outlet of existing pipeline, which may negatively affect supply reliability	

1.2. Available Literature and Comparison of Class 1 vs. Class 2 Sites

Common examples of class 2 sites include transfer nodes between pressure zones or a gravity-fed water treatment plant between a gravity source and destination supply tank (e.g., [2,5,8,10–12]). They are most often located within the water "distribution" section of the water supply chain [13]. In contrast to class 1 sites, class 2 sites offer no flexibility in the choice of flow rate, such that a turbine must be designed to efficiently operate over a broad spectrum of flow rates and corresponding available pressure heads. As the pressure may fluctuate both downstream and upstream of the turbine installation site, class 2 sites generally require more information and a more complex, customized design process. Based on a review of the scientific literature, the authors perceive a dominant focus on class 2 sites, with many researchers devising design algorithms and control solutions using microturbines and pumps-as-turbines (PATs) [7–9,13–22].

Class 1 sites are generally intermediate (such as break-pressure tanks (BPTs)) or terminal storage tanks with pressure control valves [6], typically located within the water "transmission" section of the water supply chain [13]. In the authors' experience and estimation, class 1 sites are common and represent a relatively cost-effective opportunity for renewable energy generation. They offer the water supplier flexibility (positively correlated with storage volume) in determining the rate, duration and timing of tank filling (and thus turbine operation), as this must not necessarily occur simultaneously with the downstream use in the supply zone (see Figure 1). Assuming that the inflow rate is not restricted (e.g., by the naturally occurring, seasonally varying flow from a mountain spring), the inflow regime can often be freely adjusted to maximize energy generation through a turbine, regardless of how it was managed in the past [12].

While comparatively less common, there are also studies that focus on class 1 sites. Several authors present case studies of sites at the bottom of dams [7], spring-fed pipelines [23] and storage tanks in general [24,25]. Others focus on BPTs, both isolated [6] and in series along a long-range pipeline [26,27]. However, many if not all of these authors effectively treat class 1 sites as if they were class 2 sites, assuming that the inflow regime remains unchanged and attempting to design turbines to fit this inflow regime, rather than adjusting it to allow the choice of a single, optimized turbine. Thus, they do not use the fundamental technical advantage that separates class 1 from class 2 sites [12,28]. For more background on class 1 vs. class 2 sites, see Appendix A.3.

1.3. Filling a Gap in Currently Available Design Methods for Class 1 Sites

Several other authors present methods that might be useful for class 1 sites, but the effort required to successfully adapt the methods and restrictions imposed by technical understanding and software availability for the authors' intended user group (practitioners at water supply companies and non-research water professionals) speak against pursuing this further. For instance, since the authors set out to create a tool that is free to use and based on a common tool (i.e., Microsoft Excel, which in the authors' experience is standard practice at most water supply companies), and software such as MATLAB is non-standard and requires specialized knowledge, possible adaptation of such algorithms was not considered. Furthermore, while some authors provide the basic underlying equations, the software or code in which the algorithms are implemented is not made accessible to the reader and therefore cannot be easily tested or used.

The 2016 guidelines published by the German Technical and Scientific Association for Gas and Water (DVGW) [12] identify the flexibility advantage provided by class 1 "buffered" sites, and imply that with additional data analysis a turbine can be "ideally tuned" to "balance the demands of storage management and energy generation". However, they offer no specific instructions or examples for how to translate this information into actual design parameters (flow rate and pressure drop). This is precisely the gap that the authors intend to fill with the method and accompanying Excel tool presented here. As the authors of this paper have encountered many class 1 sites and experienced an unfulfilled need for a design solution, they decided to develop their own solution. While designing turbine systems for class 2 sites represents the greater technical challenge, a practical design method for class 1 sites appears to be low-hanging fruit that has yet to be sufficiently plucked.

Given a recently resurging interest in drinking water hydropower in Germany [29–32] and elsewhere [6,7,19], the authors intend to facilitate the installation of such facilities by making it easier for small and medium water suppliers to estimate their hydropower potential and perform the basic design (determination of turbine parameters) themselves. The tool is implemented in Microsoft Excel for Windows (Office 2010 or newer), available free of cost for download at the authors' university website [33] and intended to be easily useable by practicing engineers or water professionals. If successful, it should both save time and reduce users' inhibitions about approaching this non-standard topic. To the authors' knowledge, it is the only such tool freely available on the internet.

This paper will explain the methodology behind the Excel tool, along with three other methods covered in DVGW guidelines that serve as a useful basis of comparison. These four methods will be applied to nine sites in Germany, and the results will be used to show the method in practice as well as providing a sense of the technical and economic potential that is available at class 1 drinking water hydropower sites.

2. Materials and Methods

2.1. Design Premise

The intention of this method and related Excel tool is to facilitate the implementation of microturbines in water supply systems. This is accomplished by automating the calculations required to determine the pressure head and flow rate for a hypothetical turbine that will (to the fullest extent possible with the given data) lead to the maximum electrical energy generation (kWh/annum, or kWh/a) at a given class 1 site. A basic economic analysis is also performed to provide an estimate of the project's viability.

While the Excel tool includes four calculation options to accommodate all anticipated variations on data availability for a given site (see Appendices B.1 and B.2 for a more detailed explanation), optimization of the turbine design parameters is only possible using option 1 or 2 (both slight variants on the same method), which take historical flow data into account. The primary advantage of options 1 and 2 is that there is no need to be biased by the typical inflow regime of past operations. The analysis of historical data thus allows the determination of turbine parameters from "first principles", rather than by analogy from the previous inflow rate, which may have been selected for reasons unknown to the current responsible engineer or site operator. Critically, this assumes that historic water use patterns are a reliable proxy for future water use. This paper will present the main design algorithm for calculation options 1 and 2 in its recommended form. The tool provides some additional options for the user to customize the design solution (e.g., regarding the bypass flow rate and multipliers for the outflow rate (Q_{out}) to reflect anticipated future changes), which are further described in Appendices B.3 and B.4 and the tool itself.

This method was inspired by Haakh [28], who provided a foundational and mathematically thorough treatment of turbine and pump operation in the context of water supply systems. The authors expect that the Excel tool and this article will provide a somewhat simplified, more user-friendly approach to solving a sub-set of the design challenges laid out there, with a focus on class 1 "buffered" sites. The following sections describe the essential tasks performed by the tool.

2.2. Determining the Characteristic Site Curve

Also known as the "system" curve, this describes the relationship between the exerted pressure and flow rate in the inflow pipeline as observed from a downstream reference point within the pipeline. It can also be thought of as the pressure "available" as a function of the chosen flow rate. Typically, this reference point is taken just upstream of the pressure control valve, to capture the hydraulic conditions relevant for a potential turbine. The available pressure head is equal to the maximum pressure head (h_{max}, at flow rate = 0) minus the head loss up to the point of interest (h_{loss}) and the pressure head just downstream of the control valve ($h_{downstream}$), which is necessary for the water to reach the storage tank. Since head loss increases proportionally to the square of flow rate, the curve can be well approximated (Equation (1)) with a downward-facing parabola with its vertex centered on the positive y-axis (flow rate = 0), crossing the positive x-axis at the maximum possible flow rate (pressure exerted on reference point upstream of valve = 0), at which the pressure control valve is completely open and offers no hydraulic resistance (see Figure 8 in Section 3.1 for an example). This is predicated on a simplifying assumption that assumes a constant coefficient of major pipe friction loss (defined in the U.S. as "lambda", λ), which is known to be flow rate-dependent (e.g., as captured in Moody's diagram), but is valid for conditions commonly encountered at water supply sites. The resulting second-order polynomial equation takes the following form:

$$h_{available} \, (m) \; = \; h_{max} - K_{loss} \cdot Q^2 - h_{downstream} \tag{1}$$

where K_{loss} is the head loss coefficient, which integrates the major head loss due to pipe friction and minor head loss(es) due to sources of local resistance (e.g., pipe bends, partially opened valves upstream). h_{loss} as defined in Figure A1 (Appendix A.1) is the central term in Equation (1), the product of K_{loss} with the square of the flow rate Q. While h_{max}, K_{loss} and $h_{downstream}$ could be determined analytically, the authors found empirical determination to be both more accurate and less labor-intensive. To empirically determine this parabolic curve, two points must be defined, which requires measuring five field data points:

1. Q_{1_inflow}: non-zero inflow rate at control valve, position 1 (e.g., at normal operating flow)
2. $h_{1\_upstream}$: pressure head upstream of control valve, position 1 (e.g., at normal operating flow)
3. $h_{downstream}$: pressure head just downstream of the control valve, which is necessary for the water to reach the (frequently somewhat higher) storage tank (e.g., at normal operating flow)
4. Q_{2_inflow}: inflow rate at control valve, position 2 (ideally zero flow; i.e., closed valve, but can also be a second, sufficiently different non-zero flow rate than Q_1, if closing the valve is not feasible)
5. $h_{2\_upstream}$: pressure head upstream of control valve, position 2 (ideally at zero flow; see above)

These data can usually be easily obtained, for example during a 30-min visit to a site or based on past data already acquired using a supervisory control and data acquisition (SCADA) system. If feasible, long-term measurements as well as further points along the site curve provide a more accurate picture of the real site conditions and reduce uncertainty in the resulting turbine performance. Since K_{loss} can effectively be considered a constant, it can be determined indirectly by substituting all measured values into Equation (1). Q_{max} can then be determined simply by setting Equation (1) to zero and solving for Q (Equation (2)).

$$Q_{max} = \sqrt{\frac{h_{max} - h_{downstream}}{K_{loss}}} \tag{2}$$

The reader should note that the timing of the measurement of $h_{1\_upstream}$ and $h_{2\_upstream}$ will impact the value of h_{max}, since the pressure measured depends to a small degree on the water level in the upstream tank. Ideally, these measurements should be conducted when the upstream tank is known to be at its typical lowest level, as this will provide a conservative estimate of the pressure that will at a minimum be reliably available to a potential turbine. Furthermore, the flow-dependent head losses between the control valve and storage tank are integrated into the measurement of $h_{downstream}$, if performed using a pressure gauge. The change in these losses (and therefore a change in $h_{downstream}$) due to a deviation in the turbine flow rate from Q_{1_inflow} (normal operating flow) is likely negligible, but can also be easily measured by changing the flow rate to the potential future turbine flow rate. The value for $h_{downstream}$ is usually primarily determined either by the height of the pipe outlet above the turbine (as portrayed in Figure 2) or the water level in the downstream tank (as portrayed in Figure A1), in case the pipe outlet is at the bottom of the downstream tank. Figure 2 provides a visual explanation of some of these terms for the typical case as encountered by the authors in Germany.

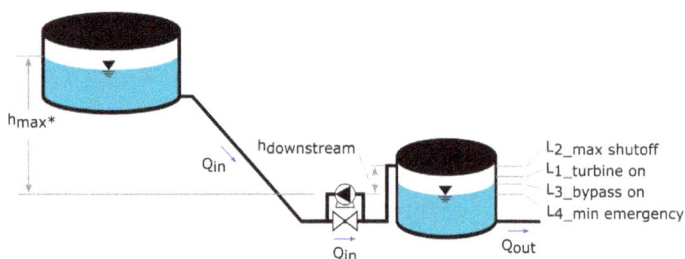

Figure 2. Visual explanation of some key parameters used in the method, as introduced in Sections 2.2 and 2.5. * Note that h_{max} depends on the current water level in the upstream tank, which is why the measurement of $h_{1\_upstream}$ and $h_{2\_upstream}$ should be conducted when the upstream tank is known to be at its typical lowest level.

2.3. Calculating the Hydraulic Power Available to the Turbine

The hydraulic power carried by the flowing, pressurized water and ultimately available to a turbine can be calculated using Equation (3):

$$P_{hydraulic} (kW) = \rho \cdot g \cdot Q \cdot h_{available} \cdot 1000 \tag{3}$$

where ρ is the density of water (kg/m^3), g is gravitational acceleration (m/s^2), Q is the flow rate (m^3/s) and $h_{available}$ is the available pressure head (m). Since ρ and g can be assumed constant (1000 kg/m^3 and 9.81 m/s^2, respectively), the equation can be simplified (Equations (4a) and (4b)) to allow the input of flow rate in more commonly used units of m^3/h or L/min:

$$P_{hydraulic} (kW) = \frac{Q \cdot h_{available}}{367}, \text{ with } Q \text{ in } m^3/h \tag{4a}$$

$$P_{hydraulic} (kW) = \frac{Q \cdot h_{available}}{6116}, \text{ with } Q \text{ in } L/min \tag{4b}$$

Given that $h_{available}$ is approximated as a second-order polynomial function of Q (Equation (3)), and this function is multiplied with Q and a constant to compute available hydraulic power, the characteristic curve of hydraulic power available to a turbine can therefore be approximated as a third-order polynomial function of Q (Equation (5)), adjusted by the appropriate constant as per Equation (4):

$$P_{available} \ (kW) = \left(h_{max}{\cdot}Q - K_{loss}{\cdot}Q^3 - h_{downstream}{\cdot}Q\right) / 367, \text{ with } Q \text{ in } m^3/h \tag{5}$$

The flow rate at which the maximum possible hydraulic power P_{max} occurs can be determined by first taking the derivative of Equation (5) with respect to flow rate Q, resulting in Equation (6):

$$P'_{available} \ (kW) = \left(h_{max} - 3{\cdot}K_{loss}{\cdot}Q^2 - h_{downstream}\right) / 367, \text{ with } Q \text{ in } m^3/h \tag{6}$$

Since Equation (5) defines a polynomial with a single maximum, setting its derivative $P'_{available}$ to zero leads to Equation (7) for the flow rate corresponding to P_{max}:

$$Q_{P\_max} \ (m^3/h) = \sqrt{\frac{h_{max} - h_{downstream}}{3{\cdot}K_{loss}}} \tag{7}$$

Substituting Q_{P_max} into Equation (5) yields the maximum possible hydraulic power for the given class 1 site. This information becomes relevant upon comparing the results of this method with the method from the 1994 DVGW guidelines [34] (see Section 2.10).

2.4. Consideration of a Bypass Pipeline Parallel to the Turbine

The tool assumes that the turbine is installed in the inflow pipeline to the storage tank and in parallel with a bypass. This ensures at a minimum that the turbine can be easily removed without interrupting supply and can also be used for the compensation of short periods of high water demand via rapid filling of the storage tank. In the event that the bypass flow rate Q_{bypass} is substantially above the turbine design flow rate $Q_{turbine}$, this can maximize total annual energy generation, by enabling $Q_{turbine}$ to be as low possible, which reduces unnecessary friction losses that would otherwise occur with a higher $Q_{turbine}$. This relationship between Q_{bypass} and $Q_{turbine}$ should become clearer in the coming sections. The recommended setting is the maximum permissible inflow rate, but this can be freely adjusted by the user.

2.5. Using a Numerical Model to Ensure Supply Reliability Based on Historical Flow Data

The inflow rate Q_{in} must at any point in time be sufficient to meet the demand placed on the storage tank by the water users in the supply zone and cannot endanger the reliability of supply by causing the storage tank to temporarily run below a minimum emergency level (e.g., for fire-fighting reserves). To determine whether this is likely to happen based on a given Q_{bypass} and $Q_{turbine}$, a simple linear, deterministic, dynamic numerical model was incorporated into the tool. The model takes a historical time series of the outflow rate from the storage tank Q_{out} as input (ideally at a maximum time interval of 15 min) and simulates Q_{in} through the turbine and bypass as well as the resulting storage tank water level L_{tank}. The initial condition for L_{tank} is 75% if no water level time series data are provided along with the time series for Q_{out}. The basic governing equation (Equation (8)) for L_{tank} at a given timestamp $t+1$ is as follows:

$$L_{tank \ @ \ t+1} \ (\% \ of \ full) = L_{tank \ @ \ t} + \frac{\left(V_{in \ @ \ T} - V_{out \ @ \ T}\right)}{V_{tank} \cdot 100} \tag{8}$$

where $L_{tank \ @ \ t+1}$ reflects the resulting tank water level at timestamp $t+1$ at the end of the time step T (between timestamps t and $t+1$), $V_{in \ @ \ T}$ and $V_{out \ @ \ T}$ are the volume of water entering and leaving the tank during the time step T (between timestamps t and $t + 1$), respectively, and V_{tank} is the volume of the storage tank at 100% capacity (m^3). To clarify, "timestamp" t refers to an instantaneous point in time (e.g., 12:00:00 (noon) on 20 June 2019), whereas "time step" T refers to the duration of time passing between the two timestamps t and $t+1$ (e.g., the 15 min between 12:00:00 and 12:15:00). V_{out} and V_{in} are obtained by multiplying the duration of the time step T ($[t + 1] - t$) with Q_{out} and Q_{in}, respectively, to yield a water volume. Q_{out} is taken from the historical time series provided by the user, whereas Q_{in} is determined by Equation (9):

$$Q_{in\ @\ t+1} = \begin{cases} Q_{bypass}, & \textit{if}\left(L_{tank\ @\ t} \le L_{2_{maxshutoff}}\ \textbf{and}\ Q_{in\ @\ t} = Q_{bypass}\right) \\ & \textit{elseif}\ L_{tank\ @\ t} \le L_{3_{bypass}on} \\ Q_{turbine}, & \textit{if}\ L_{tank\ @\ t} \le L_{1_{turbine}on} \\ 0, & \textit{if}\ L_{tank\ @\ t} > L_{2\_max\ shutoff} \end{cases} \tag{9}$$

where Q_{bypass} is the chosen bypass flow rate, $Q_{turbine}$ is the chosen turbine flow rate, $L_{2\_max\ shutoff}$ is the maximum permissible tank water level (% of full), $L_{1\_turbine\ on}$ is the threshold tank level at which the turbine is operated, $L_{3\_bypass\ on}$ is the threshold tank level at which the bypass pipe is opened, and the turbine pipe is closed. A further value, $L_{4\_min\ emergency}$, signifies the lowest permissible water level. If L_{tank} falls short of this, the $Q_{turbine}$ is excluded from the set of feasible solutions. See Figure 2 for a visual representation of these tank levels. For the method to properly function, the following relationships between values must be true (Equations (10a) and (10b)):

$$L_{2\_max\ shutoff} > L_{1\_turbine\ on} > L_{3\_bypass\ on} > L_{4\_min\ emergency} \tag{10a}$$

$$Q_{max\ inflow} \ge Q_{bypass} \ge Q_{turbine} \tag{10b}$$

where $Q_{max\ inflow}$ is the maximum permissible inflow rate into the tank, either legally according to a contract with a long-term water supplier or technically due to pipeline's natural Q_{max} (see Appendix A.1 for elaboration) or a limitation on the water source (e.g., a mountain spring). The recommended setting is for $Q_{turbine}$ to be equal to $Q_{max\ inflow}$.

Expressed in words, the system is operated such that the tank is normally filled via the turbine at $Q_{turbine}$. Only in cases of high water withdrawal (Q_{out}) in which the tank level falls very rapidly is the bypass opened (and the turbine pipe closed) to increase the inflow rate to Q_{bypass} and kept open until the maximum permissible water level is exceeded (tank is full). Then normal operation with the turbine resumes. The number of occasions on which the bypass is opened increases as $Q_{turbine}$ decreases.

By defining Q_{in} to exclusively be equal to $Q_{turbine}$ and Q_{bypass}, taking into account the impact on L_{tank}, the tool thus re-defines the inflow regime in such a way that both supply reliability can be guaranteed, and energy generation can be maximized (see Figure 1). This is the step in the tool's algorithm at which the key advantage of class 1 "buffered" sites is utilized.

2.6. Calculating Total Annual Hydraulic Energy Capture

The term "capture" refers to the annual average amount of hydraulic energy $E_{hydraulic}$ that is applied to the turbine wheel, before being converted to the mechanical energy $E_{mechanical}$ of the spinning wheel and shaft, and then further converted to electrical energy $E_{electrical}$ through a generator. $E_{hydraulic}$ is calculated using Equation (11):

$$E_{hydraulic}\left(\frac{kWh}{a}\right) = \frac{\sum_{t}^{t_{max}-1}\begin{cases} P_{hydraulic}(Q_{turbine})\cdot((t+1)-t)\cdot24\ h/d\,, & \textit{if}\ Q_{in\ @\ t} = Q_{turbine} \\ 0, & \textit{if}\ Q_{in\ @\ t} \ne Q_{turbine} \end{cases}}{\left(\frac{t_{max}-t}{365\ d/a}\right)} \tag{11}$$

where the time step ($[t + 1] - t$) is measured in units of days (standard for Microsoft Excel). The resulting Q_{in} at each timestamp t (Equation (9)) depends on the outcome of the numerical model described above. This model therefore performs two functions: (1) calculating the energy production for each time step and (2) monitoring whether the water level falls below the minimum permissible level in the storage tank and flagging such solutions as invalid.

2.7. Selection of Microturbines, Global Efficiency Curves and Calculating Total Annual Electric Energy Generation

The tool assumes a single turbine having only a narrow operating range with acceptable efficiency. The turbine's characteristic resistance curve is assumed constant, since it is fixed by the frequency of the electrical power connection (presumed to be grid electricity at 50 or 60 Hz) and the number of pole pairs in the electrical generator, which pre-determine the rotational speed of the turbine.

To provide the tool user with the calculated electrical energy generation, it is necessary to estimate the total efficiency, which is a product of the efficiency of the turbine (hydraulic to mechanical energy) and efficiency of the generator (mechanical to electrical energy). To make the tool as appealing as possible to practicing engineers, the authors chose to use data from two microturbines that are currently on the market: the in-line axial "AXENT" turbine from Stellba Hydro [35] and pumps-as-turbines (PAT) from KSB [36], companies with whom the authors have worked previously. The total efficiency η_{total} at the hydraulic best efficiency point for turbines of varying capacities was plotted relative to $P_{hydraulic}$, leading to Figure 3. For each type, AXENT and PAT (two separate models), natural log curves were fitted to the data to generate equations that enabled the estimation of η_{total} based on the input of $P_{hydraulic}$ (Equations (12a) and (12b)). The tool can be updated with new data for other types of turbines from other companies, although this cannot currently be easily done by a normal user.

$$\eta_{total} (\%) = 2.05 \cdot \ln\left(P_{hydraulic}\right) + 58.1 \ (\text{AXENT}) \tag{12a}$$

$$\eta_{total} (\%) = 2.61 \cdot \ln\left(P_{hydraulic}\right) + 57.8 \ (\text{PAT}) \tag{12b}$$

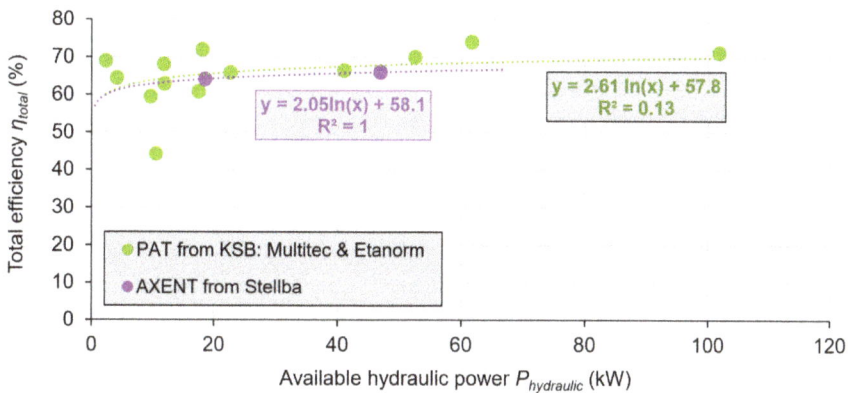

Figure 3. Global efficiency data for AXENT and pumps-as-turbines (PAT) microturbines, with best-fit natural log curves.

It is important to note that the curves in Figure 3 represent the global efficiency curve of single, optimal values of η_{total} for a hypothetical range of different turbines, not the characteristic efficiency curve of a single turbine over its operating range. Each turbine also has its own flow-dependent efficiency, such that the operating efficiency will vary from the optimal η_{total} in the event that it is operated off of its design flow rate, or if a turbine cannot be manufactured to have its peak efficiency precisely at the chosen $Q_{turbine}$. For simplicity's sake, the tool assumes that the turbine is only operated at its optimal η_{total}, at a single operating point. The flow rate must be high enough to enable the use of a microturbine with a practical size and sufficiently high efficiency, as the efficiency of turbines and generators drops rapidly with declining physical dimensions.

The total annual electrical energy generation $E_{electrical}$ for a given $Q_{turbine}$ and choice of turbine type is then determined by Equation (13):

$$E_{electrical}\left(\frac{kWh}{a}\right) = E_{hydraulic} \cdot \eta_{total} \tag{13}$$

2.8. Iteratively Determining the Optimal Turbine Design Parameters

The optimal turbine parameters are defined in the tool as the combination of Q and h that lead to the greatest annual electrical energy generation in kWh/a, based on a numerical simulation of turbine operation with the historical Q_{out} data provided by the user (see above). Generally speaking, a longer historical time series produces more reliable results by accounting for a wider range of realistic supply scenarios, but a very long time series runs the risk of using obsolete data that does not reflect the expected future supply scenarios and therefore producing sub-optimal results. In the authors' experience, 12 to 24 months is ideal.

To determine the optimal parameters, the possible solution space of values for $Q_{turbine}$ is iteratively run through the numerical model, calculating $E_{electrical}$ for each value of $Q_{turbine}$. This is first done at intervals of 5 m^3/h to determine the approximate optimal $Q_{turbine}$, and then again at intervals of 0.5 m^3/h to refine this result. The $Q_{turbine}$ leading to the greatest $E_{electrical}$ is declared optimal, and the corresponding $h_{turbine}$ calculated using Equation (1).

2.9. Estimating Economic Viability

Similar to the determination of η_{total}, the estimation of economic viability is based on available cost data for actual microturbines from the companies Stellba and KSB. These costs are based on past implemented projects, recent price quotes and the experience of collaborating engineers, and include the cost of purchase, installation and commissioning. Table 2 shows the cost items and the ranges for the two different types of turbines incorporated into the tool, as well as the total costs including the 19% value-added tax (VAT) for Germany. The peripheral costs for the AXENT turbine are generally lower, because the design elements required to install it are less complex, and it does not require protection against pressure shocks (water "hammer"), contrary to a typical PAT. Figure 4 shows the specific total cost $C_{specific}$ data for both types of turbines.

Table 2. Summary of cost parameters for AXENT turbines and PATs.

Cost Parameter / Turbine Type	Purchase Cost (Turbine and Generator)	Installation	Pipe Modi-fications	Electromechanical Control Systems	Total Incl. 19% Value-Added Tax (VAT)
AXENT (Stellba)	27,500–65,000 €	2500–4000 €	1500–3000 €	1000 €	40,500–85,700 €
PAT (KSB): Multitec and Etanorm	4400–15,700 €	5000–8000 €	5000 €	10,000 €	29,100–46,100 €
Data source (year)	Past invoice and recent price quotes (2016–2018)	Past projects (2011–2016) and engineering estimates (2016–2018)			

Equations (14a) and (14b) show the specific cost function for each turbine type. The shape of these curves corresponds approximately to data compiled from turbine projects in Switzerland [1]. The total specific costs $C_{specific}$ are calculated by entering $P_{hydraulic}$ into Equations (14a) and (14b), and the total costs C_{total} for a given site and turbine are obtained by multiplying $P_{hydraulic}$ with $C_{specific}$.

$$C_{specific}\left(€/kW_{hydraulic}\right) = 5730 \cdot P_{hydraulic}^{-0.345} \text{ (AXENT, Stellba)} \tag{14a}$$

$$C_{specific}\left(€/kW_{hydraulic}\right) = 25200 \cdot P_{hydraulic}^{-0.891} \text{ (PAT, KSB)} \tag{14b}$$

The annual benefits B_{annual} are based on user input on the applicable feed-in tariff or other electricity price at which the generated energy could be sold. In Germany, for example, the legally guaranteed feed-in tariff for turbines below a total capacity of 100 kW commissioned in 2019 is 12.27 ct. €/kWh, for a contract length of 20 years. A site is only eligible for this tariff if the water flows via gravity from its natural source to the turbine. Otherwise, the energy must normally be used on site, replacing electrical energy that would otherwise been purchased from the grid, in Germany at a price of approximately 20 ct. €/kWh. The tool combines the estimated benefits from both of these sources (Equation (15)), which can also complement each other (e.g., if there is a pump set that can be occasionally but not continuously supplied with energy by the turbine, or if the turbine is only able to cover a portion of the energy demand of the pumps).

$$B_{annual}\left(\frac{€}{a}\right) = \sum \begin{cases} E_{electrical} \cdot Price_{grid\ electricity} & \textbf{if } E_{electrical} \text{ coincides with grid energy use} \\ E_{electrical} \cdot Price_{feed-in\ tariff} & \textbf{if } E_{electrical} \text{ coincides with grid energy use} \\ & \textbf{and } site\ eligible\ for\ a\ feed-in\ tariff \end{cases} \quad (15)$$

The payback period for the project is calculated in a simple manner by comparing the initial costs and annual benefits to each other (Equation (16)), assuming the energy generation remains unchanged and without taking into account the time-dependent value of money. Depending on the tool user and decision-maker, the payback period can serve as an indicator of economic viability, if necessary complemented by consideration of C_{total}, as many projects are hindered by insufficient initial funding.

$$Payback\ period\ (a) = \frac{C_{total}\ (€)}{B_{annual}\ (€/a)} \quad (16)$$

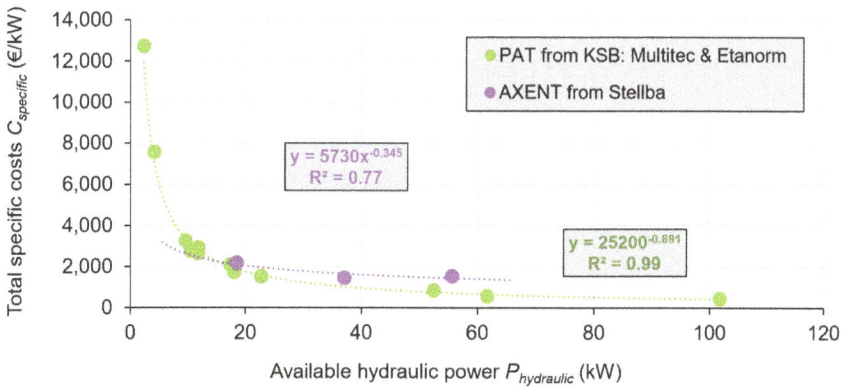

Figure 4. Specific project costs for AXENT and PATs, with best-fit power law curves.

2.10. DVGW 1994 and 2016 Methods as a Basis of Comparison

In order to provide a frame of reference for the new method, it is compared to three methods available in the 1994 and 2016 editions of the DVGW guidelines published on the topic of energy recovery through turbines in the drinking water supply. While these guidelines provide a very useful overview of the technical and operational aspects of implementing a drinking water hydropower project, these methods are not known to generate optimal design solutions for class 1 sites, and therefore only serve as a frame of reference for the new method being introduced here. The methods are compared based on the parameter $E_{electrical}$ in units of kWh/a.

The most recent edition of the DVGW guidelines [12] provides guidance in the case of class 2 "non-buffered" sites. They suggest using a frequency distribution of the occurring flow rates to determine the design flow rate of the turbine or multiple turbines. The flow rate (or mean of a range of flow rates; e.g., 52.5 m³/h representing the range from 50 to 55 m³/h) having the greatest energy density (result of Equation (17), applied to each range of flow rates) should be the design turbine flow rate. The resulting $Q_{turbine}$ is then given as input into the Excel tool to calculate $E_{electrical}$ assuming the inflow regime is modified as with the tool's optimization algorithm. In this way, this method is evaluated more generously than if it were applied as intended to class 2 sites, since the flow rates occurring above and below the chosen $Q_{turbine}$ would not contribute to energy generation, being too high or low to be efficiently processed by the turbine (see Section 3.2 and Table 4 for a concrete example).

$$E_{hydraulic}\left(\frac{kWh}{a}\right) = \frac{\sum_{t}^{t_{max}-1}\begin{cases} P_{hydraulic}(Q_{range\ mean})\cdot((t+1)-t)\cdot24\frac{h}{d},\ \textit{if}\ Q_{range\ min}\ <\ Q_{turbine}\ \le\ Q_{range\ max} \\ 0,\ \textit{if}\ Q_{turbine}\ \le\ Q_{range\ min}\ \textbf{or}\ Q_{turbine}\ >\ Q_{range\ max} \end{cases}}{\left(\frac{t_{max}-1}{365\ d/a}\right)} \tag{17}$$

This method is divided into two sub-methods, one based on the Q_{in} into the tank (approximately representing the "status quo", if the inflow regime were not modified), and the other based on Q_{out} from the tank. Historical time series are required for both in order to apply Equation (17).

The original edition of the DVGW guidelines [34] make no distinction between class 1 and 2 sites and suggests using the historically most frequently occurring flow rate to select the turbine. However, they imply that the theoretical optimal flow rate is given by Equation (18) (obtained by substituting Equation (2) into Equation (7)), at which the hydraulic power carried by the flowing water is mathematically at its maximum (see also [28,37]).

$$Q_{P\_max} = \frac{1}{\sqrt{3}}\cdot Q_{max} = 0.577\cdot Q_{max} \tag{18}$$

This method is tested on the basis of the flow rate of maximum hydraulic power. Just as for the 2016 DVGW method, the resulting $Q_{turbine}$ is given as input into the Excel tool to calculate $E_{electrical}$ assuming the inflow regime is modified as with the tool's optimization algorithm. Table 3 offers a concise summary of all methods that are compared in this paper with their corresponding short names.

Table 3. Summary of turbine design methods from the authors' perspective.

Turbine Design Method Method Characteristics	(1) HTWD [1] 2018	(2) DVGW [2] 1994	(3) DVGW 2016, a [3]	(4) DVGW 2016, b [4]
			Q_{out}	Q_{in}
Basis for design	Diverse data to determine the flow rate with the maximum annual energy generation	Q with max. hydraulic power (see Equation (18))	with historically greatest energy density (see Equation (17))	
Data requirements	Medium to high	Lowest	Medium	
Confidence of achieving max. energy generation	Highest (with high data reqs.)	Lowest	Low to medium	

[1] HTWD — Hochschule für Technik und Wirtschaft Dresden (University of Applied Sciences), the authors' home institute; [2] DVGW — Deutscher Verein des Gas- und Wasserfaches (German Technical and Scientific Association for Gas and Water), publisher of guidelines on drinking water hydropower; [3] **a**—variant of the DVGW method based on tank outflow; [4] **b**—variant of the DVGW method based on tank inflow

3. Results

3.1. Case Study of Tool Application: Break Pressure Tank (BPT) Rützengrün, Germany

The best way to understand the methodology implemented in the tool is to follow its use step-by-step for a real-world example and show the data and results as they appear to the user. A blank version of the tool as well as a version with the data for the example site BPT Rützengrün is included in the Supplementary Materials. BPT Rützengrün of the water supply utility ZWAV Plauen is located above the city of Auerbach in the Vogtland region of Saxony and receives its water (via the higher elevation BPT Vogelsgrün) from the long-range water supplier Südsachsen Wasser out of the drinking water reservoir Carlsfeld, located about 180 m above. Figure 5 shows a diagram of this supply system.

The BPT serves primarily as a pressure-reducing installation, but also has a small storage tank with a capacity of about 100 m³. Thanks to this storage, BPT Rützengrün is a class 1 site, which permits a degree of flexibility in regulating the inflow and turbine flow rate.

Figure 6 shows the first input mask, in which the user enters the minimum required information in the form of single values. Depending on the choice of calculation option (in this case option 1 and 2), those cells requiring user input turn orange. To determine the most reliable turbine parameters, the planner must not only supply the data needed to define the site curve, but also provide information about the storage tank, including volume and the feasible tank level thresholds for opening and closing the inflow valve. As per the recommended setting, the bypass flow rate Q_{bypass} is set equal to the maximum permissible inflow rate, in this case 90 m³/h.

The second input mask (not shown here) is for the time series. For BPT Rützengrün, the input time series of Q_{out} from the tank (to the downstream supply zones) ranged from 29.05.2017 to 15.02.2018 and was processed from so-called delta-event (event-based recording) raw data to have a 15-min time interval. These data are necessary to achieve reliable results. While historical time series for water level and tank inflow rate can be used by the tool to infer the storage tank volume (hence the input fields), they were neither available nor necessary, since the volume was known to be 100 m³. With the information about tank volume and the switching thresholds provided in the first input mask (Figure 6), the tank operation was simulated to determine both acceptable and optimal turbine parameters. This "simulation" consists of a simple numerical model (see Section 2.5), modifying a starting value for the tank level according to the net change in storage volume, as calculated by the difference between the inflow and outflow volumes. The planner must also choose between two types of turbines. In this case, the in-line AXENT turbine was chosen (see Section 2.7 for more details).

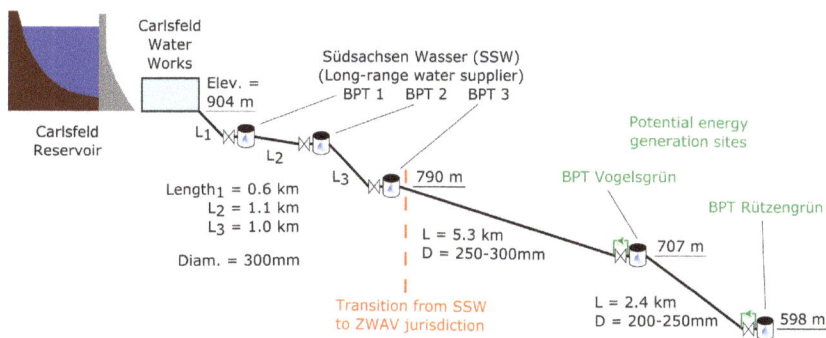

Figure 5. Diagram of the Carlsfeld supply system, with break pressure tank (BPT) Vogelsgrün and BPT Rützengrün. As illustrated in Figure 1, the potential turbines would be positioned in parallel with the existing pressure control valves.

Once all necessary input data has been provided, the Visual Basic (VBA) algorithms are run to determine the optimal turbine parameters under the given boundary conditions. The algorithms iteratively progress through all possible design flow rates (see Section 2.5 through Section 2.8), computing and graphically plotting the corresponding performance parameters annual energy generation ($E_{hydraulic}$ as well as $E_{electrical}$), power ($P_{hydraulic}$ and $P_{electrical}$) and total efficiency η_{total} for the chosen turbine type at the corresponding $P_{hydraulic}$ (for the best possible turbine, not the characteristic curve of a single turbine—see Section 2.7) (Figure 7). The optimal parameters (head and flow rate) are also displayed and added to the site curve diagram (Figure 8).

Operational Data of the Inflow Pipeline	Value	Unit	Further Explanation
Site name	BPT Rützengrün	---	results generated will be associated with this name
Q_{1_inflow}	63,1	m³/h	non-zero inflow rate at control valve, position 1 (e.g. at normal operating flow)
$h_{1\_upstream}$	10	bar	pressure head upstream of control valve, position 1 (e.g. at normal operating flow); Note: the initial reading (e.g. of the manometer) at atmospheric pressure may not be zero due to drift and must be deducted from the value under water pressure
Eligible for feed-in tariff?	yes	yes / no	e.g. in Germany the flow must be 100% gravity driven (from a higher-elevation spring, reservoir or tank), without the addition of pump energy
Is the site in Germany?	yes	yes / no	if yes, the German feed-in tariff will be automatically included in the calculation
$h_{downstream}$	0,00	bar	pressure head just downstream of the control valve, which is necessary for the water to reach the (frequently somewhat higher) storage tank (e.g. at normal operating flow)
Q_{2_inflow}	0	m³/h	inflow rate at control valve, position 2 (ideally zero flow i.e. closed valve, but can also be a second, different non-zero flow rate than Q1 if closing the valve is not feasible - the pressure head $h_{2\_upstream}$ must simply be substantially different than $h_{1\_upstream}$)
$h_{2\_upstream}$	10,7	bar	pressure head upstream of control valve, position 2 (ideally at zero flow [see above])
Q_{max_inflow}	90	m³/h	maximum permissible inflow rate into the tank, either legally according to a contract with a long-term water supplier or technically due to pipeline's natural Q_{max}
Lower threshold for Q_{out} time series		m³/h	Values below these thresholds in the time series (see next tab) are converted to zero; This is useful in the event that
Lower threshold for Q_{in} time series		m³/h	e.g. the flow measuring device erroneously reports low values even when the actual flow rate is zero.
Storage Tank Data	**Value**	**Unit**	**Further Explanation**
$L_{1\_turbine\ on}$	75	%	threshold tank level at which the turbine is put into operation; suggestion: 75%
$L_{2\_max\ shutoff}$	95	%	maximum permissible tank water level - above this value Q_{in} is set to zero such that the tank does not overflow; suggestion: 95%
$L_{3\_bypass\ on}$	60	%	threshold tank level at which the bypass pipe is opened and the turbine pipe is closed, e.g. to quickly fill the tank in times of high demand; suggestion: 60%
$L_{4\_min\ emergency}$	50	%	lowest permissible water level - any design criteria that lower the tank level below this value are excluded from the solution set; suggestion: 50%
V_{tank}	100	m³	volume of the storage tank at 100% capacity
Energy Use & Pricing Data	**Value**	**Unit**	**Further Explanation**
Electricity price on site	0,196	€/kWh	average all-in electricity price paid - the total amount billed divided by the kWh of electricity used
Feed-in tariff (if relevant)		€/kWh	can be entered insofar as known; if the site is in Germany, the feed-in tariff according to the German renewable energy law is used, regardless of the value entered here
On-site energy use		kWh/a	if available as an estimate, this can be used instead of a time series of energy use data

Figure 6. Excerpt of the first data input mask for single values, with values for BPT Rützengrün.

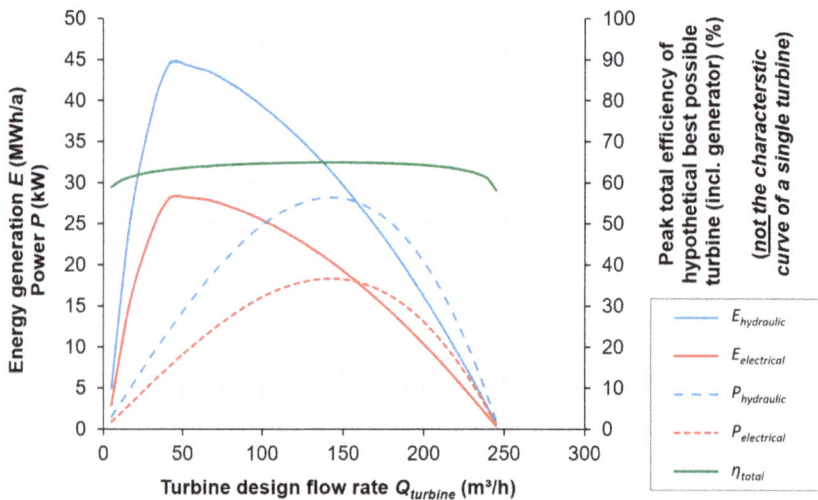

Figure 7. Calculated performance parameters of hypothetical turbines (at best efficiency point) with iteratively varied design flow rates (here from 5 to 245 m³/h), with values for BPT Rützengrün.

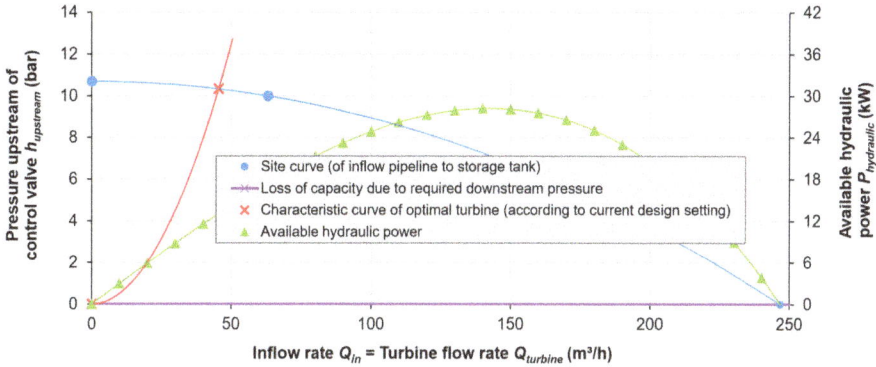

Figure 8. Site curve of BPT Rützengrün with characteristic curves of the corresponding hydraulic power as well as the (hypothetical) turbine with optimal operating parameters. 1 bar = 10.19 m of head.

After determining the optimal turbine parameters, the annual financial benefits B_{annual} are determined based on the information provided. In the same step, the total costs for purchase and installation of the turbine C_{total} are calculated on the basis of available values from the authors' experience, to roughly estimate the simple (non-inflation-adjusted) payback period (see Section 2.9 for more information). In the case of BPT Rützengrün for the input data shown here and using the recommended calculation and bypass options, $E_{electrical}$ is estimated at 26,000 kWh/a (Table 4) and the payback period at 10.0 years. According to the water supply company ZWAV Plauen, if the turbine lifetime can be assumed at 20 years, the payback period must be ≤ 10 years (less than half of the device's lifetime) to be an acceptable investment. A turbine at this site would therefore be borderline economically viable according to this assessment.

3.2. Assessing the Impact of Quality Control for Input Data

This case study also provides an opportunity to show the importance of critically assessing the data used. Figure 9 shows the time series of outflow data for BPT Rützengrün. It is clear upon inspection that the period of available data at sufficient resolution (at 15-min rather than 2-h intervals) and without gaps (which cause errors in the calculation) was not representative of the typical past operation and the expected typical future operation of BPT Rützengrün. While these data were sufficient for determining the optimal turbine parameters using the Excel tool, they estimated a much lower annual flow volume V_{annual} (144,000 m³/a) than was normal in the past and is expected for future operation (252,000 m³/a), based partially on the multi-year 2-h data. This was corroborated by the ZWAV Plauen staff, who explained that the period captured as 15-min values was unusual due to some maintenance work that was performed on the supply system. To compensate for this, $E_{electrical}$ and consequently B_{annual} were linearly increased by multiplication with the ratio between these flow volumes, a factor of 1.75. This adjustment increases $E_{electrical}$ from 26,000 to 45,600 kWh/a and lowers the payback period from 10.0 to 5.7 years (Table 4), making this project much more economically attractive than would have been the case without a careful analysis of the data. This highlights the fact that the acquisition and handling of data is not always straightforward, and ought to be done with a critical eye.

Experience with this site also demonstrated the importance of sufficiently high temporal resolution for the input data. Due to the low storage volume of 100 m³, a time interval greater than 15 min led to an unreliable simulation of the tank levels, since the incremental change in storage volume at times exceeded the tank capacity. In this case, only 2-h values were available for a longer time period (Figure 9). As these occasionally exceeded 50 m³/h, more than 100 m³ would potentially leave the tank in a single time interval. This allows no opportunity for the simulation to react by opening the inflow valve in response to the tank level falling below the switching threshold—the tank would simply be "instantly" emptied. This led the authors to embrace a rule of thumb that 15 min ought to be the maximum time interval between values for Q_{out}, even if not always strictly necessary, such as for sites with more than 1000 m^3 of storage. In general, a longer input time series is better than a shorter one, but as mentioned in Section 2.8 and for reasons made clear here, there is a trade-off between having a sufficiently long dataset for capturing the seasonally varying conditions and accidentally capturing conditions that are obsolete or unlikely to be representative of the future. A period of 12 to 24 months should be sufficient in most cases, but should also be checked for anomalies.

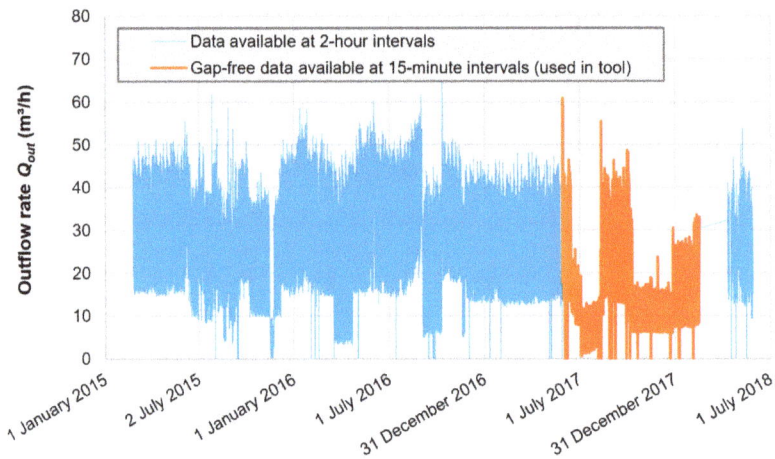

Figure 9. Outflow data series available for BPT Rützengrün, with the data used in the Excel tool highlighted in orange.

Some other authors have used long-term average values in their turbine design methods [15,25]. It is worth noting that average hourly or average daily flow data can be misleading, since the inflow rate might be much higher than the hourly average but occurring only periodically for only short periods of time (e.g., for 5 min at a time, with 15 min in between times of active flow). If this were to persist for 60 min, the hourly average flow rate would be four times lower than the actual flow rate when the inflow valve is opened, since the water volume would have flowed over 15 min rather than 60 min. The authors have encountered this situation multiple times. Designing a turbine based on the average hourly flow rate without planning to adjust the inflow rate can therefore lead to a large error, since the actual flow rate greatly deviates from the inferred flow rate based on average hourly data.

3.3. Comparison of Results Using the Newly Proposed Method with Other Methods

One purpose of this article is to compare the newly introduced method with existing methods for determining turbine parameters. Figure 10 illustrates application of method "a" suggested by the 2016 DVGW guidelines (Table 3), based on selecting the inflow rate (Q_{out}) range with the highest estimated energy generation. In this case, the Q_{out} range from 10 to 15 m^3/h is both the most frequent and most promising for highest energy generation. Thus, an average flow rate of 12.5 m^3/h emerges as the design flow rate $Q_{turbine}$. It is worth noting that the frequency of occurrence of a range of flow rates does not imply that this range will have the greatest estimated energy generation, due to the other factors that influence energy generation, such as the efficiency of the expected turbine (which declines rapidly with declining power rating), occasionally high demand that reduces energy generation at lower design flow rates (due to the necessity of bypassing more water around the turbine) and the increase in frictional head losses (with increasing flow rate). For example, the flow rate ranges 25 to

30 m^3/h and 30 to 35 m^3/h have nearly the same frequency, but the latter range has a substantially greater expected annual energy generation.

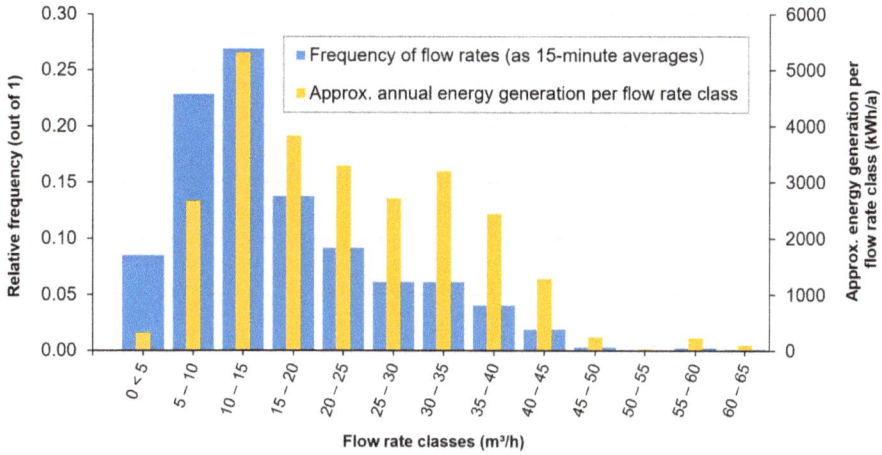

Figure 10. Frequency distribution of outflow rates and corresponding estimated annual electrical energy generation per flow rate class, at 15-min intervals for BPT Rützengrün.

The 1994 DVGW guidelines suggest that the point of maximum hydraulic power $P_{hydraulic}$ is also the optimal operating point for a turbine (Table 3). This point can be seen on the green curve of Figure 11, corresponding to a $P_{hydraulic}$ of 28.2 kW and occurring at a flow rate of 142 m^3/h. A hypothetical turbine suitable for operating at this point would have a pressure drop $h_{turbine}$ of 72.7 m, as can be seen where the fictive characteristic curve (red curve with triangle markers) of such a turbine crosses the characteristic site curve of the inflow pipeline for BPT Rützengrün.

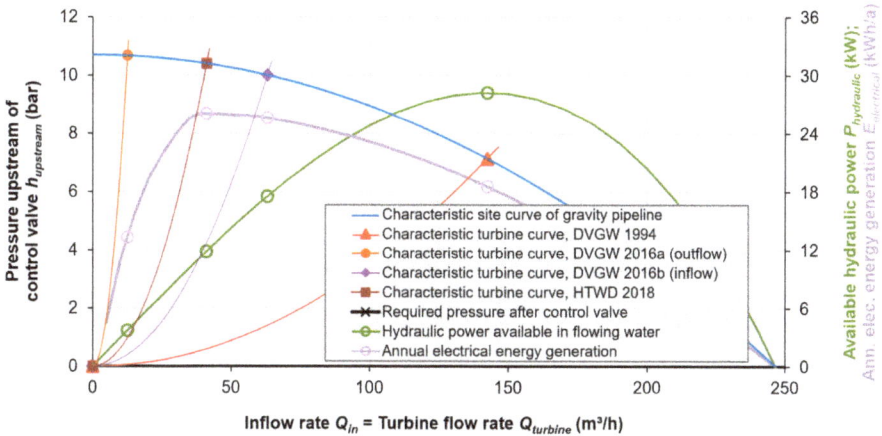

Figure 11. Site curve of BPT Rützengrün, hypothetical turbine curves for design results of all four methods, and corresponding hydraulic power as well as estimated annual electrical energy generation at each operating point.

Figure 11 also shows the results of the other three methods and their respective hypothetical turbine curves. The green and lavender curves with open circle markers respectively show the $P_{hydraulic}$ and $E_{electrical}$ corresponding to each of the turbine operating points. For the 2016 DVGW method "b" (based on inflow rate Q_{in}), no time series was available, so the single data point used to define the characteristic site curve of the inflow pipeline was taken as the most frequent inflow rate with the highest expected annual energy generation. The corresponding numbers are also summarized in Table 4. The results confirm that the 2018 University of Applied Sciences Dresden (HTWD) method yields the greatest energy generation.

It is important to reiterate (see Section 2.10) that the 2016 DVGW guidelines are explicitly intended for class 2 "non-buffered" sites, in which the system operator can "make the best" out of a hydraulic situation over which they can exercise no control. In this case, the emerging design flow rate of 12.5 m³/h is intentionally "improperly" entered into the HTWD Excel tool, which is built for class 1 "buffered" sites, and modifies the inflow regime to optimize the energy generation. In this way, the only permitted values for Q_{in} are $Q_{turbine}$ and Q_{bypass}—no other inflow rates occur. When properly applied to class 2 sites, this method yields a much lower expected energy generation, since water arriving at all flow rates not falling within a narrow range (e.g., 10 to 15 m³/h) would be bypassed around the turbine. The energy generation possible within other flow rate ranges (represented by the orange columns in Figure 10) would be lost. Using the HTWD tool for a class 1 site, these other orange columns can be almost entirely and efficiently captured by modifying the inflow rate, which is made possible by the storage tank that decouples the inflow and outflow to and from the site. This is the fundamental advantage of a class 1 site.

Table 4. Summary of the results for BPT Rützengrün using all four methods described in this article.

Parameter \ Method	(1) HTWD 2018	(2) DVGW 1994	(3i) DVGW 2016, a [1]	(3ii) DVGW 2016, a [2]	(4) DVGW 2016, b
Flow rate $Q_{turbine}$ (m³/h)	41.0	142	12.5		63.1
Pressure drop $h_{turbine}$ (m)	106	72.7	109		102
Hydraulic power $P_{hydraulic}$ (kW)	11.8	28.2	3.7		17.5
Annual energy generation $E_{electrical}$, nominal (kWh/a)	26,000	18,500	13,700	5300	25,600
Annual electrical energy generation $E_{electrical}$, corrected for flow volume (kWh/a) [3]	45,700	32,500	24,000	9300	44,800
Annual electrical energy generation $E_{electrical}$, corrected (% of result via HTWD method)	100%	71.2%	52.7%	20.4%	98.3%

[1] Method applied improperly, as if class 1 site (see Section 2.10 for elaboration); [2] method applied properly, as if class 2 site; [3] increased by a factor of 1.75 to account for expected future flow volumes (see Figure 9 and preceding discussion).

This difference between the improper and proper application of this method is shown in Table 4 (method 3i vs. 3ii), and highlights the great advantage that class 1 sites have over class 2 sites for energy generation using microturbines with narrow acceptable operating ranges (see Section 2.10 for elaboration). BPT Rützengrün as a class 1 site would have an annual electrical energy generation of about 24,000 kWh/a at the turbine flow rate of 12.5 m³/h, while it would have only 9300 kWh/a if it lacked the 100 m³ storage tank and was therefore a class 2 site—a factor of 2.6 less energy generation.

Furthermore, the use of the 2016 DVGW method "b" (based on the most frequent inflow rate with the highest expected annual energy generation) is also an improper application, since the inflow and outflow at class 2 sites are necessarily equal, as there is no storage tank. However, both of these "improper" applications of the 2016 DVGW method serve as useful comparisons, since in the absence of another method, the designer of a turbine system might choose to select a design flow rate corresponding to a known quantity about the system. As such, this flow rate also generally corresponds to the current operating state, providing a comparison with a design approach that would choose a turbine to fit the existing inflow rate. And as is evident from Table 4, using the substantially higher flow rate of 63.1 m³/h yields a total energy generation that is 98.3% of that yielded by the 2018 HTWD method—in this case hardly a significant loss. However, it cannot be assumed that this will hold true in every case, as the analysis of further sites in the following section shows. In cases where it does hold true, this method can be used to confirm this truth, which provides the designer and operator with a greater level of confidence in the ultimate design decision.

3.4. Results for Nine Sites in Germany

The four turbine design methods were applied to a total of nine sites from Saxony, Germany. The sites had a variety of characteristics and data availability and led to a range of results with varying degrees of confidence (Table 5). To highlight the hydraulic differences between the sites, Figure 12 shows the characteristic sites curves of each inflow pipeline—a kind of "finger print" for each site. The current inflow rates range from 24 to 145 m^3/h, whereas the newly proposed design flow rates range from 34 to 84 m^3/h, with an outlier of 300 m^3/h at the newly planned site Rehbocksberg. With the exception of two sites having very small (100 m^3) tanks and one site having a very large (10,000 m^3) tank, the storage capacity ranged from 1000 to 5000 m^3.

Table 5. Summary of the site characteristics, with selected design results obtained using the 2018 HTWD method.

Site	V_{tank} (m^3)	V_{annual} (m^3/a)	Q_{in} and $h_{available}$ before Turbine (Typical Operating Point)	Q_{in} and $h_{turbine}$ with Turbine	$P_{electrical}$ (kW)	Q_{bypass} (m^3/h)	Confidence in Results
Adorf-Sorge	1000	328,000	145 m^3/h 117 m	84 m^3/h 139 m	20.7	150	High
Rützengrün	100	252,000 [1]	63.1 m^3/h 102 m	41.0 m^3/h 106 m	7.5	90	High
Vogelsgrün	100	158,000	58.3 m^3/h 71.3 m	43.5 m^3/h 74.9 m	5.6	90	High
Voigtsgrün	4000	255,000	58.8 m^3/h 45.3 m	39 m^3/h 47.6 m	3.1	100	High
Chursdorf	4000	220,000	24 m^3/h 98.3 m	32.5 m^3/h 97.6 m	5.4	65	High
Mittweida	1500	260,000	55 m^3/h 19.6 m	34 m^3/h 21.8 m	2.0	150	Med.
Rochlitz	5000	207,000	61 m^3/h 63.2 m	37.5 m^3/h 68.9 m	4.4	72	Low
Neundorf	4000	292,000	100 m^3/h 39.5 m	50 m^3/h 40.4 m	3.4	130	High
Rehbocksberg	10,000	1,800,000	N.A. (new site)	300 m^3/h 55.7 m	30.0	360	High

[1] Corrected based on long-term data (see Figure 9 and preceding discussion).

For five of the nine sites it was possible to obtain sufficiently detailed data to perform the highest-quality design. In the case of Mittweida, 15-min data existed, but were saved in such a way in the (early generation) SCADA system that made them very time-consuming to retrieve. Thus one-day data were used, which were more easily accessible. For the remaining three sites no time series were available, due to one site not yet existing (the Rehbocksberg storage tank is being newly constructed) and having no SCADA data transmission. The calculation options 3 and 4 in the Excel tool are designed to handle precisely such cases, but the results are to be used with caution.

For Neundorf and Rehbocksberg, the design was based on the operating conditions planned by the respective water utilities, who were confident that they would able to maintain the design turbine flow rates. For Rochlitz, the maximum daily flow volume for any given future year was estimated and used to calculate the corresponding flow rate. This was done by first calculating the average daily flow volume in the month with the greatest flow volume (these data were available), which was 674 m^3/d. This number was then multiplied with the ratio (taken from the nearby site Chursdorf) between the maximum daily flow volume in the year and the average daily flow volume in the month with the greatest flow volume, which was 1.34. The resulting "worst-case day" had a flow volume of 900 m^3/d, which was assumed to enter the tank over a 24-h period, producing an inflow and turbine design flow rate of 37.5 m³/h.

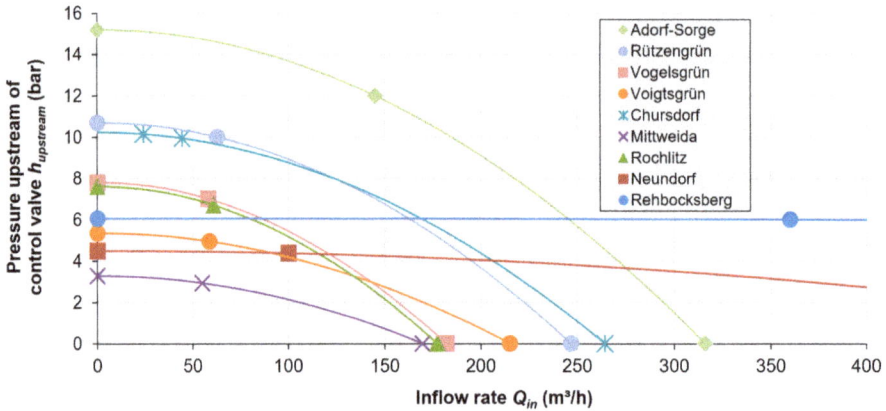

Figure 12. Overview of the hydraulic characteristic curves for all nine sites analyzed.

Tables 6 and 7 show a comparison of the technical and economic results, respectively, for all nine sites across all four methods. The HTWD method provides the greatest energy generation for every site. Using the 1994 DVGW method, the energy generation is consistently low, on average only about 71% of the optimum, ranging from 63% to 74%. The 2016 DVGW method "a" (based on the most frequent Q_{out} with the highest expected annual energy generation) produced very mixed results, ranging from 53% to 98% of the optimum, with a weighted average of 78%. The 2016 DVGW method "b" (based on Q_{in}) performed consistently better, ranging from 79% to 100% of the optimum and having a weighted average of 91%.

Table 6. Comparison of the estimated $E_{electrical}$ (in kWh/a) achieved by four different turbine design methods, in italics as percentages of the value given by method 1.

Site	(1) HTWD 2018 (New Method)			(2) DVGW 1994 kWh/a	(3) DVGW 2016, a kWh/a	(4) DVGW 2016, b kWh/a
	Data Basis: Start Date (Nr. of Days)	Calculation and Bypass Options Used	kWh/a			
Adorf-Sorge	27 April 2017 (179)	Calc. 2	80,900	59,100	78,200	69,800
		Bypass 2	*100%*	*73%*	*97%*	*86%*
Rützengrün	29 May 2018 (256)	Calc. 2	45,700	32,500	24,100	44,800
		Bypass 2	*100%*	*71%*	*53%*	*98%*
Vogelsgrün	29 May 2018 (256)	Calc. 2	20,000	14,600	19,500	19,100
		Bypass 2	*100%*	*73%*	*98%*	*96%*
Voigtsgrün	27 April 2017 (179)	Calc. 2	20,300	14,400	11,000	19,100
		Bypass 2	*100%*	*71%*	*54%*	*94%*
Chursdorf	28 October 2017 (338)	Calc. 2	36,700	23,300	32,900	29,100
		Bypass 2	*100%*	*63%*	*90%*	*79%*
Mittweida	20 July 2013 (1714)	Calc. 3	8980	6610	No data	7450
		Bypass 2	*100%*	*74%*		*83%*
Rochlitz	Single values, estimates [1]	Calc. 4	24,100	17,200	No data	22,400
			100%	*72%*		*93%*
Neundorf	Single values, future plans	Calc. 4	19,800	14,000	No data	19,800
			100%	*71%*		*100%*
Rehbocksberg	Single values, future plans	Calc. 4	180,000	126,000	No data—new storage tank	
			100%	*70%*		
Arithmetic mean	—	—	48,500	34,000	36,500	28,900
			100%	*70%*	*86%*	*91%*
Weighted mean	—	—	48,500	34,200	33,100	28,800
			100%	*71%*	*78%*	*91%*

[1] Based on the estimated maximum daily flow volume in any given year (see preceding text for clarification).

The economic results mirror the technical results. The revenue for all sites was calculated assuming 100% feeding in of the generated electricity, at the 2018 German rate of 12.33 ct. €/kWh, as all sites are eligible for the feed-in tariff under the German Renewable Energy Law. The estimated specific project costs follow a power law and ranged from 1700 €/kW$_{hydraulic}$ for the highest-capacity turbine (30.0 kW$_{electrical}$, Rehbocksberg) to 12,000 €/kW$_{hydraulic}$ for the lowest-capacity turbine (2.0 kW$_{electrical}$, Mittweida), based on past project experience from other sites with similar conditions. The estimated total project costs therefore ranged from approx. 28,000 € to 37,000 €. For the optimal case using the 2018 HTWD method, the annual revenue ranges from 1110 to 22,200 €/a, with simple payback periods from 24.2 to 1.7 years. Of the nine sites, four have payback periods less than 10 years, making them economically viable according to the standards used by the water utility ZWAV Plauen. For the remaining sites, the project costs would have to be reduced in some way to make the installation of a turbine economically viable, for example by reducing the costs of the turbine or other items (see Table 2), or acquiring financial support through state or federal grant funding for renewable energy projects.

Table 7. Comparison of the estimated annual revenue (in €/a) and simple payback period (in *a*) for the turbine parameters determined using the four different turbine design methods.

Site	(1) HTWD 2018	(2) DVGW 1994	(3) DVGW 2016, a	(4) DVGW 2016, b
Adorf-Sorge	9980 €/a	7280 €/a	9650 €/a	8600 €/a
	3.6 *a*	5.1 *a*	3.8 *a*	4.3 *a*
Rützengrün	5630 €/a	4010 €/a	2970 €/a	5530 €/a
	5.7 *a*	8.8 *a*	9.6 *a*	6.1 *a*
Vogelsgrün	2470 €/a	1800 €/a	2410 €/a	2360 €/a
	12.7 *a*	18.4 *a*	12.9 *a*	13.7 *a*
Voigtsgrün	2510 €/a	1780 €/a	1350 €/a	2360 €/a
	11.8 *a*	17.9 *a*	20.7 *a*	13.2 *a*
Chursdorf	4520 €/a	2870 €/a	4060 €/a	3590 €/a
	6.9 *a*	12.3 *a*	7.9 *a*	8.4 *a*
Mittweida	1110 €/a	815 €/a	No data	919 €/a
	24.2 *a*	34.8 *a*		30.7 *a*
Rochlitz	2980 €/a	2130 €/a	No data	2770 €/a
	10.3 *a*	15.3 *a*		11.5 *a*
Neundorf	2450 €/a	1720 €/a	No data	2450
	12.2 *a*	20.3 *a*		13.1 *a*
Rehbocksberg	22,200 €/a	15,500 €/a	No data—new storage tank	
	1.7 *a*	2.8 *a*		
Arithmetic mean	5980 €/a	4120 €a	4090 €/a	3570 €/a
	9.9 *a*	15.1 *a*	11.0 *a*	12.6 *a*
Weighted mean	3200 €/a	2300€a	2850 €/a	2540 €/a
	5.3 *a*	8.2 *a*	7.6 *a*	9.0 *a*

4. Discussion

4.1. Archetypical Sites: Handling in Excel Tool and Practical Considerations

During analysis of the nine sites presented here there were five archetypes that emerged, the closer examination of which may prove useful to the reader.

The first archetype is a typical supply tank site, with moderate storage and a more or less typical water demand pattern, reflecting a mixture of household and commercial or industrial users. Voigtsgrün, Chursdorf, Mittweida, Rochlitz, Neundorf and Rehbocksberg fall into this category.

The second archetype is a site that predominantly serves as a break pressure tank along a long-distance pipeline through mountainous terrain. Rützengrün and Vogelsgrün fall into this category. For these two sites in particular, only very small storage tanks (100 m^3) are present, which demands that temporally high-resolution outflow data is available to the Excel tool for determining both acceptable and optimal turbine flow rates. As with these two sites (see Figure 5), these kinds of sites can exist in several stages in series, which can have a simplifying cascade effect, since the downstream BPT regulates the outflow of the upstream BPT.

The third archetype is a site with a very flat characteristic site curve, resulting from very low friction losses within the desired operating range of flow rates. Normally this is a consequence of a particularly large inflow pipeline. The sites Rehbocksberg and Neundorf fall into this category (see Figure 12). In this case, the designer has much greater flexibility, because the flow rate can be made higher without risking unnecessary energy loss to pipe wall friction. For example, while the site operator ZWAV Plauen chose 50 m^3/h as the design flow rate for Neundorf, a flow rate of 100 m^3/h or even 150 m^3/h would subtract very little from the total possible energy generation. Both these sites also have high storage capacities of 5000 and 10,000 m^3 (although this is not necessarily a property of this archetype), which makes it unlikely that brief periods of high demand would endanger the security of supply by depleting tank levels. Thus, there is also less urgency for having a high-resolution time series of outflow data to verify that the chosen turbine is suitable.

The fourth archetype is a data-lean site, for which only limited information is available. If storage is large enough compared to the demand and the site can be flexibly operated due to system-level redundancy (as with Neundorf and Rehbocksberg, which both have parallel supply systems that can replace or supplement water in case of an outage), the limited data may be sufficient to produce a reliable design using the Excel tool. For cases like Rochlitz, however, more data must be gathered to see how the possible short-term spikes in demand impact the energy generation.

The fifth archetype is a site at which energy generation potential may exist, but no economically viable options are available for using the generated electricity. In the case of Adorf-Sorge, which is the second-most promising site after Rehbocksberg, the electrical power output of the turbine at the optimal operating point would be approximately 21 kW. However, the site is located deep in a forest, and is connected to the electricity grid via a 3-km-long cable, such that the maximum possible feed-in power input would be about 2.3 kW before the cable would be overloaded. This low $P_{electrical}$ could not be economically fed into the grid. There is also not sufficient energy demand at the site (no pumps or other substantial energy users), such that there is currently no known practical way to use the energy that could be generated with a turbine. Unfortunately, this removes Adorf-Sorge from the list of potential sites.

4.2. Cases in Which the Excel Tool is Not Needed or Appropriate

Pragmatic readers may ask themselves whether the data acquisition and evaluation required for using the full functionality of the Excel tool is warranted in every case. There are two main cases in which the designer of a turbine site should elect not to use the tool as recommended:

1. There is limited data available for the site and time or other constraints make it undesirable to perform new measurements. In this case the site operator can take the shortcut of using the equivalent of the 2016 DVGW method "b", and simply select a turbine that operates efficiently at the current typical Q_{in}. According to the seven sites analyzed here, this leads on average to 10% less energy generation (and annual revenue), with a risk of up to 20% less energy generation. This is the best known alternative method that removes the need for data-based work.
2. The site in question is a class 2 "non-buffered" site, for which this tool is not appropriate. Using the tool for class 2 sites will lead to gross overestimates of the potential energy generation, as shown in Table 4. This is due to the fact that the tool assumes a complete modification of the Q_{in} regime, which is not possible at a class 2 site, for which Q_{in} is necessarily equal to Q_{out}.

4.3. Limitations of the Tool

The main limitation of the Excel tool presented here is its lack of functionality for accommodating class 2 "non-buffered" sites without storage tanks. While the authors suspect that class 2 sites with economically viable energy generation potential are in the minority compared to class 1 sites, the authors already identified several sites for which a practical and technically accessible design method would be needed. Collaborators of the authors in 2013 developed a preliminary Excel tool according to this method (currently available only in German [38]), which systematically tests the historical distribution of flow rates against typical efficiency curves for different types of turbines (e.g., Francis, Pelton, pump-as-turbine) to determine both the optimal turbine type and parameters for a given site. However, the authors are not confident that this tool properly accounts for the hydraulic nature of class 2 sites. Future work could pick up where this preliminary tool left off, and perhaps incorporate its functionality directly into the Excel tool presented here. In the best case, this could also build and improve on the 2016 DVGW methods.

The remaining limitations deal with ways in which the methodology and Excel tool might not lead to optimal solutions for class 1 sites, and could be improved in the future:

1. The tool assumes the selection of a single turbine with a narrow acceptable operating range, and considers neither the possibility of turbines with wide operating ranges nor that of multiple turbines, the latter of which might lead to greater energy generation [25,27]. This was decided partly for simplicity's sake and partly out of the belief that a single turbine generally represents the most economically viable solution for class 1 sites, which is supported by one of the studies cited above [25]. Furthermore, as indicated in Section 1.2, these studies do take into account the fundamental advantage of class 1 sites, which is the ability to modify the inflow regime, instead using multiple turbines to adapt to the wide range of flow rates occurring based on the current site conditions.

2. The tool does not have a sophisticated way to support users with sites for which a feed-in tariff is either not available or not applicable (e.g., in Germany, if the water does not flow 100% via natural gradient). There is an option to enter in the total energy use on site and the percentage of which the user expects to be covered by the turbine. In the future it is planned to implement an algorithm that takes as input a time series of electricity use on site (parallel to the Q_{out} time series) and estimates how much of this energy use could be covered by the turbine, such that the user does not need to estimate this herself.

3. There is a lack of decision support in accounting for future changes in water use patterns, which other design methods seem to have accounted for [14,39,40]. However, there is a simplified factor which can be adjusted to account for possible increases or decreases in water use. In this way, an expected future water use pattern can be roughly simulated, and a turbine designed that will still be suitable for this future condition.

4. The impact of iteratively varying the threshold tank levels (see Figure 6) to activate and deactivate the turbine and bypass has not been sufficiently assessed. Sitzenfrei and Rauch [14] presented an optimization method that is similar in spirit to the one presented by the authors but applied it to a class 2 site. They pursued an optimization approach by varying parameters in a randomized fashion through 1000 simulations (Monte Carlo simulation), selecting the best solution based on the amount of energy generated over 10 years. The parameters varied in this case are the set-point water levels in the supply tank upstream of the turbine: the overflow level, the level for switching from high to low turbine flow, and the minimum level required for fire-fighting. The HTWD method introduced in this paper could be improved by implementing a similar kind of randomized (e.g., Monte Carlo) variation of the four water level thresholds used to determine when water flows through the turbine, bypass or neither. This might increase the robustness of the solution suggested by the tool and also slightly increase the total annual energy generation predicted by the tool.

5. As mentioned in Section 3.2, gaps (i.e., time intervals larger than the smallest time interval; e.g., due to missing data) in the input data time series of Q_{out} lead to an error in the calculations performed by the tool. Currently, the burden is on the user to ensure that the time series contains no gaps. In the future, this could be improved through an algorithm that automatically checks for and linearly interpolates to fill these gaps.

6.   Currently, the data from only two types of turbines from two manufacturers are incorporated into the tool. This merely reflects the authors' experience and available data until now and is not intended to imply that there are not further options. No funding links or other conflicts of interest exist between the authors and these two turbine manufacturers.

4.4. Relative Potential of Class 1 vs. Class 2 Sites

As mentioned previously, class 2 sites represent a greater technical challenge for the optimization-minded engineer than class 1 sites. Although they have not been able to confirm it, the authors suspect that class 1 sites exist in greater numbers and possess greater potential for energy generation than class 2 sites, due to their generally higher pressure heads. In particular, class 1 sites seem most relevant for small and medium-sized water suppliers with low population densities and relatively low water distribution flow rates, since their class 2 sites are unlikely to provide economically viable energy generation. It may be the case for the largest water suppliers that class 1 sites were long ago tapped for their hydropower potential, leaving predominantly class 2 sites that remain to be developed. As the largest water suppliers also tend to be more innovative, this may explain the apparently greater scientific attention given to class 2 sites (Appendix A.4). If the authors' suspicions are accurate, it would be useful to better understand the relative potential for drinking water hydropower at class 1 vs. class 2 sites, as this could provide an incentive for greater focus on supporting the design and development of class 1 sites for small and medium-sized water suppliers, which tend to make up the vast majority in the total count of a given country's water suppliers. The German water suppliers featured in this study are medium-sized.

5. Conclusions

This paper presented a novel method and accompanying Excel tool for determining the optimal parameters of a microturbine for water supply sites having a storage tank that decouples the inflow and outflow patterns, the so-called class 1 "buffered" sites. The method determines the optimal turbine flow rate based on key site characteristics and a historical time series of outflow data from the tank, simulating tank operation with a numerical model as it creates a new inflow regime. The main criterion for a viable solution is that the tank water level does not fall below a user-specified minimum. The Excel tool is not currently suitable for analyzing class 2 "non-buffered" sites without a storage tank but could conceivably be expanded to include this functionality.

The method was inspired by Haakh [28], who provided a foundational treatment of turbine and pump operation in the context of water supply systems. It fills a gap in the known methods for designing microturbines for water supply sites, which recognize the design advantages of class 1 sites, but provide no practical instructions on how to determine the optimal flow rate and corresponding pressure drop. The tool is intended to be widely accessible by being implemented in the commonly used software Microsoft Excel and is offered free of cost for download and use by any interested parties [33]. A blank version of the tool as well as a version with the data for the example site BPT Rützengrün (see Section 3.1) is included in the supplementary material.

The 2018 HTWD method presented here is compared with three other methods from two different generations of DVGW guidelines from 1994 and 2016. All four methods were used to analyze a total of nine sites located in Saxony, Germany, and the results were compared in terms of the annual energy generation. The HTWD design method estimates the greatest energy generation for each site, whereas the other three methods lead to an average of 70% to 91% of the HTWD results. If an alternative method had to be chosen, the 2016 DVGW method "b" (based on the most frequent range of inflow rates with the highest expected annual energy generation) provided the second-best results overall, ranging from 79% to 100% (with an average of 91%) of the energy generation predicted using the HTWD method (see Sections 3.3 and 3.4 for details).

Of the nine sites analyzed, four are very likely economically viable (estimated payback periods of 1.7 to 6.9 years), while four are borderline viable (payback in 10.3 to 12.7 years) and could be made viable with sufficient reduction in project costs to the operator, for example through grant funding. The remaining site (Mittweida) is very unlikely economically viable, with an estimated payback period of 24.2 years. Other factors prevent the second-most promising site (Adorf-Sorge) from being economically viable, since there is no known practical way to transport the generated electricity off site or use it on site. While a turbine has not yet been implemented at any of these nine sites, at the time of submission, the site Rehbocksberg is the furthest along in the planning, and should be installed before the end of 2019, the site Rützengrün is in an advanced planning stage and the site Chursdorf is in an early planning stage. The experience gained from analyzing all nine sites led to the description of five archetypical sites (see Section 4.1 for details), which the reader can use as points of reference when analyzing sites under his or her supervision.

Supplementary Materials: The following are available online at http://www.mdpi.com/2073-4441/11/7/1403/s1, Two appendices, and two Excel files—one containing no data (File "MicroturbineTool_HTWD_v35_blank.xlsb") and one containing data from the example site in this paper, BPT Rützengrün (File "MicroturbineTool _HTWD_v35_BPT-Ruetzengruen.xlsb").

Author Contributions: T.J.V. devised the design method and Excel tool presented in this paper, performed the analysis of the nine sites, and wrote the paper. T.G. initiated the research efforts into microturbines in the water supply in 2013, acquired the funding that enabled the work to be performed and edited the final manuscript.

Funding: The authors are grateful to the German Federal Ministry of Education and Research (BMBF) for supporting the project "Energy Efficiency in the Water Supply" (grant number 03FH018I3). Preparation of the final paper was supported by the AquaNES project (grant number 689450), which received funding from the European Union's Horizon 2020 Research and Innovation Program.

Acknowledgments: The authors are grateful to the communal water suppliers ZWA Hainichen, ZWAV Plauen and RZV Glauchau for providing data, insights from their practical experience and support during site visits. Julia Wetzel made important contributions to the development of the tool's methodology as part of her bachelor thesis work.

Conflicts of Interest: The authors declare no conflict of interest. No funding was received from the turbine manufacturers Stellba and KSB.

Appendix A. Practical Considerations for Deploying Hydropower in Water Supply Systems

Appendix A.1. Origin of Surplus Energy in Gravity-Based Water Supply Systems and Hydraulic Aspects of Their Operation

Gravity pipelines connect higher-elevation water reservoirs to lower-elevation points of water storage and water users, and are a ubiquitous feature of water supply systems. The higher-elevation reservoirs can have natural (e.g., river dams) or artificial (e.g., elevated storage tank) origins. In all gravity pipelines, the difference in potential energy (elevation head) between the upstream inlet and downstream outlet or reference point within the pipeline is fully converted into three components:

1. kinetic energy of the flowing water (velocity head),
2. pressure energy between the water molecules (pressure head) and
3. heat (and some sound) due to pipe wall (major) and local (minor) frictional resistance (head "loss") in reaction to the flowing water.

With respect to hydropower development, there are two relevant types of gravity pipelines: those with and those without pressure control at points downstream, such as transfer stations or outlets into storage tanks. In a gravity pipeline without pressure control, head loss is at a minimum, velocity head is at a maximum and pressure head fluctuates between zero and small negative or positive values along the length of the pipeline depending on the pipeline's exact downslope path (inevitably equaling zero at the outlet), such that the maximum possible flow rate is established [41]. The only remaining energy at the pipeline outlet is the velocity head of the flowing water. In a gravity pipeline with pressure control, head loss can be intentionally increased, increasing the pressure head upstream of the turbine and reducing the velocity head until the desired flow rate is established, normally using a pressure control valve at the downstream outlet (a local resistance). In cases with raw water in excess (e.g., a mountain spring during the rainy season), pressure control could also be achieved with an open basin, into which a gravity pipeline (without a pressure control valve) flows, reducing the relative pressure back to zero, and allowing excess water to spill over into natural watercourses. However, this becomes more the territory of "small" rather than "micro" hydropower, specifically a "run-of-river" scheme, which very well might precede a water supply scheme, but is not really "drinking water hydropower". These kinds of schemes are covered by many papers and by manuals (e.g., [2,42,43]).

Like a pressure control valve, a hydropower turbine also represents a local resistance. Introducing a hydropower turbine into a gravity pipeline without pressure control would therefore reduce the flow rate, which is generally undesirable for an existing water supply scheme. In a gravity pipeline with pressure control, however, a turbine can functionally replace the pressure control valve, causing no change in the flow rate. And whereas a pressure control valve converts potential energy (as elevation head) into useless head loss, a turbine converts potential energy into useful mechanical energy, and often via a generator into electrical energy. Gravity pipelines with pressure control valves can thus be thought of as possessing "surplus" potential energy, which can be "harvested" by using a turbine rather than "wasted" by using a pressure control valve, which "throttles" the flow of water.

The simplest theoretically possible gravity-fed water supply pipeline would supply water at the required flow rate without the need for any flow regulation. This would require the precise selection of the pipeline's total frictional resistance (a function of its length, diameter and inner surface material roughness), such that the maximum unregulated flow rate is equivalent to the required supply flow rate. In practice, however, a majority of existing gravity-fed supply pipelines would permit a far higher flow rate than usually required, if operated without any flow regulation. This means that these pipelines possess surplus hydraulic capacity. This arises in two main ways:

1. Intentionally, because the designer anticipates periods during which nearly the maximum flow rate will be required (e.g., evenings in a dry summer period) or expects the total demand of the supply zone to increase due to population growth and/or increase in commercial or industrial activity, or

2. Unintentionally, because the pipeline was chosen with a very generous factor of safety [14], or because demand in the supply zone is decreasing, due to declining population, increasing water use efficiency and/or cessation of commercial and industrial water use.

Both cases are common in Germany, where numerous water supply organizations have been able to construct hydropower facilities at their gravity-based systems for many decades [44–46]. Cases are also well documented in Austria [3,47], Switzerland [2] and England [48]. Regardless of the reason, in both cases flow regulation is generally employed in the form of pressure control to restrict the flow rate to the level required for supply. Pressure control is normally accomplished using a manually or remotely operated valve. While specially designed pressure-reducing valves exist, they are expensive, such that other fittings like gate or butterfly valves are frequently used, although they may not be designed for this purpose. In cases with very large altitude differences (e.g., above 200 m), pressure control can be achieved through break-pressure tanks (BPTs) over several elevation stages, to reduce the risk of pipe bursts. Here, passive pressure control can also be employed, for example using an energy-dissipating baffle wall or an in-pipe fixed-diameter orifice plate [6].

Both active and passive pressure control converts energy in the form of hydraulic pressure to energy in the form of heat, and to a lesser extent sound and vibration (at the point of resistance). It is precisely this energy that can instead be converted into useful mechanical work or electrical energy by operating a hydropower facility. In this way, the hydropower unit serves two purposes at the same time, by reducing excess pressure and simultaneously converting it to useful energy. Figure A1 illustrates the hydraulic nature of a typical potential turbine site with a pressure control valve. The turbine can be installed in a bypass pipe in parallel with the existing pressure control device or at the inflow to an existing BPT, without endangering the primary goal of supplying water [6]. Models of axial in-pipe turbines are also available, which do not necessarily require a bypass for normal operation, as they provide the same hydraulic resistance whether they are electrically connected or not, and thus pose no danger of pressure shocks (water hammer) in the case of a power outage or sudden mechanical failure.

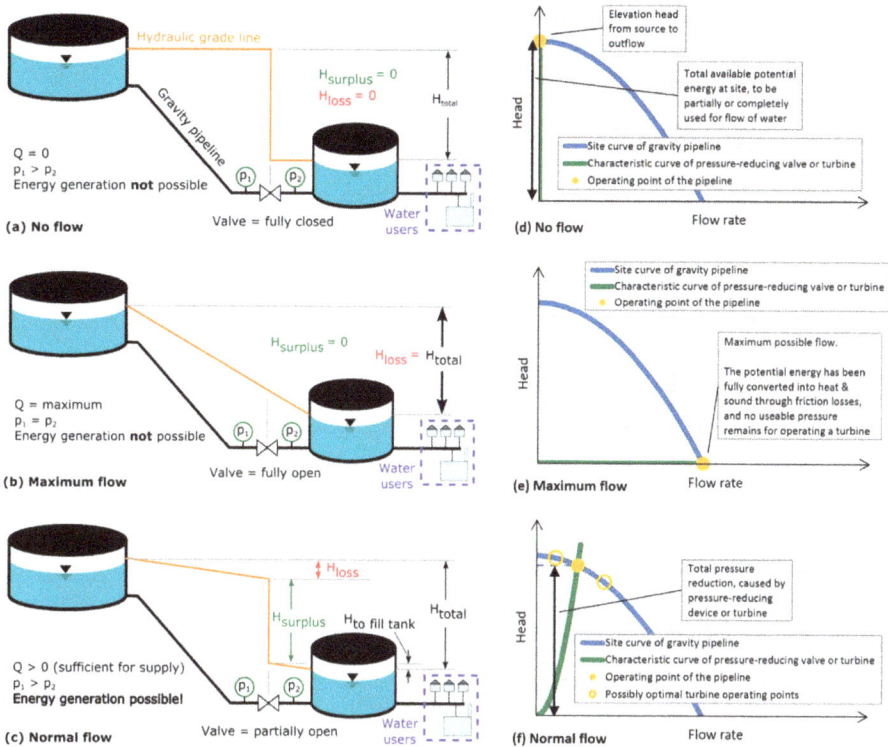

Figure A1. The basic hydraulics of a drinking water hydropower system prior to the installation of a turbine, illustrated under three conditions (no flow, maximum flow, normal flow) using the hydraulic grade line and process diagram (**a–c**) as well as a diagram of head vs. flow rate (**d–f**).

Appendix A.2. Favorable Site Characteristics for Hydropower

To maximize energy generation and economic viability (defined by initial investment, annual revenue and payback period) for both classes of sites, the following characteristics are most favorable [49]:

1. Nearly constant flow rate, either due to a site being class 1, or because the water use profile in the downstream supply zones do not fluctuate very much in the case of class 2 sites
2. Nearly constant pressure conditions
3. Existing infrastructure that can be used with only minor modifications to the piping and without any civil construction works (e.g., an easily accessible and enclosed building, control valves and pipe systems with generous amounts of space)
4. Local energy needs, such that the energy generated can most economically be used, by replacing the need to purchase energy from the grid (typically the most expensive source)
5. Conditions that meet the requirements for receiving a feed-in tariff (e.g., in Germany this is a purely natural gradient, without any pumping upstream of the turbine site)

Appendix A.3. Further Characteristics of and Implications for Turbines at Class 1 and Class 2 Sites

Appendix A.3.1. Class 1 "Buffered" Sites

Class 1 sites are generally intermediate (such as break-pressure tanks (BPTs)) or terminal storage tanks with pressure control valves [6], typically located within the water "transmission" section of the water supply chain [13]. In the authors' experience and estimation, class 1 sites represent a relatively cost-effective opportunity for renewable energy generation. They offer the water supplier flexibility (positively correlated with storage volume) in determining the rate, duration and timing of tank filling (and thus turbine operation), as this must not necessarily occur simultaneously with the downstream use in the supply zone. Assuming that the inflow rate is not restricted (e.g., by the naturally occurring flow from a mountain spring), the inflow regime can often be freely adjusted to maximize energy generation through a turbine, regardless of how it was managed in the past [12].

Pressure control at such sites is generally accomplished using a fixed resistance (e.g., orifice plate) or a simple valve (e.g., butterfly or gate), which is opened to a pre-determined set-point to reduce the inflow rate to the desired level. Since the higher-elevation water reservoir is normally a fixed point, and substantial withdrawals seldom occur en route to the storage tank (which would increase the flow rate along stretches of the pipeline, thus increasing head loss), the pressure available to a turbine generally remains constant as long as the flow rate is held constant.

This enables the selection of inexpensive turbines such as pumps-as-turbines (PATs) [36] or axial in-line turbines (such as the "AXENT" from Germany [35] or "PAM PERGA" from Spain and Germany [29]), which have a narrow effective operational spectrum and are best suited for a fixed operating point [2,42]. These also offer the greatest flexibility for installation, not normally requiring a significant structural change to the storage tank building and only minor changes to the piping. If the tank is located below the turbine installation site and the elevation head is sufficiently high, then a traditional Pelton impulse turbine can also be an option, since this reduces the pressure in the pipeline to atmospheric pressure [2,43,50]. While a so-called "counter pressure" Pelton turbine was developed by the Swiss companies Blue Water Power and Häny [51] to address precisely this limitation, it is only suitable for high-potential sites due to its very high cost, on the order of 50 times more expensive than a PAT of comparable capacity [52].

Appendix A.3.2. Class 2 "Non-Buffered" Sites

Common examples of class 2 sites include transfer nodes between pressure zones or gravity-fed water treatment plants in between gravity source and destination supply tanks (e.g., [2,5,8,11,12]). They are most often located within the water "distribution" section of the water supply chain [8]. In contrast to class 1 sites, class 2 sites offer no flexibility in the choice of flow rate, such that a turbine must be designed or chosen to efficiently operate over a broad spectrum of flow rates and corresponding available pressure heads.

Pressure control is practiced to both keep the pressure within an acceptable range for water users, to prevent pipe bursts through very high pressure, and to reduce water leakage through moderately high pressure that is not needed for maintaining satisfactory supply, for example through "pressure management" at certain times of the day (e.g., night) or in certain locations (e.g., low-elevation pressure zones). These sites are generally outfitted either with specially designed pressure-reducing valves (PRVs) or simpler gate or butterfly valves. The specially designed PRVs have an automatically adjusting variable resistance that mechanically responds to changes in pressure on both its upstream and downstream sides, to maintain a constant downstream pressure independent of flow rate. The simpler valves have a fixed characteristic curve relating their head loss coefficient to the degree of opening, and are generally opened to pre-determined set points to reduce flow rate, for example at class 1 sites or class 2 sites with a relatively constant flow rate.

At sites with sufficiently high hydropower potential (e.g., >100 kW at nearly constant operation), the best option is often a classic custom-made Francis turbine, which possesses a guide vane apparatus to flexibly react and adjust to the changing flow rates as determined by downstream water use, enabling a high efficiency over a broad operating range [2,43]. For sites with more modest hydropower potential, many scientists have devised control system designs that enable the use of inexpensive PATs to flexibly react to changing hydraulic conditions while still fulfilling the pressure control goals [7,50,53,54], in some cases accepting slightly higher leakage as a trade-off for achieving energy generation [55,56].

Appendix A.4. Review of Scientific Literature on Turbine Design Methods for Class 1 and Class 2 Sites

The potential for hydropower development in water supply schemes has been assessed by numerous scientists and engineers worldwide, and in some cases practical guidelines for implementation including design methods have been developed. However, based on a review of the scientific literature, there seems to be a dominant focus on class 2 sites, as the authors try to show in the following text.

For clarification, the term "turbine design" in this paper refers to the determination of the turbine parameters (flow rate and pressure drop across the turbine). The physical design and construction of the turbine (e.g., impeller and housing) should be carried out by a competent turbine manufacturer or research lab.

Appendix A.4.1. Studies Focusing on Class 2 Sites

Santolin et al. [57] present a method for determining the parameters of a run-of-river turbine (a variant of class 2 sites), optimizing for energy generation and economic performance. This method is similar in spirit to the method presented here for class 1 sites in water supply systems, but does not seem easily transferable.

Colombo and Kleiner [40] conduct a theoretical analysis of a turbine meant to replace a PRV for a reservoir-fed water supply system. However, this is for a class 2 site, and they also seem to make the error of associating a higher momentary power production (in kW) with a more desirable outcome, which is not necessarily true, as it often does not lead to the greatest total annual energy generation, as we show in our paper.

Carravetta et al. [13] propose a "variable operating strategy", a method for designing PAT systems to replace PRVs at class 2 sites within water distribution networks, which involves PAT in series-parallel combinations and/or with electrical rotational speed control (e.g., variable-frequency drives) to flexibly adapt to the wide range of head-flow combinations that occur. They stipulate that this complex approach is not intended for water transmission networks (i.e., class 1 sites), where there is less variability in flow conditions.

McNabola et al. [5] and Corcoran et al. [10,11] assess the potential of both class 1 and class 2 sites in Ireland and the UK, but on a conceptual and technically approximate level. They provide rough estimates of potential energy generation, without any clear guidelines on how to design a turbine for a specific site.

De Marchis et al. [53] present a hydrodynamic numerical simulation based on the "Method of Characteristics" to determine optimal parameters for PAT to be installed at class 2 sites within a water distribution network. They apply this to one of the 17 supply networks in the city of Palermo, Italy.

Sitzenfrei and Rauch [14] present a method using EPANET2 and MATLAB for determining turbine parameters at a non-conventional class 2 site in an alpine supply zone in Austria. Due to a surplus supply of natural spring water, a turbine is proposed at an outflow from the distribution network into a nearby watercourse, which serves to simultaneously generate useful energy while also flushing parts of the pipe network through increased flow rates. The amount of water released through the turbine is the surplus that remains between the amount provided naturally by the higher-elevation spring (which varies seasonally as well as daily) and the amount demanded by water users (which varies seasonally, daily and hourly).

Power et al. [15] assess the use of PAT and other turbines for four wastewater treatment plant outflows in Ireland, which are functionally equivalent to class 2 sites in drinking water supply systems.

They use an optimization algorithm in Microsoft Excel, but rely on daily average flow data "to examine turbine selection, design optimization, and economic viability".

Lima et al. [16] present a mathematical model with the software MATLAB and use the "Particle Swarm Optimization" algorithm to determine optimal parameters for turbines to replace PRVs at 2 class 2 sites with PRVs in Brazil. They also simultaneously determine the optimal sizes of the existing pipe network to optimize energy production through PAT at the former PRV sites.

Several papers mentioned here [13–16] and others not elaborated on [7,8,17–22] present methods that might be useful for class 1 sites, but the effort required to successfully adapt the methods and restrictions imposed by technical understanding and software availability for the authors' intended user group speak against pursuing this further. For instance, since the authors set out to create a tool that is free to use and based on a common tool (i.e., Microsoft Excel), and MATLAB is a non-standard software requiring specialized knowledge, possible adaptation of such algorithms [14,16] was not considered. Furthermore, while some authors provide the basic underlying equations, the software in which the algorithms are implemented is not made accessible to the reader.

Some German-language technical guidelines provide more concrete advice. The most recent 2016 edition of the German guidelines published by the German Technical and Scientific Association for Gas and Water (DVGW) [12] provides easy-to-follow guidance in the case of class 2 sites. They suggest using a frequency distribution of the occurring flow rates to determine the design flow rate of the turbine or multiple turbines. The flow rate (or group of flow rates; e.g., between 50 and 55 m^3/h) having the greatest energy density (product of total flow volume and hydraulic power) should be the design turbine flow rate.

The original 1994 edition of the DVGW guidelines [34] make no distinction between class 1 and 2 sites, and suggests using the historically most frequently occurring flow rate to select the turbine (explained in Section 2.10 above). However, they imply that the theoretical optimal flow rate is $0.577 \times Q_{max}$ (with Q_{max} defined as the greatest possible flow rate, with no pressure reduction in the pipeline), at which the hydraulic power contained in the flowing water is mathematically at its maximum [37].

Appendix A.4.2. Studies Focusing on Class 1 Sites

While comparatively less common, there are also studies that focus on class 1 sites.

Ramos et al. [7] describe an approach to design turbines for replacing PRVs at class 2 sites using the pipe network simulation tool EPANET and genetic algorithms, but also include a case study of a class 1 site in Portugal at the base of a reservoir. However, no details are provided regarding the design of the PAT proposed for this class 1 site.

Kucukali [26] analyzes a long-range water transmission pipeline with a total of 12 class 1 sites with BPTs but does not provide a concrete method regarding the ideal parameter selection of the turbines for each of the sites, presumably because the discharge remains largely constant. Since the pipeline terminates in a storage tank, however, it is conceivable that the possibility of adjusting this flow rate is neglected, which might sacrifice a possible increase in energy generation.

McNabola et al. [6] assess the potential of 10 class 1 BPT sites in Ireland, but provide only a rough estimate of the energy yield, and provide no design method for determining the optimal parameters of a turbine that would maximize the total annual energy generation.

Kougias et al. [27] present a case study with class 1 sites from northern Greece, where they have a long-range spring-fed water supply and transmission system with three BPTs (over a total of 200 vertical meters) along the route from the spring to the city 22 km further downstream. They use a so-called "Harmony Search Algorithm" (HSA) in the commercially licensed software MATLAB to determine the optimal number and dimensions of turbines for energy generation.

However, the authors prefer not to use MATLAB (as previously stated) and the HSA is not explained in detail in the paper, making it difficult to assess its usefulness. Furthermore, the flow conditions are stated to be rather constant, which makes it seem unlikely that optimization is even necessary. While not explicitly stated, this also implies the existence of a buffering storage tank at the end of the transmission pipeline, such that further optimization of energy generation might be possible as with [26].

Vilanova and Balestieri [23] present a case study in Brazil of a spring-fed supply and transmission system terminating at a water treatment plant, presumably a class 1 site with a storage tank. However, they do not provide any method for determining the optimal turbine parameters. As with other sites, the authors suspect that the inflow rate is taken as a given, rather than being seen as a parameter than can be modified to increase overall energy generation.

Novara [24] presents one simplified case study of using a PAT at a class 1 site in Italy, but provides no information about determining turbine parameters, and seems to use a very rough approximation to calculate the energy generation.

Monteiro et al. [25] introduce a method for assessing energy generation and determining the optimal number and parameters of turbines at class 1 sites, which they apply to one of three identified sites in Portugal. However, they seem to effectively treat class 1 sites as if they were class 2 sites, assuming that the inflow regime remains unchanged and attempting to design turbines to fit the inflow regime, rather than adjusting the inflow regime to allow the choice of a single, optimized turbine. Thus, they do not use the fundamental technical advantage that separates class 1 from class 2 sites.

Appendix B. User Guidelines for the Excel Tool

The following information is pertinent to users of the Excel tool, but is less relevant for judging the scientific merit of the method and results described in the article's body.

Appendix B.1. Description of the Tool

The user is only required to obtain the necessary input data and verify their accuracy, and the tool performs the calculations. The quality of the hypothetical turbine parameters depends on the amount and quality of information provided. Generally, providing more information leads to solutions that are both more reliable and more accurate.

The Excel tool has five worksheets (tabs) that are relevant for a general user:

1. Rough estimate: This sheet estimates the energy generation and economic costs and benefits based on four single input values, making the very optimistic simplifying assumption of a constant flow profile. This allows the user to determine whether it is worthwhile to continue on to the more time-intensive steps of a detailed analysis (the subsequent three sheets).
2. Single values: This sheet is for entering between 7 and 14 single values, used for generating the hydraulic site curve, calculating the economic benefits and (optionally) ensuring that the storage tank does not fall below the minimum permissible fill level due to a reduction in the flow rate (which provides the apparent "benefit" of increased energy generation).
3. Time series: This sheet is for entering time series (of the past six months to three years), used to iteratively simulate possible turbine parameters with historic data, to determine which parameters provide the greatest energy generation while still providing the daily flow volume required and (optionally) without causing unacceptable reductions in the storage tank level. Up to six time series can be entered, but generally only two are required.x
4. Turbine design: This sheet automatically determines the optimal turbine parameters based on the calculation options chosen regarding (a) level of detail and (b) choice of bypass flow. The user may then fine-tune certain design aspects before generating the technical and economic results.
5. Results: This sheet contains the results saved using a button on the previous sheet "Turbine design", providing an overview of the results obtained using various design approaches.

There are further sheets contained within the tool that perform calculations in the background, but also that display the site curve and hydraulic power curve for the current site, and display the results of the automated optimization algorithm graphically and as numbers.

Appendix B.2. Data Requirements and Corresponding Quality Criteria for Solutions

There are four main calculation options incorporated into the tool. These options are:

1. Detailed calculation with time series interval ≤15 min and consideration of storage tank levels using a historical time series of tank levels (to determine the storage capacity by inference)
2. (*recommended*) Detailed calculation with time series interval ≤15 min and consideration of storage tank levels using known or estimated useable storage tank capacity
3. Rough calculation with time series interval between 15 min and 1 d and only time series of storage tank outflow
4. Rough calculation with partially estimated single values (no time series)

The data requirements for each of these options are summarized in Table A1 below.

Table A1. Data requirements depending on desired quality of results (and corresponding calculation option): "X" = required and "(X)" = optional.

Data Type	Unit	(1) Detailed Calc., Tank Level Check via Time Series	(2) Detailed Calc., Tank Level Check via Storage Volume	(3) Rough Calc., Only Outflow Time Series	(4) Rough Calc., Estimated Single Values
		Single values (sheet 2)			
Inflow rate (Q_{in}) at control valve, position 1 (e.g., at normal flow)	m³/h	X	X	X	X
Upstream pressure ($h_{1\_upstream}$) at control valve, position 1	m; bar	X	X	X	X
Q_{in} at control valve, position 2 (e.g., at zero flow)	m³/h	X	X	X	X
$h_{2\_upstream}$ at control valve, position 2	m; bar	X	X	X	X
$h_{downstream}$ at control valve (worst case)	m; bar	X	X	X	X
Eligible for feed-in tariff? (yes/no)	—	X	X	X	X
Max. permissible Q_{in} (e.g., by contract), $Q_{max\ inflow}$	m³/h	X	X	(X)	
Min. tank level (L_{tank}) in normal operation, $L_{1\_turbine\ on}$	%	X	X		
Max. permissible tank level, $L_{2\_max\ shutoff}$	%	X	X		
Threshold for opening bypass, $L_{3\_bypass\ on}$	%	X	X		
Min. tank level in an emergency, $L_{4\_min\ emergency}$	%	X	X		
Useable storage volume, V_{tank}	m³		X		
Electricity price on site	€/kWh	X	X	X	X
Feed-in tariff (if relevant)	€/kWh	(X)	(X)	(X)	(X)
		Time series (sheet 3)—for the previous six months to three years			
Timestamp for data time series (in format TT.MM.YYYY HH:mm:ss)	—	X	X	X	
Q_{out}, from storage tank	m³/h	X	X	X	
Storage tank level, L_{tank}	%	X			
Q_{in}, to storage tank	m³/h	X	(X)		
Timestamp for Q_{in}	—	(X)	(X)		
Energy usage on site	kWh	(X)	(X)		

In general, the more data that are provided at a higher temporal resolution, the more reliable the resulting turbine parameters will be and the greater the chance of reaching the optimal energy generation. Thus calculation option 2 is recommended, with option 1 as an alternative offering a possible increase in accuracy (this was discovered after the numbering was set). These require more time and effort due to greater data requirements. If data or time are scarce, a compromise can be made on the solution quality, and a simpler calculation option 3 or 4 can be chosen. The four different options are compared on various quality criteria below (Table A2), using the rough estimate (from sheet 1 of the tool) as a benchmark.

Table A2. Quality criteria for solutions provided by the Excel tool, depending on the chosen calculation option, where 1 = lowest and 5 = highest.

Calculation Option / Quality Criteria for Solution	Rough Estimate (Sheet 1)	(1) Detailed Calc., Tank Level Check via Time Series	(2) Detailed Calc., Tank Level Check via Storage Volume	(3) Rough Calc., Only Outflow Time Series	(4) Rough Calc., Estimated Single Values
Time needed for gathering and quality-checking data (per site)	30 min. to 2 h	8 to 24 h	8 to 24 h	4 to 16 h	2 to 4 h
Confidence that tank level does not fall below min. permissible level	1	5	4	2	1
Accuracy in estimating energy generation	1	5	5	3	2
Robustness against high variation in tank outflow	1	5	5	3	1
Main advantages	Quick feedback	Best all-around solution		Faster, sometimes reliable	Small step up from rough estimate
Main disadvantages	Not reliable	Takes most time and effort		Tank levels uncertain; energy generation estimates based on daily flow volumes	

Calculation options 3 and 4 have the disadvantage that they do not account for tank levels. In order to yield reliable solutions, these options require that a turbine flow rate lies above the current typical flow rate for a given site. This assumes that the "typical" inflow rate chosen is high enough to compensate for possible short-term spikes in user demand (and outflow rate). At some sites, the inflow rate is automatically increased (above and beyond the "normal" inflow rate) in response to such events. If this behavior is known to the tool user, this higher emergency inflow rate can be chosen. If it is unknown, then a turbine could be chosen that leads to the tank level occasionally falling below the minimum permissible limit—a nonviable solution. This risk can be greatly reduced or entirely avoided by choosing calculation option 1 or 2. However, if the typical inflow rate is known to be historically constant, and the problem of low water levels has never come up, then option 3 or 4 can be acceptable, time-saving alternatives to options 1 and 2.

The primary advantage of options 1 and 2 is that there is no need to be biased by the typical inflow rate of past operations. The analysis of historical data thus allows the determination of turbine parameters from "first principles", rather than by analogy from the previous inflow rate, which may have been selected for reasons unknown to the current responsible engineer or site operator.

Appendix B.3. Choice of Bypass Flow

Since calculation options 1 and 2 may lead to the suggestion of flow rates that are lower than the current typical inflow rate, a bypass with varying flow rates is also included (as previously mentioned), as it is possible that the periodic re-routing of water around the turbine can lead to greater energy generation by selecting a relatively low turbine design flow rate. There are three options for determining the bypass flow, each with its own advantages and disadvantages:

1. Smallest possible bypass flow (minor reduction in energy generation, but smaller difference between turbine and bypass flow, which is preferable to some water supply system operators),
2. *(recommended)* Maximum permissible bypass flow (based on the user input, implies larger difference between turbine and bypass flow, but maximum energy generation and greater supply reliability) and
3. Choose the turbine flow such that in a typical situation no bypass is required (moderate reduction in energy generation, but greatest supply reliability and possibly lowest investment costs, as there is no need for electronically automated bypass valve regulation).

Appendix B.4. Assumptions/Limitations/Remarks

The following assumptions are implicit in the tool's algorithms:

- Historic water use patterns are a reliable proxy for the future. While the tool allows for some adjustment factors to account for possible future changes in both quantity and variation of user water demand, these do not aid in predicting major future trends. Therefore, it behooves the water supplier to have sufficient safeguards in place to enable manual interventions in the case that storage tank levels unexpectedly fall below permissible levels.
- Insofar as it is not already the case, the inflow rate can be kept constant during the operation of a turbine. This assumes that there are no restrictions on the inflow side, such as any put in place by the third-party operator of the reservoir or long-range supply pipeline.

The following limitations are relevant for using the tool:

- If time series are used, as recommended, having any gaps (for example, a 120-min gap in a series with otherwise 15-min intervals) leads to a false result and must be avoided by quality-checking the data.
- Only one time interval is currently possible for all data types (e.g., 15 min for both outflow and storage tank level, with the exception of the inflow rate, which has the option of a different time interval).

The following remarks may be helpful for users or potential users of the tool:

- The diameter of the pipeline plays an essential role in the availability of excess energy for hydropower generation. However, the exact diameter is not normally essential information regarding the selection of a turbine. The most reliable basis for turbine selection is the actual characteristic hydraulic site curve, derived from measurements at the storage tank flow. The turbine can normally be flexibly integrated into most pipeline systems with suitably tapered reducer and expander joints. The pipelines leading to the sites described here ranged in diameter from 150 mm at the smallest to 600 mm at the largest (for a supply main from a long-distance regional water supplier), while the pipelines in the immediate run-up to the tank typically ranged from 150 mm to 350 mm.
- When measuring pressure at field sites, one should be aware that manometers sometimes exhibit drift after years of use, such that the manometer should be separated from the water pressure and exposed to atmospheric pressure (for example, using an aeration valve) to obtain a reliable reference value corresponding to a relative pressure of 0 bar. This value can then simply be deducted from the value read when the manometer is again fully exposed to water pressure.
- Some supervisory control and data acquisition (SCADA) systems provide an option to convert the time interval of the collected data from e.g., delta-event (random, event-based time interval) to a fixed 15-min interval. If this is a feasible and reliable option, this should be used. Alternatively, the authors have developed a further Excel tool solely for the purpose of converting such delta-event time series into time series with a fixed, regular time interval. This tool can be made freely available upon request.

The maximum total annual energy generation (kWh/a) is typically achieved not at the point of maximum hydraulic power, but at the minimum feasible flow rate, limited by two factors:

1. The flow rate must at any point in time be sufficient to meet the demand placed on the storage tank by the water users in the supply zone, and cannot endanger the reliability of supply (e.g., by causing the storage tank to temporarily run empty). The bypass can, for example, be set at a higher flow rate than the turbine flow rate, in order to enable rapid filling in periods of high water withdrawal from the tank.
2. The flow rate must be high enough to enable the use of a microturbine with a practical size and sufficiently high efficiency, as the efficiency of turbines and generators drops rapidly with declining physical dimensions, and the turbine should fit into the existing infrastructure without making large structural changes in the piping network.

Exceptions to this typical case are possible, and accounted for by the Excel tool. For example, it is possible for the pipeline to have such a low friction loss-coefficient as to not exhibit a substantial drop in pressure with increasing flow rate over the conceivable operating range. The design flow rate could be made as high or low as desired, as there would be no penalty for increasing the flow rate. In this (rare) case, the site curve would resemble a horizontal line over the conceivable operating range of flow rates. As more sites are analyzed, further instances of exceptions to the typical case may be discovered.

References

1. Hintermann, M. *Electricity from Drinking Water Systems: Inventory and Feasibility Study of Drinking Water Hydropower Facilities in Switzerland (In German & French Only)*; Projektleitung DIANE Klein-Wasserkraftwerke; Bundesamt für Energiewirtschaft: Bern, Switzerland, 1994; p. 65. Available online: http://www.infrawatt.ch/sites/default/files/1994_DIANE_4df_Elektrizit%C3%A4t%20aus%20Trinkwasser-Systemen.pdf (accessed on 1 April 2019).
2. EnergieSchweiz und SVGW. *Energy in the Water Supply: Guidebook for Optimizing Energy Costs and Operation (In German Only)*; Bundesamt für Energie und SVGW: Zurich, Switzerland, 2004.
3. Aste, C.M.; Moritz, G. *TrinkHYDRO-Kärnten: Assessment of the Potential for Drinking Water Power Plants in Kärnten (In German Only)*; Technical report nr. B-EBK 9-036; ELWOG: Klagenfurt, Austria, 2009.
4. Laghari, J.A.; Mokhlis, H.; Bakar, A.H.A.; Mohammad, H. A comprehensive overview of new designs in the hydraulic, electrical equipments and controllers of mini hydropower plants making it cost effective technology. *Renew. Sustain. Energy Rev.* **2012**, *20*, 279–293. [CrossRef]
5. McNabola, A.; Coughlan, P.; Corcoran, L.; Power, C.; Williams, A.P.; Harris, I.; Gallagher, J.; Styles, D. Energy recovery in the water industry using micro-hydropower: An opportunity to improve sustainability. *Water Policy* **2013**, 1–16. [CrossRef]
6. McNabola, A.; Coughlan, P.; Williams, A.P. Energy recovery in the water industry: An assessment of the potential of micro-hydropower. *Water Environ. J.* **2013**, 1–11. [CrossRef]
7. Ramos, H.M.; Mello, M.; De, P.K. Clean power in water supply systems as a sustainable solution: From planning to practical implementation. *Water Sci. Technol. Water Supply* **2010**, *10*, 39–49. [CrossRef]
8. Carravetta, A.; Fecarotta, O.; Ramos, H.M. A new low-cost installation scheme of PATs for pico-hydropower to recover energy in residential areas. *Renew. Energy* **2018**, *2*. [CrossRef]
9. Pérez-Sánchez, M.; Sánchez-Romero, F.J.; López-Jiménez, P.A.; Ramos, H.M. PATs selection towards sustainability in irrigation networks: Simulated annealing as a water management tool. *Renew. Energy* **2017**. [CrossRef]
10. Corcoran, L.; McNabola, A.; Coughlan, P. Energy Recovery Potential of the Dublin Region Water Supply Network. In Proceedings of the World Congress on Water, Climate and Energy, Dublin, Ireland, 13–18 May 2012.
11. Corcoran, L.; Coughlan, P.; McNabola, A. Energy recovery potential using micro hydropower in water supply networks in the UK and Ireland. *Water Sci. Technol. Water Supply* **2013**, *13*, 552–560. [CrossRef]
12. DVGW. *Energy Recovery through Hydropower Facilities in the Water Supply (In German Only)*; Technical Guidelines, Worksheet W 613 (A); DVGW: Bonn, Germany, 2016.
13. Carravetta, A.; Del Giudice, G.; Fecarotta, O.; Ramos, H.M. Energy Production in Water Distribution Networks: A PAT Design Strategy. *Water Resour. Manag.* **2012**, *26*, 3947–3959. [CrossRef]
14. Sitzenfrei, R.; von Leon, J.; Rauch, W. Design and Optimization of Small Hydropower Systems in Water Distribution Networks Based on 10-Years Simulation with Epanet2. *Procedia Eng.* **2014**, *89*, 533–539. [CrossRef]
15. Power, C.; Coughlan, P.; McNabola, A. Microhydropower Energy Recovery at Wastewater-Treatment Plants: Turbine Selection and Optimization. *J. Energy Eng.* **2016**. [CrossRef]
16. Lima, G.M.; Brentan, B.M.; Luvizotto, E. Optimal design of water supply networks using an energy recovery approach. *Renew. Energy* **2017**. [CrossRef]
17. Samora, I.; Manso, P.; Franca, M.J.; Schleiss, A.J.; Ramos, H.M. *Feasibility Assessment of Micro-Hydropower for Energy Recovery in the Water Supply Network of the City of Fribourg*; Sustainable Hydraulics in the Era of Global Change; Erpicum, S., Dewals, B., Archambeau, P., Pirotton, M., Eds.; Taylor & Francis Group: London, UK, 2016; pp. 961–965. ISBN 978-1-138-02977-4.

18. Samora, I.; Franca, M.J.; Schleiss, A.J.; Ramos, H.M. Simulated Annealing in Optimization of Energy Production in a Water Supply Network. *Water Resour. Manag.* **2016**, *230*, 1533–1547. [CrossRef]

19. Carravetta, A.; del Guidice, G.; Fecarotta, O.; Ramos, H.M. PAT Design Strategy for Energy Recovery in Water Distribution Networks by Electrical Regulation. *Energies* **2013**, *6*, 411–424. [CrossRef]

20. Fecarotta, O.; Ramos, H.M.; Derakhshan, S.; Del Giudice, G.; Carravetta, A. Fine Tuning a PAT Hydropower Plant in a Water Supply Network to Improve System Effectiveness. *J. Water Resour. Plan. Manag.* **2018**, *144*. [CrossRef]

21. Novara, D.; McNabola, A. The Development of a Decision Support Software for the Design of Micro-Hydropower Schemes Utilizing a Pump as Turbine. *Proceedings* **2018**, *2*, 678. [CrossRef]

22. Pérez-Sánchez, M.; López-Jiménez, P.A.; Ramos, H.M. PATs Operating in Water Networks under Unsteady Flow Conditions: Control Valve Manoeuvre and Overspeed Effect. *Water* **2018**, *10*, 29. [CrossRef]

23. Vilanova, M.R.N.; Balestieri, J.A.P. Hydropower recovery in water supply systems: Models and case study. *Energy Convers. Manag.* **2014**, *84*, 414–426. [CrossRef]

24. Novara, D. Energy Harvesting from Municipal Water Management Systems: From Storage and Distribution to Wastewater Treatment. Extended Abstract (Not Peer-Reviewed). 2016. Available online: https://fenix.tecnico. ulisboa.pt/downloadFile/281870113703554/Extended%20Abstract%20-%20Daniele%20Novara.pdf (accessed on 9 January 2019).

25. Monteiro, L.; Delgado, J.; Covas, D.C. Improved Assessment of Energy Recovery Potential in Water Supply Systems with High Demand Variation. *Water* **2018**, *10*, 773. [CrossRef]

26. Kucukali, S. Water supply lines as a source of small hydropower in Turkey: A Case study in Edremit. In Proceedings of the World Renewable Energy Congress 2011, Hydropower Applications, Linköping, Sweden, 8–13 May 2011; pp. 1400–1407. Available online: http://www.ep.liu.se/ecp/057/vol6/004/ecp57vol6_004.pdf (accessed on 5 March 2019).

27. Kougias, I.; Patsialis, T.; Zafirakou, A.; Theodossiou, N. Exploring the potential of energy recovery using micro hydropower systems in water supply systems. *Water Util. J.* **2014**, *7*, 25–33.

28. Haakh, F. *Hydraulic Aspects of the Economic Viability of Pumps, Turbines and Pipelines in the Water Supply (In German Only)*, 1st ed.; HUSS-MEDIEN GmbH: Berlin/Oldenbourg, Germany; Industrieverlag: München, Germany, 2009; pp. 111–174. ISBN 978-3410211389.

29. Kracht, S. Out of water comes electricity—Microturbine "PAM PERGA" in water supply network (in German only). *Energie|Wasser-Praxis* **2018**, *10*, 78–81.

30. Wieprecht, S.; Kramer, M. *Investigations into the Use of Microturbines in Drinking Water Supply and Distribution Networks*; Technical report nr. 09/2012; Funded under DVGW Project W8/01/10; DVGW: Bonn, Germany, 2012.

31. Plath, M.; Wichmann, K.; Ludwig, G. *Handbook for Energy Efficiency and Energy Savings in the Water Supply (In German Only)*; DVGW & DBU: Bonn/Osnabrück, Germany, 2010.

32. Parra, S.; Krönlein, F.; Krause, S.; Günthert, F.W. Energy generation in the water distribution network through intelligent pressure management (in German only). *Energie|Wasser-Praxis* **2015**, *12*, 99–103.

33. Voltz, T.; Grischek, T. Energy management in the water supply: Excel Toolbox. Available online: https://www.htw-dresden.de/energy-in-water (accessed on 4 July 2019).

34. DVGW. *Energy Recovery through Hydropower Facilities in the Water Supply (In German Only)*; Technical guidelines, Worksheet W 613; DVGW: Eschborn, Germany, 1994.

35. Stellba Hydro: Axent. Available online: http://www.stellba-hydro.com/axent/ (accessed on 13 December 2018).

36. KSB. Application-Oriented Planning Documents for Pumps as Turbines. Available online: https://www.ksb. com/blob/52858/13564c16a6b15b3c28b1d544ae52d0e4/pat-en-data.pdf (accessed on 5 May 2019).

37. Mikus, K. Energy savings and recovery in drinking water supply (in German only). In *Mechanical and Electrical Installations in Water Works*, 1st ed.; Ebel, O.-G., Ed.; DVGW & Oldenbourg Industrieverlag: München, Germany, 1995; Volume 3, pp. 93–98. ISBN 3-486-26339-0.

38. Bahner, P.; Voltz, T.; Grischek, T. Bemessung von Pumpen als Turbinen. Available online: https://www2.htw-dresden.de/~{}wasser5/ (accessed on 5 July 2019).

39. Corcoran, L.; McNabola, A.; Coughlan, P. Predicting and quantifying the effect of variations in long-term water demand on micro-hydropower energy recovery in water supply networks. *Urban Water J.* **2016**. [CrossRef]

40. Colombo, A.; Kleiner, Y. Energy recovery in water distribution systems using microturbines. In Proceedings of the Probabilistic Methodologies in Water and Wastewater Engineering, Toronto, ON, Canada, 23–27 September 2011; pp. 1–9.

41. Brown, L. Understanding Gravity-Flow Pipelines. *Livestock Watering Factsheet*. January 2006. *British Columbia Ministry of Agriculture and Lands, Order No. 590.304–5*. Available online: https://www.itacanet.org/doc-archive-eng/water/gravity_flow_pipelines.pdf (accessed on 12 December 2018).

42. Chapallaz, J.-M.; Eichenberger, P.; Fischer, G. Manual on Pumps Used as Turbines, MHPG Series, Harnessing Water Power on a Small Scale, 11, GATE, GTZ, Eschborn. 1992. Available online: Skat.ch/book/manual-on-pumps-used-as-turbines-volume-11/ (accessed on 16 March 2019).

43. Chapallaz, J.-M.; Mombelli, H.-P.; Renaud, A. *Small Hydropower Plants: Water Turbines (In German Only)*; Impulsprogramm PACER; Bundesamt für Konjunkturfragen: Bern, Switzerland, 1995; ISBN 3-905232-54-5.

44. Jesinger, G. Possibilities and limits of energy recovery in water supply facilities (in German only). In Proceedings of the 11th Technical Water Seminar, Report Nr. 73, Munich, Germany, 22 October 1986; Bischofsberger, W., Ed.; TU Munich: Munich, Germany, 1987; pp. 185–210.

45. Heinzmann, K. Experiences with pressure-relieving turbines—Munich water works (in German only). In Proceedings of the 11th Technical Water Seminar, Report Nr. 73, Munich, Germany, 22 October 1986; Bischofsberger, W., Ed.; TU Munich: Munich, Germany, 1987; pp. 211–228.

46. Mikus, K. Experiences with pressure-relieving turbines—Stuttgart technical works (in German only). In Proceedings of the 11th Technical Water Seminar, Report Nr. 73, Munich, Germany, 22 October 1986; Bischofsberger, W., Ed.; TU Munich: Munich, Germany, 1987; pp. 237–252.

47. Schatz, J. Experiences with pressure-relieving turbines—Long-range water supply of Mühlveirtel/Austria (in German only). In Proceedings of the 11th Technical Water Seminar, Report Nr. 73, Munich, Germany, 22 October 1986; Bischofsberger, W., Ed.; TU Munich: Munich, Germany, 1987; pp. 229–236.

48. Williams, A.A.; Smith, N.P.A.; Bird, C.; Howard, M. Pumps as Turbines and the Induction Motors as Generators for Energy Recovery in Water Supply Systems. *Water Environ. J.* **1998**, *12*, 175–178. [CrossRef]

49. Voltz, T.J.; Bahner, P.; Grischek, T. Energy efficiency of pumps and small turbines—Case studies (in German only). In Proceedings of the 2nd Saxon Drinking Water Conference, Dresden, Germany, 5 September 2013; Grischek, T., Ed.; DVGW: Dresden, Germany, 2013; pp. 99–112.

50. Gaius-obaseki, T. Hydropower opportunities in the water industry. *Int. J. Environ. Sci.* **2010**, *1*, 392–402.

51. Baumann, R.; Juric, T. The counter pressure Pelton turbine as a solution for energy generation in drinking water systems (in German only). *Wasserwirtschaft* **2010**, *7–8*, 15–18.

52. Bahner, P. Deployment of Microturbines in Drinking Water Supply Networks of the FWV Elbaue-Ostharz GmbH (In German only). Diploma Thesis, Faculty of Civil Engineering, University of Applied Sciences (HTW), Dresden, Germany, 2013.

53. De Marchis, M.; Fontanazza, C.M.; Freni, G.; Messineo, A.; Milici, B.; Napoli, E.; Notaro, V.; Puleo, V.; Scopa, A. Energy recovery in water distribution networks. Implementation of pumps as turbine in a dynamic numerical model. *Procedia Eng.* **2014**, *70*, 439–448. [CrossRef]

54. Fecarotta, O.; McNabola, A. Optimal Location of Pump as Turbines (PATs) in Water Distribution Networks to Recover Energy and Reduce Leakage. *Water Resour. Manag.* **2017**, *31*, 5043–5059. [CrossRef]

55. Giugni, M.; Fontana, N.; Ranucci, A. Optimal Location of PRVs and Turbines in Water Distribution Systems. *J. Water Resour. Plan. Manag.* **2014**, *140*. [CrossRef]

56. Lima, G.M.; Luvizotto, E.; Brentan, B.M. Selection and location of pumps as turbines substituting pressure reducing valves. *Renew. Energy* **2017**. [CrossRef]

57. Santolin, A.; Cavazinni, G.; Pavesia, G.; Ardizzon, G.; Rossetti, A. Techno-economical method for the capacity sizing of a small hydropower plant. *Energy Convers. Manag.* **2011**, *52*, 2533–2541. [CrossRef]

MDPI

St. Alban-Anlage 66

4052 Basel

Switzerland

Tel. +41 61 683 77 34

Fax +41 61 302 89 18

www.mdpi.com

Water Editorial Office

E-mail: water@mdpi.com

www.mdpi.com/journal/water

www.ingramcontent.com/pod-product-compliance
Lightning Source LLC
Chambersburg PA
CBHW051711210326
41597CB00032B/5442